Sergey V. Pasechnik
Vladimir G. Chigrinov
Dina V. Shmeliova

Liquid Crystals

Related Titles

I.-C. Khoo

Liquid Crystals

2007
ISBN: 978-0-471-75153-3

I. Dierking

Textures of Liquid Crystals

2003
ISBN: 978-3-527-30725-8

E. Lueder

Liquid Crystal Displays

Addressing Schemes and Electro-Optical Effects

2001
ISBN: 978-0-471-49029-6

S.-T. Wu, D.-K. Yang

Reflective Liquid Crystal Displays

2001
ISBN: 978-0-471-49611-3

I.W. Hamley

Introduction to Soft Matter

Polymers, Colloids, Amphiphiles and Liquid Crystals

2000
ISBN: 978-0-471-89952-5

G.W. Gray, V. Vill, H.W. Spiess, D. Demus, J.W. Goodby (Eds.)

Physical Properties of Liquid Crystals

1999
ISBN: 978-3-527-29747-4

D. Demus, J.W. Goodby, G.W. Gray, H.W. Spiess, V. Vill (Eds.)

Handbook of Liquid Crystals

Four Volume Set

1998
ISBN: 978-3-527-29502-9

Sergey V. Pasechnik, Vladimir G. Chigrinov, and Dina V. Shmeliova

Liquid Crystals

Viscous and Elastic Properties

WILEY-VCH Verlag GmbH & Co. KGaA

The Authors

Prof. Sergey V. Pasechnik
Moscow State University
Laboratory of Molecular Acoustics
pasechnik@mgapi.edu

Prof. Vladimir G. Chigrinov
Hong Kong University of Science
Department of Electronic Engineering
eechigr@ust.hk

Dr. Dina V. Shmeliova
Moscow State University
Department of Physics
shmeliova@mail.ru

Cover Image
'The Gathering', created by Robert Dalgliesh and Tom Griffin, ISIS Facility, STFC Rutherford Appleton Laboratory, UK. Image copyright ISIS Facility, 2004, www.issi.rl.ac.uk. Image rendered using POV-Ray (www.povray.org) running under GridMP (www.ud.com).

Library of Congress Card No.: applied for

British Library Cataloguing-in-Publication Data
A catalogue record for this book is available from the British Library.

Bibliographic information published by the Deutsche Nationalbibliothek
The Deutsche Nationalbibliothek lists this publication in the Deutsche Nationalbibliografie; detailed bibliographic data are available on the Internet at http://dnb.d-nb.de

Printed in the Federal Republic of Germany
Printed on acid-free paper

Typesetting Thomson Digital, Noida, India
Printing Strauss GmbH, Mörlenbach
Bookbinding Litges & Dopf Buchbinderei GmbH, Heppenheim

ISBN: 978-3-527-40720-0

Contents

Liquid Crystals: Viscous and Elastic Properties
S. V. Pasechnik, V. G. Chigrinov, and D. V. Shmeliova

ISBN: 978-3-527-40720-0

Preface

It is hard to believe that only 50 years ago liquid crystals were out of practical applications and considered mostly as specific intermediate state of matter interesting for basic science. Nobody could imagine that very soon they would replace cathode-ray tubes and become the key players in display industry market. This outstanding success was achieved mainly via an enthusiasm of physicists, chemists, and engineers who not only studied unique properties of liquid crystals but also thought about their optimization providing technical privileges of liquid crystal devices.

It was realized rather soon that different physical properties played different roles in the processing of liquid crystal devices. Some characteristics such as optical index anisotropy or dielectric permittivity anisotropy were found to be most important and intensively studied. At the same time, a lot of viscoelastic parameters, except for the Frank's modules and a rotational viscosity coefficient, were considered to be of minor importance. The situation has changed when new optical modes such as OCB mode were proposed for display application. In this case, the switching (on and off) times were found to be essentially shorter than those calculated via the use of the rotational viscosity coefficient only. Such a decrease can be explained in terms of "backflow" effects arising due to an intrinsic connection between orientation and flow.

This connection can be considered a fundamental property of liquid crystals. It is responsible for a number of mechanooptical effects registered when liquid crystal layers are disturbed by shear flows. Such phenomena open good prospects for the elaboration of a new class of optical sensors, namely, liquid crystal sensors, showing extremely high sensitivity to low-frequency mechanical disturbances.

In all cases mentioned above, detailed information on viscoelastic parameters is needed to estimate technical characteristics of liquid crystal devices. Really, it is available only for a few liquid crystal compounds and materials. The situation becomes even more complicated at new applications of liquid crystals such as in photonics where strong spatial confinement essentially modifies static and dynamic behavior of liquid crystals. The near-surface layers of nanometer sizes show a very specific rheological behavior such as extremely slow rotation of an easy axis induced by strong fields. Such motion can be described in terms of specific viscoelastic

Liquid Crystals: Viscous and Elastic Properties
S. V. Pasechnik, V. G. Chigrinov, and D. V. Shmeliova

ISBN: 978-3-527-40720-0

parameters referred to as very complicated molecular processes at a boundary – liquid crystal–solid.

In the opposite case of bulk samples, the traditional hydrodynamics of incompressible nematics is not enough for describing phenomena arising from the propagation of ultrasonic waves in liquid crystal media and additional viscoelastic parameters such as anisotropic bulk elastic modules and bulk viscosities have to be introduced. Determination and analysis of temperature (frequency) dependencies of these parameters provide a unique opportunity to study critical dynamics at phase transitions of different types.

The goal of this book, which can also be called "Practical Rheology of Liquid Crystals," is to summarize the available information on rheological behavior and viscoelastic properties of liquid crystals interesting for practical applications. In accordance with this aim, we will mostly focus on experimental methods of rheological investigations omitting the microscopic picture of phenomena under consideration. The latter can be found in some brilliant books and reviews mentioned in the book. Except for phase transition problems, we restrict ourselves to nematics and ferroelectric smectic C liquid crystals. These materials can be effectively controlled by electric fields, which is of primary practical importance. We hope that this book will stimulate the progress both in traditional display science and in prospective nondisplay applications of liquid crystals such as photonics and sensors.

Sergey V. Pasechnik

1
Introduction

This book was conceived as a continuation of the series devoted to the physics and applications of liquid crystal devices [1–14]. The physical and electrooptical properties of liquid crystals (LCs) were reviewed in books written by de Gennes [1], Chandrasekhar [2], de Jeu [3], Bahadur [4], and others [5–14]. The encyclopedia of LCs collects articles on basic physical and chemical principles of LCs, as well as their application strategies in displays, thermography, and some other fields [7]. Electrooptical effects and their applications in LC devices were also discussed in books written by Blinov and Chigrinov [6, 8], Wu [11, 13], and Khoo [14]. LC optics and LC displays (LCDs) were discussed in books written by Yeh [9], Lueder [10], and Boer [12].

A deeper understanding of the basic physics of liquid crystals, including their chemical nature, macroscopic properties, and electrooptical effects, will be a considerable help to acquaint with the already existing and new liquid crystal devices. This book is oriented more toward the reader who not only like to be deeply engaged in the theory or basic physics but also is interested in applications.

This book presents the original description of rheological viscous and elastic properties of liquid crystals and shows the importance of these properties for practical applications in display and nondisplay technologies. In general, liquid crystals show quite complicated rheological behavior described in terms of a number of specific anisotropic elastic and viscous coefficients. Even in the simplest case of incompressible nematic liquid crystals, three elasticity curvature coefficients (Frank's modules) and five independent viscous-like parameters (Leslie coefficients) have to be introduced to make a proper hydrodynamic description of a number of electrooptical effects mostly used in applications. Additional viscous and elastic parameters are needed to describe extremely complicated surface dynamics of liquid crystals interacting with solids. In some specific applications, compressibility of liquid crystals is important, so additional parameters such as anisotropic bulk viscosity coefficients have to be taken into account.

The book will consider the physical nature of anisotropic viscoelasticity of liquid crystalline media, experimental methods for determination of elastic and viscous parameters important for practical applications, main directions for an optimization of existing liquid crystal displays, and physical backgrounds for the application of liquid crystals. Special attention will be paid to the properties and surface dynamics of liquid crystals described in terms of additional viscoelastic parameters such as

Liquid Crystals: Viscous and Elastic Properties
S. V. Pasechnik, V. G. Chigrinov, and D. V. Shmeliova

ISBN: 978-3-527-40720-0

anchoring strength, surface viscosity, and "gliding" viscosity. We will consider some theoretical models useful for understanding the complicated phenomena in the vicinity of liquid crystal–solid boundary, including extremely slow motion (gliding) of an easy axis, which defines a boundary orientation of LCs. The experimental methods and technique for the determination of viscous and elastic parameters of liquid crystals will also be described. We will emphasize the most suitable and reliable methods of viscosity measurements applied for studying newly synthesized liquid crystal materials of a restricted amount. Special parts of the book will be devoted to the display and photonics applications of liquid crystals where rheological properties play a key role. We will present physical backgrounds for the application of liquid crystals as sensors of mechanical perturbations (the sensors of pressure, acceleration, inclination, vibrations, etc.). Most of them are based on the intrinsic connection between shear flows and orientation of nematic liquid crystals. We will consider different types of shear flows of liquid crystals and will present linear hydrodynamic models adopted for a description of such devices. The role of viscous and elastic parameters for the optimization of the parameters of LC sensors will be analyzed in detail. We will also consider nonlinear phenomena in shear flows of liquid crystals essential for practical applications. The possibility of detecting and visualization of high-frequency acoustic fields via liquid crystals will also be discussed.

The problems of studying surface dynamics and determination of viscoelastic parameters responsible for fast and slow orientational motions will be highlighted. The aim of this book is to present the ways of optimization of liquid crystal displays by a proper choice of viscoelastic characteristics. We will focus on new types of displays where surface anchoring and surface dynamics play a key role.

We will pay special attention to describe in detail the original results obtained by the authors. For example, we will discuss the advantages of new experimental methods for shear viscosity measurements, the problem of optimization of visco-elastic properties for modern display applications, the physical backgrounds of liquid crystal sensors, the usage of ultrasound in rheological investigations, and applications of liquid crystals.

After the "Introduction," which is the first chapter of this book, we will come to Chapter 2, which is devoted to basic physical properties of LCs, such as structure and symmetry of LCs, phase transitions, and mixture preparations. Dielectric, optical, and viscoelastic properties of LCs are reviewed in this chapter, taking into account their relationship with the molecular structure and mixture content. Chapter 2 also considers the surface phenomena and cell preparations, strong and weak anchoring conditions, and the behavior of liquid crystals in magnetic and electric fields.

Chapter 3 describes anisotropic LC flows. Very complicated rheological behavior in Couette and Poiseuille LC shear flows results in a number of linear and nonlinear phenomena that have no analogues in isotropic liquids.

The optical response of LC channel at low frequencies strongly depends on LC layer thickness and boundary conditions and can be effectively controlled via electric fields, which is a key factor for LC sensor applications. Both steady LC flow regimes and flow instabilities will be considered in this chapter as key factors for LC applications as sensors and microfluid detection.

Ultrasonic methods and techniques especially developed for the characterization of LC elastic and viscous parameters are listed in Chapter 4. In particular, we have shown that such an important LC parameter as rotational viscosity coefficient can be easily extracted from LC ultrasonic data in experiments with rotating magnetic fields. We have also discussed ultrasonic methods as a unique tool for studying critical dynamics of liquid crystals at phase transitions of different types, including strongly confined systems.

Chapter 5 includes a review of experimental methods of determination of viscoelastic parameters of liquid crystals. This chapter also summarizes a very complicated behavior of LCs in near-surface layers in terms of a restricted number of parameters such as an easy axis, a surface director, an anchoring strength, and a surface viscosity. Contrary to the case of viscoelastic properties in bulk samples, these parameters reflect the interaction between LC and solid substrate. A photo-alignment technique is shown to be very effective for surface patterning with well-defined anchoring properties, which is of great practical importance. Various experimental techniques used for measuring anchoring strength such as field-off and field-on techniques are studied in detail. Near-surface layers of liquid crystals with a specific surface dynamics, such as bulk, surface, and gliding switching, are also studied.

Chapter 6 is devoted to optimal rheological properties of liquid crystals for applications in displays and photonics. We provide a general insight into various electrooptical modes in LCs with the purpose of explaining (i) the basic characteristics of the effects and their dependence on LC physical parameters and (ii) correlation of the LC rheological properties (elastic and viscosity constants, dielectric and optical anisotropy, type of LC alignment, and surface energy) with the application requirements. We consider in this chapter both active matrix (AM) and passive matrix (PM) LCD applications. Low power consumption LCD with memory effects is also highlighted. Finally, we will pay a special attention to a new trend of LC development in photonics: passive optical elements for fiber optical communication systems (DWDM components).

Extremely high sensitivity of nematic layers to the action of steady and low-frequency flows induced by a pressure gradient, which is very attractive for sensor applications, is considered in Chapter 7. The stabilizing electric fields are shown to be very effective for the optimization of technical characteristics (threshold sensitivity, dynamic range, and operating times) of LC sensors. A number of pressure gradient LC sensors (differential pressure sensors, sensors of acceleration, vibrations, inclination, and liquid and gas flows) are proposed. We have shown that LC sensors are most effective for registration of steady or low-frequency mechanical perturbations, as well as for registration, visualization, and mapping of ultrasonic fields.

The principal aims of the book are:

- to describe the practically important rheological properties of liquid crystals and preparation of liquid crystal cells most important for applications;
- to summarize the basic methods of the experimental determination of LC basic rheological parameters such as elastic and viscosity coefficients;

- to enlist LC surface interactions in terms of bulk, surface, and gliding switching as well as measurement methods for such important LC surface parameters as polar and azimuthal anchoring energy, surface, and gliding viscosity;
- to show how to control the liquid crystal behavior in electric and magnetic fields by varying its macroscopic rheological physical parameters and cell geometry;
- to compare various liquid crystal applications in displays and photonics dependent on LC rheological parameters and cell geometry;
- to present the original results of the authors in LC flow dynamics, acoustical LC phenomena, ultrasonic techniques, and LC sensors.

The book is intended for a wide range of engineers, scientists, and managers who are willing to understand the physical backgrounds of LC usage in modern display industry and nondisplay applications of liquid crystals as sensors of mechanical perturbations and as active optical elements, such as modulators, shutters, switchers, and so on. The book would be useful for students and university researchers, who specialize in the condensed matter physics and LC device development.

To the best of our knowledge, there are no books, and only a few reviews and book chapters devoted to some of the problems under considerations are available; so, we believe that our book would be of considerable interest to a relatively wider audience.

The authors are very grateful to E.P. Pozhidaev, V.M. Kozenkov, D.A. Yakovlev, A. Murauski, A. Muravsky, O. Yaroshchuk, A. Kiselev, S. Valyukh, J. Ho, X. Li, D. Huang, P. Xu, T. Du, G. Hegde, V.A. Balandin, V.A. Tsvetkov, V.I. Kireev, S.G. Ezhov, E.V. Gevorkjan, A.N. Larionov, V.I. Prokopjev, A.S. Kashitsin, E.V. Gurovich, A.V. Torchinskaya, G.I. Maksimochkin, A.G. Maksimochkin, A.V. Dubtsov, V.A. Aleshin, B.A. Shustrov, I.Sh. Nasibullayev, and A.P. Krekhov who have contributed greatly to our research program. We also owe much gratitude to H.S. Kwok, H. Takatsu, H. Takada, H. Hasebe, M. Schadt, A.S. Lagunov, D.L. Bogdanov, E.I. Kats, V.V. Lebedev, L. Kramer, S. Kralj, and N.V. Usol'tseva for the important information and many useful discussions.

References

1 de Gennes, P.G. (1974) *The Physics of Liquid Crystals*, Clarendon Press, Oxford.

2 Chandrasekhar, S. (1977) *Liquid Crystals*, Cambridge University Press, Cambridge.

3 de Jeu, W.H. (1980) *Physical Properties of Liquid Crystalline Materials*, Gordon and Breach, New York.

4 Bahadur, B. (ed.) (1991) *Liquid Crystals: Applications and Uses*, World Scientific, Singapore.

5 Vertogen, G. and de Jeu, W.H. (1988) *Thermotropic Liquid Crystals, Fundamentals*, Springer, New York.

6 Blinov, L.M. and Chigrinov, V.G. (1994) *Electrooptic Effects in Liquid Crystal Materials*, Springer, New York.

7 Demus, D., Goodby, J., Gray, G.W., Spiess, H.-W., and Vill, V. (eds) (1998) *Handbook of Liquid Crystals*, Wiley-VCH Verlag, Weinheim.

8 Chigrinov, V.G. (1999) *Liquid Crystal Devices: Physics and Applications*, Artech House, Boston.

9 Yeh, P. and Gu, C. (1999) *Optics of Liquid Crystal Displays*, John Wiley & Sons, Inc., New York.

10 Lueder, E. (2001) *Liquid Crystal Displays*, Wiley Interscience, New York.

11 Wu, S.T. and Yang, D.K. (2001) *Reflective Liquid Crystal Displays*, John Wiley & Sons, Inc., New York.

12 den Boer, W. (2005) *Active Matrix Liquid Crystal Displays: Fundamentals and Applications*, Elsevier.

13 Yang, D.-K. and Wu, S.-T. (2006) *Fundamentals of Liquid Crystal Devices*, John Wiley & Sons, Inc., New York.

14 Khoo, I.C. (2007) *Liquid Crystals*, 2nd edn, John Wiley & Sons, Inc., New York.

15 Chigrinov, V.G., Kozenkov, V.M., and Kwok, H.S. (2008) *Photoalignment of Liquid Crystalline Materials: Physics and Applications*, John Wiley & Sons, Inc., New York.

2 Physical Backgrounds for Practical Applications of Liquid Crystals

This chapter is devoted to basic physical properties of liquid crystals (LCs), such as structure and symmetry of LCs, phase transitions, and mixture preparations. Dielectric, optical, and viscoelastic properties of LCs are reviewed, taking into account their relationship with the molecular structure and mixture content. We consider the surface phenomena and cell preparations, strong and weak anchoring conditions, and the behavior of liquid crystals in magnetic and electric fields.

2.1 Anisotropy of Physical Properties of Liquid Crystals

Liquid crystals are fluids in which there occurs a certain order in the arrangement of molecules. As a result, there is an anisotropy in the mechanical, electrical, magnetic, and optical properties. The basis of the majority of specific liquid crystal electrooptical effects is found in the reorientation of the director (the axis of preferred molecular locations) in the macroscopic volume of the material under the influence of an external applied field. Anisotropy of the electrical properties of the medium (of the dielectric susceptibility and the electrical conductivity) is the origin of reorientation, whereas the dynamics of the process also depends on the viscoelastic properties and the initial orientation of the director of the mesophase relative to the field. The optical properties of the medium, its local optical anisotropy, are changed as a result of this reorientation of the director (occurring either locally or throughout the sample) in the all known electrooptic effects. We shall provide a description of the chemical structure and physical properties of liquid crystals, which are important for applications.

2.1.1 Liquid Crystal Molecules and Phases

Molecules of a specific shape form liquid crystalline phases. The most typical are rod-like molecules or rod-like molecular aggregates, which give rise to conventional nematic and smectic phases. Conventional nematic liquid crystals formed by rod-like

Liquid Crystals: Viscous and Elastic Properties
S. V. Pasechnik, V. G. Chigrinov, and D. V. Shmeliova

ISBN: 978-3-527-40720-0

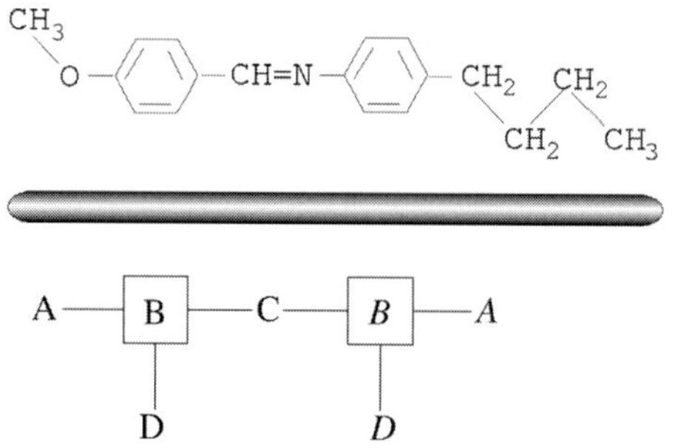

CH_3 O CH=N CH_2 CH_2 CH_2 CH_3

A B C *B* *A*

D *D*

Figure 2.1 Typical rod-like structures of nematic LC molecules.

molecules constitute a uniaxial medium, with nonpolar symmetry. The constituent molecules rotate (freely or hindered) around both their short and long axes.

A typical structure of the nematic rod-like molecule is shown in Figure 2.1. It consists of the two (less three or four) ring systems (B, *B*), sometimes with lateral substituents (D, *D*), and a linking group C between them. The groups A and *A* are the terminal groups (tails of the molecules).

Common structures of the molecular compounds, used for twist nematic (TN) and supertwist nematic (STN) liquid crystal displays (LCDs), as well as fluorinated ones, typical for active matrix (AM) LCD applications, are shown in Figures 2.2 and 2.3, respectively.

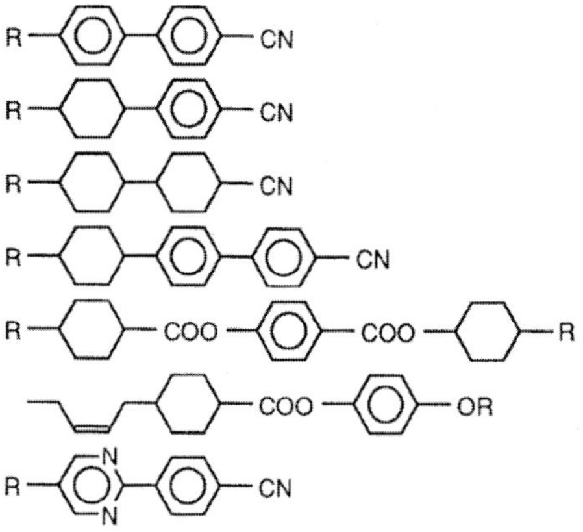

R CN

R CN

R CN

R CN

R COO COO R

COO OR

R N N CN

Figure 2.2 The structures of LC compounds for twisted nematic and supertwisted nematic display applications. The typical terminal group $R = C_nH_{2n+1}$, where $n = 3, 5, 7$ [1].

A long-range orientation order and the random disposition of the centers of gravity in individual molecules characterize nematic liquid crystals. The degree of the orientation order is characterized by the order parameter

$$S = \frac{1}{2}\langle 3\cos^2\theta - 1\rangle, \quad (2.1)$$

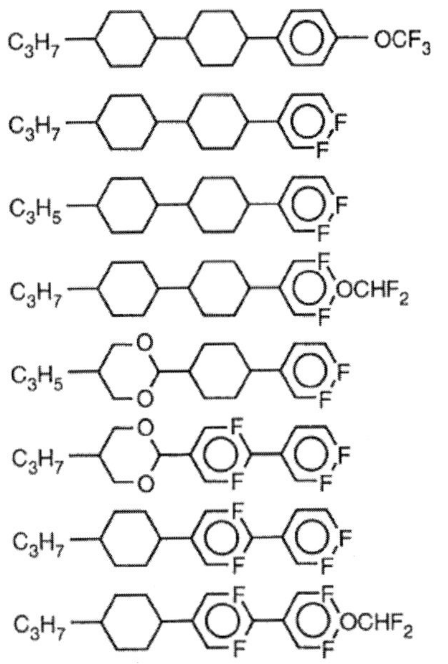

Figure 2.3 Typical structures of fluorinated nematic LC used for active matrix LCD [1].

where θ is the angle between the axis of an individual molecule and the director of the liquid crystal, and the average is taken over the complete ensemble. In a perfect crystal, $S = 1$ or $S = -1/2$, whereas in the isotropic phase, $S = 0$. For nematics, S can take, in principle, all possible values: $1 \geq S \geq -1/2$. However, for all known nematic phases formed by rod-like molecules, the order parameter is positive (negative order parameter would correspond to a nearly perpendicular location of molecular axes with respect to the director).

In principle, one compound may form several liquid crystal phases, the structure of which is defined by molecular shape (in particular, by molecular symmetry). Being displayed as a function of thermodynamic parameters such as temperature, pressure, or composition, various phases are separated from each other by phase transition lines. If a series of phase transitions in a liquid crystal occur over a temperature range, the mesophases are called thermotropic. It is also possible for mesophases to be formed from isotropic solutions of certain materials during the increase of their concentration in a suitable solvent. Such mesophases are termed lyotropic. The nematic (N) phase corresponds to the order parameter $S \sim 0.7$–0.8, which comes to zero in the isotropic (I) phase: $S = 0$. The temperature of the phase transition between N phase and I phase is called a "clearing point."

In a nematic phase, the molecules are statistically oriented along a certain preferred axis **n** called director (Figure 2.4a). The director orientation may change in space, but the characteristic distance of its variation is much longer than a molecular dimension. By external influence (e.g., through suitable treatment of the walls containing the sample), it is possible to create a uniform orientation of the

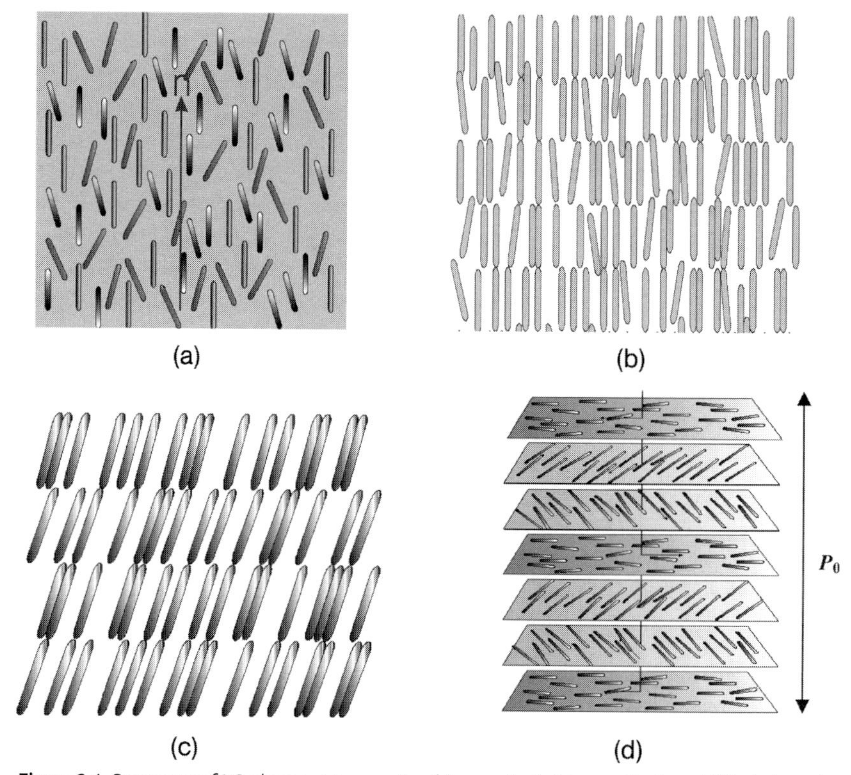

Figure 2.4 Symmetry of LC phases: (a) nematic, (b) smectic A, (c) smectic C, and (d) cholesteric.

molecular axes throughout the sample, thus obtaining a liquid monocrystal or monodomain sample.

Smectic mesophases are characterized by both the orientation and the positional order. All of them have a lamellar structure. In smectic phases, the liquid crystal molecules are arranged in layers with an average thickness d (Figure 2.4b), which is comparable to a molecular length. The layers can slide freely with respect to each other. In case the molecules are more or less perpendicular to the layer normal, the liquid crystal phase is called smectic A (Figure 2.4b). Optically, the smectic A mesophase, like the nematic one, is uniaxial with the optical axis direction coinciding with the director. If the molecules form a certain angle with a layer normal, the liquid crystal phase is recognized as smectic C, chiral or nonchiral depending on whether the rotation with respect to the normal of the molecules in neighboring layers exists or not (Figure 2.4c). On the temperature scale, smectic C phase is usually located lower than smectic A phase. The A–C phase transition is of a continuous type, that is, of the second order.

When molecules are chiral, that is, they do not possess the mirror symmetry, a variety of chiral mesophases can be observed. One of the examples of these structures called cholesteric or chiral nematic phase is shown in Figure 2.4d. Cholesteric liquid crystals are formed by optically active molecules and are characterized by the fact that

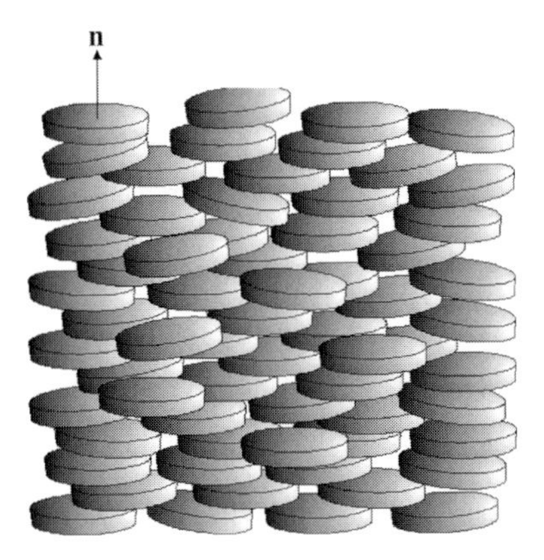

Figure 2.5 Disk-like liquid crystal.

the direction of the long molecular axes in each successive "layer" (made up of molecules that are oriented in parallel and moving freely in two directions) forms a given angle with the direction of the axes of molecules in the preceding layer. In this way, a helix is formed whose pitch (P_0) depends on the nature of the molecules (Figure 2.4d). Corresponding to the pitch P_0, the axis of orientation of the molecules (the director) rotates through an angle 2π.

Locally, like nematics, cholesterics are uniaxial. On the macroscopic scale, due to averaging, the helical structure is also uniaxial, the optical axis coinciding with the helical axis, which is always perpendicular to local (nematic) optical axes. Under a microscope, cholesterics can show the focal conic or fingerprint textures. When the helical axis is perpendicular to limiting glasses, the uniformly colored planar texture is observed. The color depends on the relative value of the pitch P with respect to a light wavelength.

Disk-like molecules form the so-called columnar mesophase (Figure 2.5). This phase is formed in materials with approximately disk-shaped molecules. The disks are packed together in columns, although their arrangement within an individual column can be either ordered or random. The columns themselves can be grouped into hexagonal or orthogonal lattices. As a rule, such a phase is optically uniaxial and negative. The optical axis coincides with the director **n**.

The chiral smectic C phase is formed by optically active molecules, which, in their racemic form, give rise to the conventional smectic C phase. The local symmetry of the C phase (C) is polar, since the plane of the molecular tilt (Figure 2.6) is no longer a mirror plane. Thus, the spontaneous polarization is allowed parallel to the layers. In the smectic C phase, each successive layer is rotated through a certain angle relative to the preceding one so that a twisted structure with the pitch P_0 is formed (Figure 2.6). The period of repetition of the physical properties in the C phase coincides with P_0. A classical example of the smectic C phase, which exhibits ferroelectric properties, is D (or L)-*p*-decyloxybenzylidene-*p*′-amino-2-methylbutyl cinnamate (DOBAMBC) (Figure 2.6). Structural classification of thermotropic liquid crystals that are most important for applications is provided in Table 2.1.

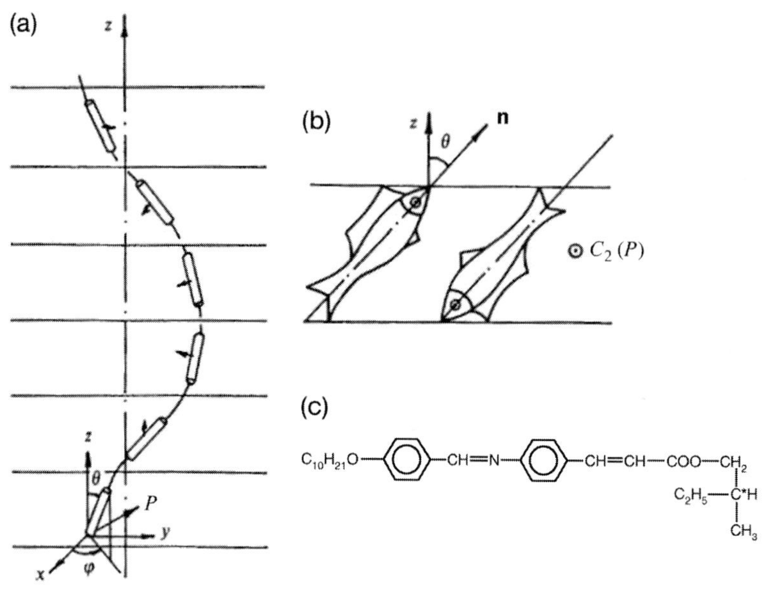

Figure 2.6 Structure and molecular arrangement of a chiral smectic C phase [1].

2.1.2
Nonliquid Crystal Compounds

The important nonliquid crystal compounds are chiral and polar additives, which are used to dope the LC material to get highly twisted or polar structures. According to general assumptions, the chiral molecule twists its nearest surroundings, thus inducing the helical rotation in the LC cell with a pitch P_0 inversely proportional

Table 2.1 Structural classification of thermotropic liquid crystals [1].[a]

Phase type	Notation	Structure
Nematic	N	Molecules are statistically oriented along a certain preferred axis **n** called director (Figure 2.4a) The symmetry is not changed by the substitution $\mathbf{n} \Rightarrow -\mathbf{n}$ (nematic is nonpolar in the bulk)
Chiral nematic or cholesteric	Ch or N^b	The axis of orientation of the molecules (the director) rotates through an angle 2π over the distance equal to the helix pitch P_0
Smectic A	SmA	Molecules are arranged in layers perpendicular to the layer normal
Smectic C	SmC	Molecules are arranged in layers oblique to the layer normal
Chiral smectic C	SmC^b	Molecules are arranged in layers oblique to the layer normal and rotated with respect to the layer normal with a period (pitch P_0). The existence of the spontaneous polarization is possible parallel to the layers

[a] The most interesting applications for liquid crystal phases are included.

to the concentration c of the chiral additive. Compounds given below are the examples of the left-handed and right-handed chiral dopants, respectively [1]:

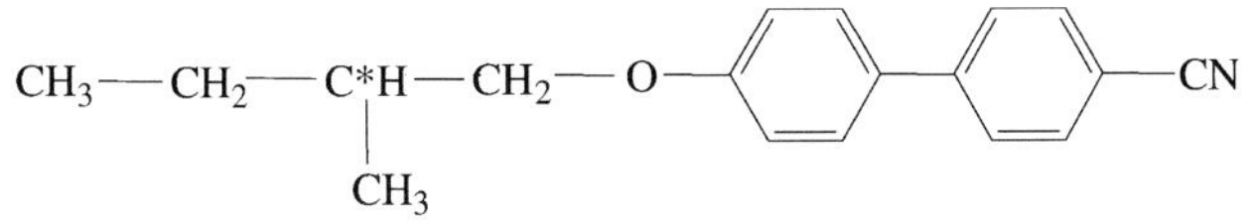

$$CH_3{-}CH_2{-}C^*H(CH_3){-}CH_2{-}O{-}C_6H_4{-}C_6H_4{-}CN$$

or

$$CH_3{-}CH_2{-}C^*H(CH_3){-}CH_2{-}C_6H_4{-}C_6H_4{-}CN$$

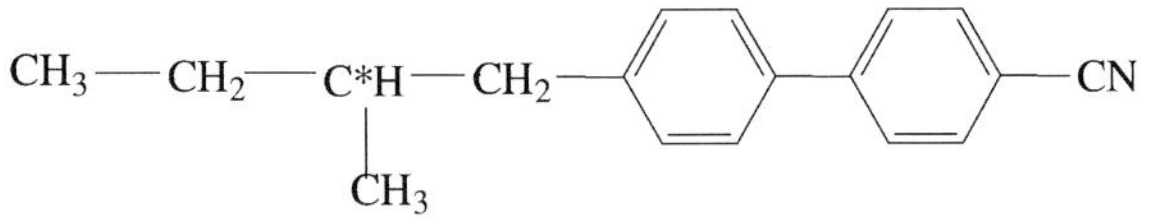

Here, the odd–even effect for chirality mentioned in the previous paragraph is used.

Very good results are obtained, using chiral dopants, based on terphenyl derivatives with the structure [2]

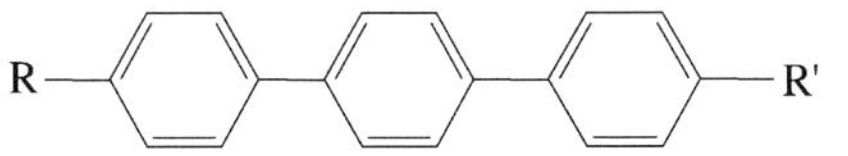

$$R{-}C_6H_4{-}C_6H_4{-}C_6H_4{-}R'$$

where R is the optically active substituent of the following type:

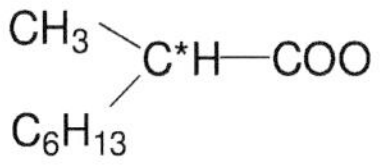

$$(CH_3)(C_6H_{13})C^*H{-}COO$$

or

$$(C_4H_9)(CH_3)C^*H{-}COO$$

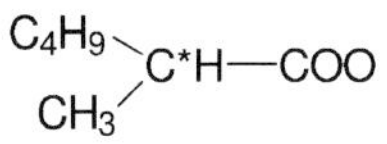

Other very useful compounds are used to form aligning layers for LCDs [1, 2]. The aligning layers must satisfy the following requirements: (i) thermal stability; (ii) solubility in suitable solvents; (iii) resistance to solubility in the LC material; (iv) proper adhesion to the substrate; (v) ability to withstand the high local temperature and mechanical effects of the rubbing or buffing process; and (vi) ability to enable the LC molecules to form pretilt angle on the substrates, which does not degrade at high temperatures. The typical substances used for the aligning layers are polyimide (PI) (Figure 2.7a) or polyvinyl alcohol (Figure 2.7b) polymers. To make the pretilt angles, sometimes the latter are doped with surfactants (Figure 2.7c).

Recently, the new "photoaligning" technique was introduced [3]. Typical materials for this nonrubbing method of the phototreatment of the substrates are dyes, dyes embedded into polymers, or pure photopolymer films, also called linear photopolymer (LPP) films. One example of LPP films is shown in Figure 2.7d.

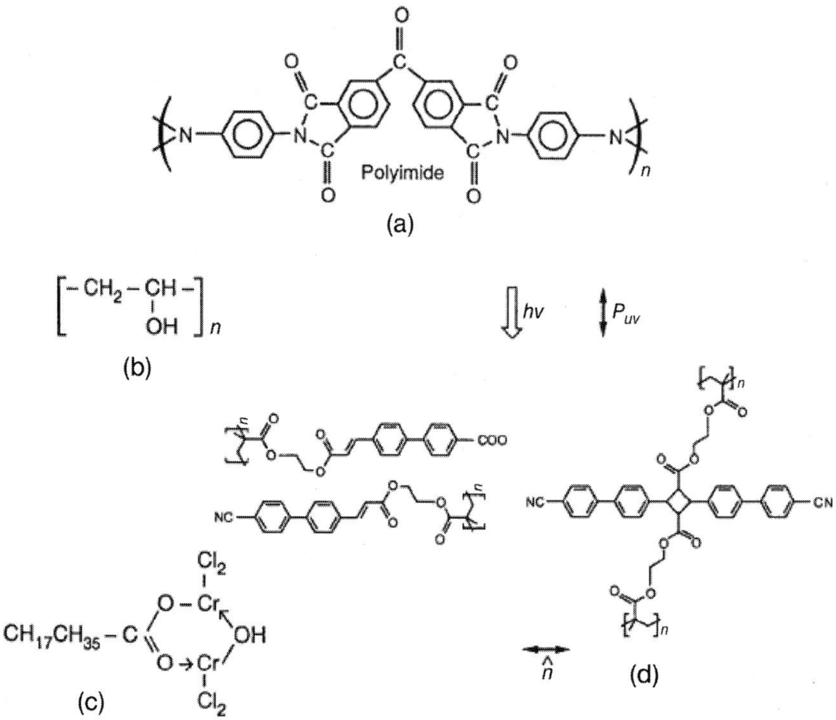

Figure 2.7 Typical liquid crystal aligning materials [1].

2.1.3
Typical Methods of Liquid Crystal Material Preparation for Various Applications

The best liquid crystalline materials for displays are, as a rule, multicomponent mixtures with a wide temperature range of operation. The general principle of constructing liquid crystalline materials for electrooptical applications is to compose multicomponent mixtures satisfying a set of necessary requirements. The requirements are varied for different applications. For example, mixtures for high informative displays must have a specially tailored ratio of elastic moduli but not an extremely high value of the dielectric anisotropy. On the contrary, materials operating at very low voltages (e.g., in image transformers based on semiconductor–liquid crystal structures) require very high values of dielectric anisotropy (for nematics) or high spontaneous polarization (for ferroelectric smectics). However, there are several common requirements for all materials. These are:

- chemical and photochemical stability;
- wide temperature range of operation;
- low viscosity;
- optimized electric and optical parameters;
- ability to be oriented by solid substrates.

In fact, the problem of stability has been solved at least for nematic materials. The majority of compounds mentioned above are chemically stable.

In order to have low viscosity and optimized electric and optical parameters, we have to carefully choose compounds from the corresponding chemical classes. For instance, low viscosity and low optical anisotropy are typical of cyclohexane derivatives, and the cyano substituent provides high dielectric anisotropy. Low melting point is achieved by composing eutectic mixtures as already discussed. In this case, the odd–even effect ought to be taken into account. To increase the clearing point, one has to dope a mixture with a substance having very high transition temperature to the isotropic phase.

This approach can be illustrated with an example of wide temperature range material ZLI-1565 worked out by "Merck." It consists of six components [1, 2] (Table 2.2). First two components provide the necessary value of the dielectric anisotropy, the next two are introduced to decrease melting point and viscosity, and the third pair increases the clearing point. As a result, the mixture has the temperature operating range from -20 to $+85\,^{\circ}$C, viscosity 19 cP (at $20\,^{\circ}$C), dielectric anisotropy $\Delta\varepsilon \approx 5$–6, and optical anisotropy $\Delta n \approx 0.13$.

A special strategy has been developed for composing ferroelectric mixtures. Such materials are based on chiral dipolar compounds that form the smectic C phase. The chemical synthesis of such compounds is extremely difficult. However, the problem can be solved step-by-step: an achiral smectic C matrix with a wide temperature range

Table 2.2 Wide temperature range material [1].

No.	Component	Weight percentage
1.	C_3H_7– ⬡ – ⬡ –CN	17%
2.	C_5H_{11}– ⬡ – ⬡ –CN	23%
3.	C_3H_7– ⬡ – ⬡ –OC_2H_5	16%
4.	C_3H_7– ⬡ – ⬡ –OC_4H_1	12%
5.	C_5H_{11}– ⬡ – ⬡ – ⬡(N, N) –C_2H_5	22%
6.	C_5H_{11}– ⬡ – ⬡ – ⬡ – ⬡ –C_3H_7	10%

may be worked out separately and then doped with a chiral dipolar additive (the chirality and dipole moment cannot be decoupled from each other as discussed).

An example of such a mixture that consists of eight components was described in Ref. [1]. A smectic C matrix contains five pyrimidine compounds with the general formula

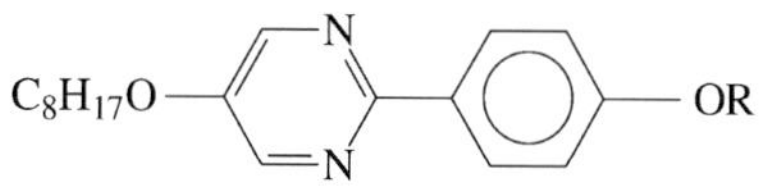

$C_8H_{17}O$ N N OR

where R = $C_{12}H_{25}$ (8.4 mol%), $C_{10}H_{21}$ (8.4 mol%), C_8H_7 (4.6 mol%), C_6H_{13} (13.7 mol %), and C_4H_9 (15.2 mol%), and 16.7 mol% of the compound

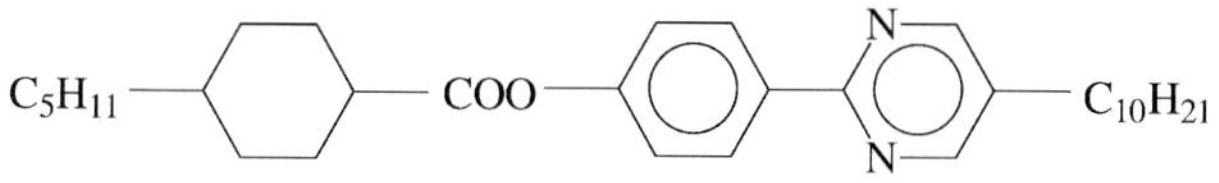

C_5H_{11} COO N N $C_{10}H_{21}$

which reduces the viscosity of the material.

The two chiral dopants with a complicated chemical structure not provided here play different roles. The first induces strong spontaneous polarization (19.7%), while the second compensates for undesirable helical structure (1.5%).

As a result, the mixture with a wide range (−7.2 to 71.6 °C) of the chiral smectic C ferroelectric phase was developed. The value of the spontaneous polarization is fairly high (41 nC/cm^2 at 25 °C).

The parameters of the liquid crystal mixtures for active matrix addressing applications include their high purity, which correlates with the value of the electrical conductivity ($\sigma < 10^{-12}$ to $10^{-14}\,\Omega^{-1}\,\mathrm{cm}^{-1}$), and a sufficiently high dielectric anisotropy to enable the proper values of the controlling voltages. The best way to fit all these requirements is to use fluorinated compounds, shown in Figure 2.3.

2.1.4 Basic Physical Properties

We will describe the most important physical parameters, which mainly determine electrooptic behavior of liquid crystal cells. According to existing phenomenological theories, we first

(1) introduce these parameters, then
(2) illustrate their dependence on the concrete molecular structure and
(3) show how to measure them and to develop new liquid crystalline mixtures, having optimal values for them.

We should note that because all the physical properties of the final mixture are interconnected and defined by the molecular structure of the components, it is impossible to arbitrarily change one liquid crystal parameter without affecting the others. This is why developing a new liquid crystal material is a delicate job, the path of certain compromises. The problem seems to be even more complicated because the list of parameters, required for application control, is not limited to the physical properties considered in this section. However, a detailed description of quality estimations of the materials is beyond the framework of this book. Sometimes, this control is even considered as the "know-how" of the producer.

2.1.4.1 Dielectric Properties

Pure organic liquids are not only dielectrics ($\sigma = 0$) but also diamagnetics, so that the magnetic susceptibility $\mu = 1 + 4\pi\chi \approx 1$, while the absorption index $n^2 = \mu\varepsilon \approx 1$.

The value of the dielectric permittivity at optical frequencies ($\varepsilon\ (\omega \Rightarrow \propto) = n^2$) is determined by the average deformation (electronic and atomic) polarizabilities of the molecule ($\langle \gamma^E \rangle$) through the Lorenz–Lorentz equation

$$\frac{n^2-1}{n^2+1} = \frac{4\pi}{3}\frac{\rho}{m} N_A \langle \gamma^E \rangle, \tag{2.2}$$

where ρ is the density of the substance, m is the molecular mass, N_A is the Avogadro number, and $\langle \gamma^E \rangle$ is the average polarizability in the electric field E.

The value of ε at low frequencies (the static dielectric permittivity) is determined in the simplest case by the Clausius–Mosotti equation

$$\frac{\varepsilon-1}{\varepsilon+2} = \frac{4\pi}{3}\frac{\rho}{m} N_A \left(\langle \gamma^E \rangle + \frac{\mu^2}{3k_B T} \right), \tag{2.3}$$

where $\mu^2/3k_B T$ is the orientation component of the average static polarizability, which depends upon the size of the dipole moment μ of the molecule. Figure 2.8 shows the variation with frequency of the dielectric permittivity for liquids with polar (curve 1) and nonpolar (curve 2) molecules.

From this must be determined the frequency f_D of its decrease or the corresponding relaxation time τ_D. The possibility of describing the relaxation of the dielectric permittivity in terms of a single time constant is based on Debye hypothesis regarding the exponential relationship governing its return to equilibrium in a constant external field (generally, this is not always satisfied). Based on the general theory of the linear response of the dielectric medium in an external field, in which a

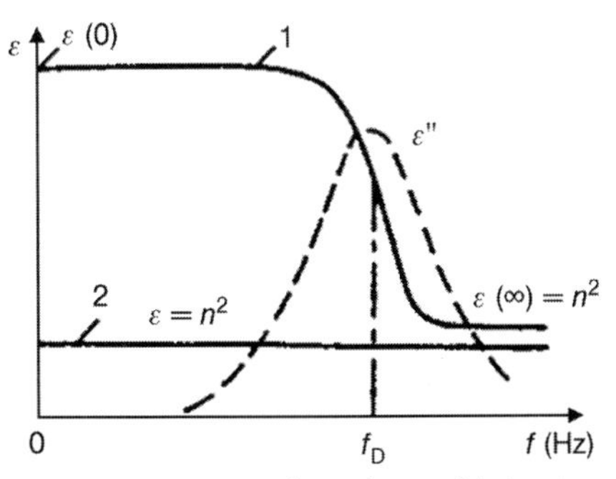

Figure 2.8 Frequency dependence of dielectric permittivities for liquid with polar (curve 1) and nonpolar (curve 2) molecules. Dielectric losses are also shown [1].

complex dielectric permittivity describes the phase lag between the displacement $D = \varepsilon^* E$ and the external field E in the dispersion region of orientation polarizability, Debye's phenomenological equations for the frequency dependence of the dielectric permittivity can be obtained:

$$\varepsilon^*(\omega) - \varepsilon(\infty) = \frac{\varepsilon(0) - \varepsilon(\infty)}{1 - i\omega\tau_D}, \tag{2.4}$$

where the real ε' and imaginary ε'' parts of the complex dielectric permittivity ε^* are given by

$$\varepsilon' = \varepsilon(\infty) + \frac{\varepsilon(0) - \varepsilon(\infty)}{1 + \omega^2\tau_D^2},$$
$$\varepsilon'' = \frac{[\varepsilon(0) - \varepsilon(\infty)]\omega\tau_D}{1 + \omega^2\tau_D^2}. \tag{2.5}$$

The dielectric losses are determined as follows:

$$\tan\varphi = \frac{\varepsilon''}{\varepsilon' - \varepsilon(\infty)} = \omega\tau_D. \tag{2.6}$$

Thus, Equation 2.6 describes the curve in Figure 2.8, which represents the frequency dependence of the real component of ε^* (the pure dielectric component), and the characteristic frequency $f_D = \omega/2\pi = (2\pi\tau_D)^{-1}$. The frequency dependence of the dielectric losses, that is, the imaginary part of ε^*, is also shown in Figure 2.8. These dielectric losses give rise to an active component of the electric current even in a purely insulating medium where there are no free charge carriers. The magnitude of the electrical conductivity caused by dielectric losses is provided by the relationship $\sigma_D = \varepsilon''\omega/4\pi$, so that the expression for the complex dielectric permittivity can also be written as

$$\varepsilon^* = \varepsilon + i4\pi\frac{\sigma_D}{\omega}. \tag{2.7}$$

The frequency dependence of the average dielectric constant measured parallel to the long molecular axis $\varepsilon_{||}$ is a characteristic of nematic liquid crystals as it corresponds to the polarization contribution related to the molecules' rotation along their short axes. The average dielectric constant perpendicular to the molecular axis $\varepsilon_{\perp}$, on the contrary, is almost independent in this frequency range as the characteristic times of the molecular rotations along a long axis are several orders of magnitude shorter.

A particularly interesting case occurs when the static value of $\varepsilon_{||}$ exceeds that of $\varepsilon_{\perp}$. In this case, as a result of the low-frequency dispersion in $\varepsilon_{||}$ at a certain frequency f_0, a change in the sign of the dielectric anisotropy of the nematic liquid crystal $\Delta\varepsilon = \varepsilon_{||} - \varepsilon_{\perp}$ can occur. Sometimes, this frequency is low, particularly in the case of nematic liquid crystals that are formed of long three-ringed molecules, for example, phenylbenzoates,

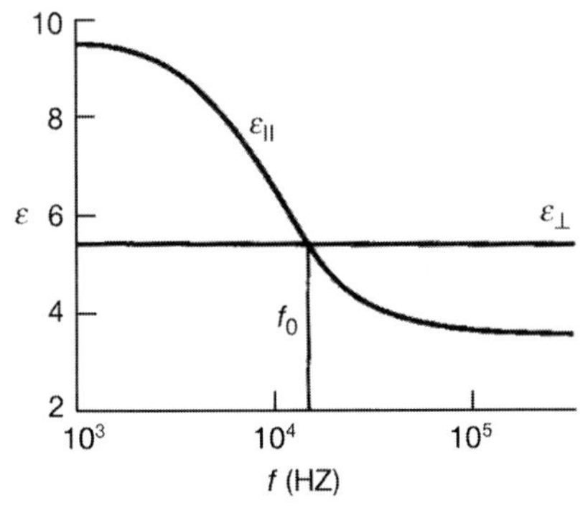

Figure 2.9 Frequency relaxation of the dielectric constant $\varepsilon_{||}$; f_0 is the dielectric inversion frequency [4].

where the barriers to rotation of the molecules around the short axes are particularly high [4]:

C_4H_9—C_6H_4—O—C(=O)—C_6H_4—O—C(=O)—C_6H_4—R

$R_1 = nC_4H_9$, $R_2 = CH_3O$

Thus, in the binary mixture of these compounds we find $f_0 \approx 20\,\text{kHz}$ at 60 °C (Figure 2.9). The dielectric sign inversion frequency f strongly depends on temperature [1, 2, 4]:

$$f_0 \approx \exp\left(-\frac{E_0}{k_B T}\right), \tag{2.8}$$

where E_0 is the corresponding activation energy. Both the values of f_0 and E_0 are defined by the value of the orientational order parameter and its temperature dependence, thus depend on the molecular structure of components in a liquid crystal mixture.

Static values of dielectric constants $\varepsilon_{||}$ and $\varepsilon_{\perp}$ are functions of liquid crystal orientational order S, the angle β between the point molecular dipole and the axis of the maximum polarizability of the molecule, the average molecular polarizability $\langle\gamma^E\rangle = (\gamma_{||}^E + 2\gamma_{\perp}^E)/3$, and its anisotropy $\Delta\gamma^E = \gamma_{||}^E - \gamma_{\perp}^E$.

In the framework of this theory, the liquid crystal dielectric anisotropy takes the form [2, 4]

$$\Delta\varepsilon = \varepsilon_{||} - \varepsilon_{\perp} = \frac{4\pi\rho}{M} N_A h F\left[\Delta\gamma^E - F\frac{\mu^2}{2k_B T}(1 - 3\cos^2\beta)\right] S, \tag{2.9}$$

where h and F are taken from the Onsager relationships valid for isotropic dielectrics as functions of the average dielectric constant $\langle\varepsilon\rangle = (\varepsilon_{||} + 2\varepsilon_{\perp})/3$ and cavity volume $a^3 = 3(\mu/4\pi)N_A\rho$ only. Thus, according to (2.9) the temperature dependence of the

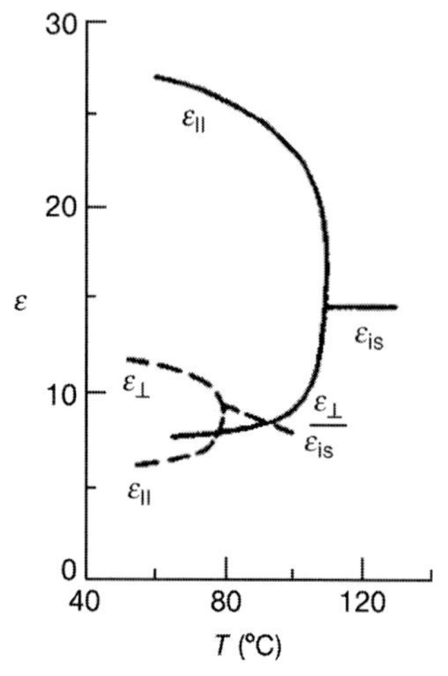

Figure 2.10 Temperature dependence of liquid crystal dielectric constants with positive (above) and negative (below) values of dielectric anisotropy $\Delta\varepsilon = \varepsilon_{\parallel} - \varepsilon_{\perp}$; ε_{is} is isotropic value of the dielectric constant.

dielectric anisotropy $\Delta\varepsilon$ is defined only by the temperature dependence of the order parameter $S(T)$ (Figure 2.10).

In case of isotropic liquid with polar molecules ($S = 0$, $\mu \neq 0$), the well-known Onsager equation applies:

$$\varepsilon_{is} = \frac{4\pi\rho}{M} N_A hF \left[\langle \gamma^E \rangle + F \frac{\mu^2}{3k_B T} \right]. \tag{2.10}$$

In view of this, for nematic liquid crystals with molecules that have a large longitudinal dipole moment we have $\Delta\varepsilon > 0$, in good agreement with the experiment. Figure 2.10 shows the temperature dependence of the $\varepsilon_{\parallel}$ and $\varepsilon_{\perp}$ components of two nematic liquid crystals [2]. In the first example, 4-butoxybenzilidene-4′-cyanoaniline, the molecules have a large (about 4–5 D) longitudinal dipole moment.

C_4H_9O — CH=N — CN

In the second example, 4-ethoxy-4′-hexyloxy-cyanostilbene, the dipole moment is approximately of the same size (because of the nitrile group), but it is directed almost normal to the long axis of the molecule.

C_2H_5O — C=CH — OC_6H_{13}
CN

In the first case, the dielectric anisotropy is positive ($\Delta\varepsilon \approx 18$, $T = 80\,°C$), but in the second case it is negative ($\Delta\varepsilon \approx -5$, $T = 60\,°C$).

In liquid crystalline mixtures, the following "additivity law" for the effective value of dielectric anisotropy is valid [1]:

$$\Delta\varepsilon_{\text{mix}} = \sum\nolimits_i C_i \Delta\varepsilon_i, \tag{2.11}$$

where C_i is molar fraction of the ith mixture component, taken at the reduced temperature $\tau = (T_{\text{NI}} - T)/T_{\text{NI}}$ (T_{NI} is the nematic to isotropic transition temperature). According to experiment, the additivity law holds for both weak polar and strong polar mixtures of liquid crystal compounds.

To obtain the desirable value of the dielectric anisotropy in a liquid crystalline material, the so-called high dipole additives are often used that possess large dipole molecular moment parallel or perpendicular to the long molecular axis. Here, two examples of such additives are given [1]:

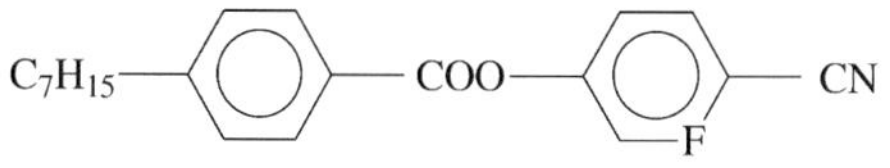

C_7H_{15} — COO — CN
F

$\Delta\varepsilon = +50$

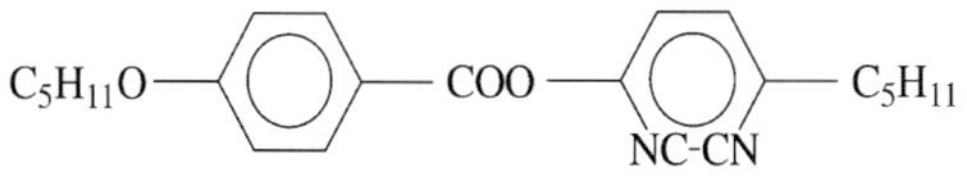

$C_5H_{11}O$ — COO — C_5H_{11}
NC-CN

$\Delta\varepsilon = -25$

In the first compound, the molecular fragments COO, CN, and F all contribute to the longitudinal dipole moment. In the second compound, two CN groups in the lateral position create a strong dipole moment perpendicular to the long molecular axis. By using high dipole additives, it is possible to develop liquid crystal mixtures with wide variation of $\Delta\varepsilon$ values (from -5 to $+25$ or even wider), required for practical applications.

2.1.4.2 **Optical Anisotropy**

If we consider the behavior of uniaxial liquid crystals at optical frequencies ($\omega \gg \omega_D \approx 10^{10}\,\text{s}^{-1}$), the orientation polarization component will not enter into the discussion. A contribution to the electric polarizability of liquid crystal molecules $\gamma_{\|}^E$ and $\gamma_{\perp}^E$ at optical frequencies is made only by the electronic and atomic parts. As a result, both parts of the complex refractive index (i.e., the refractive index n and the absorption coefficient κ) become anisotropic and each has two principal components ($n_{\|}$, $n_{\perp}$ and $\kappa_{\|}$, $\kappa_{\perp}$).

Typical temperature dependences of the principal refractive indices for three nematic liquid crystals are given in Figure 2.11 [2]. According to (2.9), the optical anisotropy $\Delta n(T) = n_{\|}(T) - n_{\perp}(T)$ temperature variation is proportional to that of the order parameter $S(T)$.

The average value of the refractive indices in the nematic phase is given by the relationship

$$\langle n^2 \rangle = \frac{1}{3}(n_{\|}^2 + 2n_{\perp}^2), \tag{2.12}$$

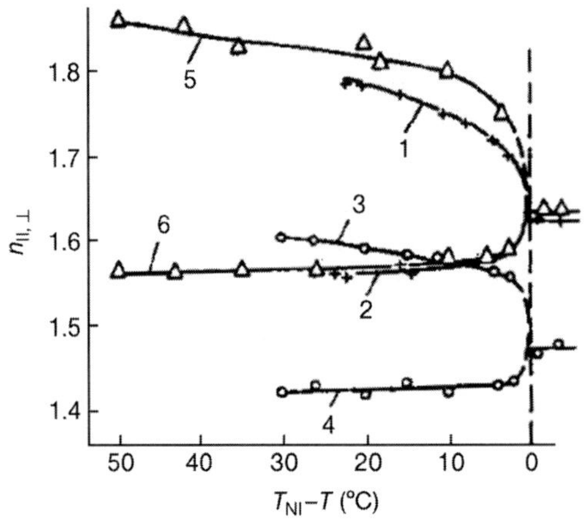

Figure 2.11 Typical dependencies of principal refractive indices $n_{\|}$ (curves 1, 3, and 5) and $n_{\perp}$ (curves 2, 4, and 6) for the three liquid crystals with different degrees of conjugation in the linkage groups. In many respects, a practical realization of oscillating flows seems to be easier than the analogous problem for the case of stationary flows described above [1, 23, 28, 32, 42, 52].

with the value $\langle n^2\rangle^{1/2}$ differing from the refractive index n_{is} in the isotropic phase because of the temperature dependence of the material density. This is due to the fact that average electrical polarizability of the molecules (as well as its components) is independent of temperature. In Figure 2.11, attention is drawn to the marked difference in the value of the refractive indices for the three materials. This is related to the differences in their molecular polarizability. In accordance with quantum mechanical theory of dispersion, the polarizability of molecule in the ground state (index 0) at frequency ω is proportional to the following sum over all possible quantum transitions from this state to higher states (k):

$$\alpha \approx \sum\nolimits_k \frac{f_{0k}}{\omega_{0k}^2-\omega^2}, \tag{2.13}$$

where f_{0k} and ω_{0k} are the oscillator strength and the frequency, respectively, of $(0 \Rightarrow k)$ transitions. If oscillator strengths of transitions to the higher levels are approximately the same, then the longest wavelength transitions will make the largest contribution to the polarizability of the molecule at a given frequency, since in this case the denominator of Equation 2.13 takes its smallest value. In the series of liquid crystal molecules with different linking or bridging groups (see Figure 2.1), the long-wave absorption band is noticeably displaced to the short-wave region of the spectrum because of the degree of conjugation in the linkage:

—N=N(→O)— > —CH=N— > —C(=O)—O—

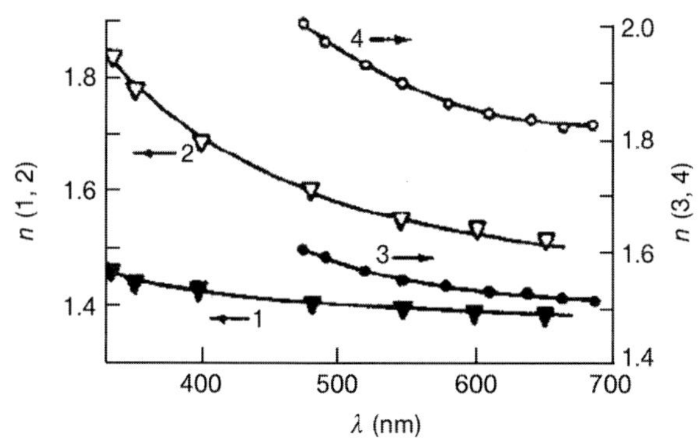

Figure 2.12 Dispersion of the refractive indices $n_{\parallel}$ (2, 4) and $n_{\perp}$ for various wavelengths. The two liquid crystals have the conjugation links COO (1, 2) and N_2O (3, 4) with the structures shown above [1, 2].

Thus, the average polarizabilities of the corresponding liquid crystal molecules decrease (Figure 2.11), and moreover, the frequency dispersion of the refractive index decreases in the same order in agreement with Equation 2.13 (Figure 2.12).

In cyclohexanecarboxylic acids, the longest wavelength absorption bands are in the far-ultraviolet region of the spectrum. Therefore, the optical anisotropy of these compounds is small ($\Delta n \approx 0.05$), and only a small dispersion of the refractive indices is observed.

According to (2.9), the optical anisotropy $\Delta n = n_{\parallel} - n_{\perp}$ is completely defined by the anisotropy of polarizability, measured parallel and perpendicular to the long molecular axis.

The values of optical anisotropy Δn increase [1, 2]

(i) with the elongation of the conjugation chain parallel to the long molecular axis;
(ii) by replacing saturated aromatic rings with the unsaturated ones;
(iii) by shortening the alkyl chain of the end molecular groups in homologue series in the form of even–odd alternation;
(iv) by increasing the values of the order parameter S(2.9).

Dependence of Δn on the molecular structure could be illustrated by Table 2.3.

Optical anisotropy of liquid crystal mixtures obeys the additivity rule for refractions:

$$\left[\frac{1}{\rho}\frac{n^2-1}{n^2+1}\right]_{\text{mix}} = \sum_i C_i \left[\frac{1}{\rho}\frac{n^2-1}{n^2+1}\right]_i, \tag{2.14}$$

where C_i is a molar fraction of ith component in the mixture. The validity of (3.14) is confirmed in experiment even for different chemical classes both for $n = n_{\parallel}$ and for $n = n_{\perp}$.

Table 2.3 Optical anisotropy correlation with the molecular structure of nematic liquid crystals [1][a].

Physical mechanism	Molecular structure	Optical anisotropy
(i)	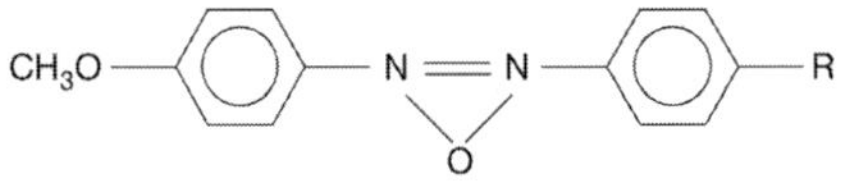 CH_3O—⟨ ⟩—N=N(O)—⟨ ⟩—R	
	$R = OCH_3$	0.26
	$R = C_4H_9$	0.21
(ii)	C_7H_{15} — AR — CN	
	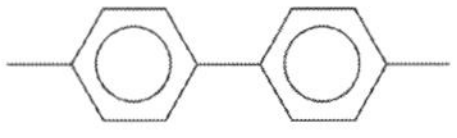	0.16
	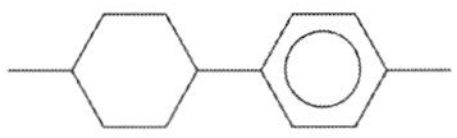	0.09
	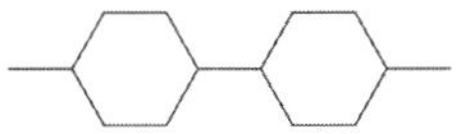	0.06

[a]Measurements were made at the wavelength $\lambda = 589$ nm.

2.1.4.3 **Viscoelastic Properties**

The viscoelastic properties of liquid crystals are very important and mainly determine the behavior of liquid crystals in external electric fields, defining such characteristics as controlling voltages, steepness of the transmission–voltage curve, response times, and so on.

We shall

(i) briefly outline the main definitions of the viscoelastic characterization of liquid crystals;
(ii) show the dependence of viscoelastic constants of nematic liquid crystals on the structure and temperature; and
(iii) discuss certain ideas how to develop new liquid crystal mixtures with given viscoelastic parameters.

We shall also describe the main methods for measurements of these parameters.

2.1.4.4 **Elasticity**

The basic difference between deformations in a liquid crystal and in a solid is that in liquid crystals there is no translational displacement of molecules on distortion of a sample. This is due to "slippage" between liquid layers. A purely shear deformation of a liquid crystal conserves elastic energy. The elasticity of an isotropic liquid is related to changes in density. In liquid crystals, variations in density can also be characterized by a suitable modulus, but the elasticity that is related to the local variation in the orientation of the director is their principal characteristic.

In the description of elasticity of a nematic liquid crystal, the following assumptions are made:

1. Director **n** reorients smoothly compared to the molecular dimension of a liquid crystal. Thus, we may conclude that the order parameter S remains constant throughout the whole volume of a liquid crystal at a fixed temperature T, while only director field **n** varies in accordance with external electric (or some other) fields.
2. The only curvature strains of the director field, which must be considered, correspond to the splay, bend, and twist distortions (Figure 2.13). Other types of deformation either do not change the elastic energy (e.g., above-mentioned pure shears) or are forbidden due to the symmetry. In nematic liquid crystals, the cylindrical symmetry of the structure as well as the absence of polarity (head to tail symmetry) must be taken into account.
3. Following the Hooke's law, only squares of the director deformations are included into the expression for the free energy.

In view of these assumptions, the density of the free volume elastic energy of nematic liquid crystal could be written as

$$g = \frac{1}{2}\left[K_{11}(\operatorname{div}\mathbf{n})^2 + K_{22}(\mathbf{n}\operatorname{curl}\mathbf{n})^2 + K_{33}(\mathbf{n}\times\operatorname{curl}\mathbf{n})^2\right]. \tag{2.15}$$

Equation 2.15 forms the basis for examining almost all electrooptical and magnetooptical phenomena in nematic liquid crystals. The first term in (2.15) describes the S deformation (splay), the second term describes the T deformation (twist), and the third term describes the B deformation (bend). These three types of deformation are illustrated in Figure 2.13. Three elastic modules K_{11}, K_{22}, and K_{33} in (2.15) characterize the corresponding values of the elastic energies.

In cholesteric (or chiral nematic) liquid crystals, the situation is very close to usual nematics. However, due to the chirality of the molecules, the lowest state of elastic energy in cholesterics does no longer correspond to the uniform director orientation

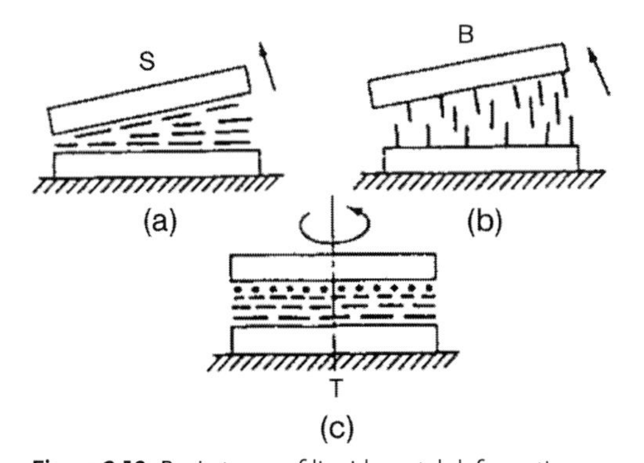

Figure 2.13 Basic types of liquid crystal deformations: (a) S deformation (splay), (b) B deformation (bend), and (c) T deformation (twist) [1].

but to the twisted one with a pitch $P_0 = 2\pi/q_0$, where q_0 is the wave vector of a cholesteric. Thus, for cholesterics the second term in expression (2.15) must be rewritten as

$$K_{22}(\mathbf{n}\,\mathrm{curl}\,\mathbf{n})^2 \Rightarrow K(\mathbf{n}\,\mathrm{curl}\,\mathbf{n} + q_0)^2, \tag{2.16}$$

where the positive and negative q_0 values correspond to the left- and right-handed helices, respectively.

The theory of elasticity of smectic liquid crystals has its own features. Deformations related to a change in the spacing between the layers are common to all smectic phases. The deformations are, in general, not related to a change in director orientation, and here an additional modulus of elasticity B occurs. In smectic A liquid crystals, the only allowed deformation is specific undulation of the smectic layers, such that interlayer distance is kept constant and director remains normal to the layer. This deformation imposes the following limitation on the director field:

$$\mathrm{curl}\,\mathbf{n} = 0, \tag{2.17}$$

and consequently, twist and bend elastic moduli diverge in the vicinity of phase transition nematic–smectic A. Equation 2.17 is not valid for smectic C (and ferroelectric smectic C) liquid crystals, where we also deal with the specific elastic constants.

Since $\mathbf{n}$ is a dimensionless quantity, the elastic constants K_{ii} must have dimensions of energy/m, that is, Newton. By dimensional arguments, these moduli should be of the order of W/a, where W is the energy of interaction of the molecules and a is their size. By assuming $W \sim 0.1$ eV (a typical value) and $a \sim 10$ Å, we derive $K \propto 10^{-11}$ N.

There are a number of experimental data on elastic moduli of liquid crystals, which are only qualitatively explained. The existing molecular approaches do not directly correspond to the real situation because molecules are considered to be spherocylinders or hard rods, far from the reality [1, 4]. For instance, the ratio K_{22}/K_{11} according to the present approaches is about 1/3, which is two times lower than the corresponding experimental range. However, the above-mentioned data are quite sufficient for developing liquid crystal mixtures with required elastic properties.

The methods of the measurements of the liquid crystal elastic moduli will be considered in detail in Chapter 5.

2.1.4.5 **Viscosity**

The dynamics of nematic liquid crystals is described by (i) director field $\mathbf{n}(\mathbf{r}, t)$ and (ii) velocities of the centers of the molecules $\mathbf{V}(\mathbf{r}, t)$. These variables in general obey the following equations:

1. The equation of continuity in incompressible liquids:

$$\mathrm{div}\,\mathbf{V} = 0. \tag{2.18}$$

2. Navier–Stokes equation in anisotropic viscous liquid:

$$\rho\left(\frac{\partial v_i}{dt} + \frac{\partial v_i}{dx_k}\right) = f_i + \frac{\partial \sigma'_{ki}}{dx}, \tag{2.19}$$

where i, k, x_i, $x_k = x, y, z$,

$$f_i = -\frac{\partial P}{dx_i} + QE_i \tag{2.20}$$

is the external force in the anisotropic liquid dielectric, σ'_{ki} is the viscous stress tensor:

$$\sigma'_{ki} = \alpha_1 n_k n_i A_{mn} n_m n_n + \alpha_2 n_k N_i + \alpha_3 n_i N_k + \alpha_4 A_{ki} + \alpha_5 n_k n_m A_{mi} + \alpha_6 n_i n_m A_{km}, \tag{2.21}$$

P is an external pressure,

$$A_{ij} = \frac{1}{2}\left(\frac{\partial v_i}{dx_j} + \frac{\partial v_j}{dx_i}\right) \tag{2.22}$$

is analogous to the viscous stress tensor for the anisotropic liquids, and

$$\mathbf{N} = \frac{d\mathbf{n}}{dt} - \frac{1}{2}[\mathbf{n} \times \text{curl}\, \mathbf{n}] \tag{2.23}$$

is the rate of motion of the director $\mathbf{n}$, which vanishes when the entire fluid is in uniform rotation with the angular velocity 1/2 curl $\mathbf{v}$.

The coefficients of the proportionality between viscous stress derivatives and the time derivative of velocity in (2.21) are called viscosity coefficients: $\alpha_1, \alpha_2, \alpha_3, \alpha_4, \alpha_5$, and α_6.

It can be shown that

$$\alpha_2 + \alpha_3 = \alpha_6 - \alpha_5,$$

that is, only five independent viscosity coefficients exist.

3. The equation of the director rotation in a nematic liquid crystal is

$$I\frac{d\Omega}{dt} = [\mathbf{n} \times \mathbf{h}] - \Gamma, \tag{2.24}$$

where $\Omega = [\mathbf{n} \times d\mathbf{n}/dt]$ is the angular velocity of the director rotation, I is the moment of inertia for the molecular reorientation, normalized to a unit volume,

$$\mathbf{h} = -\frac{\delta F}{\delta \mathbf{n}} \tag{2.25}$$

is the functional derivative of the liquid crystal volume free energy with respect to the director components $\mathbf{n}$ or the so-called molecular field, and

$$\Gamma = [\mathbf{n} \times (\gamma_1 \mathbf{N} + \gamma_2 A\mathbf{n}) \tag{2.26}$$

is the frictional torque, which is analogous to the viscous term in the Navier–Stokes equation.

Here, $\gamma_1 = \alpha_3 - \alpha_2$ and $\gamma_2 = \alpha_3 + \alpha_2$ are viscosity coefficients.

$$\gamma_1 = \alpha_3 - \alpha_2 \tag{2.27}$$

is the so-called rotational viscosity of the nematic director **n**, which characterizes the pure rotation of the nematic liquid crystal director without any movement of the centers of the molecules, called "backflow" effect. The so-called backflow effect is very important in some specific liquid crystal configurations, which we will consider in Chapter 6. The following three viscosity coefficients are important for a proper description of a backflow effect:

$$\eta_1 = \frac{1}{2}(-\alpha_2 + \alpha_4 + \alpha_5),\ \eta_2 = \frac{1}{2}(\alpha_2 + 2\alpha_3 + \alpha_4 + \alpha_5),\ \text{and}\ \eta_3 = \frac{1}{2}\alpha_4, \tag{2.28}$$

which will be considered in more detail in Chapter 5.

Let us, like in the previous case, present some known facts about viscosity of a liquid crystal.

Temperature dependence of viscosity of isotropic liquids obeys the well-known exponential law [1, 4]

$$\eta_{is} = \eta_0 \exp\left(\frac{E}{k_B T}\right), \tag{2.29}$$

where $E > 0$ is activation energy for the diffusion molecular motion and η_0 is a constant. Similar temperature dependence is observed for all nematic viscosity in the whole mesophase range, except in the vicinity of phase transition regions.

Figure 2.14 shows temperature dependence of ln η_i(2.28) on the inverse temperature $1/T$, which is very close to linear functions [1]. As seen from Figure 2.14, the "isotropic" viscosity $\eta_3 = \alpha_4/2$ does not undergo considerable change near the phase transition region, while η_1 and η_2 vanish in the isotropic phase.

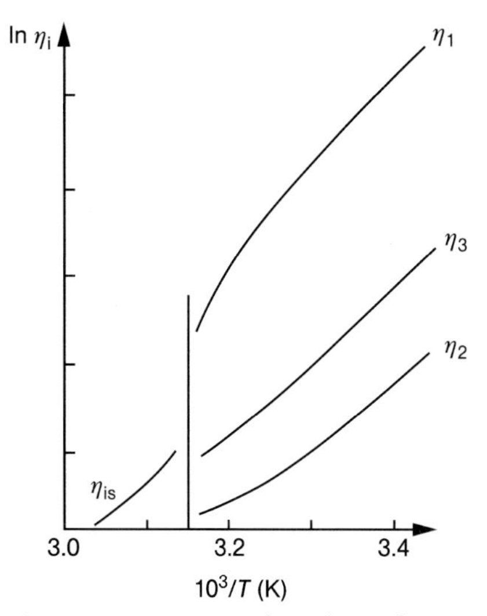

Figure 2.14 Temperature dependence of viscosity coefficients η_1, η_2, and η_3 for MBBA liquid crystals [1, 2].

The temperature dependence of the viscosity α_i (except α_4) can be expressed in terms of the order parameter $S(T)$ [1, 4]:

$$\alpha_i = a_i\, S(T) + b_i\, S^2(T). \tag{2.30}$$

As mentioned above, the most important viscosity combination of Leslie coefficients is $\gamma_1 = \alpha_3 - \alpha_2$, which defines the director response times in electrooptical effects. According to Refs [1, 4], the temperature dependence of $\gamma_1(T, S)$ could be rewritten in most general cases as

$$\gamma_1 = bS^x \exp\left(\frac{E}{k_B T}\right) \exp\left(\frac{A}{T - T_0}\right), \tag{2.31}$$

where b, A, and T_0 are temperature-independent quantities, which are functions of the molecular structure. For a nematic liquid crystal with larger values of Δn and γ_1, we have $x = 1$, while in the opposite case with low Δn, we have $x = 2$ [2]. The parameter T_0 in (2.31) indicates the so-called temperature of freezing the director motion, which is close to the glass transition temperature T_g when all the dynamic processes are frozen and the viscosity is infinite:

$$T_0 = T_g + Q, \tag{2.32}$$

where Q is about 50 K [1].

Finally, let us give typical rotational viscosity values for different chemical classes (Table 2.4). As seen from Table 2.4, including saturated fragment into the molecular structure, such as cyclohexane ring, results in a visible decrease of γ_1 values and its

Table 2.4 Rotational viscosity γ_1 for different LC chemical classes ($T = 25\,°C$) [1].

Structure of the mixture molecules[a]	Rotational viscosity, γ_1 (P)	Activation energy, E (eV)	Optical anisotropy, Δn
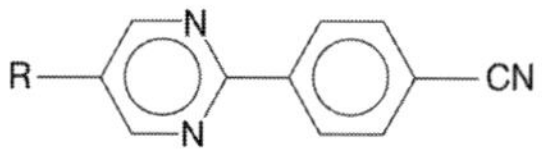 R (N, N) CN	1.9	0.565	0.174
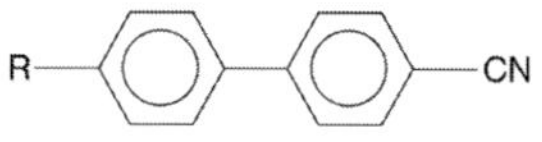 R CN	1.1	0.546	0.184
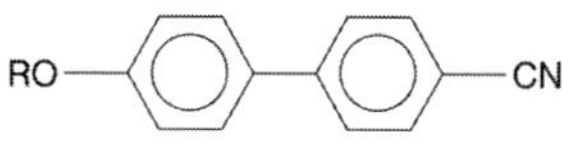 RO CN	3.3	0.496	0.177
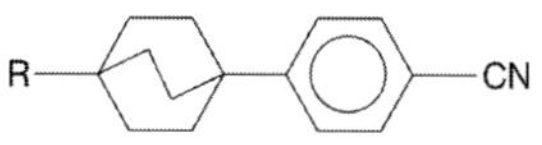 R CN	3.6	0.423	0.114
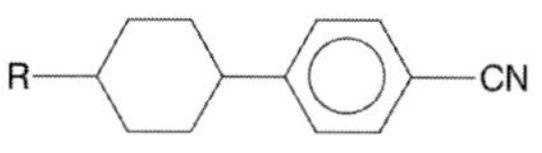 R CN	1.0	0.41	0.1

[a]Measurements were carried out in binary mixtures of fifth ($R = C_5H_{11}$) and seventh ($R = C_7H_{15}$) homologues of each compound, taken in the proportion 40 : 60.

temperature dependence; however, the Δn value also decreases due to the shortening of the conjugated bonds in the molecular structure. The behavior of viscosity in the mixtures of liquid crystal compounds is rather complicated.

The logarithmic additivity law [14]

$$(\ln \gamma_1)_{\mathrm{mix}} = x_{\mathrm{A}}(\ln \gamma_1)_{\mathrm{A}} + x_{\mathrm{B}}(\ln \gamma_1)_{\mathrm{B}} \tag{2.33}$$

seems to take place only in the mixtures of homologues in the homologous series (x_{A} and x_{B} are molar fractions of the components), while the simple additivity law ($\ln \gamma_1 \Rightarrow \gamma_1$) is not valid at all. A considerable deviation of γ_1 and E values from the relation (2.33) (depression) was observed in the mixtures of strong and weak polar compounds [1]. The physical origin of the phenomenon seems to be similar for elastic constants as discussed above.

There is no theoretical explanation of the viscosity behavior of different liquid crystal substances and their mixtures. Also, there exist only a few works where the viscosity measurements are related to the corresponding molecular structure [1, 4]. However, new liquid crystalline low-viscosity materials have been successfully developed. To make these materials, the following phenomenological rules should be remembered [1, 2]:

1. The viscosity is lower for shorter molecules. In a homologue series, even–odd alternation is observed with a marked tendency to increasing viscosity with the number of carbon atoms.
2. Alkyl end groups provide lower values of viscosity compared to alkoxy and acyloxy end groups.
3. Replacement of phenyl ring by a *trans*-cyclohexane ring results in reduced viscosity values.
4. Introducing the rings with heteroatoms increases viscosity compared to phenyl analogues (Table 2.4).
5. The most "viscous" bridging groups are the ester group $-\mathrm{COO}-$, the simple bond (as in biphenyls), and the ethane group $-\mathrm{CH}{=}\mathrm{CH}-$.

The most useful compounds for reducing viscosity in liquid crystal materials are cyclohexane derivatives due to their low viscosity, high clearing temperature, and good solubility. We will not consider here the viscous properties of cholesteric, smectic, or some other types of liquid crystals, which is rather complicated, as many unknown viscosity coefficients can be involved in our consideration due to the lower symmetry of these types of liquid crystals.

2.2 Liquid Crystal Alignment on the Surface

The interaction of liquid crystals with neighbor phases (gas, liquid, and solid) is a very interesting problem relevant to their electrooptical behavior. The structure of liquid crystalline phases close to an interface is different from that in the bulk, and this "surface structure" changes boundary conditions and influences the behavior of a

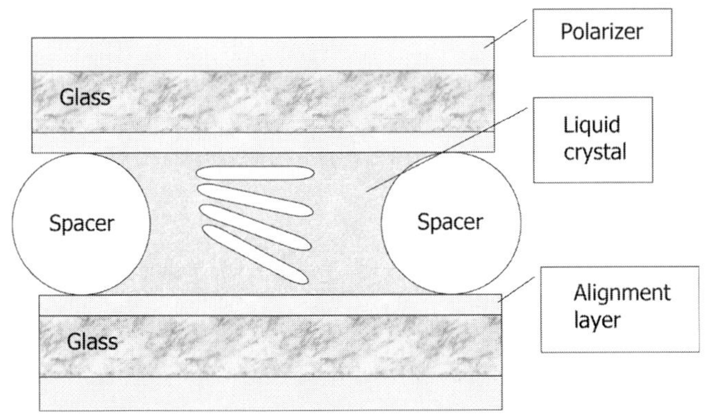

Figure 2.15 Typical liquid crystal sandwich-type cell.

liquid crystal in bulky samples. The nematic phase is of great importance from the point of view of applications in electrooptical devices; thus, in this part, we shall concentrate mainly on the surface properties of nematic liquid crystals.

2.2.1 Types of Liquid Crystal Alignment

2.2.1.1 Electrooptical Cells

In most practical applications, and when examining liquid crystals, sandwich-type cells are used (see Figure 2.15).

A flat capillary with a thickness of 1–10 μm and above is formed from two glass plates with transparent electrodes. The separation between the plates is fixed by means of an insulating spacer (mica, polyethylene, etc.). To fix a very narrow gap (about 1 μm), glass balls or pieces of glass thread of proper diameter are put between glasses. ITO (indium tin oxide) is coated onto glass usually by sputtering. ITO is conductive, so DC magnetron sputtering can be used with a high efficiency. The sputtering process is performed in a large vacuum chamber and is an in-line process. Typical thickness of the ITO layer is 20–100 nm.

In order to investigate the anisotropy of the properties of liquid crystals and the character of their electrooptical behavior, it is necessary to make a definite orientation of their molecules at the boundary walls of the cell. Molecules in the successive layers "attach" themselves to the molecules on the surface layer, and the whole sample will become a monocrystal, either ideal or deformed, depending on the orientation of these surfaces. The orientation of the molecules on the surface is characterized by two parameters: the average angle of the molecules to the plane of the surface θ_0 (preferred direction at the surface) and the anchoring energy $W = \{W_\theta, W_\varphi\}$, where W_θ and W_φ are the corresponding anchoring energies for the polar $\theta - \theta_0$ and azimuthal $\varphi - \varphi_0$ deviations of the liquid crystal director from the preferred alignment direction $\{\theta_0, \varphi_0\}$ (Figure 2.16). Using the angle θ_0, we can distinguish various orientations: homeotropic ($\theta_0 = 0$), planar ($\theta_0 = \pi/2$), and tilted ($0 \leq \theta_0 \leq \pi/2$).

Pretilt angle

Angle of local director of molecule at the surface with substrate surface

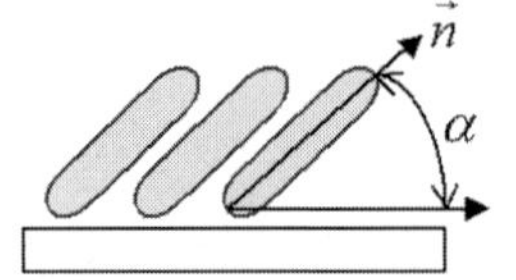

Anchoring energy

The anchoring energy consisting two parts: polar and azimuthal

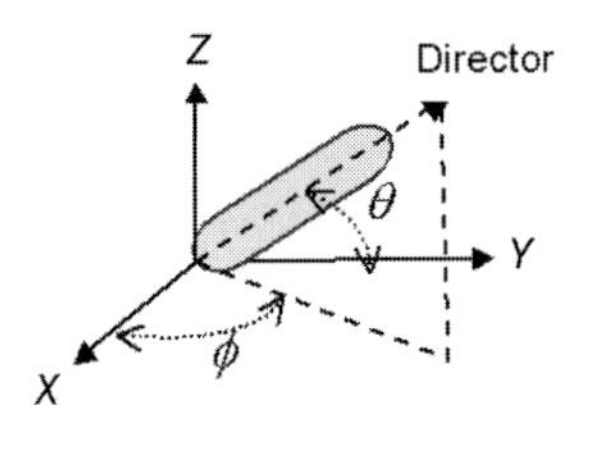

$$W(\theta) = \frac{1}{2} W_{\theta_0} \sin^2(\theta - \theta_0)$$

$$W(\phi) = \frac{1}{2} W_{\phi_0} \sin^2(\phi - \phi_0)$$

Figure 2.16 Important parameters of liquid crystal alignment: pretilt angle (above) and anchoring energy (below).

2.2.1.2 **Planar (Homogeneous) Orientation**

Most commonly, a planar orientation is produced by a mechanical rubbing of the surface of the glass with paper or cloth (Chatelain's method). Rubbing creates a microrelief in the electrode coating or glass in the form of ridges and troughs, which promotes the orientation of the molecules along these formations (Figure 2.17).

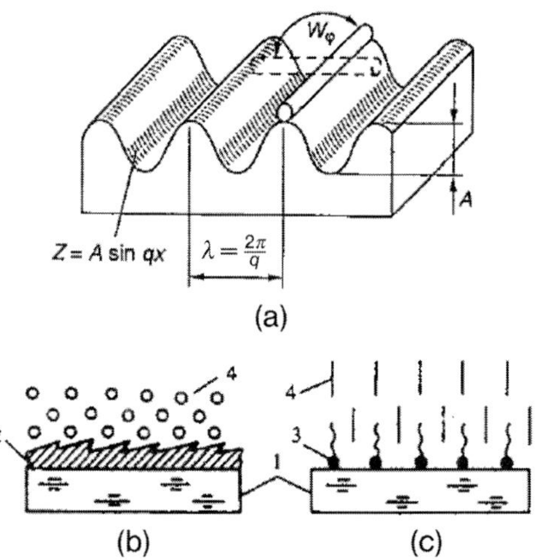

Figure 2.17 Mechanisms of planar (homogeneous) (a, b) and homeotropic (c) orientations of liquid crystals: (a) microrelief obtained by rubbing; (b) microrelief obtained by an obliquely evaporated thin film of metal; (c) orientation of liquid crystal molecules obtained by a surfactant: (1) substrate, (2) obliquely evaporated film, (3) surfactant, and (4) nematic liquid crystal. W_φ is the azimuthal anchoring energy [1].

It is very simple, but unfortunately it does not always provide a sufficiently strong anchoring of the director to the surface. Better results are given by the evaporation of metals or oxides (e.g., SiO) onto the surface at oblique incidence. The mechanism of planar orientation of nematic liquid crystals by means of an obliquely evaporated thin film of metal is illustrated in Figure 2.17b.

To explain why longitudinal ridges and troughs on the surface of the glass promote a planar orientation of a nematic liquid crystal, it is important to study the interaction of a nematic liquid crystal with such surfaces from the viewpoint of a minimum elastic energy [5]. Certain modifications in the Berreman's estimations were proposed recently, but there is still no evident experimental justification of the new approach [6], so we will not consider it in our book. For convenience, the shape of the cross section perpendicular to the ridges and troughs is taken as sinusoidal, i.e. $z = A \sin qx$:

$$\varphi(z = 0) = dz/dx = A\,q \cos qx, \tag{2.34}$$

where x is the direction perpendicular to the rubbing of the surface of the substrate, q is the wave vector of the surface structure, and A is its amplitude (Figure 2.17a). The excess of the elastic energy between the two configurations of the liquid crystal molecules parallel and perpendicular to the grooves (Figure 2.17a) can be calculated by minimizing the nematic liquid crystal elastic energy F in the half plane ($z > 0$) with the boundary conditions (2.34) in the form

$$F = \int_0^\infty g(z)\,\mathrm{d}z = \frac{1}{4} KA^2 q^3, \tag{2.35}$$

where K is elastic constant of the liquid crystal and $g(z)$ is defined from (2.15). Thus, the additional elastic energy is quadratic, depending on the depth of the relief (amplitude A) and inversely proportional to the cube of the period of the relief q.

For typical values of the depth and period of the relief ($A = 10$ Å, $2\pi/q = 200$ Å) and also of the elasticity modulus ($K = 10^{-11}$ N), the value of F is 8×10^{-5} J/m^2, which is quite reasonable. Thus, this model explains why it is more energetically favorable for the director of a nematic liquid crystal to be aligned along the ridges and troughs of a one-dimensional surface structure (planar orientation). If the relief is two dimensional (e.g., an etched surface), a similar investigation confirms the advantage of a homeotropic orientation.

Monomolecular Langmuir–Blodgett (LB) films have also been proposed as a means of achieving planar orientation [7]. The mechanism of planar orientation in this case has a "chemical" nature (dispersion and polar interaction between the molecules), although a contribution from a purely "elastic" (steric) mechanism is not excluded. Stable monolayers, in which the molecules have their long axes lying on the surface (as in certain organosilicon compounds), are also conducive to planar orientation in a nematic crystal. The quality of the LB film is high in microscopic regions, but this is difficult to obtain over macroscopic areas. This is a problem in practical production. The important point is how to decrease surface undulations. Langmuir–Blodgett films are interesting for research in surface material science, and the surfaces of the films have many unknown properties [7].

The photoaligning technique proves to possess visible advantages in comparison with the usually applied "rubbing" treatment of the substrates of the liquid crystal display cells. Photoalignment possesses obvious advantages in comparison with the usually "rubbing" treatment of the substrates of LCD cells. Possible benefits of using this technique include [8] the following:

1. Elimination of electrostatic charges and impurities as well as mechanical damage of the surface.
2. A controllable pretilt angle and anchoring energy of the liquid crystal cell, as well as its high thermal and UV stability and ionic purity.
3. Possibility of producing the structures with the required liquid crystal director alignment within the selected areas of the cell, thus allowing pixel division to enable new special LC device configurations for transflective, multidomain, 3D, and other new display types.
4. Potential increase of manufacturing yield, especially in LCDs with active matrix addressing, where fine tiny pixels of a high-resolution LCD screen are driven by thin film transistors on a silicone substrate.
5. New advanced applications of LC in fiber communications, optical data processing, holography, and other fields, where the traditional rubbing LC alignment is not possible due to the sophisticated geometry of the LC cell and/or high spatial resolution of the processing system.
6. Ability for efficient LC alignment on curved and flexible substrates.

Various mechanisms of photoalignment are described in detail in Ref. [8]. We can distinguish the following mechanisms of the photoalignment (Figure 2.18): (i) photochemical reversible *cis–trans* isomerization in azo dye-containing polymers, monolayers, and pure dye films; (ii) pure reorientation of the azo dye chromophore molecules or azo dye molecular solvates due to the diffusion under the action of polarized light; (iii) topochemical cross-linking in cinnamoyl side chain polymers; and (iv) photodegradation (photodestruction), for example, in polyimide materials. The perfect homogeneous planar and twisted liquid crystal alignment was achieved successfully in both glass and plastic substrates [8].

2.2.1.3 **Homeotropic Orientation**

There may be three mechanisms for perpendicular alignment as shown in Figures 2.17c and 2.19 [1, 7]. The first one is the amphiphilic material-assisted alignment, that is, amphiphilic material adsorbs perpendicularly to the polar surface and the liquid crystal aligns with respect to the amphiphilic material (Figures 2.17c and 2.19a). This amphiphilic material is sometimes an impurity contained in the liquid crystal or surfactant. The second mechanism is the perpendicular alignment obtained by a surface coupling agent, for example, silanes with long alkyl chains as shown in Figure 2.19b. The third mechanism is microscopic columnar structure-assisted alignment as shown in Figure 2.19c, and the alignment of the surface is obtained by SiO rotatively oblique evaporation [7].

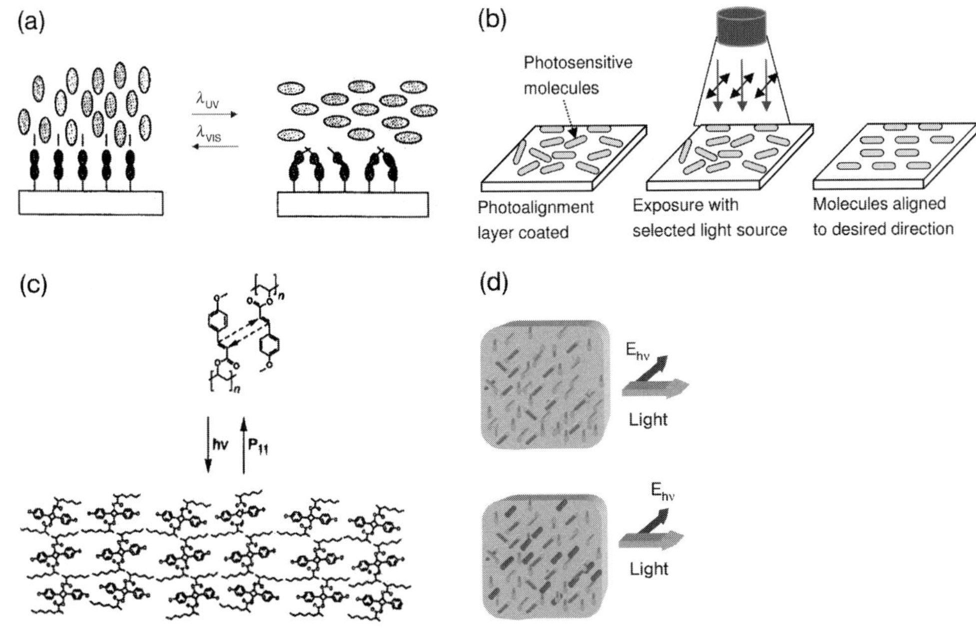

Figure 2.18 Mechanisms of liquid crystal photoalignment. (a) *cis–trans* isomerization of the "command layer" attached to the surface; (b) pure diffusion reorientation of the azo dye layer attached to the surface perpendicular to light polarization; (c) photo-cross-linking of the cinnamon fragments, leading to the preferred orientation of the aromatic rings; (d) photodegradation (photodestruction), that is, selective destruction of the molecular fragment parallel to the light polarization.

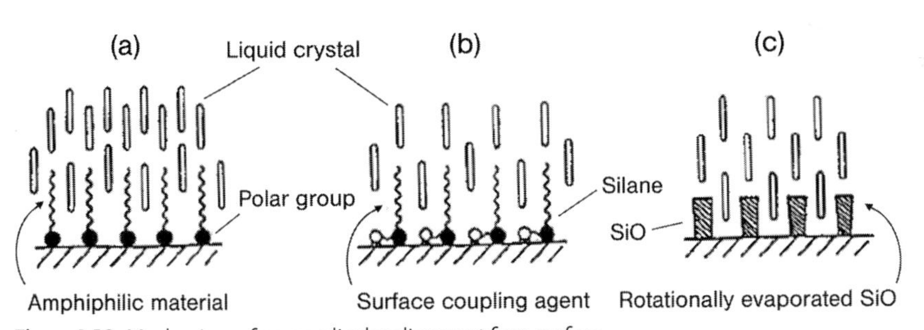

Figure 2.19 Mechanism of perpendicular alignment for a surface with strong polarity (a), for a surface treated with surface coupling agents (b), and for a SiO rotationally evaporated surface (c) [7].

Homeotropic orientation can also be produced by a photoaligning technique [8]. In this case, the sample with a photosensitive layer is illuminated by a normally incident nonpolarized light. The LC molecules tend to orient perpendicular to the light polarization vector, and consequently the only "allowed" preferred direction remains along the light propagation.

2.2.1.4 Tilted Orientation

Orientation of molecules at a given angle to the surface is achieved using layers of SiO produced by oblique evaporation at a very large angle (80–90°) between the normal to the surface and the direction to the source [7]. In this case, quasi-one-dimensional surface structure is achieved that is oriented at an angle $\alpha = \theta_0$ to the surface (Figure 2.16). This induces a tilt of the nematic liquid crystal molecules in the same direction.

Another method of pretilt angle generation includes a proper rubbing of polymer (polyimide) films. In this case, the chemical structure of the oriented nematic liquid crystals must be taken into consideration [9]. A very important thing is the presence of alkyl branches in a rubbed polyimide film [7, 9]. The absence of alkyl branches results in a low pretilt angle (~2°), polyimide films with a low density of alkyl branches lead to a medium pretilt angle (~5°), while a high pretilt angle (~20°) may be achieved for the high density of alkyl branches. The physical reason for this is evident. The presence of alkyl branches increases the tendency for a homeotropic orientation because they work on the polyimide surface in a manner similar to surfactants (see Figure 2.17c). Tilted orientation of the nematic liquid crystal molecules can also be achieved by using surfactants [1, 2]. The pretilt angle irreversibly increases with temperature due to the increase in the flexibility of the polymer side chains. The effect is more pronounced in case of larger dielectric constants of liquid crystals [7].

A high pretilt alignment (~20°) of liquid crystal molecules can be achieved by an oblique evaporation technique of SiO (Figure 2.17b). At present, the tilt angle produced by practical rubbing methods is about few degrees. Hence, a technique that induces higher tilt angles is required.

The idea of nanodomains and the mixing of vertical and homogeneous alignment polyimide materials together to achieve high pretilt liquid crystal angles was developed [10]. Instead of trying to achieve a homogeneous mixture of two different alignment materials, an inhomogeneous alignment layer consisting of homogeneous and homeotropic domains can be used to achieve the same goal (Figure 2.19). Since it is difficult to fabricate homogeneous mixtures of two different materials, the two materials were simply let to segregate and form domains. As long as the domains are small enough, the result will be a uniform pretilt angle for the liquid crystal molecules. It turned out that reliable high pretilt angles could be obtained (Figure 2.20).

Low and high liquid crystal pretilt angles can also be obtained by a photoalignment technique [8], in particular using a nonpolarized obliquely incident light (Figure 2.21).

Azo dye photoalignment materials can be used for the purpose. A subsequent thermopolymerization of the azo dye layers is necessary to stabilize liquid crystal alignment [8]. A high reliable pretilt angle can also be made by a nanostructured surface, similar to the one shown in Figure 2.20, where the rubbing operation is replaced by obliquely incident UV exposure to arrange a photodegradation process in a homogeneous polyimide. A reliable temperature-stable pretilt angle between 0° and 90° was demonstrated.

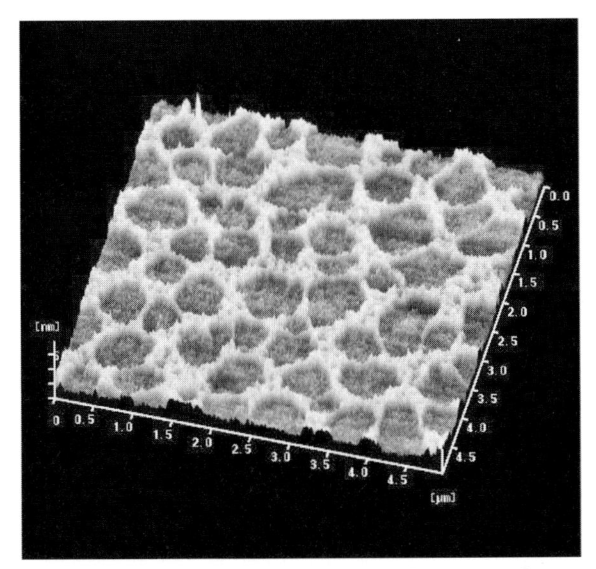

Figure 2.20 Nanostructured liquid crystal surface. Atomic force micrograph of the nanostructured surface. Light (dark) colored region corresponds to vertical (horizontal) polyimide, respectively [10].

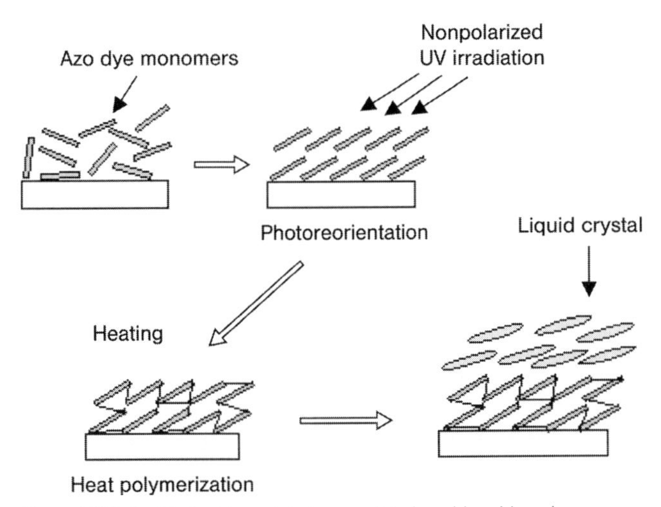

Figure 2.21 Pretilt liquid crystal alignment induced by obliquely incident nonpolarized light [8]. The thermopolymerization stabilized the photoalignment produced by azo dye layers.

2.2.1.5 **Other Types of Liquid Crystal Alignment**

Different types of liquid crystal orientations are shown in Figure 2.22 for nematic liquid crystals and in Figure 2.23 for cholesteric and smectic liquid crystals.

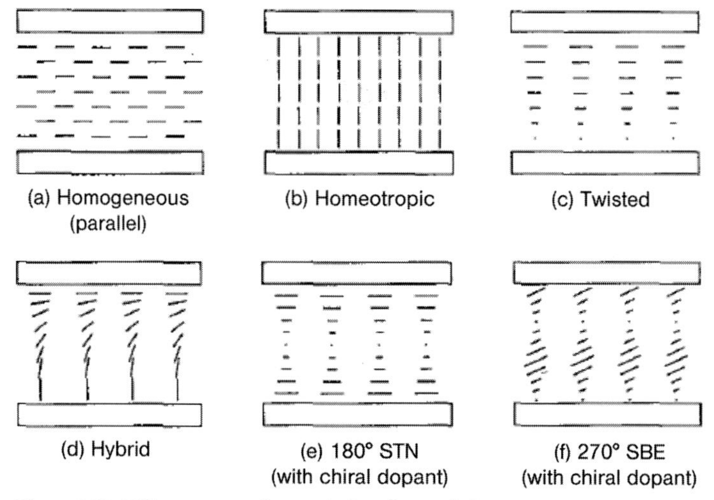

Figure 2.22 Different types of nematic liquid crystal alignments [1].

We have already described the planar (homogeneous), homeotropic, and tilted orientations. Other liquid crystal alignments interesting for applications are of a hybrid (planar to homeotropic transitions between the two substrates) or twisted type (Figure 2.22). We should know that it is impossible to create a twisted nematic liquid crystal orientation with a rotation angle more than 90°. Thus, a certain amount of a chiral dopant has to be added to provide the necessary twist angle of the nematic configuration (180° or 270° is most suitable for applications).

The alignment configurations of cholesteric liquid crystals are shown in Figure 2.23. When the director of a cholesteric liquid crystal is oriented along the surfaces of a cell,

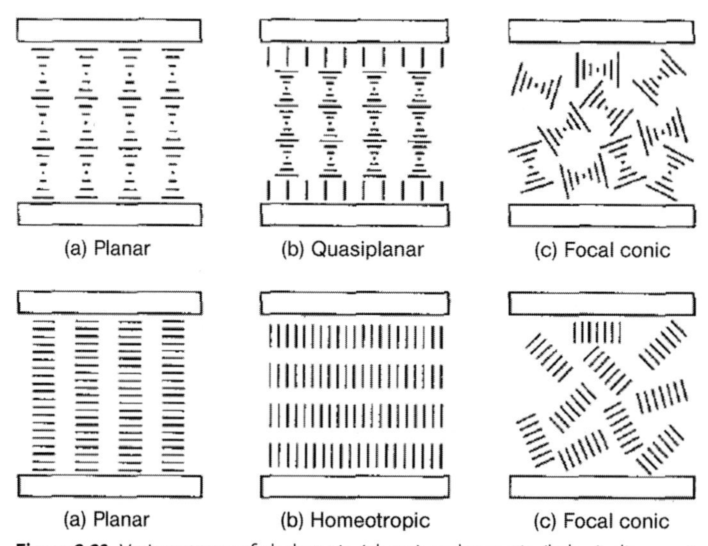

Figure 2.23 Various types of cholesteric (above) and smectic (below) alignments [1].

by special treatment, the axis of the helix is perpendicular to the glass surfaces. In this case, a transparent, optically active planar texture is obtained. When, on the other hand, the molecules of the cholesteric are oriented normal to the glass surfaces, the axis of the helix is forced to become parallel to the glass, although it can have an azimuthal orientation. In this case, the quasiplanar or "fingerprint" texture takes place. In the more general case, when directions of the helical axis in various parts of the cell are random along the azimuth and are not parallel to the glass surfaces, we have a polycrystalline sample, or the so-called focal conic texture (Figure 2.23). When opposite walls are treated in a different way (with the planar and homeotropic orientation at opposite boundaries), a hybrid cholesteric texture is obtained.

An alignment of ferroelectric liquid crystals (FLCs) and antiferroelectric liquid crystals (AFLCs) can be obtained by special means [1]. The most important factors to consider in order to realize a bistable switching in FLCs are (i) to avoid zigzag defects due to the smectic layers' tilt in two opposite directions; (ii) to produce an appropriate anchoring energy by rubbing the polymer aligning layer (small energy results in surface defects, while strong anchoring prevents the effect of bistable switching in an electric field [1, 2]); (iii) to obtain sufficiently high pretilt angles on the substrates to avoid the degeneration of FLC switching trajectory; and (iv) to prevent the accumulation of the surface charge near the boundaries, which diminish the effective switching field. The last problem was solved by orienting an FLC on the photoaligned substrates to diminish the number of the surface traps near the substrates [8]. Perfect electrooptical performance of the photoaligned FLC display with a memorized gray scale was demonstrated [8].

In AFLC cells, the optical axis is located nearly along the rubbing direction. Thus, the alignment is stable and is more easy to obtain than in the case of FLCs. The absence of the macroscopic polarization in the AFLC state, as well as a possibility to avoid surface defects, makes the surface AFLC orientation more simple to produce for display applications [1]. The polyimide photoaligned films were applied for the first time to align AFLC molecules, which needed a planar alignment [12]. Liquid crystal molecules were aligned perpendicular to the polarization direction of UV light. The AFLC cell showed some defects; however, the AFLC cell alignment was enhanced by controlling the cooling rate and post-UV treatment.

2.2.2 Surface Energy

The energy needed to deviate the liquid crystal molecules (the director) from the preferred orientation at the surface is called anchoring energy. The anchoring energy should not be mistaken with the energy of adhesion of the liquid crystal with the solid surface or the surface energy of the liquid crystal–solid interface, which are of the order of 0.2–0.4 J/m^2, that is, several orders of magnitude higher than the anchoring energy of the director reorientation at the surface (10^{-6}–10^{-3} J/m^2) [1]. We will define the anchoring energy in more detail.

Assume the interface is in the x, y-plane (see Figure 2.16) and the equilibrium position of the director (the easy direction) is defined by the polar θ_0 and azimuthal φ_0

angles. The easy direction at the nematic–solid interface is predetermined by a specific treatment of the solid surface. We can distinguish the homeotropic ($\theta_0 = \pi/2$), planar ($\theta_0 = 0$), and tilted ($0 < \theta_0 < \pi/2$) orientations. In turn, the planar orientation can be homogeneous (with a unique angle φ_0), heterogeneous (when several easy directions are possible at the crystalline substrate), or degenerate (all angles are equally probable and the cylindrical symmetry with respect to the surface normal exists). The same is true for the x–y projection of the director for the tilted orientation.

The anchoring energy shows how much energy one has to spend to deviate the director from the easy direction (where it is anchored). In the first approximation, the anchoring energy is expressed by the Rapini potential (see Figure 2.16) [1, 2]

$$W = -\frac{1}{2} W_0(\mathbf{n} \cdot \mathbf{d}). \tag{2.36}$$

As the director deviation in the polar and azimuthal plane came to be characterized by different anchoring energies, the terms "polar" and "azimuthal" anchoring energies were introduced. The corresponding contributions to the anchoring were written as (Figure 2.16)

$$W_\theta = \frac{1}{2} W_{\theta 0} \sin^2(\theta - \theta_0), \quad W_\varphi = \frac{1}{2} W_{\varphi 0} \sin^2(\varphi - \varphi_0). \tag{2.37}$$

However, (2.37) does not provide an adequate description of the anchoring phenomenon in the case when the liquid crystal alignment angle is high; so, instead of (2.37), the more general expression is preferred in the form of the special series, for example, Legendre polynomials [13]:

$$W(\theta) = -\frac{1}{2}\sum_{n=1}^{\infty} A_{2n} P_{2n}(\cos\theta) = -\frac{1}{2} W_0 \cos^2\theta - \frac{1}{2}\sum_{n=2}^{\infty} W_{2n} \cos^{2n}\theta. \tag{2.38}$$

The shape or the Rapini anchoring potential is shown in Figure 2.24. However, some experimental data may be fit better with other shapes of the surface potential. For example, the elliptic sine shape

$$W = \frac{1}{2} W_0 \sin^2(\theta^{-}\theta_0, k), \quad 0 \le k \le 1 \tag{2.39}$$

has been predicted in Ref. [14], which is reduced to the Rapini shape only for a special value ($k = 0$) of the modulus of the elliptic function ($0 \le k \le 1$). This function can describe a very sharp minimum of the surface potential near the easy direction (see Figure 2.24, curve 2). Usually, the second term in the series (2.38) proves to be sufficient for the adequate description of the experimental phenomena [15]. In Figure 2.24, a particular form of the potential (2.38) is also shown (curve 3). In any case, once the form $W(\theta)$ is chosen, the problem of searching the director field in a layer of a finite thickness can be solved at the macroscopic level by solving the variation problem of the minimization of a total value of the liquid crystal bulk and surface energy [1, 2]. More detail description of the methods for the determination of the liquid crystal anchoring energy will be provided in Chapter 5.

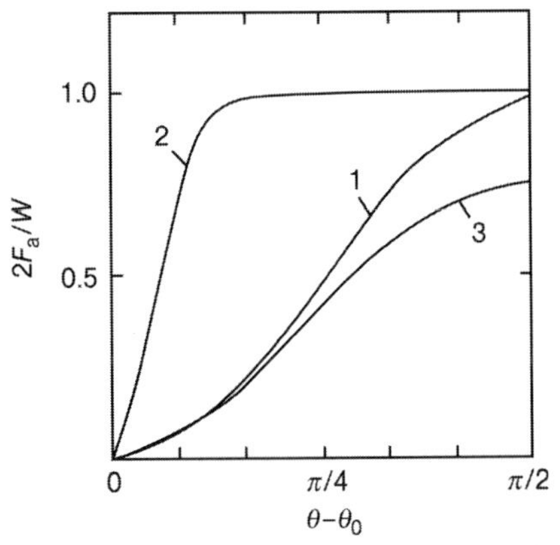

Figure 2.24 Surface anchoring potential as a function of the deviation angle (from the preferred orientation at the surface: (1) "Rapini" potential (2.37), (2) the elliptic sine potential (2.39), and (3) Legendre expansion (2.38) with the two first terms [15].

2.3
Liquid Crystals Under Magnetic and Electric Fields

The action of the electric field on the liquid crystal layer results in the deformation of the initial molecular (director) distribution and a subsequent variation of the LC cell optical properties [1, 2]. Director **n** reorients in an electric field under the action of the dielectric torque, which is proportional to the dielectric anisotropy $\Delta\varepsilon$ (2.8). The corresponding contribution g_ε to the density of the nematic free energy (2.15) gives [1]

$$g_\varepsilon = -\frac{\mathbf{DE}}{2} = -\frac{\varepsilon_0\varepsilon_\perp E^2}{2} - \frac{\varepsilon_0\Delta\varepsilon(\mathbf{En})^2}{2}, \tag{2.40}$$

that is, director **n** tends to align itself along the field ($\mathbf{n}||\mathbf{E}$) if $\Delta\varepsilon > 0$ and is perpendicular to it ($\mathbf{n}\perp\mathbf{E}$), provided that $\Delta\varepsilon < 0$.

The effect of a magnetic field on the director orientation comes from the corresponding interaction energy density:

$$g_h = -\frac{1}{2}\mu^{-1}\Delta\chi(\mathbf{nB})^2, \tag{2.41}$$

where $\Delta\chi = \chi_{||} - \chi_\perp$ is the diamagnetic anisotropy ($\Delta\chi > 0$ in typical nematics).

The elastic torque supports the initial director orientation, fixed by the boundary conditions on the surface (S), which in case of a strong surface anchoring is

$$\mathbf{n}|_S = \mathbf{no}.$$

As a result, a compromised director profile appears that satisfies the condition of the minimum free energy:

$$F_V = \int_V (g_k + g_\varepsilon + g_H)\mathrm{d}\tau, \qquad \mathbf{n}|_S = \mathbf{no}, \tag{2.42}$$

where $g_k = 1/2[K_{11}(\text{div}\ \mathbf{n})^2 + K_{22}(\mathbf{n}\ \text{rot}\ \mathbf{n})^2 + K_{33}(\mathbf{n} \times \text{rot}\,\mathbf{n})^2]$ is the elastic energy density and K_{ii} denote the elastic moduli; $q_0 = 2\pi/R_0$ characterizes the "natural chirality" and is equal to zero in pure nematics.

In a more general case of a finite director anchoring at the boundaries, we have to write the total energy F of liquid crystals as follows:

$$F = F_v + F_s, \tag{2.43}$$

where F_s is the surface energy, $F_s = W$ (2.36)–(2.39). The electrooptic response in case of a weak boundary anchoring is a special case. By attaining the minimum of the nematic free energy (2.42) or (2.43), it is possible to derive the equilibrium director distribution in a static case. To find the response times, we have to solve the equations of nematodynamics in the electric field. The corresponding analysis shows that the director reorientation is always accompanied by the macroscopic flow, the so-called backflow. The only exclusion is the pure twist rotation of the director (see Chapter 6). Backflow considerably affects the characteristic times of the electrooptical effects in uniform structures, especially in the case of strong deformations of the initial director orientation [1, 2].

On the basis of the director distribution, we can derive the electrooptical response of a nematic liquid crystal cell (such as birefringence), rotation of the polarization plane of the incident light, total internal reflection, absorption, or some other important characteristics of the cell. Basic electrooptical liquid crystal modes and the role of viscoelastic parameters for the optimization of flat panel display configurations will be discussed in Chapter 6.

References

1 Chigrinov, V.G. (1999) *Liquid Crystal Devices: Physics and Applications*, Artech House.

2 Blinov, L.M. and Chigrinov, V.G. (1994) *Electrooptical Effects in Liquid Crystalline Materials*, Springer, New York.

3 Chigrinov, V.G., Hasebe, H., Takatsu, H., Takada, H., and Kwok, H.S., (2008) Liquid-crystal photoaligning by azo dyes. *J. SID* **16/9**, pp. 897–904.

4 de Jeu, W.H. (1980) *Physical Properties of Liquid Crystalline Materials*, Gordon and Breach, New York.

5 Berreman, D.W. (1972) Solid surface shape and the alignment of an adjacent nematic LC. *Phys. Rev. Lett.*, **28**, 1683.

6 Fukuda, J., Yoneya, M., and Yokoyama, H. (2007) Surface-groove-induced azimuthal anchoring of a nematic liquid crystal: Berreman's model reexamined. *Phys. Rev. Lett.*, **98**, 187803.

7 Uchida, T. and Seki, H. (1991) Surface alignment of liquid crystals, in *Liquid Crystal Applications and Uses* (ed. B. Bahadur), World Scientific, Singapore.

8 Chigrinov, V.G., Kozenkov, V.M., and Kwok, H.S. (2008) *Photoalignment of Liquid Crystalline Materials: Physics and Applications*, John Wiley & Sons, Inc., New York.

9 Myrvold, B. and Komdo, K. (1995) The relationship between chemical structure of nematic liquid crystals and their pretilt angles. *Liq. Cryst.*, **18**, 271.

10 Kwok, H.S. and Yeung, F.S.Y. (2008) Nano-structured liquid crystal alignment layers. *J. Soc. Inform. Display*, **16**, 911.

11 Ho, J.Y.L., Chigrinov, V.G., and Kwok, H.S. (2007) Variable liquid crystal pretilt angles generated by photoalignment of a mixed polyimide alignment layer. *Appl. Phys. Lett.*, **90**, 243506.

12 Hyun, S.-Y., Moon, J.-H., and Shin, D.-M. (2004) Application of photoalignment technologies to antiferroelectric liquid crystal cell. *Mol. Cryst. Liq. Cryst.*, **412**, 369.

13 Barbero, G., Madhusudana, N.V., and Durand, G. (1984) Anchoring energy of nematic liquid crystals: an analysis of the proposed forms. *Z. Naturforsch.*, **39**, 1066.

14 Barnik, M.I., Blinov, L.M., Korkishko, T.V., Umanskiy, B.A., and Chigrinov, V.G. (1983) Investigation of NLC director orientational deformations in electric field for different boundary conditions. *Mol. Cryst. Liq. Cryst.*, **99**, 53.

15 Yokoyama, H. and van Sprang, H.A. (1985) A novel method for determining the anchoring energy function at a nematic liquid crystal wall from director distortions at high fields. *J. Appl. Phys.*, **57**, 4520.

3
Flows of Anisotropic Liquids

In this chapter, we will focus on linear and nonlinear phenomena arising from liquid crystal flows. In accordance with the global idea of the book, we will pay attention mainly to the effects that are interesting from a viewpoint of practical applications. We will start with a general description of Couette and Poiseuille shear flows of liquid crystals and emphasize their difference from the similar isotropic liquid flows. After that, a rheological behavior in steady shear flows will be described in detail. The main attention will be paid to hydrodynamic instabilities that have no analogues in isotropic liquid flows. We will show the key role of boundary orientation at nonlinear phenomena in liquid crystal flows. In particular, the specific primary instability, an escape of a director from the flow plane, which arises in homeotropic samples of nematics under Poiseuille flows, will be described in detail. The interaction of shear flows and electric fields will be under consideration too. After that, we will go to the more complicated case of oscillatory flows. We will present a description of the experimental setups and results obtained for linear and nonlinear regimes of a director motion. Specific phenomena such as secondary roll instability and long-living domains induced by intensive shear flows will also be considered. Finally, we will analyze the influence of weak surface anchoring described in terms of an anchoring strength and surface viscosity on linear and nonlinear flow phenomena in liquid crystals. The information presented below will be used in Chapter 7 elaborating on liquid crystal sensors of mechanical perturbations.

3.1
Couette and Poiseuille Flows in Isotropic Liquids and Liquid Crystals

Nematic liquid crystals can be considered as *anisotropic fluids* with the same mechanism of translational molecular motions as in isotropic *liquids*. So such property as high *fluidity* referred to zero value of static shear elastic modules is similar in both cases. The main difference arises due to a long-range orientational order described in terms of a tensor order parameter Q_{ij}. Usually, in nematic phase, the tensor order parameter is expressed by two independent characteristics, the scalar order parameter S and the director $\mathbf{n}$. The latter as an additional *hydrodynamic parameter* has to be incorporated into hydrodynamic equations of nematic liquid

Liquid Crystals: Viscous and Elastic Properties
S. V. Pasechnik, V. G. Chigrinov, and D. V. Shmeliova

ISBN: 978-3-527-40720-0

crystals, presented in Chapter 2. The connection between translation motions and the orientational ones described accordingly in terms of a velocity field **v**(**r**, *t*) and a director field **n**(**r**, *t*) can be considered as a *fundamental physical property of liquid crystals*. It is responsible for a number of specific phenomena existing in liquid crystal media. Some of them, such as various *instabilities* induced by electric fields in thin layers of nematic liquid crystals or backflow effects arising at the turning off electric fields are of great practical importance (Chapter 6).

The connection between **v**(**r**, *t*) and **n**(**r**, *t*) also results in a number of linear and nonlinear phenomena in flows of liquid crystals. In general, liquid crystals show non-Newtonian rheological behavior. It means, for example, that apparent shear viscosity depends on the shear rate. At the same time, strong magnetic (electric) fields allow to stabilize orientational structure. In thin layers of LC, the proper surface treatment can also provide the given orientation (e.g., planar or homeotropic, see Chapter 2). Liquid crystal with a stabilized orientational structure can be considered a *conventional Newtonian fluid* with shear viscosity determined by a field direction [1] or by a surface-induced direction. Such rheological behavior has to be taken into account at viscometric studies of liquid crystals. It will be described in detail in Chapter 5.

A number of interesting flow-induced phenomena in smectic A and C phases are out of scope of this book. A reader can find description of such effects in Oswald and Pieranski [2] and in original papers [3, 4]. We also do not consider rheological behavior of polymeric liquid crystals, lyotropic liquid crystals, and active liquid crystals. The last two classes of LC materials are of interest for understanding very complicated phenomena existing in living nature. In particular, active liquid crystals show unusual rheological properties that demand a special consideration [5–7]. Nevertheless, they are of rare technical application in modern industry.

In most cases considered here, the set of standard hydrodynamic equations of incompressible nematics presented in Chapter 2 will be enough to describe various phenomena arising from liquid crystal flows. Nevertheless, such hydrodynamic description fails in some particular cases, such as the formation of defect structures under intensive flows. To solve such problems, generalized mesoscopic or semimicroscopic theories were proposed in the past decade. The former introduce into hydrodynamics the tensor order parameter [8, 9], or the changeable scalar order parameter [10–12], instead of, or addition to, a director. A Fokker–Planck equation is used at semimicroscopic approach derived by Hess [13, 14]. We will mention some of the approaches at description of particular problems.

Two basic types of shear flows, the Couette flow and the Poiseuille flow, are usually considered in rheology of liquid crystals. They are illustrated in Figure 3.1 for the two dimensional geometry.

From the point of view of general rheology (see Ref. [15]), these flows belong to the so-called controllable (Couette) and partly controllable (Poiseuille) viscometric flows. All viscometric flows can be considered as a number of sublayers moving steadily as rigid bodies and occupying the same position in space at all times (a laminar motion). The simplest cases of plane flows are shown in Figure 3.1. These sublayers are plane, at least at low shear rates $s = dv_x/dz$. Usually, a hypothesis of nonslipping ($v_x = 0$ at $z = 0$ and h) is applied for a velocity on a solid surface. It results in velocity profiles

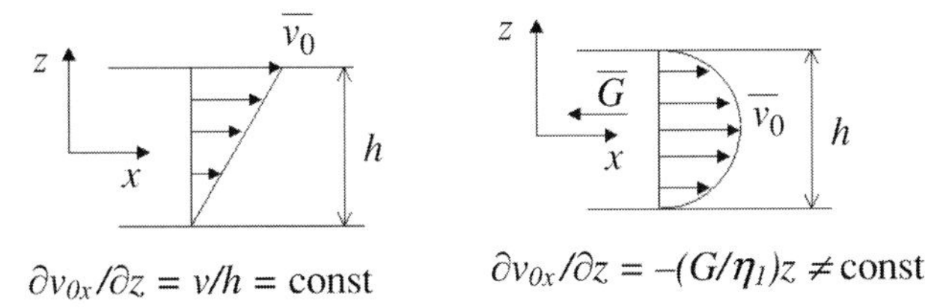

(a) Couette flow (b) Poiseuille flow

Figure 3.1 Basic types of shear flows.

shown in Figure 3.1. Relative motion of sublayers is accompanied by the energy dissipation described in terms of friction forces F_s acting between adjusting layers. In the simplest case, the friction force is proportional to the velocity gradient and is governed by Newton's law:

$$F_s = \eta S \frac{dv_x}{dz}, \tag{3.1}$$

where S is the square of contacted surfaces and η is the shear viscosity coefficient. It is worthwhile to note that on a microscopic scale the friction forces in liquids arise due to the transport of the momentum by diffusing molecules in the direction normal to the sliding planes. It is quite different from molecular mechanisms responsible for the energy dissipation at relative motion of solids.

The fluids that follow the law (3.1) are called Newton's liquids. The shear viscosity coefficient of such objects remains constant at variation of the shear rate at least for laminar flows of different types. A lot of simple isotropic and homogeneous liquids such as water, alcohol, glycerol, and so on show Newtonian behavior. At the same time, it does not hold for complex liquids, such as polymer solutions, emulsions, suspensions, and so on, where the measured viscosity was found to depend on the shear rate. A number of rheological phenomena such as "shear thinning" (decreasing of an effective viscosity with a shear rate) were established for flows of non-Newtonian liquids. A detailed description of such effects is out of scope of this book and can be found elsewhere (see, for example, Ref. [15]).

Liquid crystals also belong to the general class of non-Newtonian liquids. So some common rheological effects, "shear thinning," for example, were established for them too. At the same time, the most phenomena found in linear and nonlinear shear flows of liquid crystals are very specific. In particular, the mechanisms of the emergence of hydrodynamic instabilities are quite different for isotropic and anisotropic liquids and are described in terms of two different dimensionless parameters – the Reynolds number *Re* (isotropic Newtonian liquids) and the Ericksen number *Er* (nematic liquid crystals). These parameters are defined as [16]

$$\mathrm{Re} = \frac{\rho l v}{\eta} \tag{3.2}$$

$$Er = \frac{\eta l}{\rho D}, \tag{3.3}$$

where ρ is the liquid density, l is the characteristic length for the given flow, and v is the characteristic velocity of the flow, and the orientation diffusivity

$$D = \frac{K}{\gamma_1}, \tag{3.4}$$

The Reynolds number is referred to the well-known *hydrodynamic instability*, namely, to the *turbulence*, which is characterized by a number of vortexes and an additional viscous dissipation of energy.

The Erickson number defines the flow velocity needed to produce orientational distortions in liquid crystals layer. The estimates made for typical values of material constants of LCs [16] have shown that this velocity is about six order smaller than the velocity at which the turbulence occurs. So the latter phenomenon does not play any role in the description of hydrodynamic instabilities of liquid crystals presented below.

3.2 Hydrodynamic Instabilities in Couette and Poiseuille Steady Shear Flows

One can find a lot of original papers and some reviews [16, 17] with a description on hydrodynamic instabilities in liquid crystals. We will shortly describe the most interesting results for Couette and Poiseuille flows. The special attention will be paid to the specific instabilities arising under Poiseuille flows at homeotropic boundary orientation, which is important for LC sensor elaboration (see Chapter 7).

The general classification of instabilities in accordance with time and spatial characteristics includes:

1. Homogeneous instabilities arising as spatially homogeneous changes of orientation throughout the entire layer of a liquid crystal.
2. Spatially periodical instabilities with a regular dependence of orientation on coordinates that are visualized as domain structure with the unique direction of domains and with a period close to the layer thickness (or smaller).
3. Long scaled instabilities with characteristic sizes of nonhomogeneous regions essentially exceeding the layer thickness.
4. Nonstationary instabilities referred to the relatively slow and periodic time variations of nonequilibrium structure.

Below we will focus on the two first types of instabilities.

Usually, the theoretical analysis of the problem depends essentially on the type of *basic state* describing the nondisturbed fields of hydrodynamic parameters. The simplest case of basic state corresponds to the uniform orientation n_0 of a liquid crystal. It can be achieved by two ways.

It is well known that in most cases intensive stationary flows (with high values of a shear rate s) produce the stable orientation of a director in the flow plane at the angle

$$\theta_{\text{flow}} = \arctan\sqrt{\alpha_3/\alpha_2} \tag{3.5}$$

relative to the flow velocity **v** [16]. It holds for particular flows of different types [18] and all over the layer excluding for the thin boundary layers with the thickness $\xi(s)$ defined as [19]

$$\xi^2(s) = \frac{K}{2s\sqrt{\alpha_2\alpha_3}}, \tag{3.6}$$

where K and η are the effective values of a Frank elastic module and a shear viscosity. For the typical values of $K \sim 10^{-11}$ N, $\eta \sim 0.1$ Pa s we will obtain $\xi(s) \sim 10\,\mu$m at $s \sim 1\,\mathrm{s}^{-1}$. It means that the rheological behavior of relatively thick layers of liquid crystals aligned by intensive flows does not depend on the boundary conditions, which is in accordance with the data of shear viscosity measurements. The rare exclusions correspond to the change of α_3 sign (the so-called nonaligned nematics) that takes place in the vicinity of nematic–smectic A phase transition.

3.3 Steady Flows of Liquid Crystals

3.3.1 Homogeneous Instability at Initial Planar Orientation Normal to the Flow Plane

The second possibility to get the uniform orientation n_0 in the basic state is the usage of particular type of boundary orientation, namely, the planar orientation directed normally to the flow plane. In this case, the basic state at small values of the shear rate s corresponds to the linear (parabolic) velocity profile for Couette (Poiseuille) flow with homogeneous planar orientation throughout the sample. At the shear rate increasing up to some critical value s_c, the *homogeneous instability* connected with the reorientation of a director in the plane of the layer was observed both for Couette and for Poiseuille flows [20–23].

The physical mechanism for the explanation of such an effect proposed by Guyon and Pieranski [20, 21] is illustrated in Figure 3.2.

In accordance with it, the time variation of some spontaneous initial fluctuation of a director $\delta \mathbf{n}_z = \mathbf{n}_1 - \mathbf{n}_0$ can be increased or decreased by a flow depending on the signs of Leslie coefficients α_2 and α_3. For a particular situation shown in Figure 3.2, the basic flow with a constant shear rate $s = dv_x/dz$ induces the viscous torque $\Gamma_z = -\alpha_2 s \delta n_z$ acted on the director. This torque is proportional to δn_z and is positive for $\alpha_2 < 0$ (which is correct due to thermodynamic arguments). The torque Γ_z results in the rotation of a director around Z-axis to a new position $\mathbf{n}_2 = \mathbf{n}_1 + \delta \mathbf{n}_x$ as shown in Figure 3.2. As a result, the second viscous torque $\Gamma_x = -\alpha_3 s \delta n_x$ induces a rotation of a director around X-axis. For $\alpha_3 < 0$ (flow aligned nematics), this rotation leads to an increase in the initial director fluctuation that means the arising of instability. For $\alpha_3 > 0$ (nonaligned nematics), the initial fluctuation is suppressed and the system becomes stable relative to the mechanism mentioned above. This mechanism is valid for a Poiseuille flow too [22].

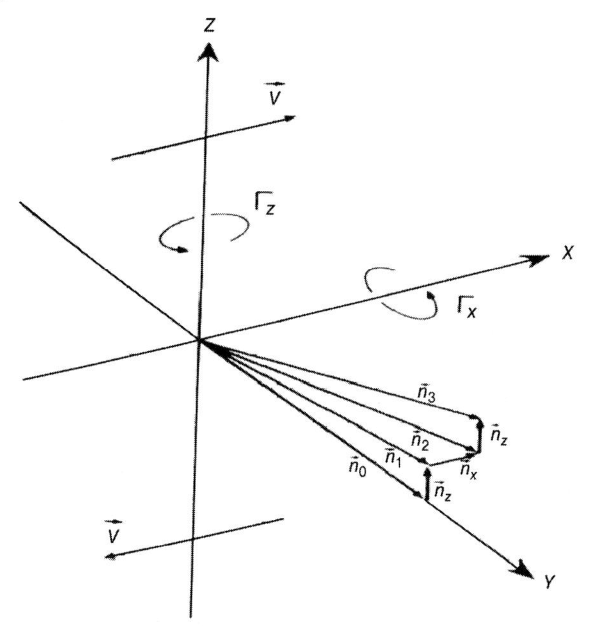

Figure 3.2 Mechanism for homogeneous instability in Couette flow. (After Guyon and Pieranski [20, 21].)

Analytic solutions for a threshold shear rate s_c corresponding to the instability under consideration can be obtained in the framework of simplified model referred to a steady flow with a constant shear rate s [21]. It can be considered a good approximation for a quasisteady reversal flow realized in the experiment. The corresponding equations are expressed as:

$$-\mu_0\,\Delta\chi H^2 n_z + K_{11}\frac{d^2 n_z}{dz^2} = \alpha_3 s n_y \tag{3.7}$$

$$-\mu_0\,\Delta\chi H^2 n_y + K_{22}\frac{d^2 n_y}{dz^2} = \alpha_2 s n_z, \tag{3.8}$$

where n_y and n_z are the components of director fluctuations. The solution with the minimal rate of variation, which corresponds to the threshold, is written as

$$\frac{n_z}{n_{z0}} = \frac{n_y}{n_{y0}} = \cos\left(\frac{\pi z}{d}\right). \tag{3.9}$$

Substituting (3.9) into Equations 3.7 and 3.8 gives rise to the system of uniform algebraic equations that has nontrivial solutions at zero value of the determinant. It results in the next equation for the critical shear rate s_c corresponding to the hydrodynamic instability:

$$\alpha_3\alpha_2 s_c^2 = \left(\frac{K_{11}\pi^2}{h^2} + \mu_0\,\Delta\chi H^2\right)\left(\frac{K_{22}\pi^2}{h^2} + \mu_0\,\Delta\chi H^2\right). \tag{3.10}$$

The expressions for weak and strong fields can be easily obtained from (3.10) as

$$s_c = \frac{\pi^2}{h^2}\sqrt{\frac{K_{11}K_{22}}{\alpha_2\alpha_3}}, \qquad H = 0, \tag{3.11}$$

$$s_c = \frac{\mu_0\,\Delta\chi H^2}{\sqrt{\alpha_2\alpha_3}}, \qquad H \gg \frac{\pi}{h}\sqrt{\frac{K_{ii}}{\mu_0\,\Delta\chi}}. \tag{3.12}$$

The estimates made in accordance with (3.10)–(3.12) for 200 μm layer of MBBA were found to correspond reasonably to the experimental data. In particular, the calculated threshold shear rate s_c for zero field (0.05 s^{-1}) was close to the experimental value.

One of the essential consequences of Equation 3.11 is a decrease in the critical shear rate on thickness. This result is valid for a Poiseuille flow too and makes possible to use wedge-like cells for the visualization of hydrodynamic instabilities, as it will be shown here. On the contrary, for electrically induced instabilities, such as Fréedericksz transitions, the critical voltage does not depend on the thickness at least for strongly anchoring surfaces.

A detailed hydrodynamic description of a homogeneous instability for the particular boundary conditions mentioned above was provided by Manneville and Dubois-Violette both for Couette [24] and for Poiseuille [25, 26] flows. The obtained results induce some corrections in preliminary estimates [21] for threshold value s_c of the shear rate. In particular, this parameter for Couette flow in the absence of external (magnetic or electric) fields can be expressed by the approximate expression

$$s_c h^2 = \pi^2\sqrt{\frac{K_{11}K_{22}\eta_1}{\alpha_2\alpha_3\eta_3}}. \tag{3.13}$$

This expression predicts the decrease in the critical shear rate with a layer thickness as h^{-2}.

A deep insight into the problem including the account of nonlinear effects just above the threshold of the emergence of a homogeneous instability was provided by Manneville [27]. In particular, he pointed out the strong parallel between the Landau–Hopf (L–H) bifurcation theory traditionally used for the description of hydrodynamic instabilities and more universal Landau–Khalatnikov (L–H) theory of time-dependent critical phenomena taking place at phase transitions. The latter is based on the motion equation for the order parameter η

$$\frac{d\eta}{dt} = -\Lambda\frac{\partial F}{\partial \eta}, \tag{3.14}$$

where Λ is a kinetic coefficient, F is the free energy that can be expressed as a power expansion relative to the order parameter. In particular, in the vicinity of the second order transition, it takes the form

$$F = F_0(T_0) + \alpha(T-T_0)\frac{\eta^2}{2} + \gamma\frac{\eta^4}{4}, \tag{3.15}$$

where T_0 is the critical temperature, α and γ denote the parameters that are considered to be independent of temperature. It is simple to show that the presented kinetic equation for η has the same form as the phenomenological motion equation for the amplitude of the parameter A referred to as the distorted state in L–B theory:

$$\frac{\mathrm{d}A}{\mathrm{d}T} = \alpha A - \gamma A^3. \tag{3.16}$$

To adopt the Landau–Khalatnikov theory for the description of the homogeneous hydrodynamic instability, one has to replace the thermodynamic parameter (temperature T) in the expansion (3.16) with the hydrodynamic parameter (shear rate s) and the order parameter η with the angle φ of a director rotation. It makes possible to use the well-known results previously obtained from the theory of critical phenomena. In particular, the dependence of the order parameter (azimuthal angle φ) on the shear rate just above the transition follows the power law $\varphi \sim (s - s_c)^\beta$ with a mean field value of the critical exponent $\beta = 1/2$. Such behavior holds everywhere except the extremely narrow critical region where fluctuations play a dominant role. It is worthwhile to note that the square-root behavior mentioned above is in accordance with experimental results, although it is restricted to a relatively narrow neighborhood of the threshold ($s < 1.5s_c$).

Magnetic (H) field applied along the initial planar orientation (n_0) or electric field (E) oriented along the layer normal provide the shift of the homogeneous instability threshold. In this case, the critical shear rate can be estimated approximately by expression [24]

$$(K_{11}q_z^2 + \mu_0\,\Delta\chi H^2 - \varepsilon_0\,\Delta\varepsilon E^2)(K_{22}q_z^2 + \mu_0\,\Delta\chi H^2) = s^2\alpha_2\alpha_3\eta_3/\eta_1, \tag{3.17}$$

where $q_z = \pi/h$ is the wave number. For nematics with a negative value of dielectric permittivity anisotropy, both magnetic and electric fields have to suppress the instability. According to (3.17), in the case of strong fields, the critical shear rate does not depend on the thickness and is proportional to the squared field strength. These conclusions were also confirmed by experiments [28] with a Couette flow.

The homogeneous instability at the same boundary conditions was also observed for Poiseuille flow in the experiments of Guyon and Pieranski [22] and Janossy *et al.* [23]. In this case, four possible types of orientational distortions shown in Figure 3.3 were identified [22].

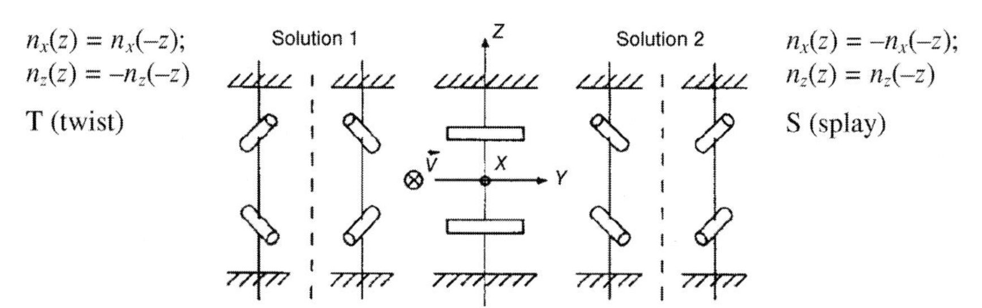

Figure 3.3 Four possible types of orientational distortions under Poiseuille flow. (After Guyon and Pieranski [22].)

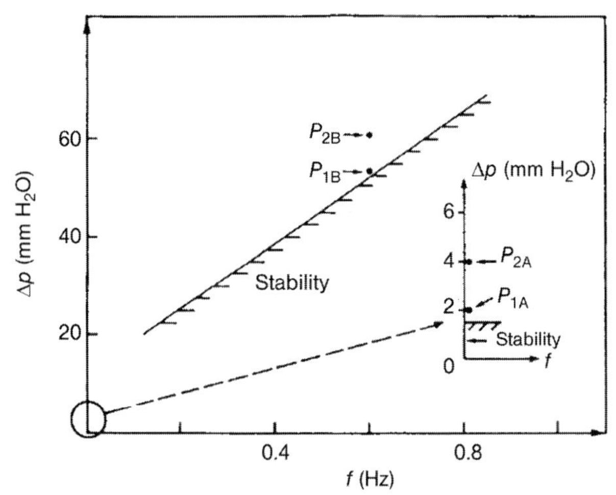

Figure 3.4 Critical difference of applied pressure ΔP_c as function of the inversion frequency; P_{1A}, P_{1B} – the threshold for a homogeneous instability, P_{2A}, P_{2B} – the threshold for a nonhomogeneous periodical instability. (After Guyon and Pieranski [22].)

A quasistationary Poiseuille flow was realized in the capillary (200 μm of thickness and 80 mm in length) via a rather complicated system that provided low-frequency (10^{-3}–10 Hz) reversals of the direction of a flow while the module of pressure difference producing a flow was maintained constantly during intervals between reversals. The stability diagram for this flow is shown in Figure 3.4.

It is worthwhile to note that a homogeneous instability was observed as a number of big size domains separated by domain walls orthogonally oriented to the flow direction. It was found that the application of a stabilizing magnetic field suppressed the instability. The same effect was observed at an increase in the inversion frequency. So the threshold pressure difference could be increased 10–50 times.

At the preliminary theoretical estimates of the presented results, the shear flow was considered as the stationary one [22, 23]. In the absence of fields, the threshold for a homogeneous instability is defined by the dimensionless Ericksen number [18]

$$Er = \frac{\Delta P/\Delta x}{\eta_3}\sqrt{\frac{\alpha_2\alpha_3}{K_{11}K_{22}}}\left(\frac{h}{2}\right)^3 = \bar{s}_c\sqrt{\frac{\alpha_2\alpha_3}{K_{11}K_{22}}}\left(\frac{h}{2}\right)^2, \tag{3.18}$$

where $G = \Delta P/\Delta x$ is the value of a pressure gradient, $\bar{s}_c$ is the averaged value of the shear rate module. The estimates made for a case of MBBA had shown that corresponding critical values of the Ericksen number (Er_c) and the pressure gradient (G_c) are equal to 12.8 and 16.5 kPa/m. The latter value is slightly lower than the experimental one (24.5 kPa/m). More rigorous theoretical solution of the problem was performed by Dubois-Violette and Manneville [24]. The theoretical value for $Er_c = 17.3$ turned out to be much closer to the experimental result. The threshold for T-deformation was found lower than that for S-deformation. So, the first one had to be realized under an increasing pressure gradient that is also in agreement with experimental results.

The influence of magnetic field on the threshold of a homogeneous instability was considered in Ref. [26]. In the case of a strong stabilizing field, it can be estimated from the expression

$$Er_c = F + bF^{2/3}, \tag{3.19}$$

where $F = (h/2)^2 \mu_0 \, \Delta\chi H^2/K$, $b = 2.55$. The estimates made via (3.19) showed a good agreement with the experimental results [23]. Contrary to the case of a simple shear flow, the nonhomogeneous long scaled periodic instability was not induced by a magnetic field at low-frequency reversals of the flow direction.

3.3.2
Periodic Instability at Initial Planar Orientation Normal to the Flow Plane

The homogeneous instability described above was not found for Couette flow under strong stabilizing magnetic fields. Instead, the periodic instability was observed as a system of rolls with long axes parallel to the flow direction and with a period close to the layer thickness [21]. The expression for a threshold shear rate was obtained for one constant approximation by Galerkin method as

$$s^2_{\text{roll}} = \frac{(\mu_0 \, \Delta\chi H^2 + Kk_x^2 + Kk_z^2)}{\alpha_3\alpha_2(1+g)} = s^2_{\text{gom}} \frac{[1 + Kk_x^2/(\mu_0 \, \Delta\chi H^2 + Kk_z^2)]}{1+g}, \tag{3.20}$$

where g is a parameter defined by Leslie coefficients and estimated as 3–3.5, k_i is the wave vector component. Expression (3.20) predicts a decrease in the ratio $r = s_{\text{roll}}/s_{\text{gom}}$ of threshold shear rates for the periodic (s_{roll}) and homogeneous (s_{gom}) instabilities with an increase in the magnetic field strength H. At $r < 1$, the periodic instability becomes preferable, which was observed experimentally. Numerical calculations performed by Dubois-Violette and Manneville [25, 28] have shown a good agreement with the experimental data [21].

For quasistationary Poiseuille flow, the system of periodic rolls with a period two to three times bigger than the layer thickness was observed [22] at a frequency of the flow inversion f higher than 0.5×10^{-3} Hz (see Figure 3.4). At the same time, the essentially higher value (0.15 Hz) was reported for the periodic instability induced by an oscillating flow [23].

3.3.3
Instability at Initial Planar Orientation in the Flow Plane

A theoretical analysis of LC stability relative to a steady simple shear flow for the case of an initial planar orientation coinciding with the flow plane was done by Pikin [29, 30]. It was found that hydrodynamic instability can arise only in the case $\alpha_3/\alpha_2 < 0$ when the Ericksen number $Er = \gamma_2 sh/K_{11}$ exceeds the critical value Er_c defined as

$$Er_c = \frac{4,8}{\sqrt{|\alpha_3/\gamma_2|}}. \tag{3.21}$$

This result is true also for Poiseuille flow [31]. Theoretical analysis done in [32] showed that in the latter case at an increase in the pressure gradient, the maximal angle of a director approached asymptotically the θ_{flow}, which is in accordance with the result previously obtained by Leslie [33]. The existence of a stable flow-induced angle was used by Gähwiller [34] to determine of a complete set of Leslie coefficients (in the case of MBBA $\theta \rightarrow \theta_{flow} = 7.1°$).

3.3.4 Hydrodynamic Instabilities at Initial Homeotropic Orientation

The case of a homeotropic surface orientation is essentially different from those (the initial planar orientation normal or parallel to the flow plane) described above. Even for relatively weak flows, there is a nonzero torque $\Gamma_z = -\alpha_2 s$ acting on a director that is not compensated by an elastic torque. It results in finite distortions of the initial homogeneous orientation without any threshold that is different from the case $n_0 \perp v, \nabla v$ (see Section 3.3.1). Usually, at the same shear rate, the value of this torque is essentially higher than the reciprocal value $\Gamma_z = -\alpha_3 s$ for the initial planar orientation in the flow plane (see Section 3.3.3). So homeotropic samples of LCs have to show the maximal sensitivity to the velocity gradients that plays a key role for LC sensors (Chapter 7).

It is obvious that a theoretical analysis of possible hydrodynamic instabilities for homeotropic boundary orientation is more complicated (compared to the case of planar orientation) as the basic linear state corresponds to the inhomogeneous initial orientation. So computer simulations are usually used for this purpose.

The flow-induced distortions of the initially homeotropic orientation under the action of steady shear flows were first studied by Wahl and Fisher [35, 36] and Gähwiller [34].

Different scenarios of nonlinear behavior were realized depending on the type of the shear flow and the kind of the liquid crystal.

The simplest result was obtained for flow-aligned nematics under simple steady shear flow by Wahl and Fisher [35, 36]. Increase in shear rate in this case led to the monotonic inclination of a director *in a flow plane* and changes in the polar angle up to saturation value θ_{flow} at high shear rate. The same scenario could be achieved for nonaligned nematics by applying a stabilizing electric field [37]. Such experiments allowed determination of the ratio α_3/α_2 for Leslie coefficients and will be described in more detail in Chapter 5. Investigations of nonaligned nematics under simple shear flow in the absence of fields gave contradictory results. Cladis and Torza [38] (CBOOA) observed the "tumble" of a director to the new steady position in the flow plane at the critical shear rate s_c (at lower shear rates inhomogeneous cellular flow was registered). At the same time, Pieranski and Guyon [39] (HBAB) did not found any stationary state at high shear rates. At moderate shear rates (about 1 s^{-1}), they observed the escape of a director from the shear plane whereas the secondary roll instability (with the axes of rolls parallel to the flow direction) was detected at higher shear rates. The theoretical analysis of such phenomena can be found in [40, 41]. The case of a steady Poiseuille flow is essentially different and will be considered below in detail.

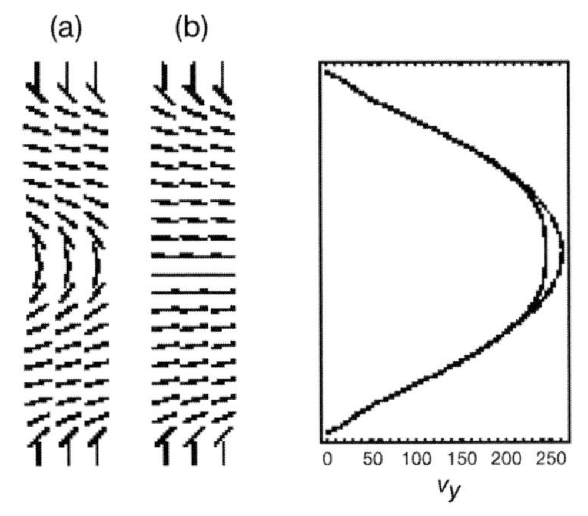

Figure 3.5 Two different director configurations for steady state of LC layer for Poiseuille flow and the corresponding flow velocity. The solid and dotted lines are for the director configurations (a) and (b), respectively. (After Deniston *et al.* [8].)

The nonlinear behavior of nematic liquid crystals under Poiseuille flow is of special interest due to a number of specific phenomena that do not exist in a simple shear flow. Such phenomena have to be taken into account while measuring shear viscosity as will be shown in Chapter 5. It is also important for sensor applications of liquid crystals based on the usage of homeotropic layers showing extremely high sensitivity to a pressure gradient (see Chapter 7).

In general, it was recognized that different scenarios accompanied by arising specific instabilities could be realized with an increase in pressure gradient applied to LC layer. For example, Deniston *et al.* [8] conducted a direct computer simulation of the dynamic behavior of the initially homeotropic sample in the framework of the generalized hydrodynamic theory. They found two possible steady states of the system, shown in Figure 3.5.

Configuration (a) corresponds to the traditional case that will be discussed below. The second configuration (b) is obtained by computer simulation of a nematic subjected to a flow in isotropic phase with cooling afterward. This configuration is similar to the equilibrium configuration of a director in a cylindrical tube with normal anchoring.

The strict theoretical description of nonlinear phenomena in the case under consideration is not complete yet; essential progress was achieved only for the primary homogeneous instability connected with the escape of a director from the flow plane. Below this case will be considered in detail.

The first experimental results for instabilities of homeotropic samples of a flow-aligned nematic (MBBA) under a pressure gradient were obtained by Hiltrop and Fisher [42]. They studied the specific radial Poiseuille flow in the layer of LC confined by two circular glass plates. The flow was induced by a pressure drop from the center to outside parts of the cell. Three types of orientational structures were registered depending on the pressure gradient. At low- and high-pressure gradients, orientational distortions were concentrated at the flow plane. They also observed the escape of the director from the flow plane for moderate pressure gradient.

Two different approaches were applied to explain complicated nonlinear phenomena taking place in homeotropic samples under Poiseuille flow [8, 40].

In the first approach, the standard hydrodynamic theory of Leslie with a director as a unique additional parameter was used to conduct stability analysis. In particular, Zuniga and Leslie [40] found two principal modes – a bow (B) mode and peak (P) mode.

The B-mode correspondent to the geometry shown in Figure 3.5a (the same polar angles $\theta(\pm h/2) = \pi/2; \theta(0) = \pi/2$ in the middle and at boundaries of LC layer) become unstable relative to the fluctuations of an azimuthal angle at some threshold pressure gradient G_c. The correspondent critical value A_c of the dimensionless pressure gradient

$$A_c = \frac{G_c h^3}{8K_1} \rightarrow G_c \sim \frac{1}{h^3} \tag{3.22}$$

was estimated as 20 for material parameters of MBBA. In this case, the maximal angle of inclination from the initial homeotropic orientation was equal to 42°. The peak mode ($\theta(\pm h/2) = \pi/2;\ \theta(0) = 0$) correspondent to the structure in Figure 3.5b was found to be stable relative to the transition of a director from the flow plane.

The second approach mentioned above is based on generalized hydrodynamic equations in which the director is added (or replaced) by a variable scalar order parameter S (or by a variable tensorial order parameter Q_{ij}). Such approach seemed to be natural at description of nanoscaled phenomena such as defects arising under intensive flows or flow-induced orientation in isotropic phase of liquid crystals. In particular, the transition from B to P mode can be assigned to the sharp variations of the order parameter shown in Figure 3.6.

According to Figure 3.6, the scenario of B–P transition includes thinning of the central layer with an increase in pressure gradient and it is "unzipping" that leads to the stable peak mode. This scenario can be realized without the escape of a director from the flow plane. It is worthwhile to note the essential simplification used in the theory that ignores the anisotropy of viscoelastic properties of LCs (e.g., shear viscosity).

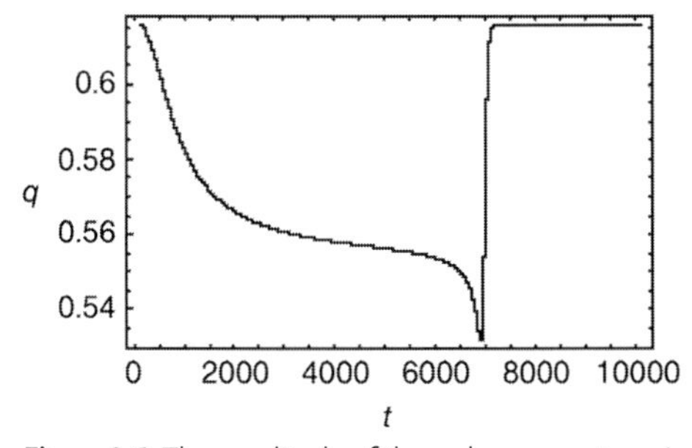

Figure 3.6 The amplitude of the order parameter q in the center of the channel as a function of a dimensionless time. (After Deniston *et al.* [8].)

The alternative description of such transformation in terms of a director includes the emergence of a primer homogeneous instability from the basic B-mode connected with the escape of a director from the flow plane. Such mechanism will be considered below.

3.3.5
Orientational Instability in a Nematic Liquid Crystal in a Decaying Poiseuille Flow

A detailed experimental study of the homogeneous "out-off-plane" instability was performed using a specific quasistationary decay flow in a wedge-like cell. The main idea of experiment is based on Equation 3.22 that predicts rather strong ($G_c \sim h^{-3}$) dependence of the threshold pressure gradient on the thickness of an LC layer. It paves the way for visualizing the hydrodynamic instabilities, as described below. The experimental cell is shown schematically in Figure 3.7 [43]. A capillary with a wedge gap was formed by glass plates with inner surfaces coated by a thin conducting SnO_2 layer, which made it possible to apply an electric field to the liquid crystal layer. Treating the surfaces with chromolane ensured a homeotropic orientation (perpendicular to the surface) of the nematic on the substrates. The main feature of the cell

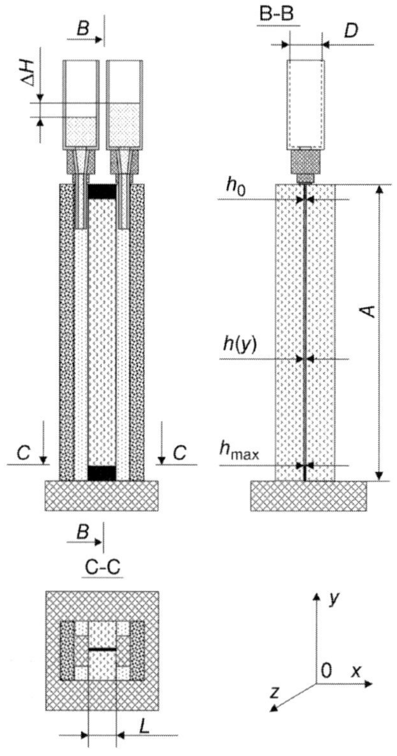

Figure 3.7 Geometry of a wedge-shaped cell: $A = 10$ cm, $L = 1$ cm, $h_{max} = 210\,\mu m$, $h_0 = 33\,\mu m$, and $D = 1.5$ cm.

was wedge-shaped with a local layer width h varying along the y-axis. The linearity of the $h(y)$ dependence and the absolute values of local width h were monitored from the variation of the phase difference between the ordinary and the extraordinary rays caused by a decrease in the alternating voltage ($U_0 = 45$ V, a frequency of 5 kHz) applied to the MBBA layer to zero. The absolute error in determining the local width was approximately 2–3 μm. Prior to the experiment, the cell was mounted vertically and was filled with the liquid crystal so that the material filled the capillary, the filling channels, and a part of expansion vessels (cylindrical pipes of diameter D). Decaying Poiseuille flow (along the x-axis) was produced because of introduction of the crystal into one of expansion vessels. The initial pressure drop ΔP_0 created in this case and proportional to the initial difference ΔH_0 in the levels of the liquid crystal was calculated to within 5% from the mass of the crystal introduced in the cell and the diameter of the expansion vessels. The experiments were carried out at temperature $T = 22 \pm 0.5\,°\mathrm{C}$.

For a small wedging, the capillary can be treated as a set of channels having different widths and parallel to the x-axis, to which the same pressure gradient $\Delta P/L$ is applied. In addition, in view of the large aspect ratio of the cell

$$\frac{A}{(h_{\max} + h_0)/2} \approx 800,$$

$$\frac{L}{(h_{\max} + h_0)/2} \approx 80.$$

We can expect that a plane Poiseuille flow along the x-axis is formed in the capillary (except for the boundary regions at the ends of the cell); this is confirmed by observing the movement of small impure particles (2–4 μm in diameter) added to the nematic. The intensity $I(t)$ of light with a wavelength of 628 nm (He–Ne laser) transmitted (along the z-axis) through the capillary was detected from an area of diameter about 0.3 mm by a photodiode and recorded in digital form (with the help of A to D converter) on the hard disk of a computer. Two versions of positions of the polarizer and the analyzer were used in the experiment: crossed polaroids oriented at an angle of $\alpha = 45°$ to the direction of the flow (geometry a) and at an angle of $\alpha = 0°$ (geometry b). Geometry b made it possible to detect the escape of the director from the x–z plane of the flow. Shadow images of the cell in crossed polaroids in geometries a and b were recorded simultaneously with the help of a digital camera.

For small initial pressure drops ($\Delta P_0 \leq 6$ Pa) in the entire range of local cell thicknesses $h_0 < h < h_{\max}$, no changes in the intensity of transmitted light were detected in b geometry; consequently, the director preserved its orientation in the plane of the flow. In a geometry, this regime corresponds to the shadow image of the cell, consisting of dark and bright fringes arranged along the direction of the flow and formed as a result of interference of the ordinary and extraordinary rays. The phase delay δ appears as a result of a change in the refractive index, which is in turn associated with a deviation of the director from the initial homeotropic orientation. The dynamics of the interference fringes is as follows: the formation of the fringe structure in the region of large local thicknesses begins immediately after the

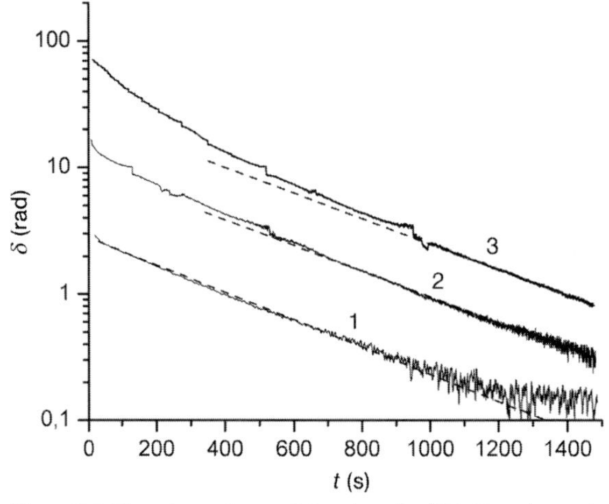

Figure 3.8 Time dependence of the phase lag $\delta(t)$. Experimental data (solid curves) and results of simulation (dashed curves): $h = 139\,\mu\text{m}$, $\Delta P_0 = 0.4$ Pa (1), $\Delta P_0 = 0.8$ Pa (2), $\Delta P_0 = 2.8$ Pa (3).

emergence of the initial pressure difference; after this, the system of fringes moves. Recording locally the intensity of transmitted light $I = I_0 \sin^2(\delta/2)$ in *a* geometry, where I_0 is the input intensity, it was found that the phase lag $\delta(t)$ decreased exponentially with time (curve 1 in Figure 3.8). The analysis of such type of dependence can be fulfilled in the framework of a simple linearized model of a quasistationary decay flow taking into account small changes of a polar angle at a constant value of the effective viscosity coefficient. It provides a new method of measuring of shear viscosity coefficients described in detail in Chapter 5. In particular, the fitting curve 1 shown in Figure 3.2 corresponds to the value of shear viscosity $\eta_{\text{hom}} = 0.16 \pm 0.02$ Pa s obtained for MBBA that is in good agreement with the results of independent measurements [44] for maximal Miesowicz viscosity. The time dependence of a light intensity in a linear regime of a director motion was used to extract the instant values of a pressure gradient drop ΔP according to the linearized model $\delta(t) \sim \Delta P^2$, described in Chapter 5. It can be fulfilled even for higher values of the initial pressure drop ΔP_0 producing nonlinear distortions in the thicker part of a wedge-like cell used in experiments. In this case, $\delta(t)$ dependences obtained in the thinner part of the cell do not follow the simple exponential low (curves 2 and 3 in Figure 3.8) due to overall changes of an effective viscosity connected with nonlinear orientational distortions. The modified exponential dependence with a relaxation time proportional to a slowly varying effective viscosity was found to be in accordance with experimental results (Figure 3.8).

With increasing initial pressure drop ($\Delta P_0 > 6$ Pa), the signal intensity $I(t)$ of transmitted light in *b* geometry, which is recorded in the range of large thicknesses of the cell, exhibits two peaks (curve 2 in Figure 3.9b), indicating the escape of the director from the plane of the flow. Figure 3.9a shows the theoretical dependences $I(t)$ in geometries *a* and *b* as well as angle $\phi_m(t)$ of deviation of the director from the plane

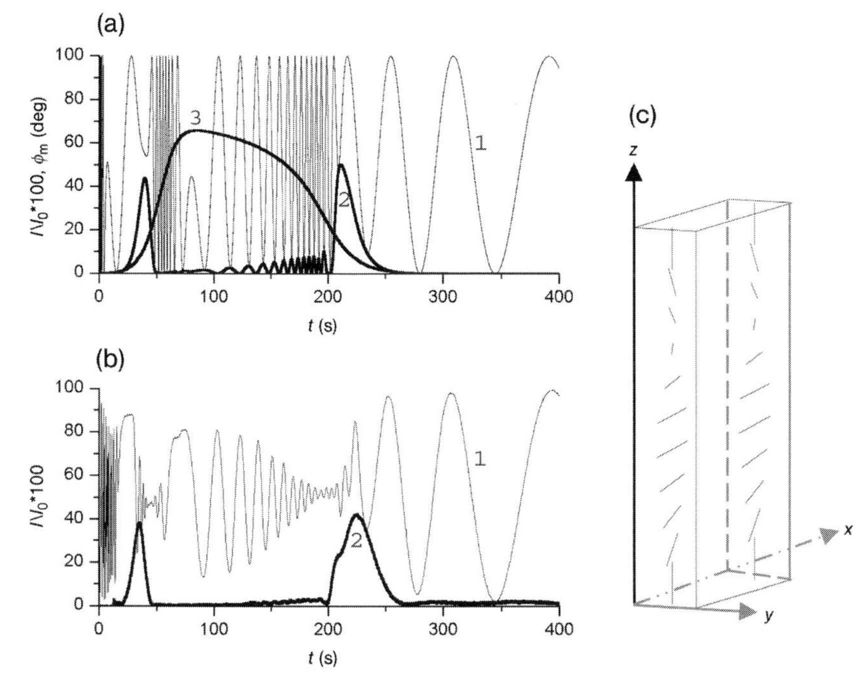

Figure 3.9 Time dependences of the intensity of transmitted light, $I(t)$, in geometry *a* (curves 1) and *b* (curves 2) and of the angle of deviation of the director from the flow plane at the center of the layer, $\phi_m(t)$ (curve 3): (a) theoretical calculations: $\Delta P_0 = 21$ Pa, $h = 90\,\mu$m; (b) experimental data: $\Delta P_0 = 20$ Pa, $h = 86\,\mu$m; (c) schematic diagram of orientation of the director at the instant corresponding to $\phi_m = 60°$.

of the flow at the center of the layer. The nonlinear nematodynamics equations [45] for a planar layer, when the director and the velocity are functions of coordinate *z* and time *t* [46], were solved numerically using the material parameters for MBBA. The intensity of transmitted light was calculated using the Jones matrix method [47]. Angle ϕ_m characterizes the orientation of the director at the center of the layer:

$$\mathbf{n}_m = (0, \sin\phi_m, \cos\phi_m).$$

The director distribution at the instant corresponding to $\phi_m = 60°$ is shown in Figure 3.9c. The first peak of the $I(t)$ signal in *b* geometry is associated with the escape of the director from the plane of the flow in the case of a large initial pressure drop. As the pressure drop $\Delta P(t)$ decreases below the threshold value, the director returns to the plane of the flow (second peak on the $I(t)$ curve in *b* geometry) and relaxes over long periods to the uniform homeotropic orientation.

The transition associated with the escape of the director from the plane of the flow is observed most clearly in the shadow image of the cell (Figure 3.10). In *b* geometry (Figure 3.10b), the shadow image is (in the increasing order of the local layer thickness) dark field I in the range of smaller thicknesses, light fringe II, and the

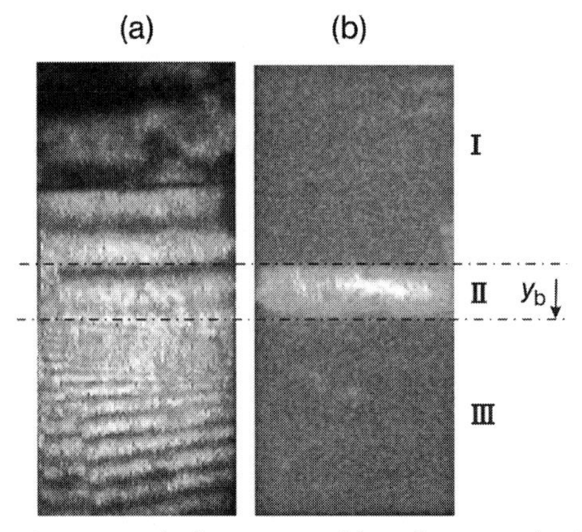

Figure 3.10 Shadow images of the cell in crossed polaroids at instant $t = 30$ s for $\Delta P_0 = 15.5$ Pa: in geometry *a* (left) and in geometry *b* (right).

low-intensity region III. In *a* geometry (Figure 3.10a), two regions can be clearly distinguished on the shadow image, region I + II, corresponding to relatively small thicknesses, in which wide interference fringes parallel to the direction of the flow are observed, and region III, corresponding to large thicknesses, where narrow interference fringes are transformed into wide fringes.

The polarization and optical analysis, as well as a comparison of microphotographs (Figure 3.10) with time dependences of transmitted light (Figure 3.9b), make it possible to unambiguously identify all regions on the shadow image of the cell: region I (the director is in the plane of the flow, the azimuthal angle ϕ_m of deviation of the director at the center of the layer is zero), region II (the director emerges from the plane of the flow ($0° < \phi_m < 20°$), and region III (the director is oriented almost perpendicular to the plane of the flow, $\phi_m \rightarrow 90°$). The recording of shadow images and the intensity of transmitted light in *b* geometry made it possible to clarify the nature of the formation of the region corresponding to the escape of the director from the plane of the flow. Immediately after the application of the initial pressure gradient (for 10–15 s) in the range of large values of local thicknesses of the nematic layer, light fringe II is formed, which subsequently moves along the y-axis toward smaller thicknesses over a time of approximately 30–40 s. After approximately 50–60 s, the position of the light fringe stabilizes, and the fringe begins to move slowly in the direction of increasing layer thickness. This stage corresponds to a steady-state decaying Poiseuille flow. The boundary y_b between regions II and III can be seen most clearly, while the boundary between regions I and II becomes less clear as fringe II moves toward large thicknesses. The width of region II attains its minimal value when it begins its reverse motion and increases as fringe II moves toward large local thicknesses. The presence of two peaks in the $I(t)$ signal in *b* geometry (Figure 3.9b) is due to the fact that light fringe II passes twice through the point of observation. At the stage corresponding to a steady-state decaying Poiseuille flow, the time dependence $y_b(t)$ of the position of the boundary between regions II and III was recorded since this boundary remains the clearest during the entire experiment. The simultaneous recording of the phase lag $\delta(t)$ in *a* geometry in the range of small local widths of the liquid crystal layer makes it possible to reconstruct the time dependence of the pressure drop, $\Delta P(t)$. The connection between $\delta(t)$ and $\Delta P(t)$ in the regime of small deformation will be considered in Chapter 5.

Thus, using the data on time dependence $y_b(t)$ of the position of the boundary and the dependence $\Delta P(t)$. Reconstructed from $\delta(t)$, we can associate the value of pressure drop with the position of boundary y_b recorded in the experiment. The latter is connected with the local thickness h_b of the layer by a simple linear dependence. So, it is possible to extract the dependence of the critical pressure gradient ΔP_c, corresponding to the escape of the director from the plane of the flow, on local thickness h_b of the liquid crystal layer. Figure 3.11 (curves marked by *a*) shows the $\Delta P_c(h_b)$ dependence obtained for various values of the initial pressure drop ΔP_0. It can be seen from the figure that for large values of *h*, the curves obtained for different values of ΔP_0 almost coincide. This is due to the fact that large thicknesses correspond to large time intervals following the application of the initial pressure drop, when a quasistationary flow sets in the cell. The velocity varies slowly with time and the director can follow the variation of pressure. Accordingly, the position of the boundary y_b in such a flow regime is determined only by the current value of ΔP and does not depend on ΔP_0.

The results of the analysis of the effect of an electric field applied along the *z*-axis on the threshold for the escape of the director from the plane of the flow are given by curves marked by *b* in Figure 3.11. The reduction of the critical value $\Delta P_c(h_b)$ is due to

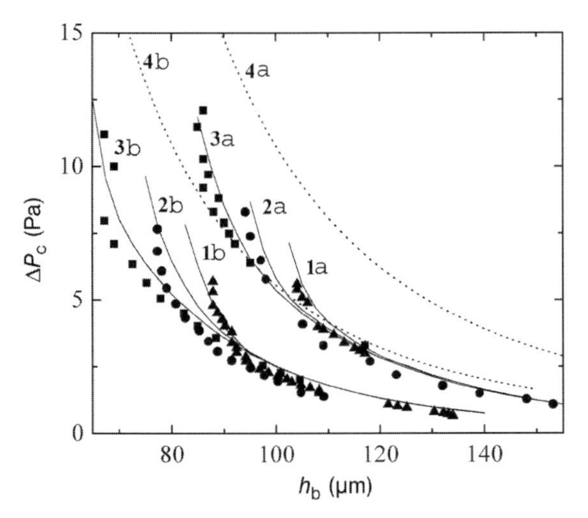

Figure 3.11 The threshold pressure drop ΔP_c corresponding to the escape of the director from the flow plane as a function of the local layer thickness h_b. Experimental data are shown by symbols and the results of calculations are given by curves. $U = 0$ V: $\Delta P_0 = 9.4$ Pa (▲ and 1a), $\Delta P_0 = 12.8$ Pa (● and 2a), $\Delta P_0 = 15.5$ Pa (■ and 3a); $U = 3$ V: $\Delta P_0 = 8.7$ Pa (▲ and 1b), $\Delta P_0 = 10.8$ Pa (● and 2b), and $\Delta P_0 = 14.1$ Pa (■ and 3b). Curves 4a and 4b correspond to calculations for the case of a steady-state Poiseuille flow for $U = 0$ and 3 V, respectively.

the fact that the electric field exerts a destabilizing effect (in addition to the flow) on the homeotropically oriented MBBA layer (negative anisotropy of permittivity).

Figure 3.11 also shows the theoretical $\Delta P_c(h_b)$ dependences obtained from the results of simulation of the nonlinear nematodynamics equations [45] for the case when the director and velocity are functions of coordinate z and time t [46] using the material parameters of MBBA. The results are in good agreement with the experimental data considering that the wedge-shape cell was simulated in numerical calculations by a set of planar capillaries, taking into account experimental errors in determining ΔP_c and h_b.

Figure 3.11 also shows for comparison the dependences $\Delta P_c^{st}(h)$ of the critical pressure drop corresponding to the escape of the director from the plane of the flow, which were calculated for a steady-state Poiseuille flow (curves 4). The critical value of the pressure drop for a steady-state Poiseuille flow $\Delta P_c^{st} \sim 1/h^3$ systematically exceeds the corresponding values of ΔP_c for a decaying flow. This is due to the fact that the return of the director to the plane of the decaying flow occurs upon a decrease in pressure $\Delta P(t)$ below ΔP_c^{st} over a finite time (on the order of the director relaxation time), during which the pressure continues to decrease.

3.3.6
Influence of a Decay Flow on Electrohydrodynamic Instability in Liquid Crystals

As it is shown above, weak decay flows with $\Delta P < \Delta P_c^{st}$ induced the "in-plane" homogeneous motion of a director. Nevertheless, it is of interest to find out some influence of such flows on the structures induced by other factors. Below we describe one particular case concerning the interaction between a weak decay flow and electrohydrodynamic (EHD) roll instability induced in a homeotropic LC layer by a low-frequency electric field.

It is well known that the initial orientation of LC samples stabilized by surfaces can be destabilized by action of electric or magnetic fields [45]. The primary instability, Fréedericksz transition, arises at voltages higher than some threshold value U_F and can be assigned to dielectric (magnetic) permittivity anisotropy. The existence of the secondary EHD instability is connected with a bulk charge originating from the periodic ion vibrations in LC media with anisotropy of an electric conductivity (Helfrich–Kerr mechanism). The amplitude of such vibrations grows with decreasing frequency. So, relatively low frequencies are needed to produce EHD roll instability. In experiments, the latter is visualized as a number of electroconvective rolls.

In initially planar samples, a threshold voltage V_{roll} does not depend on the layer thickness h and the spatial period of such structure is comparable to d. It is well known that just above the threshold voltage, the roll instability is visualized as a system of regular light and dark stripes – stationary Williams's rolls. The axes of rolls are oriented normally to the initial surface direction determined by a proper surface treatment [48, 49]. This property was used, for example, to control the direction of the flow-induced surface orientation emerging upon filling of the cells with LCs in case of a weak surface anchoring [50].

At increasing voltage, the rolls become distorted and are transformed into a number of moving vortexes strongly scattering light (dynamic light scattering mode).

It is worthwhile to note that at high frequencies, the specific roll structure (shevrons) with a period dependent on voltage is realized. The mechanism responsible for such structure essentially differs from Helfrich–Kerr mechanism mentioned above.

The emergence of EHD instability in homeotropic samples is essentially a complicated phenomenon [48, 51]. It is well known that the initial homeotropic orientation of LC sample with a negative sign of dielectric permittivity anisotropy can be destabilized by the action of an electric field. The primary instability – Fréedericksz transition – arises at voltages higher than some threshold value U_F. The secondary roll instability arising at higher voltages looks rather chaotic contrary to the well-oriented rolls observed in planar cells. It is quite understandable as there is no privilege direction defined by surfaces and degeneration relative to an azimuthal angle takes place for both the primary instability and for the secondary one.

The existence of four types of roll structures, namely, normal rolls, inclined rolls, abnormal rolls, and defect rolls was established in experiments [51]. The experimental data described below were obtained at a voltage 13.6 V and a frequency $f_U = 240$ Hz, which is lower than the Lifshits' point for MBBA ($f_U = 270$ Hz in the absence of magnetic field), correspond to the inclined rolls. The applied voltage was just above the threshold value (13.1 V). The estimates made in Ref. [51] show that at these parameters the inclination angle (the angle between the axes of rolls and the local director) is lower than 15°.

The result of the action of a decay flow on the EHD structure is shown in Figure 3.12. It is of interest that the degeneration is removed at an extremely low pressure drop (of order 1 Pa) that is essentially lower than thresholds of flow-induced instabilities described below. The long axes of rolls are oriented in the direction of a pressure gradient that is in accordance with the flow-induced roll instability in the planar samples [22, 23]. It is of interest that the roll structure is distorted when the action of a decay flow becomes negligible. It means that the latter is more favorable in comparison with a regular roll structure. In many respects, the action of a weak flow is similar to that registered in the case of weak magnetic fields.

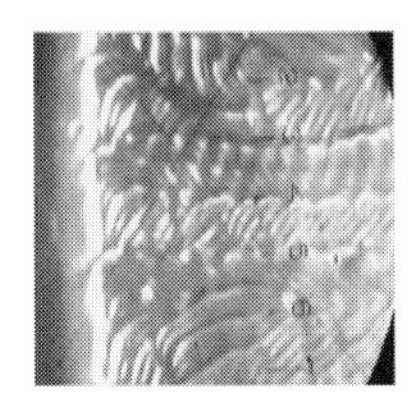
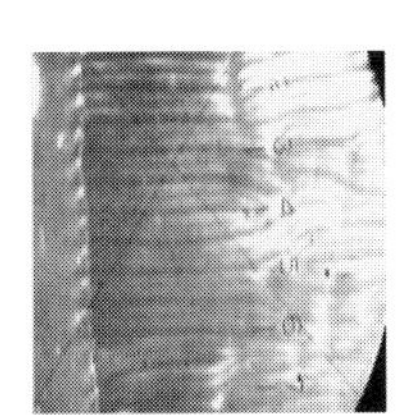
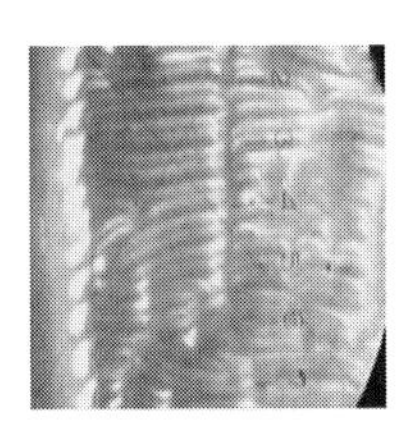
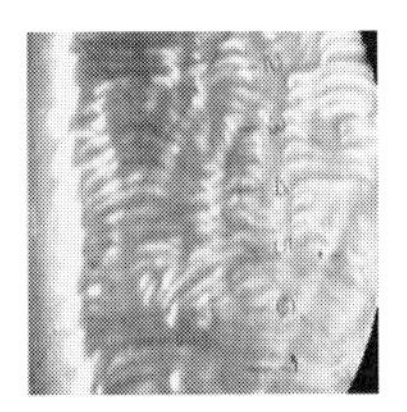

The time after start of a decay flow:

0 s 180 s 420 s 930 s

Figure 3.12 Influence of a decay flow on the EHD structure arising due to low-frequency electric field. $\Delta P(0) = 5$ Pa; $U = 13.6$ V; $f_U = 240$ Hz; the thickness of the layer in the center of the images $\langle h \rangle = 120\,\mu$m, the vertical size of the images -6.4 mm.

To explain the presented results, one has to take into account the destabilizing torque that stimulates the above-mentioned escape of a director from the flow plane. It is reasonable to consider the axes of rolls to be normal to the local director. It means that such mechanism provides the orientation of the roll axes in the flow direction as it was observed in experiments.

It is important that the influence of a weak decay flow on EHD instability is registered in nonpolarized light that makes such effect interesting for sensor applications of liquid crystals.

3.4 Hydrodynamic Instabilities Under Oscillating Flows

In many respects, a practical realization of oscillating flows seems to be easier than the analogous problem for the case of stationary flows described above. Moreover, in most experiments [23, 60] devoted to the study of the simplest plane stationary Couette and Poiseuille flows, the latter were replaced by the oscillating flows with a periodic inversion of a flow direction. Indeed, at low frequencies of inversion (e.g., 10^{-3} Hz [23]), such approximation seems to be reasonable. Nevertheless, it can be broken for relatively thick (about 200 μm) LC layers often used in experiments [23]. So, some results reported for the latter case are possibly connected with a nonstationary rheological behavior.

It is important to realize in experiments the harmonic oscillating flow. It makes possible to get results applicable for a comparison with the theoretical description operating mainly with such type of motion.

The harmonically varied shear motion needed for a Couette flow is relatively simple for practical realization. For example, one can use the LC cell that consists of the motionless plate and the moving one with the LC layer fixed by spacers. The moving plate connected mechanically with a loudspeaker transmits a simple shear deformation to the LC layer under investigation. A special attention has to be paid to avoid a transverse plate motion that can drastically modify LC behavior [53].

The realization of harmonic variations of a pressure difference applied to LC layer that induces an oscillating Poiseuille flow is an essentially complicated problem, as it will be shown below.

That is why the oscillating Couette flows were studied in more detail compared to the Poiseuille ones.

3.4.1 Oscillating Coutte Flow

The nonlinear behavior of LC layer under an oscillating Couette flow and in the presence of magnetic (electric) fields was studied in [21] for the initial planar orientation normal to the shear rate. It was found that a number of dynamic regimes for the roll instability controlled by both the frequency of oscillations and the external field strength (Figure 3.13).

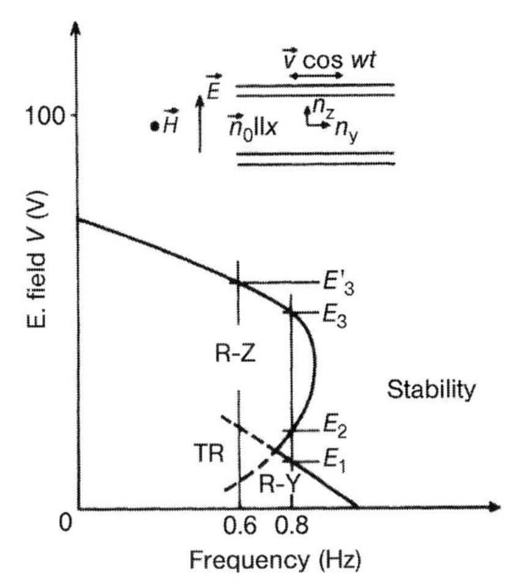

Figure 3.13 Threshold diagram for different regimes of roll instability. The ratio of oscillations amplitude to the layer thickness is equal to 0.148; R-Z: the rolls are visualized as the inversion of dark and light stripes every half of the period, which corresponds to the change of a sign for n_z projection at a constant sign for n_y; R-Y: there is no inversion of dark and light stripes that correspond to the change of a sign for n_y projection at a constant sign for n_z; TR: intermediate regime with an exotic texture. (After Ref. [21].)

Different spatial structures controlled by the frequency of oscillations in Couette flow are also realized at the homeotropic boundary orientation.

The small values of the amplitude of oscillations correspond to the linear regime of a director motion in the shear plane. Experimentally, this case was studied in detail [54] for the layer of 5 CB under the action of vibrations with amplitudes 0–20 μm and at a frequency variation in the wide range (0.01–200 Hz). The authors have found an excellent agreement between the results of numerical calculations and the experimental data including time dependences of the intensity of light passed through the cell, amplitude, and phase parameters of optical response.

An increase in the amplitude of vibrations results in the direct emergence of a nonhomogeneous roll instability [54, 55] that is quite different from the case of oscillating Poiseuille flow where the primary homogeneous instability takes place. The latter will be considered below. A theoretical analysis of the roll instability in a Couette flow can be found in a number of papers [54, 56–58]. The obtained analytic expressions for the threshold amplitudes of oscillations at low (high) frequencies are only in qualitative agreement with the experimental results. The quantitative agreement was obtained only for a numerical solution of the basic equations at relatively low frequencies (<100 Hz). At higher frequencies, one has to take into account the inertial properties of LCs [58]. This case corresponds to the shear waves propagating in a liquid crystal layer and will be considered below in more detail. A number of acoustooptical effects were studied for the sonic and ultrasonic frequency ranges [59, 60]. A lot of results of such types were summarized by Kapustina [53].

Among other possible variants of boundary conditions, special attention was paid to the case where the initial orientation coincides with the flow-induced direction described by the angle θ_{flow} (see above). This orientation is quite stable for a steady shear flow. At the same time, it was shown [61] that oscillating flow produces the specific orientational instability with the threshold amplitude of oscillations dependent on frequency. Experimentally, such possibility was not studied up to now, although today's technologies provide the boundary orientation with a controlled pretilt angle [62, 63].

3.4.2
Oscillating Poiseuille Flow: Planar Orientation

The first experimental study of nematic liquid crystals under an oscillating Poiseuille flow was performed for the simplest case of a boundary orientation, namely, for a planar orientation normal to the flow plane [22, 23]. Indeed, the pressure gradient applied to the layer was constant during a half of the period followed by the abrupt change of the flow direction. So, it was difficult to wait for an exact agreement with the theory considered in terms of harmonically varied pressure gradient. Nevertheless, it explained the main features of the original experimental results. In particular, it was found that in the absence of fields, a threshold of the homogeneous instability described above increased with the frequency of oscillations. Moreover, this instability was replaced by the roll instability (with axes of rolls oriented along the flow direction) for frequencies higher than 0.15 Hz (see Figure 3.14). For a low pressure difference ΔP, the spatial period of the rolls was close to the layer thickness and decreased as ΔP^{-2} for high pressure difference. It points out the formation of two systems of convective rolls near boundaries.

Additional application of stabilized magnetic and electric fields makes the frequency dependence of the threshold pressure gradient to be very complicated, as shown in Figure 3.15.

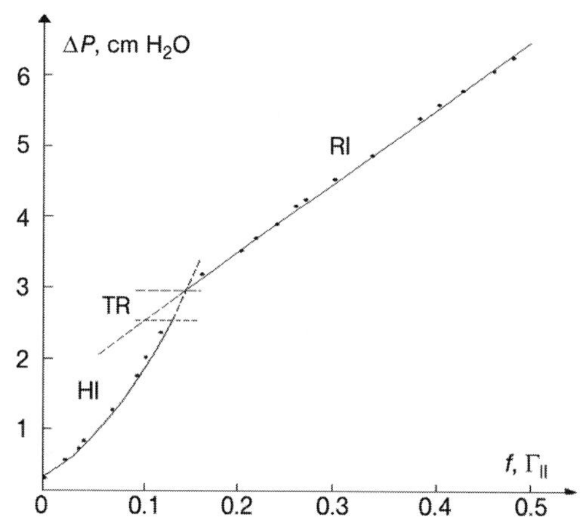

Figure 3.14 Frequency dependencies of the threshold pressure differences for homogeneous (HI) and roll (RI) instabilities in MBBA in the absence of external fields; TR: transient regime [23].

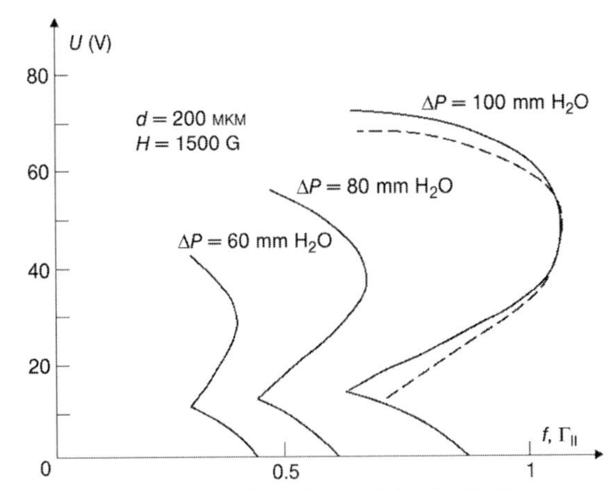

Figure 3.15 Frequency dependence of the threshold corresponding to roll instability for MBBA in the presence of electric field [23].

Theoretical aspects of the structure formation under oscillating Poiseuille flow for the given geometry were considered in a number of publications [26, 57, 64]. It was shown [26] that theoretical backgrounds, derived for an analysis of hydrodynamic instabilities arising under steady flow, could be applied with some modifications to the case of oscillatory flows. In particular, in both cases, a fluctuating parameter $w(r, t)$ is represented by the unique expression

$$w(r,t) = A\exp(\sigma t). \tag{3.23}$$

For oscillating flows, the parameter A depends both on spatial coordinates and time, whereas the latter dependence is absent for a steady flow. For the basic state oscillating with a period T, this parameter is also represented by the function with the same period (the Fluke's theorem):

$$A = A(r,t) = A(r,t+T) \tag{3.24}$$

The unique condition for emergence of a homogeneous instability is expressed as $Re\{\sigma\} \geq 0$.

This inequality corresponds to the monotonic increase of the initial fluctuation, and the threshold of the homogeneous instability can be found from corresponding equality. It is possible to use for this purpose the Fourier expansion of the function $w(r, t)$ and time-dependent coefficients in the initial hydrodynamic equations. After that, one has to extract the identical harmonic components. It transforms the initial hydrodynamic equations into the infinite system of the ordinary differential equations. The threshold of the homogeneous instability is determined via a solution (using numerical methods) of the corresponding boundary-value problem for this system. In practice, the calculations are rather cumbrous and demand a powerful computer. The results obtained for low frequencies essentially depend on the time profile of pressure gradients, which means that a lot of harmonics have to be used to analyze experimental data mentioned above [23].

The account of the additional time coordinate leads to the expansion of the mode classification. In particular, for the oscillating flow with the period T one can extract the so-called Z mode defined by the expression

$$n_z\left(t+\frac{T}{2}\right) = -n_z(t); \qquad n_y\left(t+\frac{T}{2}\right) = n_y(t) \tag{3.25}$$

and Y mode

$$n_z\left(t+\frac{T}{2}\right) = n_z(t); \qquad n_y\left(t+\frac{T}{2}\right) = -n_y(t). \tag{3.26}$$

Upon using the Fourier expansion, such solutions correspond to the even and odd harmonics. So, for Poiseuille flow four regimes, Y-T, Z-T, Y-S, Z-S with different symmetry properties relatively to the spatial and time coordinates are possible.

The alternative way for the theoretical analysis of the problem is connected with the direct numerical integration of the initial equations. This procedure is relatively simple for oscillating flows with inversion of a flow direction at the constant module of a pressure gradient. The frequency dependence of the threshold Ericksen number for the homogeneous instability in Z-T regime obtained in such manner [26] was found to be in accordance with the experimental results [23]. The obtained increase (approximately linearly) in the threshold pressure gradient (proportional to the Ericksen number) with frequency means that at high frequencies the near boundary distortions have no enough time (before the inversion of the flow direction) to spread over the entire LC layer.

Later Tarasov *et al.* [57] have shown via direct computer simulation that oscillating Poiseuille flow (contrary to the Couette flow) results in emergence of the homogeneous instability with distortions corresponding only to T-type symmetry. At the same time, the even instability existing in the low-frequency regime is replaced by the odd instability at increasing frequency. The quantitative correspondence of experimental [23] and theoretical results was obtained by taking into account the time profile of low-frequency pressure difference used in the experiment.

3.4.3 Oscillating Poiseuille Flow: Homeotropic Orientation

The response of the initially homeotropic layers of LCs on the action of the oscillating Poiseuille flows is of special interest. As it was mentioned above, there is no threshold for linear distortions of a homeotropic sample, and the effective torque

$$\Gamma = \alpha_2 \frac{\partial v_x}{\partial z} \tag{3.27}$$

acted upon a director via velocity gradient is maximal for this particular orientation. The optical response of a homeotropic layer on small orientational distortions can be made very pronounced by using polarized light and optimal geometry of experiments. It makes the use of homeotropic layers the main element of LC sensors described in Chapter 7.

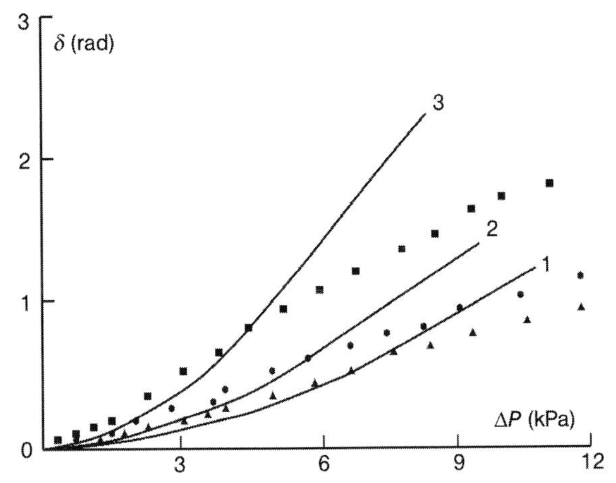

Figure 3.16 Experimental [65] and theoretical [66] dependences of the maximal flow-induced phase difference on the amplitude of a pressure difference: (1) $h = 9\,\mu\text{m}$, (2) $h = 17\,\mu\text{m}$, (3) $h = 32\,\mu\text{m}$; $f = 57\,\text{Hz}$.

The nonlinear phenomena such as the homogeneous instability described above for a steady Poiseuille flow is also interesting for theoretical description and experimental study.

One of the first investigations of the linear response of homeotropic layer to the oscillating Poiseuille flows was conducted by Blinov *et al.* [65]. The experiments were performed with the thin layer (5–60 μm) of 5CB at sound frequency (57 Hz). The authors have registered the linear deformation of the initial structure by using the transmitted and reflected polarized light. They have studied and analyzed some interesting effects such as the damping of low-frequency acoustic wave, a modulation via the flow the of the director in the surface layer, and so on. The dependences of the maximal phase difference δ_m on the amplitude of an applied pressure ΔP_m are shown in Figure 3.16. They correspond, at least, for relatively small values of ΔP_m to the calculations performed in the framework of the linear model [66].

The used cell did not show a very high sensitivity to the pressure oscillations (the maximal applied pressure difference was about 12 kPa). It can be explained by a relatively small thickness of the layer as the sensitivity of a homeotropic sample drastically depends on the latter parameter.

More systematic (and interesting for practical applications) results were reported [67–70] in infrasonic frequency range where the sensitivity of homeotropic samples becomes extremely high. It makes it possible to study both linear and nonlinear phenomena in oscillating Poiseuille flows. Below we will describe such experiments and discuss obtained results.

3.4.3.1 Experimental Setup for Low-Frequency Poiseuille Flow

The general scheme of the experimental setup is shown in Figure 3.17. The low-frequency (<1 Hz) variable pressure difference applied to the LC layer was generated

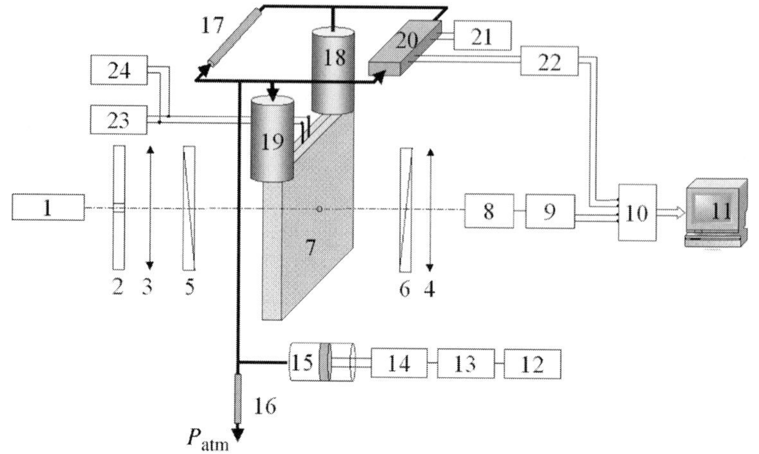

Figure 3.17 Experimental setup: (1) He–Ne laser; (2) diaphragm; (3, 4) lenses; (5, 6) polarizes; (7) LC cell; (8) photodiode; (9) amplifier; (10) AD converter; (11) computer; (12–15) mechanical system for air pressure supply; (16, 17) capillaries; (18, 19) tubes partly filled with LC; (20–22) pressure sensor with electric supply and amplifier; (23) ac generator; (24) voltmeter.

by the special mechanical system [21] based on the regulated compression of air in the cylinder 15.

The mechanical system includes the DC motor 13 with a rotation velocity regulated by the variation of the electrical current provided by electric supply unit 12. The rotational motion is slowed and transformed into the translational motion (in x-direction) of a piston compressing the air inside cylinder 15 via a special unit 14 including a worm gear and an oscillating crank gear. Such decision provides extremely slow (0.01–1 Hz) motion of a piston $x = r\sin(\omega t)$ and variations of a pressure (proportional to x) in accordance with a harmonic law; $\Delta P = \Delta P_0 \sin \omega t$. It makes possible the correct comparison between theoretical calculations and experimental results, which is under question for the case of reversal motion used in earlier experiments [23] and described above. The amplitude of pressure variations ΔP_0 applied to the LC cell can be changed in the range (1–1000 Pa) both by replacing the radius r of the crank or by using capillaries 16, 17 of different length and diameter, which connected both sides of the cell and the entire aerodynamic system with the environment. The latter makes it possible to minimize extremely slow (<0.01 Hz) variations of the pressure difference applied to the LC layer and induced by thermal instabilities of different parts of the system. Moreover, such decision excluded the procedure of a precise choice of the null position of the mechanical system, which plays a key role in the analogous system without capillaries [71]. The wide range of variations of the controlled parameters (ΔP_0, ω, h) provides a detailed study of both linear and nonlinear phenomena in LC under oscillating Poiseuille flows.

Tubes 18, 19 shown in Figure 3.17 provide both filling of the cell with LC and minimization of errors arising due to periodical changes of the difference ΔH in levels of meniscues in the tubes. The minimal diameter D_{min} of the tubes sufficient for the latter purpose can be estimated by comparing the maximal hydrostatic pressure $\Delta P_H = \rho g\ \Delta H_{max}$ with the amplitude of the air pressure difference ΔP_0 applied to the cell. It is clear from the geometry of the experiment that $\Delta H_{max} = V_{T/2}/(\pi D^2)$, where ΔH_{max} is the maximal (for a period) change of the level difference, $V_{T/2}$ is the volume flowing through the cell for half the period. The latter can be found by the integration of the instant volume flow rate (dV/dt) defined for the wedge-like cell under consideration as follows:

$$
\begin{aligned}
-\frac{dV}{dt} &= \sum_{i=1}^{n} \frac{d\Delta V}{dt} = \int_0^A \frac{\Delta P(t)}{12\cdot\eta_3\cdot L}\left(h_{max} - \frac{(h_{max}-h_0)}{A}y\right)^3 dy \\
&= \frac{\Delta P(t)A}{12\cdot\eta_3\cdot L}\left(h_{max}^3 - \frac{3h_{max}^2(h_{max}-h_0)}{2} + h_{max}(h_{max}-h_0)^2 - \frac{(h_{max}-h_0)^3}{4}\right) \\
&= \frac{\Delta P(t)A}{12\cdot\eta_3\cdot L}\frac{(h_{max}+h_0)(h_{max}^2+h_0^2)}{4}.
\end{aligned}
\tag{3.28}
$$

Expression (3.28) was obtained by considering the wedge-like channel as a number of parallel channels of a constant thickness and by putting the effective viscosity equal to the intermediate Miesowicz coefficient η_3. The final result for $V_{T/2}$ is expressed as

$$
\begin{aligned}
V_{T/2} &= \int_0^{T/2} dV = \int_0^{T/2} \frac{A(h_{max}+h_0)(h_{max}^2+h_0^2)}{4\cdot 12\cdot\eta_3\cdot L}\Delta P_0 \sin\left(\frac{2\pi\cdot t}{T}\right) dt \\
&= \frac{A(h_{max}+h_0)(h_{max}^2+h_0^2)}{4\cdot 12\cdot\eta_3\cdot L}\frac{P_0 T}{\pi}.
\end{aligned}
\tag{3.29}
$$

So, the inequality $\Delta P_H \ll \Delta P_0$, leads to the next criteria for a minimal diameter of the tubes:

$$
D \gg \sqrt{\frac{\rho g A(h_{max}+h_0)(h_{max}^2+h_0^2)}{4\cdot 12\cdot\eta_3\cdot L}\frac{T}{\pi^2}}.
\tag{3.30}
$$

The estimated value of D_{min} for the maximal period $T = 100$ s, realized in the experiments is equal to 2.6 mm, which is essentially lower than the real diameter of the tubes ($D = 13.5$ mm).

The optical part of the experimental setup was traditional for a polarization–optical investigation. It was registered simultaneously the time variations of the intensity $I(t)$ of polarized light passed through the cell and the time dependences $\Delta P(t)$ of the pressure difference applied to a liquid crystal layer. The cell was placed between crossed polarizers at the angle $\beta = 45°$ with respect to the flow direction (geometry *a*). In this case, the maximal amplitude of the flow induced light intensity is observed. Lenses 3 and 4 are used to minimize the scanned area dimensions up to the limit when the local thickness can be considered a constant value. It was possible to apply additionally high-frequency (3 kHz) electric field to study combined effects of flow and field.

3.4.3.2 Linear In-Plane Motion of a Director Under Oscillating Poiseuille Flow

In this section, we describe experimental results corresponding to the motion of LC director in flow plane. The visual observations and measurements were performed in crossed polarizers oriented at 45° relative to the flow direction (geometry *a*). Geometry *b* was found to be insensitive to the "in-plane" motion of a director.

Visual observations performed at low values of a pressure gradient show the picture of interference stripes parallel to the flow direction and similar to that observed in a decay flow (Figure 3.10). They arose at an increase of an instant pressure difference and moved upward and downward in accordance with pressure oscillations. It is clear that such motion reflects the periodic flow-induced variations of the polar angle θ and corresponding birefringence that depends both on θ and h. Examples of local optical response are shown in Figure 3.18.

One can see the rising additional local extremes (corresponding to the motion of the interference stripes) on $I(t)$ dependences at an increase of both the pressure difference amplitude ΔP_0 and the local thickness h.

The presented dependences provide the calculation of the maximal phase difference δ_m in accordance with

$$I = I_0 \sin^2(\delta/2) \tag{3.31}$$

realized for each halve of the period of pressure oscillations. The latter parameter can be easily connected with the amplitude of the polar angle variations in the framework of linearized models [66–68].

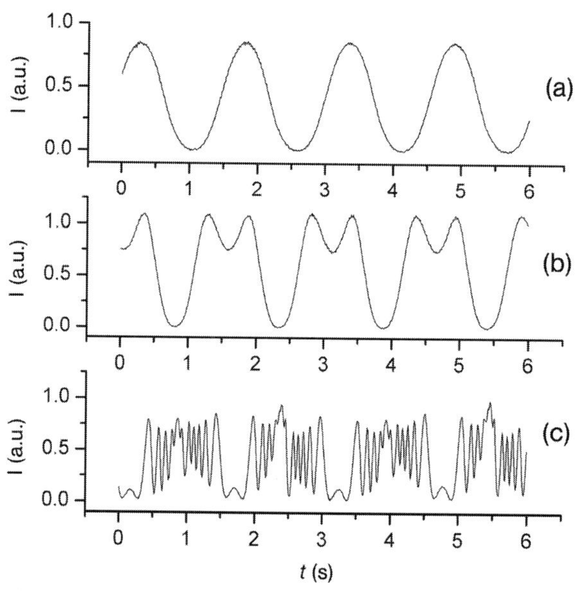

Figure 3.18 Time dependences of light intensity $I(t)$ obtained in geometry "*a*" at different experimental parameters: $T = 3.1$ s, $h = 112\,\mu$m, (a) $\Delta P_0 = 16.4$ Pa, $\delta_m = 2.18$ rad; (b) $\Delta P_0 = 23.0$ Pa, $\delta_m = 4.3$ rad; (c) $\Delta P_0 = 133.5$ Pa. MBBA 22 °C.

In particular, for the case of additional action of electric field E on the homeotropic layer of LC, the system of coupled hydrodynamic equations is written as [68]

$$\rho \frac{\partial v_X}{\partial t} = -\frac{\partial P}{\partial x} + \eta_1 \frac{\partial^2 v_X}{\partial z^2} + \alpha_2 \frac{\partial^2 \theta}{\partial z \partial t} \tag{3.32}$$

$$k_{33} \frac{\partial^2 \theta}{\partial z^2} = \alpha_2 \frac{\partial v_X}{\partial z} + \gamma_1 \frac{\partial \theta}{\partial t} + \varepsilon_0 \, \Delta\varepsilon E^2 \theta, \tag{3.33}$$

where ρ, η_1, $\Delta\varepsilon$, K_{33} are the density, the maximal shear viscosity, the dielectric permittivity anisotropy, and the Frank's constant of LC. This system describes a time evolution of the velocity $V_x(z, t)$ and orientation $\theta(z, t)$ fields for a flow plane (here θ is the angle of declination from the initial homeotropic orientation).

By introducing dimensionless parameters $\tilde{t} = \omega \cdot t$, $\tilde{z} = z/h$, the general solution of this system can be represented as follows:

$$\theta = \theta_0 + \bar{\theta} \tag{3.34}$$

$$\bar{\theta}(\tilde{z}, \tilde{t}) = \theta_r(\tilde{z}) \cos \tilde{t} + \theta_i(\tilde{z}) \sin \tilde{t}. \tag{3.35}$$

Functions θ_r andθ_i are defined by the substitution of expression (3.35) into the system (3.32) and (3.33):

$$\theta_r(\tilde{z}) = -\frac{\Delta P_0}{\Delta x} \frac{h}{\eta_2 \omega \left[\left(\frac{\omega_E}{\omega} \right)^2 + m^2 \right]} \left\{ \left(\frac{\omega_E}{\omega} \right) \times \left[\tilde{z} + \frac{1}{2} \frac{\cos k_2 \left(\tilde{z} + \frac{1}{2} \right) \cosh k_1 \left(\tilde{z} - \frac{1}{2} \right) - \cos k_2 \left(\tilde{z} - \frac{1}{2} \right) \cosh k_1 \left(\tilde{z} + \frac{1}{2} \right)}{(\cosh k_1 - \cos k_2)} \right] - m \left[\frac{1}{2} \frac{\sin k_2 \left(\tilde{z} - \frac{1}{2} \right) \sinh k_1 \left(\tilde{z} + \frac{1}{2} \right) - \sin k_2 \left(\tilde{z} + \frac{1}{2} \right) \sinh k_1 \left(\tilde{z} - \frac{1}{2} \right)}{(\cosh k_1 - \cos k_2)} \right] \right\} \tag{3.36}$$

$$\theta_i(\tilde{z}) = -\frac{\Delta P_0}{\Delta x} \frac{h}{\eta_2 \omega \left[\left(\frac{\omega_E}{\omega} \right)^2 + m^2 \right]} \left\{ \left(\frac{\omega_E}{\omega} \right) \times \left[\frac{1}{2} \frac{\sin k_2 \left(\tilde{z} - \frac{1}{2} \right) \sinh k_1 \left(\tilde{z} + \frac{1}{2} \right) - \sin k_2 \left(\tilde{z} + \frac{1}{2} \right) \sinh k_1 \left(\tilde{z} - \frac{1}{2} \right)}{(\cosh k_1 - \cos k_2)} \right] + m \left[\tilde{z} + \frac{1}{2} \frac{\cos k_2 \left(\tilde{z} + \frac{1}{2} \right) \cosh k_1 \left(\tilde{z} - \frac{1}{2} \right) - \cos k_2 \left(\tilde{z} - \frac{1}{2} \right) \cosh k_1 \left(\tilde{z} + \frac{1}{2} \right)}{(\cosh k_1 - \cos k_2)} \right] \right\}, \tag{3.37}$$

where the frequency dependence of wave numbers k_1 and k_2 is expressed as

$$k_1 = \sqrt{\frac{1}{2}\left[\frac{\omega_E}{\omega_0} + \sqrt{\left(\frac{\omega_E}{\omega_0}\right)^2 + m^2\left(\frac{\omega}{\omega_0}\right)^2}\right]} \tag{3.38}$$

$$k_2 = \frac{m\left(\frac{\omega}{\omega_0}\right)}{\sqrt{2\left[\frac{\omega_E}{\omega_0} + \sqrt{\left(\frac{\omega_E}{\omega_0}\right)^2 + m^2\left(\frac{\omega}{\omega_0}\right)^2}\right]}} \tag{3.39}$$

$$m = 1 - \frac{(-\alpha_2)}{\eta_2}\frac{1}{1-\lambda}. \tag{3.40}$$

The obtained expressions for k_1 and k_2 depend on the following three frequencies:

ω – the frequency of the applied pressure gradient,

$\omega_0 = \frac{K_{33}}{h^2\gamma_1}$ – the characteristic frequency of the director motion in the absence of electric field, and

$\omega_E = \frac{\varepsilon_0\,\Delta\varepsilon E^2}{\gamma_1}$ – the analogous frequency in the presence of strong electric field.

In the case of weak fields ($\omega_E \ll \omega_0$), two wave numbers, describing space variations of orientation become equal:

$$k_1 \approx k_2 \approx k \approx \sqrt{\frac{1}{2}m\frac{\omega}{\omega_0}}. \tag{3.41}$$

The expression (3.41) previously obtained in [66] means a decrease in the boundary layer thickness with frequency ω.

At the same time, for strong electric fields ($\omega_E \ll m\omega$)

$$k_2 \cong 0, \qquad k_1 \approx \sqrt{\omega_E/\omega_0} \approx (U/U_F)^2,$$

where $U_F = \pi\sqrt{K_{33}/\varepsilon_0\,\Delta\varepsilon}$ is the threshold voltage for a Fréedericksz transition. So, the character of space distortions and the thickness of the boundary layers do not depend on frequency for this case.

The analysis of the obtained expressions shows that the amplitude of orientational oscillations decreases with frequency. The analogous decrease takes place via application of electric field for liquid crystals with a positive value of $\Delta\varepsilon$. In the latter case, increase in electric field strength also results in decrease in the imaginary part $\theta_i(z)$ and corresponding phase shift β between a director motion and oscillations of a pressure gradient:

$$tg\beta = \frac{\theta_i}{\theta_r}. \tag{3.42}$$

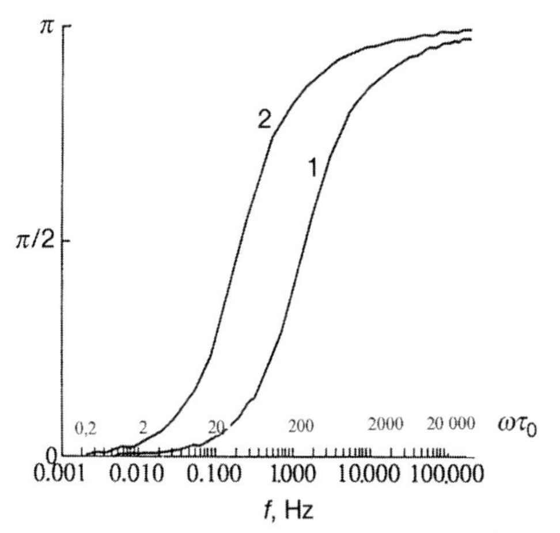

Figure 3.19 Theoretical frequency dependences of the phase shift between a director motion and pressure gradient oscillations for the layer of 5 CB ($h = 50\,\mu m$) confined by surfaces with a homeotropic (1) and a planar orientation (2). (After Ref. [66].)

In the absence of electric field, the phase shift depends on the frequency of oscillations. For low frequencies

$$\omega \ll \omega_0/m \tag{3.43}$$

$\beta \to 0$, so the director oscillates in a phase with pressure oscillations, whereas at high frequencies $\beta \to \pi$ and the orientational motion is in the opposite phase with a pressure variation. The frequency dependence of the phase shift is shown in Figure 3.19.

It is convenient to compare the presented theoretical expressions with the experimental results in terms of the optical phase delay δ. For small distortions from the initial homeotropic orientation, the maximal variations of this parameter δ_m can be expressed as

$$\delta_m \cong \delta_1 \times (\Delta P_0)^2, \tag{3.44}$$

where $\delta_1(T, h) = \delta_m(\Delta P_0 = 1\ \mathrm{Pa})$ – the reduced phase delay.

The experimental dependences of the maximal phase delay δ_m on the squared amplitude of a pressure difference are shown in Figures 3.20 and 3.21. They are in good agreement with the general prediction of the linear model (3.44).

The parameter δ_1 defines the sensitivity of LC layer to the pressure gradient that plays a key role in sensor applications of liquid crystals (Chapter 7). Its change with the frequency of oscillations is reflected through a decrease in the inclination of lines in Figure 3.20 for shorter periods. It is worthwhile to note an extremely strong dependence of δ_1 on the local thickness h shown in Figure 3.21. This fact is in qualitative accordance with the results of simplified quasistationary model [67] that predicts $\delta_1 \sim h^7$ in the absence of fields. It is obvious that a quasistationary regime is realized at low values of a local thickness where parameter δ_1 does not depend on the period of oscillations (see Figure 3.22).

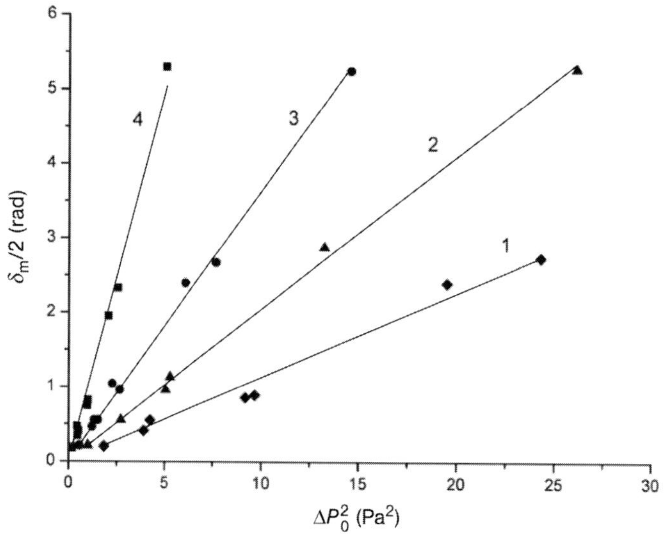

Figure 3.20 Dependencies of phase difference δ_m on the squared pressure amplitude ΔP_0 in MBBA 22 °C for different periods. Symbols: experimental results, lines: approximation by linearization. $h = 107\,\mu m$; (1) $T = 28$ s, (2) $T = 12.5$ s, (3) $T = 5.6$ s, (4) $T = 3$ s.

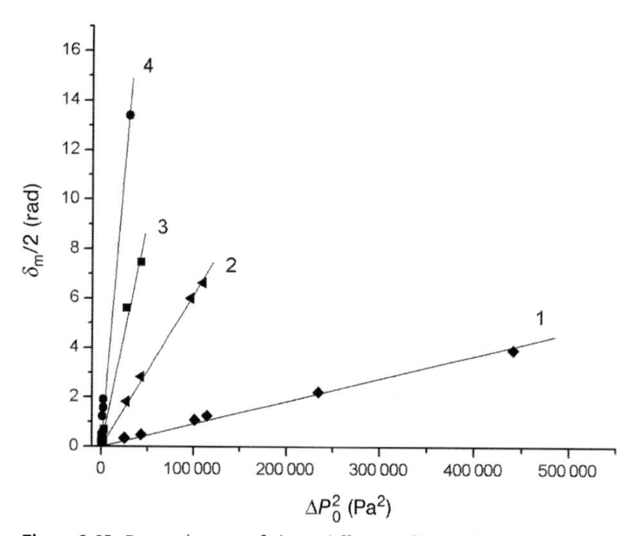

Figure 3.21 Dependencies of phase difference δ_m on the squared pressure amplitude ΔP_0 in MBBA for a different layer thickness. Symbols: experimental results, lines: approximation by linearization. $T = 3.2$ s; (1) $h = 25.4\,\mu m$, (2) $h = 29.6\,\mu m$, (3) $h = 32.8\,\mu m$, (4) $h = 37.6\,\mu m$.

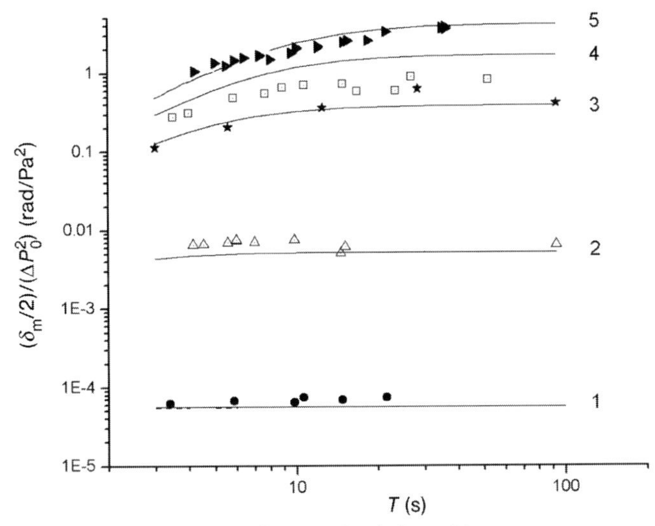

Figure 3.22 Dependencies of a normalized phase delay in a homeotropic layer of MBBA on pressure oscillations period for different values of a local thickness. Symbols: experimental data, lines: theoretical curves: (1) $h = 30\,\mu\text{m}$; (2) $h = 58\,\mu\text{m}$; (3) $h = 107\,\mu\text{m}$; (4) $h = 114\,\mu\text{m}$; (5) $h = 150\,\mu\text{m}$.

At the same time, such dependence takes place at higher thicknesses. It means that a quasistationary approximation does not hold and the general solution (3.35)–(3.37) has to be applied for a comparison between the experiment and the theory. The results of such a comparison for MBBA with available material parameters are presented in Figures 3.23 and 3.24. The reasonable quantitative agreement makes strong arguments for the possibility of a priori estimation of a linear response of LC layer on the time-dependent pressure gradient. It paves the way for optimization of the technical parameters of LC cells used in sensor applications. In particular, it can be done via additional usage of electric fields. In the case of MBBA ($\Delta\varepsilon < 0$), the maximal increase in the phase difference δ_m induced by field is rather moderate (see Figure 3.24) that can be explained by a critical slowing down of director fluctuations at approaching Fréedericksz transition from below.

Previously, this fact was confirmed by acoustooptical investigations of LC with a negative sign of $\Delta\varepsilon$ [72]. In the vicinity of this transition, the dependence of the relaxation time on voltage is described by the power law with critical exponent equal to 1:

$$\tau \sim \left(\frac{U_F - U}{U_F}\right)^{-1}, \tag{3.45}$$

where U_F is the voltage of Fréedericksz transition.

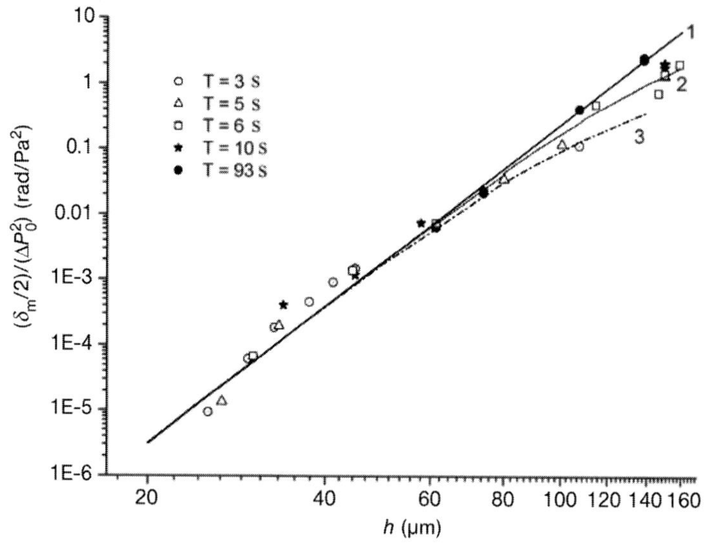

Figure 3.23 Dependence of the reduced phase delay on the layer thickness at different periods of pressure oscillations. Symbols: experimental data, (1) a stationary approximation, (2 and 3) a general solution for $T = 6$ and 3 s. MBBA 22 °C.

The critical increase in the relaxation time breaks the low-frequency regime for the entire frequency range, so the general expressions (3.35)–(3.37) have to be applied to compare the theory and the experiment. The result of such a comparison shown in Figure 3.24 confirms the possibility of a quantitative description of a

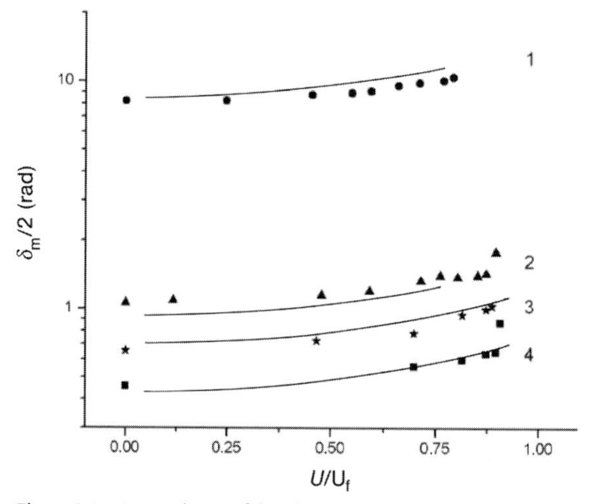

Figure 3.24 Dependence of the phase delay on electric voltage U. Symbols: experimental data, lines: results of numerical calculations. (1) $h = 61\,\mu m$, $T = 6$ s, $\Delta P_0 = 33.2$ Pa; (2) $h = 61\,\mu m$, $T = 6$ s, $\Delta P_0 = 11$ Pa; (3) $h = 38\,\mu m$, $T = 4.7$ s, $\Delta P_0 = 47.6$ Pa; (4) $h = 44\,\mu m$, $T = 15.8$ s, $\Delta P_0 = 22.5$ Pa. MBBA 22 °C.

linear regime in the presence of electric field. It is important for the elaboration of electrically controlled sensors based on the usage of LC with high positive value of dielectric permittivity anisotropy. The results of such type will be presented and discussed in Chapter 7.

3.4.3.3 **Hydrodynamic Instabilities Under Oscillating Poiseuille Flows**

The problem of hydrodynamic instabilities arising in a homeotropic layer of NLC under oscillating Poiseuille flow can be considered as the most complicated in comparison with the particular cases described above. It is connected with the nature of a basic state that is presented by the linear regime of an oscillating motion in the flow plane (see Section 3.3.2). Nevertheless, the main reason, namely, the space dependence of the shear rate, which provides the primary homogeneous instability is common for both steady and oscillating Poiseuille flows. The theoretical prediction of such instability was made by Krekhov and Kramer [40]. Up to now, only rare experiments [71, 74] were performed to study this case. In particular, an escape of a director from the flow plane referred to the homogeneous instability was reported in Ref. [71]. The experiments were conducted with a sample of MBBA placed into a plane capillary with a length 15 mm and a gap of 23 μm. The special mechanical system was used to apply the harmonically varying pressure difference. The measurements were performed in a rather narrow frequency range (5–20 Hz) correspondent to the high-frequency limit ($\omega\tau_0 \gg 1$). The dynamic changes in orientation were registered in polarized light when one of the axes of crossed polarizers was parallel to the flow direction. Such optical geometry, described above for a case of a decay flow, is insensitive to the motion of a director in a plane flow. So, an escape of a director from the flow plane was easily detected as a double peak on $I(t)$ dependence (Figure 3.25). It was found that the threshold of this effect increased approximately proportionally with a frequency (Figure 3.26). The theoretical calculations showed a good agreement with the experimental dependence. The corresponding profiles of orientations are shown in Figure 3.27.

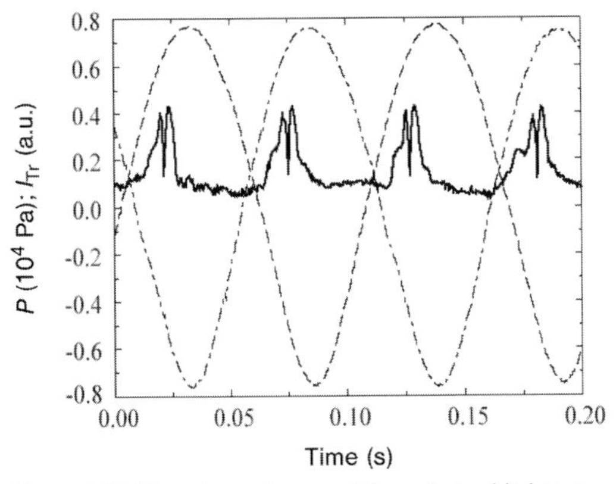

Figure 3.25 Time dependences of the polarized light intensity (solid curve) and the pressure difference (dotted curve). (After Ref. [71].)

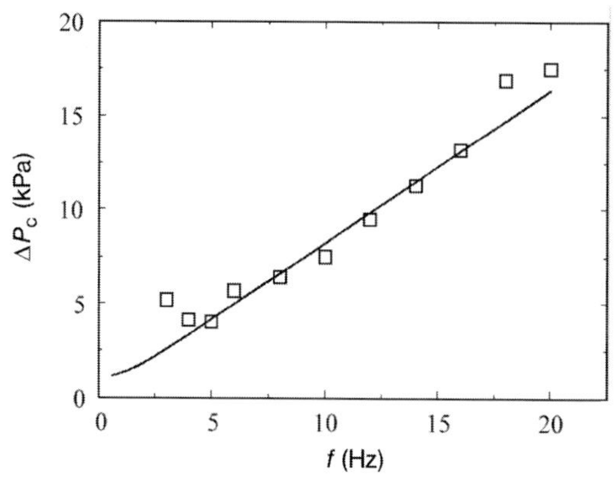

Figure 3.26 Frequency dependence of the threshold pressure difference (symbols: experimental results, line: theoretical calculations).

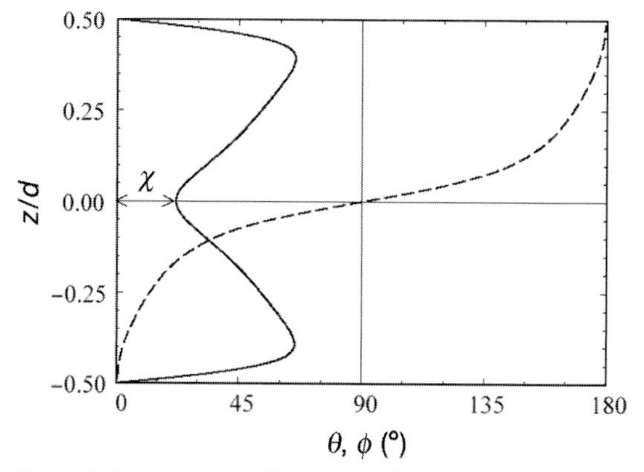

Figure 3.27 Director profiles for out-of-plane transition: polar angle $\theta(z)$ (solid line), azimuthal angle $\phi(z)$ (dashed line); $\omega t = \pi/2$, $f = 10$ Hz, $h = 23\,\mu$m. MBBA 22 °C.

The theoretical analysis of a homogeneous instability was performed in Ref. [75]. It took into account different possible types of symmetry for a basic state (Table 3.1). The numerical calculations were made for the minimal threshold in the absence of electric field that corresponded to the first type of symmetry (Table 3.2) referred to as the fluctuating parameters.

Table 3.1 Symmetry of the basic state.

	n_{0x}	n_{0z}	v_{0x}
t	Odd	Even	Odd
z	Odd	Even	Even

Table 3.2 Symmetry of fluctuations for hydrodynamic parameters.

	$n_{1y}(t)$	$v_{1y}(t)$	$n_{1y}(z)$	$v_{1y}(z)$
I	Even	Even	Even	Odd
II	Even	Even	Odd	Even
III	Odd	Odd	Even	Odd
IV	Odd	Odd	Odd	Even

It was found that the dimensionless pressure gradient,

$$a_c = \frac{G_0 h}{\omega(-\alpha_2)}, \tag{3.46}$$

did not depend on the frequency at high frequencies and varied as ω^{-1} in low-frequency limit. Universal frequency dependence for a pressure gradient is shown in Figure 3.28.

The case of additional action of electric field was considered too. It was shown that electric field led to increase (decrease) in the threshold pressure gradient for positive (negative) signs of dielectric permittivity anisotropy. In the latter case, instability can arise at extremely weak flows due to a well-known mechanism responsible for a Fréedericksz transition and taking place at critical field E_F (Figure 3.29). The role of the flow in this case is to remove the azimuthal degeneration intrinsic to a B-effect.

Below we will present some experimental confirmation of the theoretical conclusions mentioned above.

Detailed experimental studies of "out-of-plane" motion at low frequencies of pressure oscillations [70] were performed by using the wedge-like cell mentioned above. It made possible to study simultaneously both linear and nonlinear regimes of a director motion.

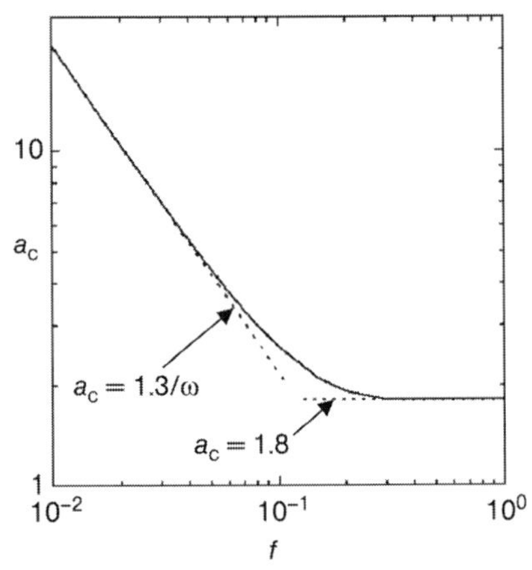

Figure 3.28 The frequency dependence of the dimensionless threshold pressure gradient a_c; MBBA 25 °C, $E_0 = 0$ V, $h = 100$ μm; the solid curve: an exact numerical solution, the dashed curve: an approximate analytic solution.

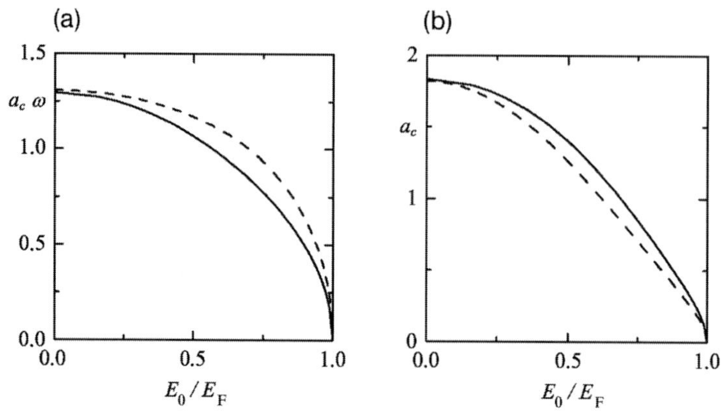

Figure 3.29 The dependences of the dimensionless threshold pressure gradient a_c on the relative value E_0/E_F of the electric field strength. The solid (dashed) lines correspond to a numerical calculation (an approximate analytic solution); $h = 100\,\mu m$, (a) $f = 0.01$ Hz, (b) $f = 1$ Hz. MBBA 22 °C.

The visual observations of the cell revealed flow-induced structures analogous in many respects to those described above for a case of a decay quasistationary flow though some difference was also noticed.

In particular, the sharp boundary between two regions (I and II) corresponding to "in-plane" (I) and linear "out-of-plane" (II) regimes of a director motion was visualized in both cases. The time of formation of this boundary in the case of an oscillatory flow exceeded by order the period of oscillations. It is important to note that after this time the position of boundary y_b became fixed at least for the periods shorter than 30 s. Slight periodical motion of the boundary relative to the averaged position was registered only at very low frequencies ($T > 30$ s). In general, the width of the intermediate region (II) is smaller than for the case of decay flow and equal to 1–2 mm. It corresponds to the thickness variation of about 2–4 μm.

The region (III) corresponding to strongly nonlinear changes of an azimuthal angle is well visualized in both optical geometries "*a*" and "*b*". The picture obtained in geometry "*a*" represents a number of narrow interference stripes moving in accordance with a pressure difference variation. The contrast of this picture becomes maximal at the moment when the instant pressure difference is close to zero. The narrow bright stripe (region II) similar to that observed in a decay flow was instantly visualized for an oscillating flow too (Figure 3.10). At the same time, the region (III) was rather dark in all phases of oscillations. The analogous picture observed at higher frequencies (>1 Hz) was explained previously by inversion of the direction of "out-of-plane" transition [71]. The region (I) corresponding to "in-plane" motion of a director remained dark at observation in geometry "*b*".

The drastic changes in a dynamic optical response due to crossing the boundary was also obtained at the standard local study of light intensity variations for the both geometries (Figure 3.30).

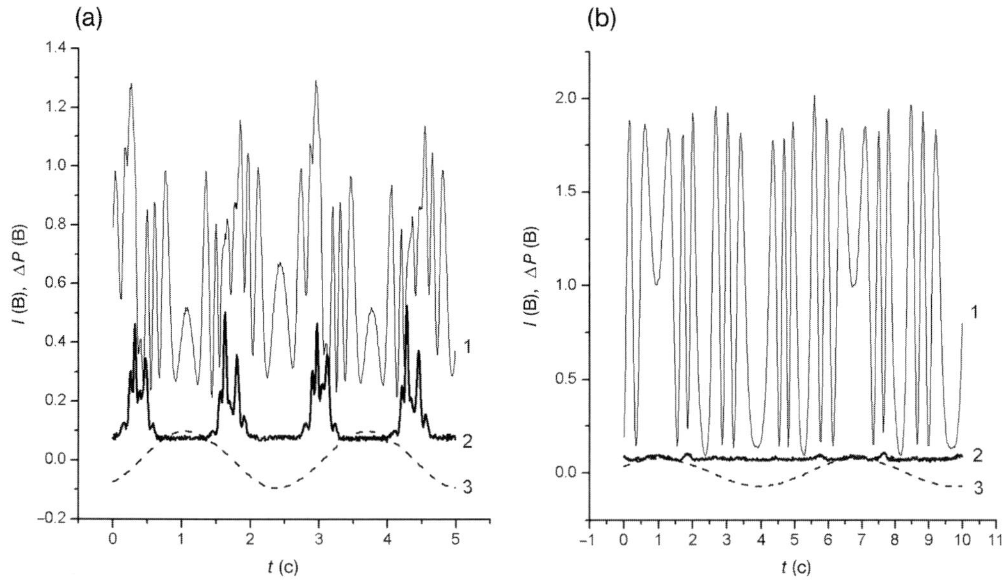

Figure 3.30 Time dependences of light intensity $I(t)$ in geometry "*a*" (1), b (2); (3) time dependence of the pressure; (a) $\Delta P_0 = 536$ Pa, $h = 130\,\mu$m ($h_b = 128\,\mu$m); (b) $\Delta P_0 = 398$ Pa, $h = 119\,\mu$m ($h_b = 121\,\mu$m). MBBA 23 °C.

It is worthwhile to note that the boundary described above becomes not so sharp and shows some inclination at higher frequencies. It can be connected with a slight surface inhomogeneity to the cell that did not play an essential role in a linear regime described above. These effects induce some additional (up to $\pm 5\,\mu$m) but no dramatic errors in the determination of a local thickness h_b corresponding to the position y_b of the boundary. The threshold local thickness h_c can be considered as the most important from the point of view of the experimental checking of the theoretical predictions concerning a homogeneous instability.

In particular, it was established that the variations of both the amplitude and the frequency of the pressure applied to the cell induced a shift of the boundary to the new position corresponding to the new value of the local layer thickness. It can be used to restore the inverse dependence, namely, $\Delta P_c(h)$ that in some respects is more convenient for a comparison with the theory as it usually deals with a threshold amplitude ΔP_c. So, the wedge-like cell provides both a visualization of the homogeneous instability and a quantitative information about threshold parameters (h_c or ΔP_c).

The example of $\Delta P_c(h)$ obtained at extremely low frequencies is shown in Figure 3.31. The threshold amplitudes of the pressure difference is well described by the power law $\Delta Pc \sim h^{-3}$ that is in accordance with the estimates made for a steady flow [31]. It confirms the realization of low-frequency limit in the described experiments.

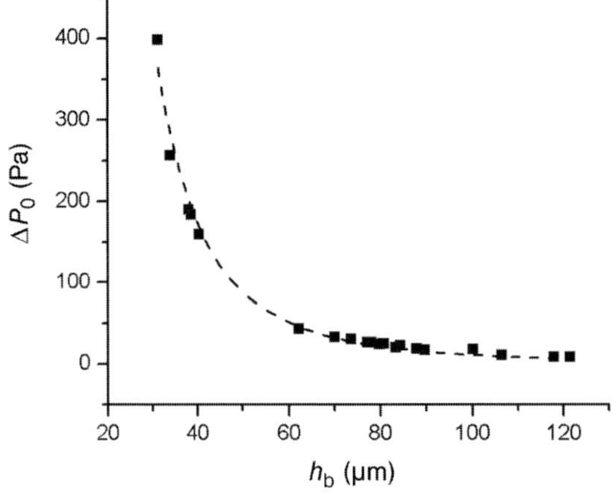

Figure 3.31 Amplitude of the threshold pressure difference ΔP_0^c corresponding to the escape of a director from the flow plane as a function of a local layer thickness h_b. The experimental points are approximated by the power law (dashed curve) $\Delta P_0^c = A \cdot h^B$ with $B = -3.22 \pm 0.13$. MBBA 23 °C.

The theoretical analysis of the problem has shown the existence of the universal dependence of a dimensionless pressure difference $a_c = (\Delta P_0/L)h/[\omega(-\alpha_2)]$ on the parameter $\omega\tau_0$. The experimental data are presented in accordance with this dependence in Figure 3.32. They confirm the existence of the two

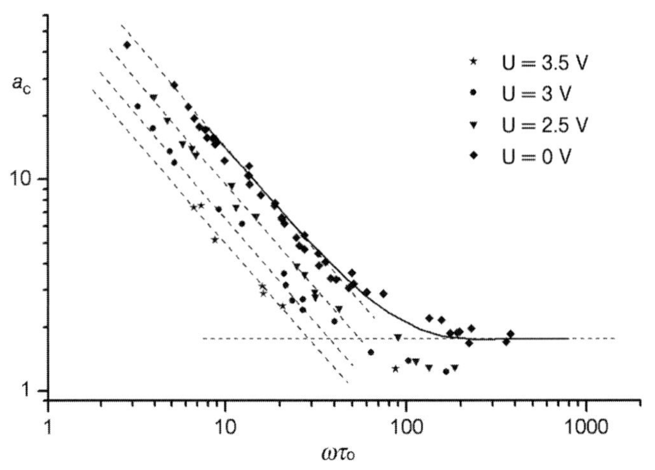

Figure 3.32 Universal dependences for a threshold of the homogeneous instability at different voltages. The symbols correspond to the experimental data; theoretical dependences are presented by dotted curves (for the asymptotic regimes) and by a solid curve (results of a complete numerical calculations). MBBA 23 °C.

(low-frequency and high-frequency) asymptotic regimes predicted by the theory (see Figure 3.28).

Experimental data at low frequencies are in good quantitative agreement with the numerical calculations though some difference (about 20%) was noticed for high frequencies. The latter can be attributed to the additional errors in determination of the inclined boundary mentioned above.

One of the most interesting problems solved in described experiments was the combined action of shear flow and electric field on the hydrodynamic instabilities arising in shear flows of NLC. In the case of negative sign of dielectric permittivity anisotropy (e.g., MBBA), both electric fields and shear flows induce destabilization of the initial orientation. In the absence of a shear flow, the field-induced distortions are degenerated to the azimuthal angle. One can wait to see that even slight flow can remove this degeneration and is able to modify the total picture and threshold values for a Fréedericksz transition. Moreover, this effect holds even for the highly dissipative structures such as convective rolls induced by low-frequency electric field (see Section 3.3.6). From the point of view of the flow-induced homogeneous instability, the additional action of destabilizing electric field has to stimulate the escape of a director from the flow plane with a decrease in the threshold pressure gradient. This effect is seen well in Figure 3.32.

At low frequencies ($\omega\tau_0 \ll 20$), the influence of destabilizing electric field can be described by universal dependence shown in Figure 3.33, where the threshold pressure difference at the given voltage $\Delta P_0(U)$ is normalized on the similar parameter in the absence of field. One can see that the application of electric field in the vicinity of Fréedericksz transition provides a critical decrease in the threshold pressure difference (up to zero). It may be of practical importance for elaboration of highly sensitive LC sensors of low-frequency vibrations.

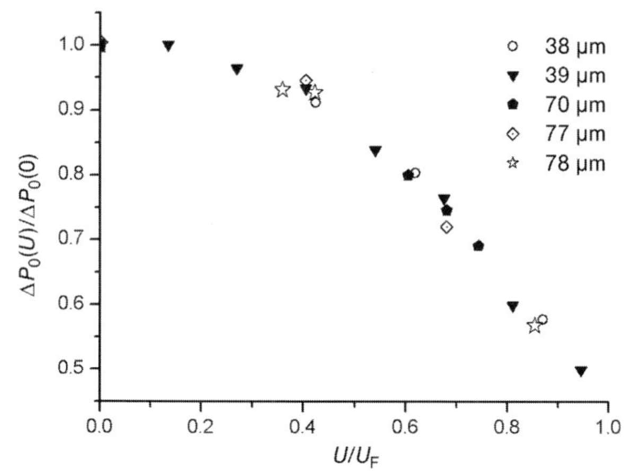

Figure 3.33 Universal dependence of the threshold amplitude of the pressure difference ΔP_0 on the electric voltage U for different values of a local layer thickness h. MBBA 23 °C.

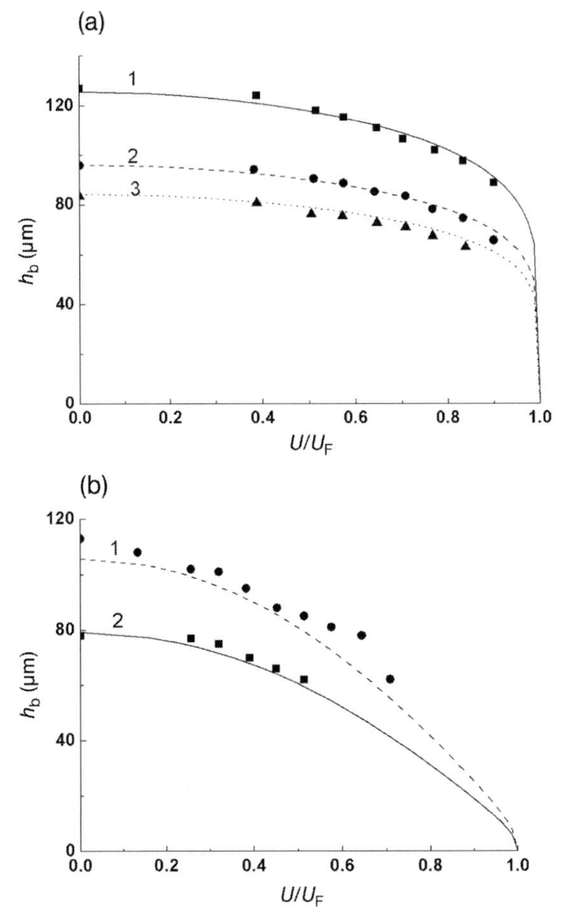

Figure 3.34 Dependencies of the threshold local thickness h_b on the applied voltage U normalized on the Fréedericksz transition voltage U_F for low frequencies (a) and high frequencies (b); the solid curves correspond to the analytic solutions of the low-frequency ($\omega\tau_0 \ll 1$) and the high-frequency ($\omega\tau_0 \gg 1$) regimes, the dashed curves represent the results of numerical simulations [75]; (a) (1) $T = 40.6$ s, $\Delta P_0 = 6.7$ Pa, (2) $T = 40.6$ s, $\Delta P_0 = 15$ Pa, (3) $T = 14$ s, $\Delta P_0 = 22.3$ Pa; (b) (1) $T = 5$ s, $\Delta P_0 = 25.9$ Pa, (2) $T = 3.3$ s, $\Delta P_0 = 52.4$ Pa. MBBA 23 °C.

It is possible to obtain the analytic solutions of the problem [75, 76] for both the low-frequency ($\omega\tau_0 \ll 1$) and the high-frequency ($\omega\tau_0 \gg 1$) regimes. Such solutions are in good accordance with the experimental data, as shown in Figure 3.34a. The agreement between the theory (numerical simulations) and the experiments was also obtained for intermediate frequencies, which is demonstrated by Figure 3.34b.

The same is true for liquid crystals with a positive value of the dielectric permittivity anisotropy. In this case, application of electric field stimulates stabilization of the linear "in-plane" regime of a director motion and a shift of the boundary to the region of smaller thickness (Figure 3.35).

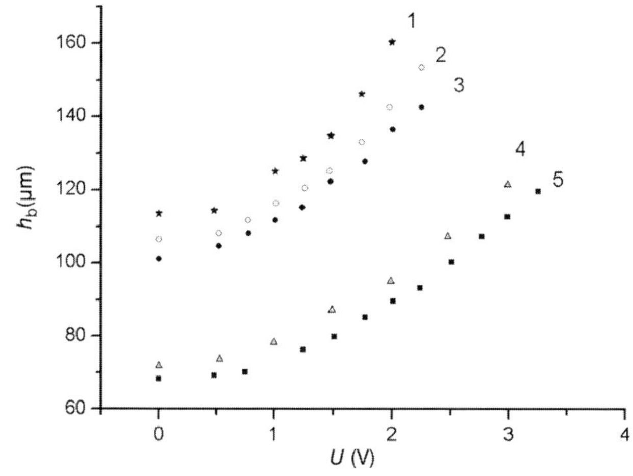

Figure 3.35 Dependence of the threshold layer thickness h_b on the applied voltage U. (1) $T = 67$ s, $\Delta P_0 = 10$ Pa; (2) $T = 49$ s, $\Delta P_0 = 11.4$ Pa; (3) $T = 49$ s, $\Delta P_0 = 12.7$ Pa; (4) $T = 16.1$ s, $\Delta P_0 = 33.1$ Pa; (5) $T = 15.5$ s, $\Delta P_0 = 44$ Pa. ZhK616 24 °C.

This effect is also quantitatively described in terms of universal dependences as shown in Figure 3.36. Contrary to the case of $\Delta\varepsilon < 0$, the range of stabilizing applied voltages is rather wide (0–100 V for $h > 50\,\mu m$). It makes the electric field very effective at the control of both linear and nonlinear regimes of a director motion that is important for practical applications (Chapter 7).

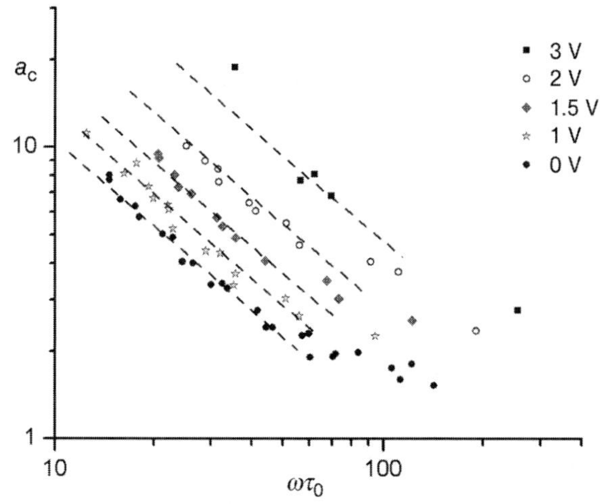

Figure 3.36 Dependencies of the dimensionless threshold amplitude a_c on the parameter $\omega\tau_0$ at different voltages. Dashed lines represent the results of theoretical calculations in a low-frequency range. ZhK616 24 °C.

3.5
Secondary Instabilities in Poiseuille Flows

There are some interesting effects induced by decay and oscillating flows in the regions of wedge-like cell with the thickness exceeding the threshold thickness for a homogeneous instability. These *secondary* effects arise in the conditions far from equilibrium when the basic state corresponds to the *primary* instability connected with the escape of a director from the flow plane. It is worthwhile to note that existence of dissipative structures is a general property of the systems with strongly irreversible processes. In our case, the next structures of such type were experimentally observed.

3.5.1
Domain Walls

The domain walls separate the regions with the opposite signs of y-projections of a director ($\mathbf{n}_y$ and $-\mathbf{n}_y$). Such structure was induced by oscillating flows [40, 70, 71]. At low frequencies, the flow stimulates not only the rise domain walls but also their slow transformation (Figure 3.37). In the final stage, the walls are oriented in the flow direction. These walls disappeared for 1–5 min after the flow stopped. This time is comparable to the characteristic time τ_0 of a director motion introduced above. Previously, the domain walls were observed for an initially planar orientation [22]. But in the latter case, they were oriented normally in the flow direction and obviously arose due to the defects on the channel edges.

3.5.2
Secondary Roll Instability in Oscillating Flow

The use of the wedge-like cell provided the first observation of the secondary roll instability in a homeotropic nematic layer under oscillatory Poiseuille flow. It arose in

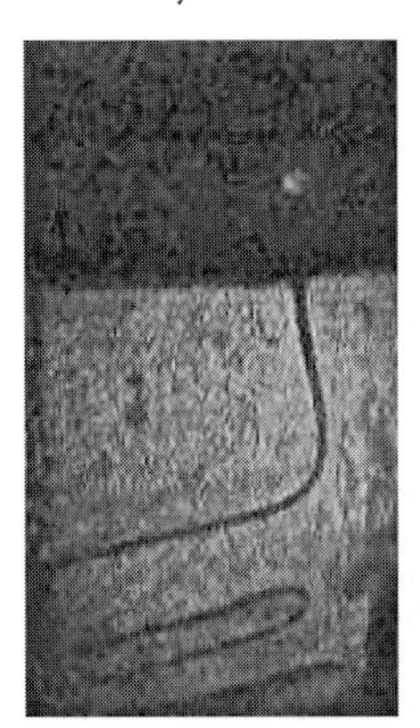
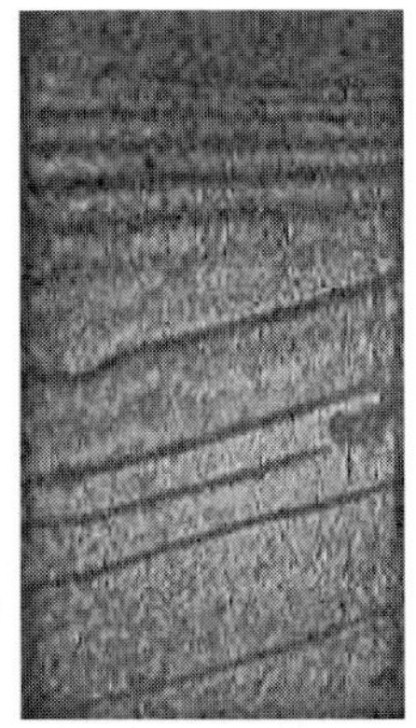

Figure 3.37 Evolution of domain walls induced by an oscillating Poiseuille flow; snapshots correspond to 85 min (left) and 120 min (right) after beginning of the flow; geometry "*a*", $\Delta P_0 = 26.5$ Pa, $T = 33.7$ s. MBBA 23 °C.

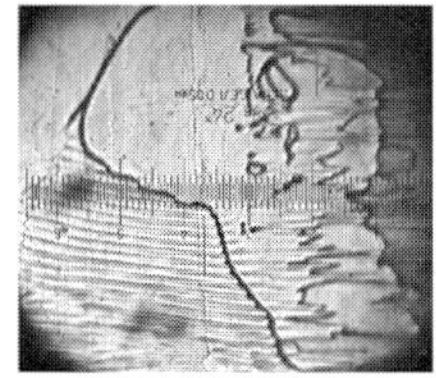

Roll structure near the threshold of arise

$T = 12.5$ s

$\Delta P_0 = 126$ Pa

$<h> = 140$ μm

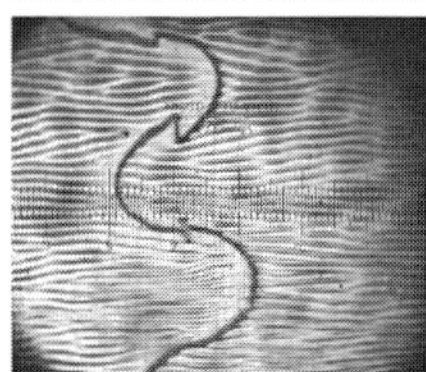

Roll structure above the threshold of arise

$T = 12.5$ s

$\Delta P_0 = 126$ Pa

$<h> = 160$ μm

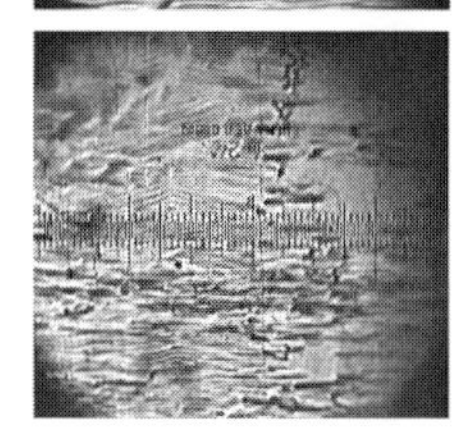

Disappearance of the roll structure. Chaos

$T = 12.5$ s

$\Delta P_0 = 177$ Pa

$<h> = 160$ μm

Figure 3.38 Roll's instability in oscillatory Poiseuille flow. MBBA 22 °C.

the region of the first homogeneous instability connected with the escape of a director from the flow plane. Rolls were dislocated into domains and moved in accordance with overall flow motion. Roll's axes were oriented exactly along the flow direction in the region near the threshold of appearance and were inclined far from the threshold. In general, the angle of inclination increased with the layer thickness and showed a slow precession. Increasing thickness also led to the development of a number of defects, which are transformed into a chaos-like structure showing strong light scattering. The evolution of the roll structure into chaos is demonstrated by photos in Figure 3.38.

The space period of the rolls near the threshold q_{flow} was obtained via light diffraction. The ratio q_{flow}/h and the analogous ratio for electroconvective rolls oriented by flow (see Section 3.3.6) are shown in Figure 3.39 as functions of the layer thickness h. The relative period of the secondary flow instability decreases with h, contrary to the case of electrically induced rolls where the period is approximately equal to the layer thickness.

Presumably, the secondary roll instability can be produced by the same destabilizing mechanisms as those considered previously [22, 23] for the initial planar orientation perpendicular to the flow plane (see Section 3.3.1). Analysis presented in Section 3.3.5 for a homogeneous instability decay flow shows that such type of orientation is also realized in the central part of the cell with a homeotropic surface orientation. In this case, the period of the secondary instability has to be slightly smaller (due to boundary layers) than the layer thickness. Obviously, this situation really takes place in the thinner part of the cell. At the same time, it is well known that in intensive flows the overall orientation of LC is close to the flow direction. Such

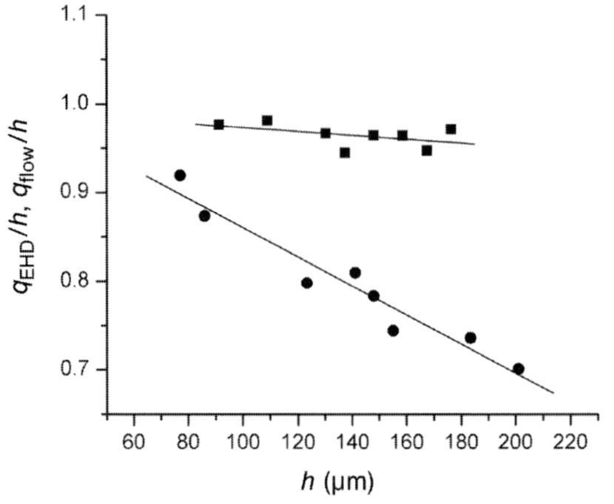

Figure 3.39 Comparison of roll structures periods: (●) rolls, induced by oscillating flow; (■) electroconvective rolls stabilized by oscillating Poiseuille flow. MBBA 22 °C.

mechanism that is most favorable in the thicker part of the cell can suppress the perpendicular orientation in the central part of the layer. It has to decrease the size of the central part and the corresponding period of the flow-induced roll structure that is in accordance with the data presented in Figure 3.39.

The frequency dependences of the thresholds for homogeneous and roll instabilities and also for chaos-like structure are presented in Figure 3.40 using dimensionless coordinates a_c and $\omega\tau_d$. The two latter thresholds were registered due to the appearance of the light diffraction produced by the periodical pace structure and it is suppressed by the chaos-like structure. In general, the threshold for the second roll instability in the case of homeotropic surface alignment is higher

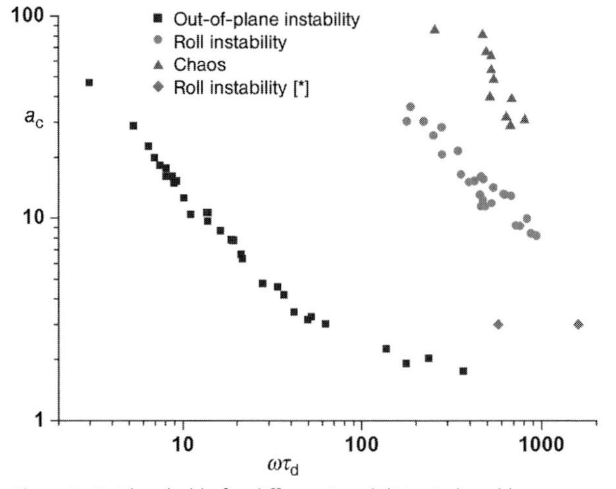

Figure 3.40 Thresholds for different instabilities induced by oscillatory Poiseuille flows; * – a threshold for a case with planar surface orientation [23]. $U = 0$ V, MBBA 22 °C.

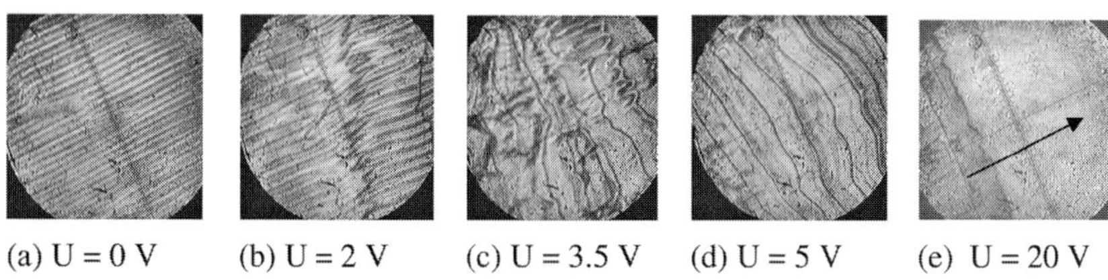

Figure 3.41 Electrically induced transformation of the secondary roll instability into the system of disclination lines. ZhK616 ($\Delta\varepsilon = +3.4$), $\Delta P_0 = 71$ Pa, $T = 16.7$ s, $\langle h \rangle = 162\,\mu$m, $\emptyset$ 6.4 mm, 22 °C; the flow direction is pointed by the arrow.

than that established for the simpler case of the planar surface orientation normal to the flow plane. Nevertheless, the thresholds for these cases are of the same order of magnitude, so the similar destabilizing mechanisms can act in both cases.

The investigation of the secondary roll instability in a liquid crystal mixture (ZhK616) with a positive sign of $\Delta\varepsilon$ under electric field revealed a nontrivial behavior. The experiments were conducted in the cell where one electrode was separated into two parts. It stimulated the emergence of defects in the vicinity of intraelectrode gap. In the absence of electric field, the roll structure looked similar to that observed for MBBA (Figure 3.41a). It would be reasonable to wait for suppressing the roll instability under applied electric field. The real scenario at increasing voltage was rather complicated; at relatively low voltages, the roll structure is deformed and the linear defect oriented perpendicular to the flow director was formed in the intraelectrode gap at the pressure inversion (Figure 3.41b). At the next moment, this line was moved by flow. The same phenomena took place every half of a period. So, a number of pair defects relatively symmetrical to the intraelectrode gap were observed after some periods of oscillations (Figure 3.41c). Increasing electric field results in suppressing the roll structure with a simultaneous transformation of defects in linear disclinations (Figure 3.41d). The defects were totally suppressed at voltages higher than 10 V (Figure 3.41e). In the latter case, the intraelectrode gap was well visualized under combined action of flow and field. So, the described effect can be used to produce well-oriented linear defects. It is important for physics of defects in liquid crystals. Such phenomena can be responsible for the formation of large-scale domains described in [22].

3.5.3 Long-Living Domains Produced by Flows

The memory-like effects in nematic liquid crystals are of especial interest from the viewpoint of basic science and practical applications. Some of them induced by electric fields will be described in Chapter 7. Below we present the experimental results concerning the long-living structures induced in the wedge-like cell both by the decay flow and the oscillatory one.

For the decay flow, it was established that macroscopic domains of different sizes and shapes arose in the thicker part of the cell at Ericksen numbers bigger than the

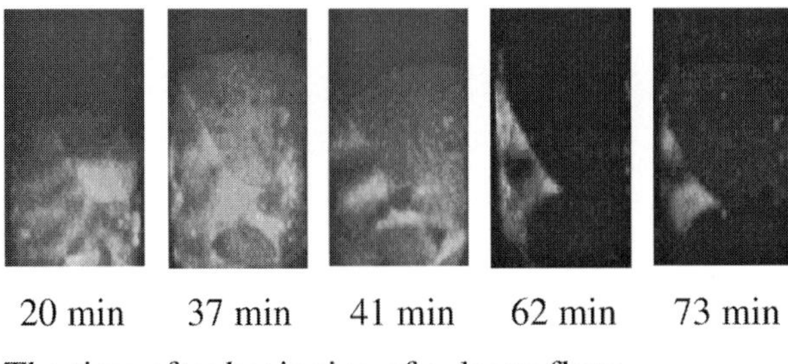

Figure 3.42 Relaxation of a long-living domain induced by a decay flow; MBBA, the initial pressure difference $\Delta P(0) = 37$ Pa, the range of local thicknesses corresponding to the image $h = (120, 175)$ μm; geometry "*a*".

critical value for out-of-plane transition. These regions were characterized by extremely long lifetime after the flow stopped. For example, it required more than 3 h for relaxation of the domain structure (induced by pressure gradient $G = 2500$ Pa/m and observed at local thickness about 130 μm) to the initial homogeneous homeotropic orientation (dark state in crossed polarizers). This time was essentially more than the decay time of a flow (about 10 min). At longer times, the slow relaxation process was well visualized in crossed polarizers oriented at 45° relative to the flow direction (geometry "*a*"), as shown in Figure 3.42. The observations in this geometry revealed a number of inclined interference stripes slowly moving inside domains, while the surrounding was already dark. In geometry "*b*" (one of the crossed polarizers was oriented along the flow direction), the domains were visualized as more bright regions separated from dark homeotropic surrounding by narrow bright boundary. It means that the mean orientation inside domains differs from the orientation in the flow plane.

The sizes of domains increased with the increase in pressure gradient and unique region of abnormal orientation could arise. After stopping the flow, the square of domains was slowly decreased due to the motion of domain walls (with a velocity of order 1 μm/s).

The analogous long-living regions were induced also by low-frequency oscillating flows. They arose after an exposure time t_{exp} (typically some minutes) in the regions with strong anisotropic scattering of light. The overall size of the regions increased with both the amplitude of pressure oscillations and the exposure time. In particular, it was possible to produce big unique regions that occupied the most area of the cell. The dynamics of such regions after stopping the flow is similar to that of a decay flow.

The nature of such regions with abnormally long lifetime is not well understood yet. One can propose that they are connected with local breaking of the nematic order mentioned above. The generalized theory [11] with a variable order parameter predicts the emergence of a number of sublayers parallel to the layer plane separated by regions with sharp change of the order degree. Such structure can show metastable behavior.

There is no doubt that the long-living structures described above arise also upon filling of the cells with LCs in the case of homeotropic orientation. So, such effects are of practical importance.

3.6
Shear Flows at Weak Anchoring

The theoretical analysis of the flow-induced phenomena described above was mostly carried out under the assumption of strong surface anchoring. It is obvious that the effects of weak anchoring (see Chapter 5) have to modify both static and dynamic behavior of liquid crystals as in the case of electric fields. The theoretical description taking into account weak anchoring was developed in the past decade [69, 77–82]. It includes some particular cases corresponding to linear and nonlinear regimes of a director motion. In the case of a linear regime, the attention was mainly focused on oscillating flows where weak anchoring effects described in terms of an anchoring strength and a surface viscosity (see Chapter 5) can modify the overall dynamic response of LC layer on the oscillating shear stress (Couette flow) or the pressure gradient (Poiseuille flow). Only rarely were experimental studies of flow-induced behavior of liquid crystals at weak anchoring [69, 77] performed so far.

3.6.1
Linear Oscillating Flows at Weak Anchoring

In many respects, the interest in linear flows in LC layers confined to weakly anchored surfaces was stimulated by the possibility of extracting information about the so-called surface viscosity coefficient predicted by theory. In the proposed theoretical models [83–85], it is referred to as the existence of near-surface layers described by a surface director $\mathbf{n}_s$ that moves slower in comparison with the bulk one. It produces additional viscous moments acting on the bulk director from the boundary layers, which modifies the dynamical response of LC on the external force, applied to the layer. The surface viscosity defines the value of this moment. It denotes the phase shift between bulk and boundary layers. Estimates made [86] have shown that this shift lies in microsecond range and can hardly be detected directly. So some experimental technique [87–89] was proposed to get information about fast surface dynamics and values of surface viscosity coefficients. Nevertheless, up to now the reliable estimates of the latter parameter were rarely reported [89].

Recently, the optical study of oscillating shear flows was considered as the alternative method for the registration of fast surface dynamics at weak anchoring [77, 78]. One can wait to see that the use of mechanically induced oscillations of liquid crystal makes it possible to exclude the electric properties of liquid crystals from the problem under consideration. So, such technique can be rather universal for LC materials and surfaces of different types.

In spite of essential theoretical progress in this direction, there are only two experimental works [69, 77] devoted to the special study of weak anchoring effects on shear flows of liquid crystals. In both cases, a combined homeoplanar orientation was used as the nondisturbed state of LC layer. This state is more complicated for a theoretical analysis than the homogeneous one, which was described in detail [78, 79]. Nevertheless, the former is more suitable for practical realization of a weak polar anchoring.

Below we will follow the general theoretical description of the problem presented in Ref. [69] for the case of Poiseuille shear flow.

3.6.1.1 General Equations

Let us consider a nematic layer with thickness h confined between two solid substrates. The origin of the Cartesian coordinate system is chosen in the layer center with z being perpendicular to the layer. The oscillatory Poiseuille flow induced by pressure gradient $(\Delta P/\Delta x)\cos(\omega\tau)$ applied along the x-axis.

The solution of the nematodynamics equations for the velocity $\mathbf{v}$ and director $\mathbf{n}$ corresponding to the motion of a director in the flow plane is expressed as

$$\mathbf{v} = (v_z, 0, 0), \qquad \mathbf{n} = (n_x, 0, n_z). \tag{3.47}$$

Velocity component v_z is equal to zero due to incompressibility condition div $\mathbf{v} = 0$.

It is convenient to use for numerical calculations the dimensionless variables t' and z' introduced as $t' = t\omega^{-1}$ and $z' = z/h$.

The complete set of dimensionless dynamic nematic equations is

$$\varepsilon(n_x n_{z,zz} - k_{31} n_z n_{x,zzx,zz} + F n_x n_z) + (n_z n_{x,t} - n_x n_{z,t}) + \frac{1}{\lambda - 1}(n_z^2 + \lambda n_x^2) v_{x,z} = 0 \tag{3.48}$$

$$-a_p \cos t + \partial_z \left\{ -n_z n_{x,t} - \lambda n_x n_{z,t} + \frac{\alpha_4 + (\alpha_3 + \alpha_6) n_x^2 + (\alpha_5 - \alpha_2 + 2\alpha_1 n_x^2) n_z^2}{-2\alpha_2} v_{x,z} \right\} = 0 \tag{3.49}$$

$$n_x^2 + n_z^2 = 1, \tag{3.50}$$

where $\varepsilon = 1/\tau_h\omega$, $\tau_h = \gamma_1 h^2/K_{11}$, $a_p = (\Delta P/\Delta x)(h/(-\alpha_2)\omega)$, $k_{31} = K_{33}/K_{11}$, $\lambda = \alpha_3/\alpha_2$, α_i is the Leslie viscosity, and K_{ii} the elastic constants.

By introducing new variable θ, the angle between director orientation and x-axis, one can rewrite (3.47) in the following form:

$$n_x = \cos\theta(z,t), \quad n_y = 0, \quad n_z = \sin\theta(z,t), \quad v_x = v_x(z,t), \quad v_y = 0, \quad v_z = 0. \tag{3.51}$$

Thus, the normalization condition is satisfied automatically and Equations (3.48) and (3.49) take form

$$\theta_{,t} - K(\theta) v_{x,z} = \varepsilon(P(\theta)\theta_{,zz} + 1/2 P'(\theta)\theta_{,z}^2 + F\cos\theta\sin\theta), \tag{3.52}$$

$$0 = -a_p \cos(t) + \partial_z\{-(1-\lambda)K(\theta)\theta_{,t} + Q(\theta) v_{x,z}\}, \tag{3.53}$$

with

$$K(\theta) = \frac{\lambda\cos^2\theta - \sin^2\theta}{1-\lambda}, \qquad P(\theta) = \cos^2\theta + k_{31}\sin^2\theta,$$

$$Q(\theta) = \frac{\alpha_4 + (\alpha_5 - \alpha_2)\sin^2\theta + (\alpha_3 + \alpha_6 + 2\alpha_1\sin^2\theta)\cos^2\theta}{2(-\alpha_2)}, \tag{3.54}$$

where $F = \pi^2 E^2/E_F^2$, $E_F = (\pi/h)\sqrt{K_{11}/(\varepsilon_0\,\Delta\varepsilon)}$, and $\Delta\varepsilon$ is the dielectric permittivity anisotropy.

Boundary condition for the velocity field (no-slip):

$$v(z = \pm 1/2) = 0. \tag{3.55}$$

There are two possible types of the director boundary conditions: symmetrical (weak homeotropic or planar boundary alignment on both substrates) and hybrid (one substrate with strong homeotropical anchoring and another one with weak planar anchoring).

The boundary condition for weak anchoring surfaces can be written as [86]

$$\pm P(\theta)\theta_{,z} + \frac{1}{2}\frac{Wh\partial f_s}{K_{11}\partial\theta} + \frac{\omega h}{K_{11}}\eta_s\frac{\partial\theta}{\partial t} = 0, \tag{3.56}$$

where W is the polar anchoring strength and $f_s(\theta - \theta_0) = \sin^2(\theta - \theta_0)$ is the function entering into the Rapini potential for specific surface energy per unit area:

$$F_s = (1/2)Wf_s(\theta - \theta_0), \tag{3.57}$$

$\eta_s = \gamma_1 \cdot l_{\gamma_1}$ is the so-called surface viscosity, which determines the viscous losses in near-boundary layer of thickness $l_{\gamma 1}$. $l_{\gamma 1}$ has a dimension of length and corresponds to characteristic viscosity length (a boundary layer where rotation viscosity γ_1 has experienced the influence by the solid surface). In practice, the particular type of boundary condition is determined by a proper surface treatment. For a strong anchoring, we have $\theta = const$ on the boundary (equal to zero for the planar alignment and $\pi/2$ for the homeotropic alignment).

Symmetrical boundary conditions are represented as

$$\pm P(\theta)\theta_{,z} + \frac{hW}{2K_{11}}\frac{\partial f_s}{\partial\theta} + \frac{h\omega}{K_{11}}\eta_s\frac{\partial\theta}{\partial t} = 0, \quad \text{“}-\text{” for } z = +1/2 \quad \text{and} \quad \text{“}+\text{” for } z = -1/2. \tag{3.58}$$

In the hybrid cell, the boundary conditions are nonsymmetric:

$$\theta = \pi/2|_{z=+1/2} \quad \text{(strong homeotropic anchoring)} \tag{3.59}$$

$$-\cos\theta \cdot \theta_{,z} + W_h\sin\theta + \eta_h\cos\theta\cdot\theta_{,t} = 0|_{z=-1/2} \quad \text{(weak planar anchoring)}, \tag{3.60}$$

where $W_h = hW/(P_0K_{11})$, $\eta_h = h\gamma_1 l_{\gamma_1}\omega/(P_0K_{11})$, $W > 0$.

3.6.1.2 Linear Oscillating Flow at Symmetrical Boundary Conditions

The set of Equations 3.52 and 3.53 with symmetrical boundary conditions (3.55) and (3.58) with a constant average angle on the boundaries. For the small flow amplitudes a_p, one can introduce the small perturbations $\tilde{\theta}$, $\tilde{U}$ of hydrodynamic parameters relative to the initial state θ_0, v_{0x}:

$$\theta = \theta_0 + \tilde{\theta}, \qquad v_x = v_{x0} + U, \qquad |\tilde{\theta}| \ll 1, \qquad |U| \ll 1, \qquad \partial f_s/\partial\theta = 2\tilde{\theta}, \tag{3.61}$$

with boundary conditions

$$\begin{aligned} &U(z=\pm 1/2) = 0, \quad \tilde{\theta}_{,z} - E\tilde{\theta} - G\tilde{\theta}_{,t} = 0|_{z=-1/2}, \quad \tilde{\theta}_{,z} + E\tilde{\theta} + G\tilde{\theta}_{,t} = 0|_{z=+1/2}, \\ &E = hW/(P_0 K_{11}), \quad G = h\gamma_1 l_{\gamma_1}\omega/(P_0 K_{11}), \quad W > 0. \end{aligned} \tag{3.62}$$

So, the set of the linear differential equations for small perturbations $\tilde{\theta}$ and U is presented as

$$\tilde{\theta}_{,t} - K_0 U_{,z} = \varepsilon(P_0\tilde{\theta}_{,zz} + F\tilde{\theta}), \qquad -(1-\lambda)K_0\tilde{\theta}_{,tz} + Q_0 U_{,zz} = a_p\cos(t). \tag{3.63}$$

The analytic solution of the set of Equation (3.63)

$$\tilde{\theta}(z,t) = T_1(z)\cos(t) + T_2(z)\sin(t), \qquad U(z,t) = U_1(z)\cos(t) + U_2(z)\sin(t). \tag{3.64}$$

with

$$\begin{aligned} T_1(z) &= -a_p M\left(\frac{(c_1 + M_f c_2)f_1(z) - (c_2 - M_f c_1)f_2(z)}{(1+M_f^2)(c_1^2+c_2^2)} - 2M_f z\right) \\ M_f &= \frac{\varepsilon F Q_0}{Q_0 - (1-\lambda)K_0^2} \\ T_2(z) &= -a_p M\left(\frac{(c_1 + M_f c_2)f_2(z) + (c_2 - M_f c_1)f_1(z)}{(1+M_f^2)(c_1^2+c_2^2)} - 2z\right) \\ M &= \frac{K_0}{2(Q_0 - (1-\lambda)K_0^2)}. \end{aligned} \tag{3.65}$$

Here, c_i denotes functions of z, W_h, and η_h.

In the case of strong surface anchoring ($E \gg 1$), the boundary conditions (3.62) become quite simple:

$$\tilde{\theta} = 0, \qquad z = \pm 1/2. \tag{3.66}$$

In this case, the functions $\tilde{T}_i(z)$ depend on the frequency of oscillations through the dependence of the wave number $k \sim \omega^{1/2}$. At high enough frequency, the latter parameter defines the distortions of a velocity and of an orientation in near-surface layers of LCs. It is also true for a weak anchoring if viscous contribution into the boundary condition is essentially smaller than the elastic one

$$\nu = G/E = \eta_s\omega/W \ll 1. \quad (3.67)$$

It is easy to estimate the frequency $f_m = \omega_m/2\pi$ at which surface viscosity effects can play the most important role by putting $\nu = 1$. For $h = 20 \times 10^{-6}$ m, $\eta_s = 10^{-8}$ Pa s m f_m varies from 15 to 150 Hz in the range of anchoring strength 10^{-6}–10^{-5} J/m^2. Of course, the thickness of the cell has to be decreased with increasing anchoring strength to detect the effects of weak anchoring on the orientational dynamics of LC layer.

It was shown by computer modeling of the dynamic behavior of a homeotropic layer [78] that the effects connected with the surface viscosity can be detected by analyzing the frequency dependence of optical response at frequencies comparable to f_m. Electric fields can also be useful for such study as they provide control of the spectra of a director motion [68]. The maximum value of the light intensity is shown in Figure 3.43 as a function of the flow frequency (Figure 3.43a) and the external electric field intensity (Figure 3.43b) for the different values of the surface viscosity. The influence of the surface viscosity is very strong in some regions of the flow frequency (f = 10–1000 Hz), and this parameter can be controlled by the external stabilizing electric field (for the fixed flow frequency). Note that the field dependence is more accurate in comparison to the first curve and can be easily reproduced in experiments.

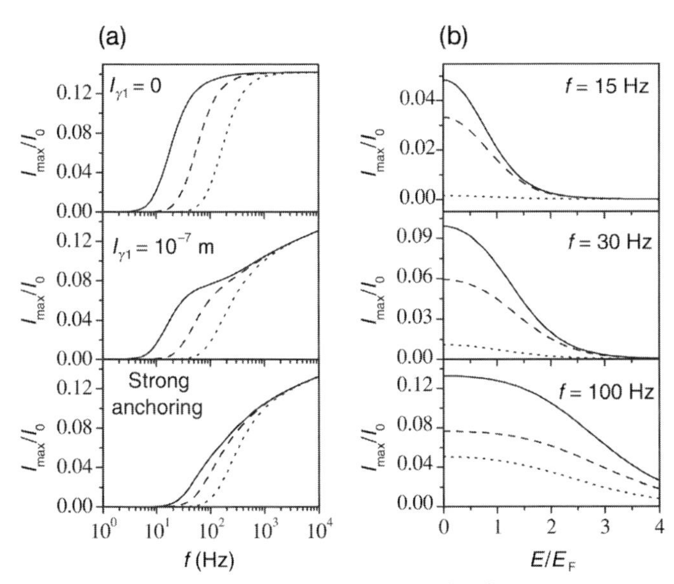

Figure 3.43 Maximum of light intensity versus (a) flow frequency and (b) external electric field intensity. $a_p = 0.2$, $h = 10\,\mu$m, $W = 10^{-6}$ J/m^2; (a) E/E_F: (—–) 0, (- - -) 2, (···) 4; (b) $l_{\gamma1}$ (m): (—–) 0, (- - -) 10^{-7}, (···) strong anchoring. Calculations were done for 5CB.

3.6.1.3 Linear Oscillating Flow at Hybrid Boundary Conditions

First, we have to define the influence of a weak anchoring on the initial orientational structure and optical properties of a homeoplanar layer.

For a stationary case in the absence of electric field, Equation 3.52 for a director motion becomes relatively simple:

$$P(\theta)\theta_{,zz} + (1/2)P'(\theta)\theta_{,z}^{2} = 0. \quad (3.68)$$

This equation has very simple solution in the case of one constant approximation ($K_{11} = K_{33}$):

$$\theta = C_1 z + C_2, \quad (3.69)$$

where C_1 and C_2 are the constants to be determined from boundary conditions. For strong anchoring at both surfaces:

$$C_1 = \pi/2, \qquad C_2 = \pi/4. \quad (3.70)$$

In the case of a weak anchoring on the planar substrate, these constants have to be determined from the next expressions

$$C_2 = \pi/2 - C_1/2 \quad (3.71)$$

$$C_1 - (1/2)\chi^{-1}\sin(2C_1 h) = 0, \quad (3.72)$$

where the parameter

$$\chi = K_{11}/Wh \quad (3.73)$$

can be considered a small parameter for the surfaces with a moderate anchoring. For example, for $K_{11} = 8.5 \times 10^{-12}$ N, $h = 20 \times 10^{-6}$ m (typical parameters of our experiments), χ varies from 0.04 for $W = 10^{-5}$ J/m^2 to 0.4 for $W = 10^{-6}$ J/m^2. In this case, the expressions for C_1 and C_2 can be obtained due to a first-order expansion of (3.72)

$$C_1 = \pi/2(1-\chi), \qquad C_2 = \pi/4(1+\chi). \quad (3.74)$$

It results in a nonzero value of the polar angle on the weak anchoring plate:

$$\theta(z = -1/2) = \pi\chi/2. \quad (3.75)$$

The difference between the weak anchoring surface and the strong one can be detected by optical measurements.

Using (3.69) and (3.74), we can write down the expression for δ as

$$\delta_0 = \frac{2\pi h}{\lambda}\Delta n\left[\frac{(1/2)-\chi}{1-\chi}\right]. \quad (3.76)$$

For strong anchoring surface expression, (3.76) results in

$$\delta = \frac{\pi}{\lambda} h\,\Delta n, \quad (3.77)$$

which means that the phase delay for homeoplanar orientation is twice smaller than for a planar one.

The existence of nonhomogeneous orientation as an initial stationary state of the cell with nonsymmetric boundary conditions makes the problem essentially more complicated for an analytic solution. Nevertheless, some useful results can be obtained without exact solution of the problem. First, it is quite reasonable to consider that the flow-induced small declinations $\tilde{\theta}$ from the initial nonhomogeneous state, approximately defined by expression (3.69), are proportional to the amplitude a_p of the driven force independent of the type of surface anchoring as in the case of a homeotropic alignment. For small deviations from the initial state, the optical phase delay can be expressed according to (3.76) as

$$\delta = (2\pi h\,\Delta n/\lambda)\langle\cos^2(\theta_0(z)+\tilde{\theta}(z,t))\rangle \approx (2\pi h\,\Delta n/\lambda)\langle\cos^2\theta_0(z)-\tilde{\theta}(z,t)\sin 2\theta_0(z)\rangle, \tag{3.78}$$

where $\langle\ \rangle$ is the mean value and $\theta_0(z)$ is the initial state determined by (3.69), and we consider the case $\tilde{\theta} \ll \theta_0$. In this case, the flow-induced difference $\tilde{\delta}$ between the maximal δ_{max} and the minimal δ_{min} values of δ has to be proportional to a_p contrary to the quadratic dependence for the case of a homeotropic sample. So, one can wait to see that for a hybrid cell, $\tilde{\delta}$ has to be proportional to the amplitude a_p of the driven force

$$\tilde{\delta} \sim a_p \sim \left(\frac{\Delta P}{\Delta x}\right)T. \tag{3.79}$$

This result can be checked experimentally by analyzing the intensity of polarized light passing through a hybrid cell. The coefficient of proportionality can depend on the frequency as in the case of symmetrical homeotropic anchoring. At high enough frequencies ($\varepsilon \gg 1$), the boundary layers become very thin and do not contribute essentially to the phase delay changes. In the low-frequency limit ($\varepsilon \ll 1$), the phase delay changes do not depend on frequency and include only elastic surface contribution. So, the intermediate frequency range seems to be most useful to extract information about surface viscosity.

3.6.1.4 **Experimental Technique and Results**

In many features, experiments conducted with flow-induced orientational structure in a homeoplanar cell [69] are similar to those described in Section 3.3.3 for the case of a homeotropic layer.

The construction of the cell with homeoplanar alignment differs from that of homeotropic cells (Figure 3.7) mostly by the type of surface treatment. The central channel of the hybrid cell with a constant gap was formed by two glass plates with inner surfaces treated in a special manner to provide a hybrid orientation. One of the surfaces was coated by a chrome di-stearyl film to get strong homeotropic anchoring. The opposite surface was spincoated by a 0.5% dye solution (SD1) in DMF and illuminated by UV polarized light ($\lambda = 360$ nm, $J = 0.8$ mW/cm^2) in a standard manner. The cell was separated into different zones that were treated by UV with different exposure times (t_{ex}). Usually, such procedure provides a different degree of an orientational order in a photosensitive layer and different anchoring strength (at least, as far as azimuthal anchoring is concerned [6]). The geometry of the experiment is close to describing above for an exception of the inhomogeneous basic state.

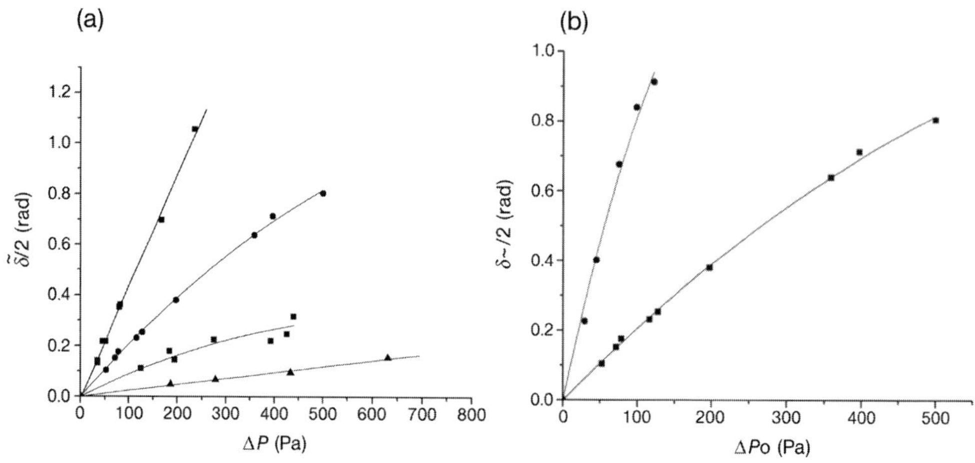

Figure 3.44 Dependencies of flow-induced variations of phase delay $\tilde{\delta}$ in a hybrid cell on the pressure difference amplitude ΔP_0; (a) ZhK616, $h = 20\,\mu m$, $t_{ex} = 10$ min; (1) ■, $T = 0.14$ s; (2) ● $T = 0.10$ s; (3) ■, $T = 0.064$ s; (4) ▲, $T = 0.04$ s; (b) ZhK616, $T = 0.10$ s; (1) ● $h = 42.3\,\mu m$; (2) ■, $h = 20\,\mu m$; solid lines: polynomial approximation of the second order.

The experimental setup, shown in Figure 3.17, was also applied for a study under consideration. Some modification was made only in the initial stage of pressure generation to provide higher frequencies. Experiments were done in geometry *a* useful for a study of in-plane motion of a director. In principle, the out-of-plane motion is possible due to the same destabilizing mechanisms as those that were considered above for the case of homeotropic orientation. So, to be sure that the director moves in the plane of the flow, the optical response of the cells in geometry "*b*" was controlled, too.

The existence of the initially deformed structure in the homeoplanar cell results in nonsymmetric response. In this case, the difference $\tilde{\delta}$ between maximal δ_{max} and minimal δ_{min} values of phase delay (instead of δ_{max} for a homeotropic cell) can be used as the main informative parameter connected with linear deformations of the orientational structure. Contrary to the case of a homeotropic orientation, this parameter linearly depends on the amplitude of the pressure difference applied to the cell (Figure 3.44a and b) at least at low values of the amplitude and frequency. Some nonlinear declinations take place at high enough values of frequencies. The sensitivity of such dependencies to the thickness variations (Figure 3.44b) at a frequency about 10 Hz is not as strong as for low-frequency data obtained for strong homeotropic anchoring Figure 3.21.

According to theoretical conclusions presented above, the frequency dependences of a phase difference δ (for homeotropic samples) and of a phase delay difference $\tilde{\delta}$ (for homeoplanar samples) are most sensitive to the variations of surface anchoring energy and surface viscosity. As already mentioned, low-frequency data obtained for homeotropic samples of MBBA can be quantitatively described in the framework of a model with strong anchoring. For homeoplanar samples with photoalingment treatment, one can wait to see the inclinations of the

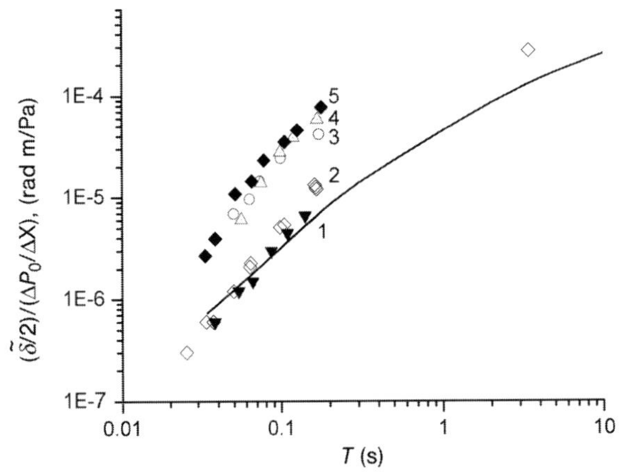

Figure 3.45 Normalized differences of phase delay as functions of pressure oscillations period: $h = 20\,\mu m$: (1) $t_{ex} = 20$ min, (2) $t_{ex} = 10$ min; $h = 45\,\mu m$: (3) $t_{ex} = 10$ min; (4) $t_{ex} = 20$ min; (5) $t_{ex} = 40$ min. ZhK616. Symbols: experimental data, line: theoretical curve for $l_{\gamma 1} = 10^{-7}$ m, $W_p = 10^{-7}$ J/m^2 MBBA.

frequency dependences from those predicted by such type of model especially for the case of relatively low anchoring energy. The examples of frequency dependencies of normalized phase delay difference $\tilde{\delta}/(\Delta P_0)$ at different experimental conditions are shown in Figure 3.45. One can see that there is no essential difference for the curves obtained at different exposure times of UV treatment.

So, the fast surface dynamics is insensitive to this parameter, at least at the control parameters (frequency and thickness) realized in experiments. There is some tendency of a decrease in the mean slope of the curves with decreasing layer thickness. It can be considered as a weak influence of surface viscosity, as it has to be more pronounced at lower values of thickness. Nevertheless, the comparison between experimental results and calculated frequency dependencies does not allow one to extract the exact value of the surface viscosity coefficient. Estimates have shown that the thickness of the boundary layer corresponding to the surface viscosity does not exceed 10^{-6} m. To extract more precise information about surface viscosity coefficient for surfaces with an intermediate anchoring (more than 10^{-6} J/m^2 as in the case described), one has to increase the frequency of oscillations with decreasing thickness of the layer. Moreover, it is needed to make more precise measurements of polar anchoring strength to extract reasonable values of the surface viscosity coefficient. Besides, independent measurements of polar anchoring strength seem to be important too.

3.6.2 Hydrodynamic Instabilities at Weak Anchoring

Recently, the influence of weak anchoring on nonlinear phenomena induced by shear flows in nematic liquid crystals was considered in some publications [80–82]. The developed theoretical description is restricted to the simplest type of instabilities arising under steady flows (plane *Couette* or *Poiseuille* flows) in the samples with the initial planar orientation normal to the flow plane (see Sections 3.3.1 and 3.3.2). The solution of the nematodynamic equations is essentially simplified for this case due to a homogeneous orientation and linear (Couette flow) or parabolic (Poiseuille flow) velocity profile corresponding to the linear basic state. It was shown by the linear stability analysis [80–82] that weak anchoring results in a decrease in the threshold shear rate (or pressure gradient) corresponding to the homogeneous instability. Moreover, in the case of Poiseuille flow, the variation of the anchoring strength can produce a crossover between homogeneous or spatially periodic (roll) instabilities, at least for MBBA. Such crossover was not found for Couette flow of MBBA at the same set of material parameters, though this effect has to be rather sensitive to slight variations in material constants [81]. A detailed analysis of the problem in the case of additional action of magnetic (electric) fields can be found in [82].

It is obvious that the destabilizing mechanism shown in Figure 3.2 involves changes of both azimuthal and polar angles. So one can wait to see the influence of corresponding azimuthal and polar anchoring on the threshold of a homogeneous instability.

Indeed, the computer simulation of the problem [80] confirmed that this conclusion was valid for both Couette and Poiseuille steady flows. It is demonstrated by Figures 3.46 and 3.47 [80], where dimensionless threshold shear rates a_c^2, corresponding to the emergence of homogeneous instability, are presented as functions of dimensionless anchoring strengths w_a and w_p. Here, the parameter a_c^2 is proportional to the shear rate (pressure gradient) for a Couette (Poiseuille) flow and depends on a number of viscoelastic parameters of LC, whereas the parameters w_a and w_p are proportional to the azimuthal (W_a) and polar (W_p) anchoring strengths ($w_a = W_a h / K_{22}$, $w_p = W_p h / K_{11}$).

There is no experimental confirmation of the results presented above. Meanwhile, it would be interesting to conduct such experiments with a well-controlled surface

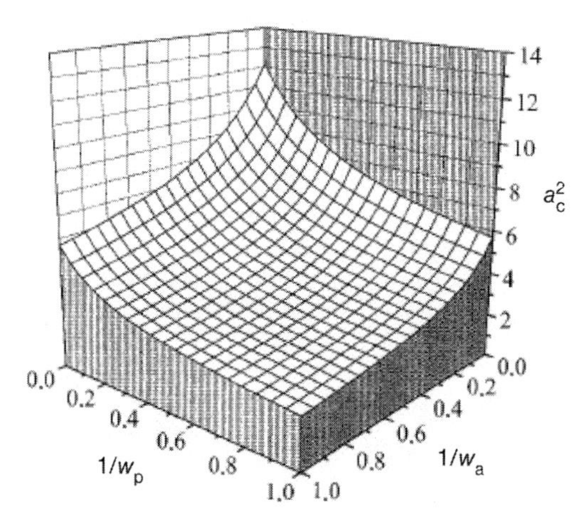

Figure 3.46 Critical shear rate a_c of Couette flow versus anchoring strengths: MBBA. (After Tarasov and Krekhov [80].)

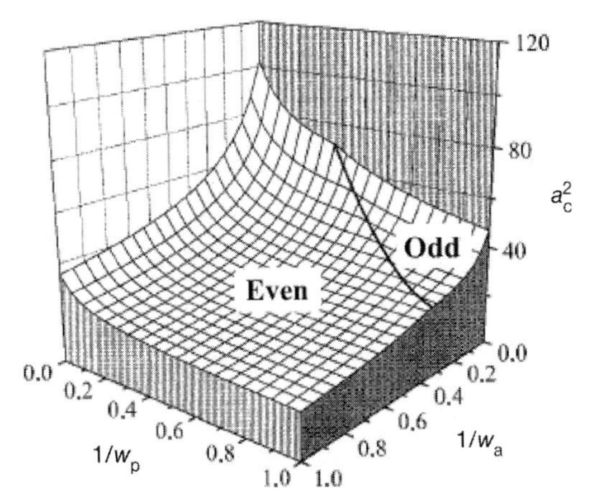

Figure 3.47 Critical shear rate a_c of Poiseuille flow versus anchoring strengths: MBBA. (After Tarasov and Krekhov [80].)

anchoring (e.g., provided by photoalignment technique). The proposed idea of using flows as an alternative technique for the determination of anchoring strengths [80] seems to be rather attractive as such procedure is applicable to different types of surfaces.

Similar information on surface anchoring can also be obtained from the study of nonlinear behavior of LC at a slightly supercritical shear rate. Theoretical analysis [81] has shown that for weaker anchoring, the director rotates more strongly and the relaxation time of the amplitude of the basic state perturbations significantly increases. The azimuthal and polar angles in this regime also strongly depend on anchoring strengths.

We conclude that experimental studies of linear and nonlinear phenomena in liquid crystals under shear flows can be considered as new tools for understanding weak anchoring surfaces that are of practical importance.

The general summary of the results presented above in this chapter important for practical applications of liquid crystals can be formulated as follows:

1. Liquid crystals show a very complicated rheological behavior in Couette and Poiseuille shear flows due to an intrinsic connection between translation motion (velocity field) and the collective rotation of long molecular axes described in terms of a director. This connection results in a number of linear and nonlinear phenomena that have no analogues in isotropic liquids.

2. The response of liquid crystal to the applied shear (Couette flow) or pressure gradient (Poiseuille flow) drastically depends on the initial boundary orientation.

3. The simplest case studied in detail corresponds to the initial planar orientation normal to the flow plane. In this case, the changes in orientational structure take place at some threshold flow amplitude. In contrast, there is no threshold flow amplitude for the initial homeotropic structure.

4. In the latter case, a liquid crystal layer shows maximal sensitivity to the action of shear flows. It is important for the elaboration of highly sensitive LC sensors of mechanical motion (see Chapter 7).
5. The optical response of a homeotropic layer on the action of steady and oscillating shear flows is well described in the case of small orientational distortions corresponding to the director motion in flow plane. At low frequencies, the sensitivity of liquid crystal cells strongly depends on the layer thickness and can be effectively controlled via electric fields. It is of key importance for possible practical applications, which will be described in Chapter 7.
6. For a Poiseuille flow, such linear regime breaks due to the emergence of a homogeneous instability connected with the escape of a director from the flow plane. The secondary instabilities of different types were also observed in intensive Poiseuille flows. Some of them show memory-like behavior that has to be taken into account at the application of liquid crystals.
7. Shear flows of liquid crystals can provide new information on surface anchoring properties that are of practical importance (see Chapter 5). More experiments are strongly needed to prove this conclusion.

References

1 Stewart, I.W. (2004) *The Static and Dynamic Continuum Theory of Liquid Crystals: A Mathematical Introduction*, Tailor and Francis.

2 Oswald, P. and Pieranski, P. (2005) *Smectic and Columnar Liquid Crystals*, Taylor and Francis, London.

3 Auernhammer, G.K., Brand, H.R., and Pleiner, H. (2002) Shear-induced instabilities in layered liquids. *Phys. Rev. E*, **66**, 061707.

4 Pleiner, H. and Brand, H.R. (1999) Nonlinear hydrodynamics of strongly deformed smectic C and C* liquid crystals. *Physica A*, **265**.

5 Ziebert, F. and Aranson, I.S. (2008) Rheological and structural properties of dilute active filament solutions. *Phys. Rev. E*, **77**, 011918.

6 Stannarius, R., Nemeş, A., and Eremin, A. (2005) Plucking a liquid chord: mechanical response of a liquid crystal filament. *Phys. Rev. E*, **72**, 020702(R).

7 Eremin, A., Nemeş, A., Stannarius, R., Schulz, M., Nádasi, H., and Weissflog, W. (2005) Structure and mechanical properties of liquid crystalline filaments. *Phys. Rev. E*, **71**, 031705.

8 Deniston, C., Orlandini, E., and Yeomans, J.M. (2001) Simulations of liquid crystals in Poiseuille flow. *Comput. Theor. Polym. Sci.*, **11**, 389.

9 Das, M., Chakrabarti, B., Dasgupta, C., Ramaswamy, S., and Sood, A.K. (2005) Routes to spatiotemporal chaos in the rheology of nematogenic fluids. *Phys. Rev. E*, **71**, 021707.

10 Calderer, M.C. and Mukherjee, B. (1997) On Poiseuille flow of liquid crystals. *Liq. Cryst.*, **22**, 121.

11 Calderer, M.C. and Liu, C. (2000) Poiseuille flow of nematic liquid crystals. *Int. J. Eng. Sci.*, **38**, 1007.

12 Mukherjee, B., Mazumder, S., and Calderer, M.C. (2001) Poiseuille flow of liquid crystals: highly oscillatory regimes. *J. Non-Newt. Fluid Mech.*, **99**, 37.

13 Hess, S. (2002) Complex fluid behavior: coupling of the shear stress with order parameter tensors of ranks two and three

in nematic liquid crystals and in tetradic fluids. *Physica A*, **314**, 310.

14 Rienacker, G. and Hess, S. (1999) Orientational dynamics of nematic liquid crystals under shear flow. *Physica A*, **267**, 294.

15 Tanner, R.I. (2000) *Engineering Rheology*, Oxford University Press.

16 Dubois-Violette, E. and Manneville, P. (1996) Flow instabilities in nematics, in *Pattern Formation in Liquid Crystals* (eds A. Buka and L. Kramer), Springer, Berlin.

17 Oswald, P. and Pieranski, P. (2005) *Nematic and Cholesteric Liquid Crystals*, Taylor and Francis, London.

18 Currie, P.K. (1979) Apparent viscosity during viscometric flow of nematic liquid crystals. *J. Physique*, **40**, 501.

19 Stephen, M.J. and Straley, J.P. (1974) Physics of liquid crystals. *Rev. Mod. Phys.*, **46**, 617.

20 Pieranski, P. and Guyon, E. (1973) Shear-flow induced transition in nematics. *Solid State Commun.*, **13**, 435.

21 Pieranski, P. and Guyon, E. (1974) Instability of certain shear flows in nematics. *Phys. Rev. A*, **9**, 404.

22 Guyon, E. and Pieranski, P. (1975) Poiseuille flow instabilities in nematics. *J. Physique Colloques*, **36**, Cl-203.

23 Janossy, I., Pieranski, P., and Guyon, E. (1976) Poiseuille flow in nematics: experimental study of the instabilities. *J. Physique*, **37**, 1105.

24 Manneville, P. and Dubois-Violette, E. (1976) Shear flow instability in nematic: theory of steady simple shear flows. *J. Physique*, **37**, 285.

25 Manneville, P. and Dubois-Violette, E. (1976) Steady Poiseuille flow in nematics: theory of the uniform instability. *J. Physique*, **37**, 1124.

26 Manneville, P. (1979) Theoretical analysis of Poiseuille flow instabilities in nematics. *J. Physique*, **40**, 713.

27 Manneville, P. (1978) Non-linearities and fluctuations at the threshold of a hydrodinamic instability in nematic liquid crystals. *J. Physique*, **39**, 911.

28 Dubois-Violette, E., Guyon, E., Janossy, I., Pieranski, P., and Manneville, P. (1977) Theory and experiment on plane shear flow instabilities in nematics. *J. Mecanique*, **16**, 733.

29 Pikin, S.A. (1974) Couette flow of a nematic liquid. *Sov. Phys. JETP*, **38**, 1246.

30 Pikin, S.A. (1991) *Structural Transformations in Liquid Crystals*, Gordon and Breach Science Publishers, NY.

31 Zuniga, I. and Leslie, F.M. (1989) Orientational instabilities in plane Poiseuille flow of certain nematic liquid crystals. *J. Non-Newt. Fluid Mech.*, **33**, 123.

32 Zuniga, I. (1990) Orientational instabilities in Couette flow of non-flow-aligning nematic liquid crystals. *Phys. Rev. A*, **41**, 2050.

33 Leslie, F.M. (1968) Some constitutive equations for liquid crystals. *Arch. Ration. Mech. Anal.*, **28**, 265.

34 Gähwiller, Ch. (1972) Temperature dependence of flow alignment in nematic liquid crystals. *Phys. Rev. Lett.*, **28**, 1554.

35 Wahl, J. and Fischer, F. (1972) A new optical method for studying the viscoelastic behaviour of nematic liquid crystals. *Opt. Commun.*, **5**, 341.

36 Wahl, J. and Fischer, F. (1973) Elastic and viscosity constants of nematic liquid crystals from a new optical method. *Mol. Cryst. Liq. Cryst.*, **22**, 359.

37 Skarp, K., Carlsson, T., Lagerwall, S.T., and Stebler, B. (1981) Flow properties of nematic 8 CB. *Mol. Cryst. Liq. Cryst.*, **66**, 199.

38 Cladis, P.E. and Torza, S. (1975) Stability of nematic liquid crystals in Couette flow. *Phys. Rev. Lett.*, **35**, 1283.

39 Pieranski, P. and Guyon, E. (1974) Two shear-flow regimes in nematic *p*-*n*-hexylo-benzilidene-*p*′-aminobenzonitrile. *Phys. Rev. Lett.*, **32**, 924.

40 Zuniga, I. and Leslie, F.M. (1989) Shear flow instabilities in nonaligning nematic liquid crystals. *Europhys. Lett.*, **9**, 689.

41 Zuniga, I. and Leslie, F.M. (1989) Shear flow instabilities in non-flow-aligning nematic liquid crystals. *Liq. Cryst.*, **5**, 725.

42 Hiltrop, K. and Fischer, F. (1976) Radial Poiseuille flow of a homeotropic nematic LC layer. *Z. Naturforsch.*, **31**, 800.

43 Pasechnik, S.V., Krekhov, A.P., Shmeleva, D.V., Nasibullaev, I.Sh., and Tsvetkov, V.A. (2005) Orientational instability in a nematic liquid crystal in a decaying Poiseuille flow. *J. Exp. Theor. Phys.*, **100**, 804.

44 Kneppe, H. and Shneider, F. (1981) Determination of the viscosity coefficients of the liquid crystal MBBA. *Mol. Cryst. Liq. Cryst.*, **65**, 3.

45 de Gennes, P.-G. and Prost, J. (1993) *The Physics of Liquid Crystals*, 2nd edn, Clarendon Press, Oxford.

46 Krekhov, A.P. and Kramer, L. (1996) Flow-alignment instability and slow director oscillations in nematic liquid crystals under oscillatory flow. *Phys. Rev. E*, **53**, 4925.

47 Yeh, P. and Gu, C. (1999) *Optics of Liquid Crystal Displays*, John Wiley and Sons, Inc., New York.

48 Hertrich, A., Decker, W., Pesch, W., and Kramer, L. (1992) The electrohydrodynamic instability in homeotropic nematic layers. *J. Phys. II France*, **2**, 1915.

49 Kramer, L. and Pesch, W. (1996) Electrohydrodynamic Instabilities in Nematic Liquid Crystals, in *Pattern Formation in Liquid Crystals* (eds A. Buka and L. Kramer), Springer, Berlin.

50 Bock, T.M., Blasing, J., Frette, V., and Rehberg, I. (2000) Alignment visualization using electroconvection of planar nematics. *Rev. Sci. Instrum.*, **71**, 2800.

51 Buka, A., Toth, P., Eber, N., and Kramer, L. (2000) Electroconvection in homeotropically alligned nematics. *Phys. Rep.*, **337**, 157.

52 Porter, R.S. and Johnson, J.F. (1966) Some flow characteristics of mesophase types. *J. Chem. Phys.*, **45**, 1452.

53 Kapustina, O.A. (1984) *Acoustooptical Phenomena in Liquid Crystals*, Gordon and Breach, NY.

54 Börzsönyi, T., Buka, A., Krekhov, A.P., and Kramer, L. (1998) Response of a homeotropic nematic liquid crystal to rectilinear oscillatory shear. *Phys. Rev. E*, **58**, 7419.

55 Belova, G.P. and Remizova, E.I. (1985) Peculiarities of ocoustooptic interaction in a homeotropically aligned layer of a nematic liquid crystal under periodical shear deformation. *Acoust. Zh.*, **31**, 289 (in Russian).

56 Kozhevnikov, E.N. (1986) Domain structure in a normally oriented liquid crystal layer under action of low-frequency shear. *Sov. Phys. JETP*, **64**, 793.

57 Tarasov, O.S., Krekhov, A.P., and Kramer, L. (1999) Nematic liquid crystal under plane oscillatory flows. *Mol. Cryst. Liq. Cryst.*, **328**, 573.

58 Kozhevnikov, E.N. (2002) Domain structure in a layer of nematic liquid crystal under oscillating Couette flow. *Bulletin Perm. St. Uni. Phys.*, **1**, 63 (in Russian).

59 Scudieri, F., Bertolotti, V., and Mellone, S. (1976) Acoustohydrodynamic instability in nematic liquid crystals. *Appl. Phys.*, **47**, 3781.

60 Scudieri, F. (1976) High-frequency shear instability in nematic liquid crystal. *Appl. Phys. Lett.*, **29**, 398.

61 Tarasov, O.S. and Krekhov, A.P. (1999) Orientational instability of a nematic liquid crystal under oscillating shear flow. *Kristallografia*, **44**, 1121 (in Russian).

62 Takatoch, K., Hazegawa, M., Koden, M., Itoh, N., Hazekawa, R., and Sakamoto, M. (2005) *Alignment Technologies and Applications of Liquid Crystal Devices*, Taylor and Francis, London.

63 Chigrinov, V.G., Kozenkov, V.M., and Kwok, H.S. (2008) *Photoaligning: Physics and Applications in Liquid Crystal Devices*, Wiley-VCH Verlag GmbH, Weinheim.

64 Pikin, S.A. and Chigrinov, V.G. (1974) Instability of Poiseuille flow of a nematic fluid. *Zh. Eksp. Teor. Fiz.*, **67**, 2280 (in Russian).

65 Blinov, L.M., Dadivanjan, S.A., Reshetov, V.D., Subachus, D.B., and Jablonskii, S.V. (1990) Peculiarities of Poiseuille flow in flat capillaries on the example of the acoustically disturbed liquid crystal. *Zh. Eksp. Teor. Fiz.*, **97**, 1597 (in Russian).

66 Tarasov, O.S. and Krekhov, A.P. (1998) Nematic liquid crystal under oscillating Poiseuille flow. *Kristallografia*, **43**, 516 (in Russian).

67 Pasechnik, S.V. and Torchinskaya, A.V. (1999) Behaviour of nematic layer oriented by electric field and pressure gradient in the striped liquid crystal cell. *Mol. Cryst. Liq. Cryst.*, **331**, 341.

68 Pasechnik, S.V., Torchinskaya, A.V., Shustrov, B.A., and Urmanova, T.N. (2001) Nonlinear optical response of nematic liquid crystal on varying pressure difference in the presence of electric field. *Mol. Cryst. Liq. Cryst.*, **367**, 727.

69 Pasechnik, S.V., Nasibullayev, I.Sh., Shmeliova, D.V., Tsvetkov, V.A., and Chigrinov, V.G. (2006) Oscillating Poiseuille flow in the hybrid cell with UV activated anchoring surface. *Liq. Cryst.*, **33** (10), 1153–1165.

70 Shmeliova, D.V. (2005) Investigation of LC structures induced by shear flow and electric field, PhD thesis, Moscow State Academy of Instrument Engineering and Computer Science.

71 Toth, P., Krekhov, A.P., Kramer, L., and Peinke, J. (2000) Orientational transition in nematic liquid crystals under oscillatory Poiseuille flow. *Europhys. Lett.*, **51**, 48.

72 Ezhov, S.G., Pasechnik, S.V., and Balandin, V.A. (1984) Effect of an electric field on the temporal characteristics of the acoustooptic effect in nematic liquid crystals. *Sov. Tech. Phys. Lett. (USA)*, **10**, 202.

73 Krekhov, A.P. and Kramer, L. (1996) Flow-alignment instability and slow director oscillations in nematic liquid crystals under oscillatory flow. *Phys. Rev. E*, **53**, 4925.

74 Krekhov, A.P., Borzsonyi, T., and Toth, P. *et al.* (2000) Nematic liquid crystals under oscillatory shear flow. *Phys. Rep.*, **337**, 171.

75 Nasibullaev, I.Sh. (2004) Orientational instabilities in the nematic liquid crystals in the flow, PhD thesis, IFMK, UNTS RAS, Ufa.

76 Nasibullayev, I.Sh. and Krekhov, A.P. (2003) Instabilities in nematic liquid crystal under combine action of the oscillatory Poiseuille flow and electric field. Proceedings of 7th ECLC, p. 43.

77 Khazimullin, M.V., Börzsönyi, T., Krekhov, A.P., and Lebedev, Yu.A. (1999) Orientational transition in nematic liquid crystal with hybrid alignment under oscillatory shear. *Mol. Cryst. Liq. Cryst.*, **329**, 247–254.

78 Nasibullaev, I.Sh. and Krekhov, A.P. (2001) Behavior of a nematic liquid crystal in oscillatory flow at weak surface anchoring. *Crystallogr. Rep.*, **46**, 495.

79 Nasibullaev, I.Sh., Krekhov, A.P., and Khazimulin, M.V. (2000) Dynamics of nematic liquid crystal under oscillatory flow: influence of surface viscosity. *Mol. Cryst. Liq. Cryst.*, **351**, 395.

80 Tarasov, O.S., Krekhov, A.P., and Kramer, L. (2001) Influence of weak anchoring on flow instabilities in nematic liquid crystals. *Liq. Cryst.*, **28**, 833.

81 Tarasov, O.S. (2004) Shear-induced rotations in a weakly anchored nematic liquid crystal. *Liq. Cryst.*, **31**, 1235.

82 Nasibullayev, I.Sh., Tarasov, O.S., Krekhov, A.P., and Kramer, L. (2005) Orientational instabilities in nematic liquid crystals with weak anchoring under combined action of steady flow and external fields. *Phys. Rev. E*, **72**, 051706.

83 Vorflusev, V.P., Kitzerow, H.S., and Chigrinov, V.G. (1997) Azimuthal surface gliding of a nematic liquid crystal. *Appl. Phys. Lett.*, **70**, 3359.

84 Tsoy, V.I. (1995) Freedericksz transition dynamics in a nematic layer with a surface viscosity. *Mol. Cryst. Liq. Cryst.*, **264**, 51.

85 Durand, G.E. and Virga, E.G. (1999) Hydrodynamic model for surface nematic viscosity. *Phys. Rev. E*, **59**, 4137.

86 Kedney, P.J. and Leslie, F.M. (1998) Switching in a simple bistable nematic cell. *Liq. Cryst.*, **24**, 613.

87 Oliveira, E.A., Figueiredo, A.M., and Durand, G. (1991) Gliding anchoring of lyotropic nematic liquid crystals on amorphous glass surfaces. *Phys. Rev. A*, **44**, R825.

88 Petrov, A.G., Ionescu, A.Th., Versache, C., and Scaramuzza, N. (1995) Investigation of flexoelectric properties of a palladium-containing nematic liquid crystal, Azpac, and its mixtures with MBBA. *Liq. Cryst.*, **19**, 169.

89 Vilfan, M., Olenik, I.D., Mertelj, A., and Čopič, M. (2001) Aging of surface anchoring and surface viscosity of a nematic liquid crystal on photoaligning poly-.vinyl-cinnamate. *Phys. Rev. E*, **63**, 061709.

4
Ultrasound in Liquid Crystals

This chapter is devoted to the application of ultrasonic methods in rheology of liquid crystals. First, we will describe the methods based on anisotropic propagation of longitudinal waves through the compressible liquid crystalline media. The construction of acoustic chambers to realize pulse and resonator methods will be presented, as well as the specific peculiarities of measurements in liquid crystals. A special attention will be paid to physical backgrounds that make it possible to use ultrasound for studying viscoelastic properties of nematic liquid crystals (NLCs). In particular, ultrasonic investigations of the rotational viscosity coefficient at variations of temperature and pressure will be considered in detail. After it, we will provide a short excursion into ultrasonic methods based on shear waves in nematic liquid crystals. A detailed description of ultrasonic investigations performed at different phase transitions (nematic–isotropic liquid, nematic–smectic A, and nematic–smectic C) will compose the next section of the chapter. We will show that ultrasonic methods provide a unique possibility for an experimental study of critical dynamics and viscoelastic properties of liquid crystals in the vicinity of phase transitions. Finally, the ultrasonic investigations in confined liquid crystal samples will be described. The presented methods can be effectively used in rheological studies of synthesis of modern liquid crystal materials. The possibility of ultrasonic study of phase diagrams and critical dynamics of liquid crystals under strong confinement will be demonstrated too.

4.1
Methods and Technique of Ultrasonic Investigations of Liquids and Liquid Crystals: Longitudinal Waves

Ultrasonic methods are widely used to study rheological properties of condensed matter including isotropic liquids, polymers, and solids [1]. In spite of the fact that first ultrasonic investigations of liquid crystals were performed about 50 years ago (see, e.g., the review by Natale [2]), there are some problems that have not yet been solved. That is why ultrasonic technique is of relatively limited use except for some cases discussed below.

Liquid Crystals: Viscous and Elastic Properties
S. V. Pasechnik, V. G. Chigrinov, and D. V. Shmeliova

ISBN: 978-3-527-40720-0

Most of the ultrasonic experiments, described in literature, are based on the use of two main types of ultrasonic waves – longitudinal waves and the shear waves [3]. They differ in the direction of oscillating motion of particles relative to the direction of wave propagation. In the case of longitudinal waves, both directions coincide while they are normal to each other in the case of shear waves. It is well known that shear waves may easily propagate in solids but quickly decay in isotropic liquids where they correspond to *overdamped modes*. The latter is also true for nematic liquid crystals. Contrarily, longitudinal waves show relatively weak attenuation at propagation in isotropic liquids and liquid crystals, especially at low frequencies. So, they correspond to *propagating modes*. Two types of waves are also quite different from the viewpoint of experimental setups and obtained parameters. In this section, we will consider the ultrasonic methods based on longitudinal waves.

The principle of ultrasonic measurements by using longitudinal waves is illustrated by the scheme of the simple setup shown in Figure 4.1. Usually, ultrasonic longitudinal waves are formed in an acoustic chamber filled with a liquid under investigation [4]. The setup presented in Figure 4.1 can be used to study the propagation of acoustic pulses through liquid media. It includes three generators (1–3) needed to form high-frequency radio pulses of definite duration and amplitude by mixing radio frequency continuous oscillations and video pulses.

Two piezoelectric transducers (4 and 5) served for the direct transformation of the input high-frequency radio signal into mechanical oscillations of liquid particles and for an inverse transformation. Oscilloscope 6 is used to control the time of pulse propagation and also the amplitude of output pulses. It provides measurements of the ultrasonic velocity and the attenuation. In some experimental setups, the first transducer is used for both direct and inverse transformations. In this case, the second transducer is replaced by a passive reflector of ultrasonic waves. Two kinds of piezoelectric transducers, monocrystal and polycrystalline ferroelectric plates, are used in ultrasonic range of frequencies.

Monocrystalline plates are usually fabricated from crystals of quartz or lithium niobate ($LiNbO_3$) to provide maximal efficiency of an electroacoustic transformation. For example, a quartz plate has to be oriented normally to Z-axis of a crystal to obtain

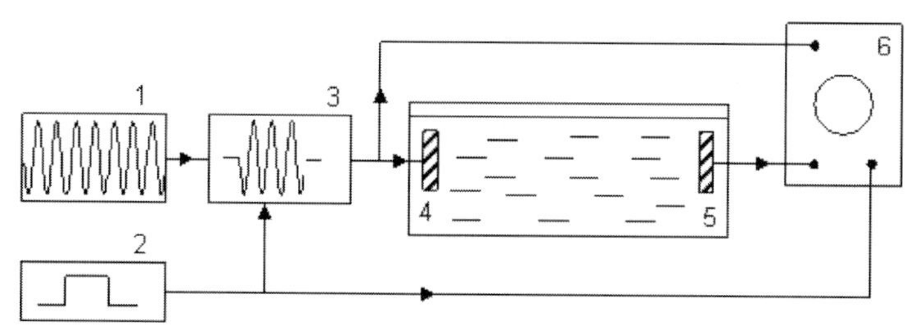

Figure 4.1 Typical scheme of ultrasonic setup for the study of liquids via longitudinal waves: (1) generator of high-frequency continuous oscillations; (2) generator of video pulses; (3) generator of radio pulses; (4, 5) piezoelectric transducers; (6) oscilloscope.

maximal longitudinal oscillations of the plate in Z-direction. One can induce acoustic oscillations in a plate by applying high-frequency electric voltage to the electrodes evaporated on the opposite planes of the plate. Ultrasonic measurements are usually performed at frequencies close to the resonance frequencies f_n of an element, expressed as

$$f_n = (2n+1)(C/d), \tag{4.1}$$

where C is the velocity of longitudinal ultrasonic waves traveling in a plate in Z-direction, d is the thickness of a plate, and n is the number of resonance oscillations. The basic resonance frequency f_0 corresponding to $n=0$ is about 1.5 MHz for a quartz plate of 1 mm thickness.

One of the main advantages of monocrystalline elements is the possibility of an effective electroacoustic transformation not only at basic resonance frequency but also at higher resonance frequencies. So, they can be used for ultrasonic measurements up to a frequency of about 100 MHz. At higher frequencies, the electromechanical transformation is achieved by placing the piezo element into a cavity of a special resonator of electromagnetic waves. This makes it possible to reach extremely high ultrasonic frequencies (up to some gigahertz).

Polycrystalline transducers are fabricated by agglomeration and polarization of special ceramics. The most popular materials of such type, the so-called PZT ceramics, are composed from the solid solutions $PbTiO_3$–$PbZiO_3$. There is no problem in getting such elements of different sizes and shapes with a high-value coefficient of an electromechanical transformation. Nevertheless, such elements show high efficiency only at the basic resonance frequency and practically are useless at frequencies higher than 20 MHz. More details of an ultrasonic technique can be found in Refs [3–6].

In liquids, mechanical vibrations of a transducer result in periodic mechanical oscillations of molecules and local changes in pressure and other thermodynamic parameters. Such changes can be considered as equilibrium changes only at relatively low frequencies. In this case, a wave propagates through a liquid with low attenuation and with a velocity c determined by an adiabatic compressibility of a liquid β_a:

$$\beta_a = 1/(\rho c^2). \tag{4.2}$$

In general, it takes some finite time τ (the time of relaxation) for a system to reach an equilibrium state, and therefore the sound propagation is an irreversible process accompanied by dissipation of energy of an ultrasonic wave. There are a number of possible relaxation mechanisms even in the case of simple liquids [5, 7]. They can be connected, for example, to the transfer of energy from translational degrees of freedom to the internal ones (intermolecular relaxation) or to structural transformation of a liquid (structural relaxation). Both these processes contribute to the total absorption coefficient (α) of ultrasound, which determines the change of amplitude A of an ultrasonic wave with a distance x by an equation

$$A = A_0 \exp(-\alpha x), \tag{4.3}$$

where $A_0 = A(x=0)$ is the initial amplitude.

There is a background contribution (Stroke's absorption) α_0 proportional to the frequency squared that is connected with shear viscosity η_s:

$$\alpha_0 = (2\pi^2/pc^3)(4\eta_s/3)f^2. \tag{4.4}$$

For simple liquids, this contribution becomes dominant at extremely high frequencies (0.1–1 GHz). At moderate ultrasonic frequencies (1–10 MHz), the measured value of the absorption coefficient exceeds essentially the value calculated from Equation 4.4. The difference $\alpha_v = \alpha - \alpha_0$ is often described in terms of the so-called bulk viscosity η_v, which reflects the dissipation of energy via relaxation phenomena that take place at local density changing. For simple liquids, the contribution α_v of bulk viscosity to the absorption coefficient is expressed as

$$\alpha_v = (2\pi^2/pc^3)\eta_v f^2. \tag{4.5}$$

The main difference between the bulk viscosity and the shear viscosity is the frequency dependence of these parameters. The shear viscosity is frequency dependent when frequencies are extremely high (higher than 1 GHz [5, 8]). At the same time, the bulk viscosity can show the essential frequency dependence in the region of relatively low ultrasonic frequencies (1–100 MHz) corresponding to the frequency range usually realized at ultrasonic measurements [3–5]. Such dependence may be rather complicated in the case of a media characterized by a number of relaxation processes. It is important to note that relaxation processes contribute not only to the absorption coefficient but also to the velocity of ultrasound. So, the measurements and analysis of acoustic parameters (c and α) at different frequencies (acoustic spectroscopy) provide a lot of information about viscoelastic properties and relaxation processes in liquids.

The experimental technique for ultrasonic measurements is elaborated to a very high level [3–6]. For example, the typical experimental errors at measurements of an ultrasonic velocity to an absolute value are of order 0.1%, which can be considered as very high precision. At the same time, the typical errors at measurements of an absorption coefficient are in the range 1–10%. Nevertheless, the latter parameter is more sensitive to relaxation processes and is of great importance in ultrasonic studies.

In spite of a variety of particular experimental setups used for ultrasonic studies of liquids, most of them are based on two kinds of methods: (i) an impulse method and (ii) a resonator method.

4.1.1 Impulse Method

The example of instrument realization of such method was described above (Figure 4.1). In this case, an input electric radio impulse is transferred to an acoustical high-frequency impulse by the first piezoelectric transducer that acts as a piston. The acoustic impulse propagates through a liquid to a distance l (an acoustic way) and is transformed into an output electric radio impulse.

In the simplest modification of the method, one can measure the time shift between input and output pulses τ, which is proportional to the velocity of the sound:

$$\tau = c/l. \tag{4.6}$$

This parameter can be obtained by processing two radio impulses, which helps determine ultrasonic velocity c if acoustic way l is known.

The absorption coefficient can be determined by measuring the decrease in amplitude of an output radio impulse induced by attenuation of sound passing through the acoustic chamber. The decrease in this parameter imposed by the double transformation of signals in piezoelectric transducers can be excluded by calibration or by variation of an acoustic length.

In spite of simplicity of the basic idea of the method, there are a number of factors that have to be taken into account to minimize possible errors.

First, the acoustic field formed by oscillating piston differs from the ideal case of a plane wave, which is usually used in calculations, due to the diffraction phenomena. This difference depends on the ratio $r = \lambda/D$, where λ is the wavelength ($\lambda = c/f$), and D is the diameter of a piston. The plane wave regime is realized in the case

$$r \ll 1. \tag{4.7}$$

At the typical value of ultrasonic velocity in liquids ($c \approx 1.5 \times 10^3$ m/s), it corresponds to

$$D \gg 1\ \text{mm} \tag{4.8}$$

at a frequency $f = 1.5$ MHz.

It means that the diameter of a piezoelectric transducer has to be large enough (especially for low-frequency measurements) to avoid errors via diffraction effects (usually $D \geq 20\lambda$ can be considered a good approximation). So, at frequency 1.5 MHz it corresponds to $D \geq 20$ mm.

The second factor under consideration is the time duration T of an input radio impulse. First, it has to be long enough to identify the wave envelope with a continuous harmonic wave of the basic frequency f. Usually, the radio impulse has to include at least 10 high-frequency oscillations to match the condition mentioned above. It means that the minimal duration $T_{\min}$ of the impulse is expressed as

$$T_{\min} \geq 10/f. \tag{4.9}$$

At the same time, the total space length of an acoustic impulse propagating in liquids does not have to exceed the double acoustic way to avoid an interference with the secondary impulse reflected from transducers (see Figure 4.1). It results in the restriction of the maximal duration $T_{\max}$ of an impulse:

$$T \leq 2\tau. \tag{4.10}$$

To fulfill both conditions, one has to optimize the dimensions of an acoustic camera. For example, at frequency $f = 1.5$ MHz, the acoustic way l longer than 5 mm is needed to correctly measure the ultrasonic velocity.

Additional restrictions on the geometrical size of an acoustic chamber arise when ultrasonic measurements of an absorption coefficient are conducted. In accordance with expression (4.3) in the case

$$\alpha l \gg 1 \tag{4.11}$$

the output signal will be too small to be registered. Such situation can occur, for example, at high-frequency measurements (as an absorption coefficient increases with frequency) or at ultrasonic investigations of phase transitions points where critical increase in absorption coefficient takes place.

The opposite case

$$\alpha l = 1 \tag{4.12}$$

is undesirable too as the resolution of the method can be insufficient to obtain reliable data. Usually, it occurs at low-frequency measurements. So, the optimal conditions of measurements correspond to the intermediate case

$$\alpha l \approx 1. \tag{4.13}$$

The latter is difficult to realize for liquids characterized by essential changes in ultrasonic absorption, in particular for liquid crystals. So, different variations of the method based on chambers with two fixed lengths or on systems with variable acoustic ways were successfully used [3–5]. In the latter case, the pulse-phase method of a variable distance, the ultrasonic velocity and the absorption coefficient can be determined due to the interference of input and the first output radio pulses. Such interference takes place within the range of impulse duration defined by the next inequality:

$$\tau < T < 2\tau. \tag{4.14}$$

The right side of this inequality analogous to expression (4.10) means that reflected pulses traveling into a chamber do not contribute to the interference mentioned above.

The experimental procedure includes changing the acoustic base from l to $l + \Delta l$ and counting the exact number N of interference minima induced by this motion. In addition, the amplitude of the input impulse is changed from value A to $A - \Delta A$ via an attenuator to compensate for the changes in the amplitude of the output impulse produced by the absorption of ultrasound in liquids. It helps determine both the sound velocity and the absorption coefficient via simple expressions:

$$c = f\,\Delta l/N \tag{4.15}$$

$$\alpha = \Delta A/\Delta l. \tag{4.16}$$

The pulse phase method of a variable frequency, another modification of the pulse phase method, can be used in cases where the motion of the transducer is undesirable (e.g., in smectic liquid crystal, as such motion destroys the layered structure). Such technique provides measurements of changes in the ultrasonic velocity (Δc) and the absorption coefficient $\Delta\alpha$ induced by the variation of external parameters (e.g., temperature). These changes are compensated for by the variation of both the frequency (Δf) and the amplitude (ΔA) of the input impulse at the fixed acoustic base (l) to restore the minimum of an interference picture. In this case, the frequency dependence of phase shift of the signals inside transducers has to be taken into account. The next expressions, obtained in Ref. [9], can be used in calculations:

$$\Delta C/C = \Delta f/[(f_2 - f_1)\tau], \tag{4.17}$$

$$\Delta\alpha = \Delta A/l, \tag{4.18}$$

where τ is defined by expression (4.6), f_1 and f_2 are the frequencies corresponding to the neighbor interference minima.

On the basis of the estimates and remarks made above, one can conclude that precise ultrasonic measurements at low frequencies can be provided only for the bulk samples of typical volumes 5–50 cm^3. Although it does not play an essential role in most cases, this restriction is critical for ultrasonic studies of newly synthesized materials, such as liquid crystals.

4.1.2
Resonator Method

The main idea of the method is to use standing acoustic waves instead of the propagating ones as in the case of impulse method. It allows one to obtain reliable information at frequencies lower than 1 MHz at relatively a small amount of liquid.

In the simplest case, such method can be realized by using two piezoelectric transducers, as shown in Figure 4.1. The first transducer transforms high-frequency harmonic oscillations of electric voltage into acoustic oscillations that form a standing ultrasonic wave between transducers. This regime is realized at a number of resonance frequencies f_k, which are defined by

$$f_k = kC/(2l). \tag{4.19}$$

The amplitude of output signal U_k generated by the second transducer depends on the absorption coefficient of the liquid. For an ideal resonator with a negligible thickness of transducers and a complete reflection of an acoustics wave from the boundary liquid–transducer it can be expressed as [10]

$$U_k = U_{k,\mathrm{m}}\left[1 + \frac{\sin^2(\pi f/f_k)}{sh^2(\alpha l)}\right], \tag{4.20}$$

where $U_{k,\mathrm{m}}$ is the maximal value of the amplitude taking place at f_k. This expression is valid only in the vicinity of a resonance peak. For a low attenuation ($\alpha l \ll 1$), which usually holds at low frequencies, expression (4.20) can be linearized to obtain

$$\alpha\lambda/\pi = 1/Q_k = \Delta f_k/f_k, \tag{4.21}$$

where Q_k is an acoustic quality of a resonator connected with losses in a liquid and Δf_k is the width of the resonance peak at $U_k = 0.707\,U_{k,\max}$.

There are additional losses in a real resonator determined by a number of factors (diffraction and scattering of ultrasonic waves, nonideal reflection of waves from piezoelectric transducer, mechanical damping of the piezoelectric transducers at fixing in a chamber, etc.). Such losses can be summarized in the additional parameter Q_0. So, the corrected expression for an absorption coefficient will be written as

$$\alpha = \frac{\pi}{\lambda}(Q_k^{-1} - Q_0^{-1}). \tag{4.22}$$

By using Equations 4.19 to 4.22, one can calculate the value of sound velocity and absorption coefficient. Usually, this method demands a careful calibration using well-studied liquids with low ultrasonic absorption (e.g., solution: ethanol–water) to identify the proper resonance peaks, to determine the parameter Q_0, and to exclude diffraction errors in sound velocity measurements. It can be shown [11] that in ultrasonic resonator, the relative contribution of diffraction losses increases with decreasing frequency. That is why one has to use piezoelectric transducers of a big diameter and increase the acoustic way for minimizing this contribution. For example, the plane transducers with diameter $D = 58$ mm placed at distance 10 mm were used to measure ultrasonic parameters of liquid crystals in the frequency range 0.1–1 MHz [12]. It corresponds to the volume of the sample about 30 cm^3. In practice, the construction of acoustical chambers is rather complicated. In particular, it has to provide strictly parallel installation of both transducers. The latter can be simplified by using transducers with surfaces of definite curvature [13]. In this case, diffraction losses can be effectively decreased, which provides the usage of relatively moderate amount of liquids under acoustic investigations. An example of such resonator will be described below.

4.1.3
Ultrasonic Technique for the Study of Liquid Crystals

4.1.3.1 Peculiarities of Ultrasonic Investigations of Liquid Crystals

One of the main advantages of such studies is the use of bulk samples where the influence of surfaces is negligible. It essentially simplifies the theoretical description of structural changes in liquid crystals induced by external fields in comparison with the case of thin optically transparent layers usually considered in optical studies of LC. At the same time, a thorough account of difference between liquid crystals from isotropic liquids is important.

Firstly, liquid crystals are essentially anisotropic objects. It means that ultrasonic measurements, which can provide detailed information about viscous and elastic properties of different liquid crystal phases, have to be done in oriented samples. The most proper method for obtaining such samples for ultrasonic experiments is the use of relatively strong magnetic fields (with induction in the range 0.1–1 T). Usually, it leads to relevant complications of the experimental setups in comparison with those used for isotropic liquids. In particular, the linear sizes of acoustic cameras are limited, which makes it difficult to carry out measurements at relatively low frequencies. Second, it is well known that the most phase transitions in liquid crystals can be of a weak first order (nematic–isotropic liquids (N–I) and nematic–smectic C (N–C)), of the second order (smectic A–smectic C (A–C)), or show the intermediate behavior (nematic–smectic A (N–A)). In all cases, they are accompanied by additional relaxation processes, such as relaxation of order parameters and of critical fluctuations of order parameters. The critical slowing down of these processes plays an essential role in ultrasonic measurements in the vicinities of phase transitions. At low-frequency measurements, it results in a critical increase in an ultrasonic attenuation (up to some decades) and corresponding decrease in the amplitude of an acoustic pulse upon approaching phase transition temperatures. So, it is not simple to make a proper choice of an optimal acoustic base (the distance of propagations of acoustic pulses). It is important to note that critical dynamic properties in the vicinities of phase transitions in liquid crystals are anisotropic too and even can be described (in the case of N–A) transition by different kinetic mechanisms for different orientations [14, 15].

Now, we describe some examples of a particular experimental technique used for the study of oriented samples of liquid crystals. We focus on different acoustic chambers as the most original parts of experimental setups.

4.1.3.2 **Ultrasonic Chambers for the Study of Liquid Crystals**

Construction of a many-channel chamber [16, 17], which allows one to realize simultaneously the pulse phase methods of a variable distance and of a variable frequency (see above), is presented in Figure 4.2a. The main parts of the chamber are the cassettes with four pairs of piezo transducers of different basic resonance frequencies. One of the cassettes is fixed while the other can be moved via a special wedge – like a mechanism that transforms a vertical movement of a micrometric screw into horizontal motion of the cassette. Such decision solves two problems mentioned above. First, it provides measurements in horizontal magnetic field at

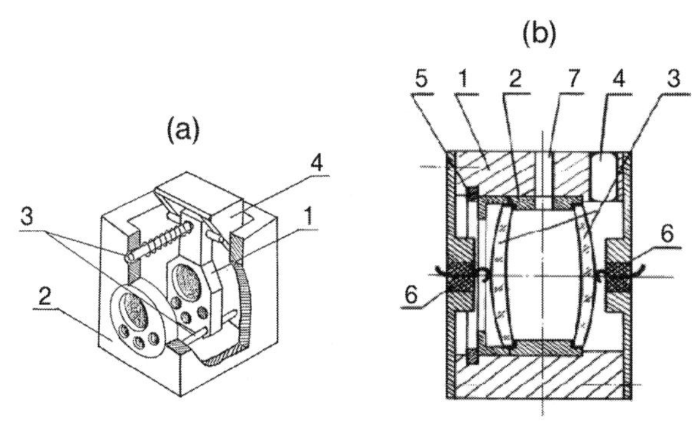

Figure 4.2 The construction of the acoustic cameras for investigations of liquid crystals by the pulse phase (a) and resonator (b) methods; (a) 1 – the cassettes with four pairs of piezo transformers; 2 – the camera body; 3 and 4 – the director rails and the wedge needed to move one of the cassettes; (b) 1 – the duralumin casing; 2 – the stainless ring; 3 – the concave–convex quartz transducers, 4 – the quartz thermometer; 5 – the fixing nut; 6 – the contact electrodes; 7 – the filling hole.

different angles θ between the vector of magnetic induction **B** and the wave vector ***q***. Second, one can adjust the acoustic way to most suitable value correspondent to the absorption coefficient. After that, one can measure relative changes in ultrasonic absorption and velocity by the pulse phase method of a variable frequency. It is of especial importance for ultrasonic studies of smectic phases where any motion can destroy layered structure. At the same time, measurement of absolute values of acoustical parameters in isotropic and nematic phases is possible via the pulse phase method of a variable distance. It was possible to perform such measurements in the frequency range 3–46 MHz by using the acoustic camera shown in Figure 4.2a. A sample of liquid crystal (about 10 cm^3) was oriented by a static magnetic field of induction **B** varying in the range 0–0.64 T. It was possible to get the angular dependencies of acoustic parameters by rotating a camera relative to **B** (nematic and smectic C phases) or by cooling a sample from a nematic phase to smectic phases at different angles between **B** and ***q***. It is worthwhile to note that a similar decision was described in Ref. [18].

The acoustic camera (acoustic resonator) that was used at low-frequency ultrasonic investigations of liquid crystals [10, 19] by a resonator method is shown in Figure 4.2b. The main original decision is the use of the concave–convex transducers to minimize diffraction losses arising at low frequencies. Transducers are made of monocrystalline plates of lithium niobate (yx 36° cut). The radii of curvature of opposite surfaces of the transducers are equal to +75 mm (r_1) and −400 mm (r_2) at diameter 30 mm. Two silver electrodes were put on the surfaces of each transducer by electrolysis. Transducers were symmetrically fixed to the body of a chamber with a clue to achieve a high acoustic Q factor of a resonator. The distance between the transducers was about 5 mm, which provides ultrasonic data of a moderate amount of LC (about 4 cm^3). The camera was thermally stabilized and placed in a rotational magnetic field (0.3 T) to get detailed information on viscous properties of LC (see below).

The ultrasonic chamber with two distances is shown in Figure 4.3. It was used [20] to measure (by the pulse method) the complex of acoustic parameters (ultrasound

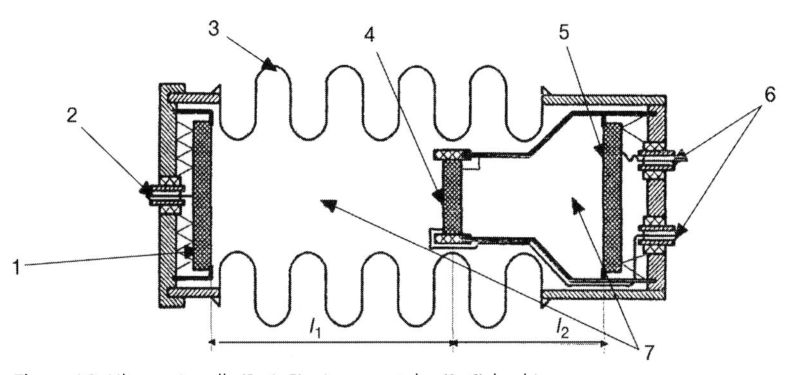

Figure 4.3 Ultrasonic cell: (1, 4, 5) piezo crystals; (2, 6) lead-in; (3) separating rolling diaphragm; (7) liquid crystal. (After Larionov *et al.* [20].)

velocity c and attenuation α at different angles θ between LC director and wave vector) and the density ρ of LC under varying pressure and temperature. So, the dependencies of viscous and elastic parameters of liquid crystals on PVT-thermodynamic-state parameters have been studied.

The ultrasonic cell contained three electroacoustic transducers. A radio frequency pulse of amplitude A_0 was applied to transducer 4 immersed in liquid–crystal sample 7. The sound pulse traveled to the transducers 1 and 5, where it was transformed into the radio frequency pulses of amplitude A_1 and A_2 accordingly:

$$A_1 = k_1 \cdot A_0 \cdot \exp(-\alpha l_1), \qquad A_2 = k_2 \cdot A_0 \cdot \exp(-\alpha l_2),$$

where k_1 and k_2 are the sound transmission coefficients of acoustic channels, l_1 is the variable acoustic distance, which is the function of temperature and pressure, and l_2 is the distance between piezo crystals 4 and 5, which is constant. The velocity and the absorption coefficient of ultrasound were measured at frequencies 2.67 MHz and 500 kHz, respectively. The accuracy of measurements was 0.02 and 0.08% for a velocity at frequencies 2.67 MHz and 500 kHz, respectively, and 2% for an absorption coefficient. The density of liquid crystal under changeable pressure and temperature can be determined by comparing the time needed for sound pulses to propagate the distances l_1 and l_2. After filling and sealing, the acoustic cell was inserted into a pressure vessel made of diamagnetic steel. It is important to note that the total amount of a sample under consideration was about 1 cm^3, which makes such a cell very attractive for rheological studies of liquid crystals. Experiments were performed in the pressure range 1–600 bar and in the temperature range 290–365 K. The uniform orientation of the nematic director was obtained in the usual way [21, 22] using the static magnetic field and the rotating one (of induction 0.15 T). In the latter case, the ratio $\gamma_1/\Delta\chi$ can be determined from the analysis of the time dependencies of an ultrasonic attenuation [21, 22], as will be shown below.

4.2 Ultrasonic Viscosimetry of Nematic Liquid Crystals

4.2.1 Theoretical Background

Anisotropic propagation of ultrasound in NLC was considered by a number of authors in the framework of hydrodynamic approximation [23–25]. Though these theories are essentially different, they lead to similar results relative to linear longitudinal waves. That is why we will consider this problem using Leslie–Ericksen hydrodynamic theory, which is of main practical use.

Let us consider an ultrasonic wave with a wave vector $\boldsymbol{q}$ oriented at the angle θ relative to the initial position of a nematic director and X_3-axis.

The system of linearized hydrodynamic equations in this case derived from general equations [25] can be expressed as [26]

$$\rho_0 \frac{\partial v_1}{\partial t} = -\frac{\partial P}{\partial x_1} + (\mu_1 + \alpha_4)\frac{\partial^2 v_1}{\partial x_1^2} + \frac{1}{2}(\alpha_4 + \alpha_5 - \alpha_2)\frac{\partial^2 v_1}{\partial x_1^2}$$

$$+ \left(\mu_1 + \mu_2 + \frac{\alpha_4 + \alpha_5 + \alpha_2}{2}\right)\frac{\partial^2 v_3}{\partial x_1 \partial x_3} + \alpha_2 \frac{\partial^2 \theta^*}{\partial t \partial x_3} \tag{4.23}$$

$$\rho_0 \frac{\partial v_3}{\partial t} = -\frac{\partial P}{\partial x_3} + \frac{1}{2}(\alpha_3 + \alpha_4 + \alpha_6)\frac{\partial^2 v_3}{\partial x_1^2}$$

$$+ (\mu_1 + \mu_2 + \mu_3 + \alpha_1 + \alpha_4 + \alpha_5 + \alpha_6)\frac{\partial^2 v_3}{\partial x_1^2}$$

$$+ \left(\mu_1 + \mu_2 + \frac{\alpha_4 + \alpha_6 - \alpha_3}{2}\right)\frac{\partial^2 v_1}{\partial x_1 \partial x_3} + \alpha_3 \frac{\partial^2 \theta^*}{\partial t \partial x_1} \tag{4.24}$$

$$\gamma_1 \frac{\partial \theta^*}{\partial t} - K_{11}\frac{\partial^2 \theta^*}{\partial x_1^2} - K_{33}\frac{\partial^2 \theta^*}{\partial x_3^2} + \alpha_2 \frac{\partial v_1}{\partial x_3} + \alpha_3 \frac{\partial v_3}{\partial x_1} = 0 \tag{4.25}$$

$$\frac{\partial P}{\partial t} = -\rho_0 C_0^2 \left(\frac{\partial v_1}{\partial x_1} + \frac{\partial v_3}{\partial x_3}\right). \tag{4.26}$$

In these equations, θ^* refers to deviation of the director from its initial orientation imposed by an ultrasonic wave, α_i to the Leslie coefficients, introduced for an incompressible LC media (Chapters 2 and 3), and μ_i to dissipative parameters arising due to compressibility of nematic, which can be called bulk viscosity coefficients. The solution of this system is expressed in a standard manner:

$$\mathrm{v}_1, \mathrm{v}_2, \theta^* \sim \exp i(\omega t - q_1 x_1 - q_3 x_3), \tag{4.27}$$

where

$$q_1 = q \sin\theta, \qquad q_3 = q \cos\theta \tag{4.28}$$

are the components of the wave vector $\boldsymbol{q}$, for a wave propagating in θ direction. By taking into account (4.27) and (4.28), one can easily get the expression for an absorption coefficient (α) corresponding to the imaginary part of the wave vector $\boldsymbol{q}$ of a longitudinal ultrasonic wave:

$$\frac{\alpha}{f^2} = \frac{\alpha^\perp}{f^2} + a_\alpha \cos^2\theta + b_\alpha \cos^4\theta, \tag{4.29}$$

where

$$\frac{\alpha^\perp}{f^2} = \frac{2\pi^2}{\rho_0 C_0^3}(\mu_1 + \alpha_4) \tag{4.30}$$

$$a_\alpha = \frac{2\pi^2}{\rho_0 C_0^3}(\mu_2 + \mu_3 + \alpha_5 + \alpha_3 - \gamma_1 \lambda^2) \tag{4.31}$$

$$b_\alpha = \frac{2\pi^2}{\rho_0 C_0^3}(\alpha_1 + \gamma_1 \lambda^2). \tag{4.32}$$

The obtained expressions are close to those presented in Ref. [25] except the combination ($\gamma_1\ \lambda^2$) entering into expressions (4.31) and (4.32). This combination results from an account of director oscillations in an ultrasonic wave, which were ignored in the work mentioned above. Moreover, the expressions (4.29) to (4.32) are analogous to well-known results, obtained by Forster *et al.* [23] and frequently used for the analysis of acoustic properties of liquid crystals [6, 27]. Both the general expression (4.29) and (4.31) and (4.32) are valid in the latter case at the next substitutions:

$$\mu_1 + \alpha_4 = \nu_4 + \nu_2 \tag{4.33}$$

$$\mu_2 + \mu_3 + \alpha_5 + \alpha_3 - \gamma_1 \lambda^2 = 2\nu_3 - \nu_2 + \nu_5 - \nu_4 \tag{4.34}$$

$$\alpha_1 + \gamma_1 \lambda^2 = 2(\nu_1 + \nu_2 - 2\nu_3), \tag{4.35}$$

where ν_1, ν_2, ν_3 are the shear viscosity coefficients, while ν_4, ν_5 are the bulk viscosity coefficients in Forster *et al.* notation [23].

Results obtained show that some useful information about both shear and bulk viscosities can be obtained by analyzing angular dependencies of an ultrasonic absorption coefficient. In particular, in accordance with (4.32) the frequency-independent parameter b_α does not include bulk viscosities, and in this way one can extract information about a combination ($\alpha_1 + \gamma_1 \lambda^2$) of dissipative coefficients of incompressible NLCs.

Indeed, this problem turns out to be not so trivial. First, it was shown earlier [28] that the standard hydrodynamics was insufficient for the description of ultrasound propagation in liquid crystals. In particular, it predicts an isotropic ultrasonic velocity that contradicts the experimental data. There were some attempts both on phenomenological [28] and on microscopic levels [29] to take into account possible relaxation processes in nematics. In particular, in the framework of Fokker–Planck model [29], it was shown that the relaxation of a nematic order parameter resulted in critical contribution E_i in a complex elastic modulus of a liquid crystal:

$$E_i = B\frac{\left(\frac{\partial S_0}{\partial U_{\alpha\alpha}}\right)^2 - 3i\omega\tau x\left(\frac{\partial S_0}{\partial U_{\alpha\alpha}} + \frac{3}{4}x\right)}{1 - i\omega\tau} - 9Bx\left(\frac{\partial S_0}{\partial U_{\alpha\alpha}} + \frac{3}{2}x\right)\cos^2\theta \frac{-i\omega\tau}{1 - i\omega\tau}$$
$$+ \frac{81}{4}Bx^2\cos^4\theta \frac{-i\omega\tau}{1 - i\omega\tau}, \tag{4.36}$$

where B denotes generalized susceptibility of NLCs that can be extracted from an absolute value of an absorption coefficient, $U_{\alpha\alpha}$ is the volume variation in an ultrasonic wave, $\tau = \gamma/B$ is the relaxation time of the order parameter S ($S = S_0$ at equilibrium), γ is the kinetic coefficient,

$$x = \langle \cos^2\theta_i - \cos^4\theta_i \rangle = \frac{2}{15}\left(1 + \frac{5}{7}\langle P_2 \rangle - \frac{12}{7}\langle P_4 \rangle\right), \tag{4.37}$$

a parameter that includes the second ($\langle P_2 \rangle = \langle P_2(\cos^2\theta_i)\rangle = S_0$) and the fourth ($\langle P_4 \rangle = \langle P_4(\cos^2\theta_i)\rangle$) momentum of an equilibrium distribution function, P_i – the Legendre polynomials.

Expressions for critical contributions to an anisotropy of the sound velocity (ΔC) and the absorption coefficient ($\Delta\alpha$) can be easily obtained from real and imaginary parts of a complex modulus E (4.36):

$$\Delta C(\theta) = \mathrm{Re}\left[\frac{E^c(\theta) - E^c(0)}{2\rho_0 C_0}\right] = -\frac{9Bx}{2\rho_0 C_0}\left[\left(\frac{\partial S_0}{\partial U_{\alpha\alpha}} + \frac{3}{2}x\right)\cos^2\theta - \frac{9}{4}x\cos^4\theta\right]\frac{\omega^2\tau^2}{1+\omega^2\tau^2} \tag{4.38}$$

$$\Delta\alpha(\theta) = \mathrm{Im}\left[\frac{E^c(\theta) - E^c(0)}{2\rho_0 C_0^3}\right] = -\frac{9Bx}{2\rho_0 C_0^3}\left[\left(\frac{\partial S_0}{\partial U_{\alpha\alpha}} + \frac{3}{2}x\right)\cos^2\theta - \frac{9}{4}x\cos^4\theta\right]\frac{\omega^2\tau}{1+\omega^2\tau^2}. \tag{4.39}$$

Comparison of expressions (4.29) and (4.39) shows that a critical relaxation process can contribute to both parameters a_α and b_α describing the anisotropic absorption coefficient. Moreover, it results in the emergence of the frequency-dependent anisotropy of an ultrasonic velocity vanishing in low-frequency limit. It is interesting that there is a simple connection between the critical contributions defined by (4.38) and (4.39):

$$\Delta\alpha(\theta) = (\tau/C_0)\frac{\Delta C(\theta)}{C_0}, \tag{4.40}$$

which is valid in the case of the unique critical process. Taking into account the second relaxation process (e.g., intramolecular relaxation [30]) makes the connection between $\Delta\alpha$ and ΔC more complicated. In any case, a thorough analysis of experimental data on both the anisotropic absorption coefficient and the anisotropy of ultrasound velocity at different frequencies is needed to reveal the real possibility of ultrasonic methods in the rheology of liquid crystals.

Nevertheless, at least in one case one can ignore most problems described above. Let us consider the propagation of ultrasound through a nematic sample described by a director which slow varies with time (a characteristic time τ_n is in the range 0.1–100 s). The propagation time of ultrasound defined by (4.6) at a reasonable size of an acoustic camera is of order 10^{-5} s and is essentially smaller than τ_n. So, each ultrasonic pulse propagates in liquid crystal as in a quasistationary media. It means that the expression for static ultrasonic absorption coefficients (4.29) is valid for slow director motion, which contributes to this parameter via dependence $\theta(t)$. This simple idea was applied for analyzing time dependences of ultrasonic absorption coefficients induced by shear flow [31], different types of magnetic fields (pulsed [6], rotating [10, 20–22]), and a combined action of electric and magnetic fields [32]. From the viewpoint of practical applications, ultrasonic studies of LC in a rotating magnetic field are of special interest as they lead to a simple and effective method of determining a rotational viscosity coefficient γ_1. The most simple analysis of the problem can be made in the case of synchronic regime of a director rotation established first in Zwetkoff's classical experiments [33], which will be described in detail in Chapter 5. In this case, there is a phase gap between field and director depending on the twist viscosity coefficient γ_1 so the latter can be calculated from such types of experiments by expression (see Chapter 5):

$$\gamma_1 = \frac{H^2 \cdot \mu_0 \, \Delta\chi \cdot \sin 2\varphi}{2\omega_H}, \tag{4.41}$$

where H and ω_H denote the strength and angular velocity of magnetic field, respectively, and $\Delta\chi$ is diamagnetic susceptible anisotropy.

In ultrasonic experiments, the phase gap and the ratio $\gamma_1/\Delta\chi$ can be determined from the comparison of the dynamic angular dependencies of an attenuation coefficient and the static ones. It is obvious that in this case fast relaxation processes mentioned above do not play any role as they contribute to both dependencies. It allows to calculate the twist viscosity coefficient γ_1 knowing the value of $\Delta\chi$. One can estimate the critical value ω_H^c of the angular velocity corresponding to the change in regimes (from the synchronic regime to the asynchronous one) by putting $\sin 2\varphi = 1$ in expression (4.41). It gives

$$\omega_H^c = \frac{\mu_0 \, \Delta\chi H^2}{2\gamma_1}. \tag{4.42}$$

For reasonable values of $\Delta\chi \approx 10^{-6}$, $\gamma_1 \approx 1$ P, it results in $\omega_H^c \approx 0.5$ rad/s at $H \approx$ 3 kOe. Such relatively low value of a critical angular velocity can be realized by mechanical rotation of a permanent magnet that makes experimental setups rather cumbrous (e.g., a mass of 3 kOe permanent magnet is about 70 kg). The use of weak magnetic fields is restricted by existence of minimal strength of saturation H_s (about 1 kOe) needed for a perfect orientation of bulk samples of liquid crystals. A decrease in the field strength down to H_s results in deviation of experimental dependences $\alpha(t)$ from those predicted by the simple model mentioned above. It also prevents correct determination of the ratio $\gamma_1/\Delta\chi$ due to measurements of critical angular velocity ω_H^c that is possible to realize at $H \gg H_s$. It is worthwhile to note that the time dependencies of an ultrasonic absorption coefficient obtained at asynchronous regime of rotation cannot be explained by a simple hydrodynamic model of homogenous sample [34] and demand an alternative description [22].

Nonetheless, one can consider ultrasonic measurement in rotating magnetic field as a reliable method for determining the twist viscosity coefficient. A number of investigations confirm this conclusion [6, 20–22]. It is also worthwhile to mention the attempt [32] to exclude the rotation of magnetic field by using a combined action of electric and magnetic fields on a bulk sample of a liquid crystal. In the latter case, one can apply a simple hydrodynamic description of the problem. It is important to note that both parameters $\Delta\chi$ and γ_1 can be ascertained by analyzing experimental data.

4.2.2 Experimental Confirmation

4.2.2.1 Static Regime

Experimentally, the anisotropy of ultrasonic parameters in nematic liquid crystals, especially of an absorption coefficient, was established about 40 years ago [35, 36]. Since then, a lot of experiments with liquid crystals oriented by magnetic fields have been performed. The results of ultrasonic studies were summarized in some reviews and books (see, for example, Refs [2, 6]). In all the cases, the general expression (4.29) was enough to describe the angular dependence of an ultrasonic absorption coefficient.

Typical angular dependences of the low-frequency anisotropy of an absorption coefficient,

$$\frac{\Delta\alpha}{f^2}(\theta) = \frac{\alpha}{f^2}(\theta) - \frac{\alpha^{\perp}}{f^2}, \tag{4.43}$$

measured by a resonator method [10] in BBBA (*p*-*n*-butoxybenzilidene-*n*-butylaniline) for different temperatures, are shown in Figure 4.4.

The results are well approximated (solid lines) by expression (4.29) to extract parameters a_α and b_α. A comparison [37] of low-frequency values of b_α with those extracted from measurement at essentially higher frequencies [9, 37] did not reveal an essential frequency dependence of this parameter, at least out of N–I and N–S_A phase transition regions (Figure 4.5).

It means that the possible relaxation contribution to this parameter, mentioned above (see 4.39, is negligible and one can use b_α for determining dissipative parameters of an incompressible NLCs. So, the temperature dependence of b_α, shown in Figure 4.6, must reflect the analogous dependence of a combination $(\alpha_1 + \gamma_1\lambda^2)$. Contrary to b_α, the parameter a_α essentially depends on frequency (Figure 4.7).

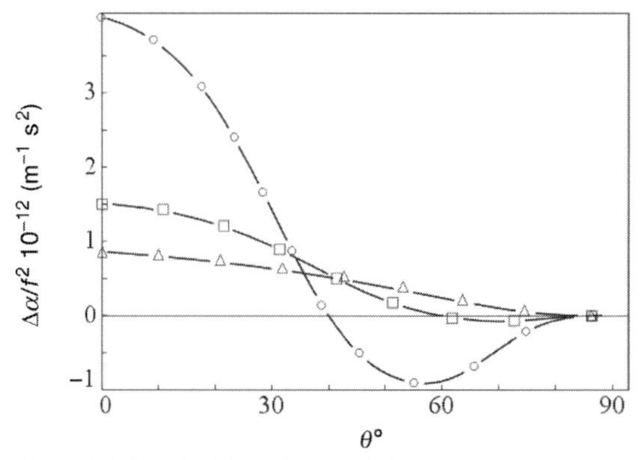

Figure 4.4 Angular dependences of ultrasonic attenuation anisotropy $\Delta\alpha/f^2$ in nematic phase of BBBA at different temperatures: (O) 315.8 K, (□) 317.0 K, (△) 328.3 K; $f = 0.5$ MHz. (After Prokopjev [10].)

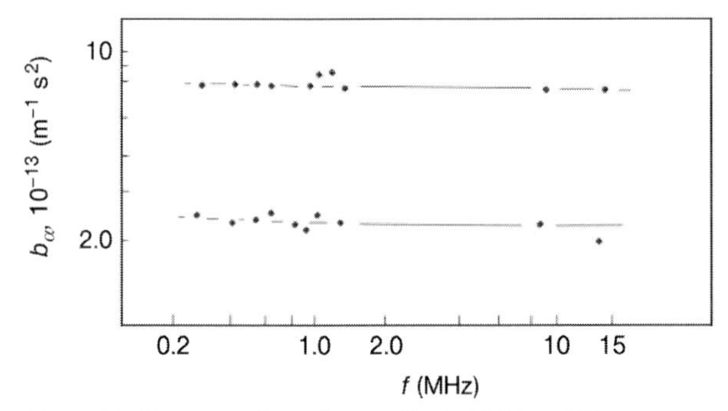

Figure 4.5 Frequency dependence of b_α in BBBA at different temperatures: (■) 320.3 K; (●) 332.3 K [37]. Data at frequencies higher than 3 MHz are taken from Ref. [9].

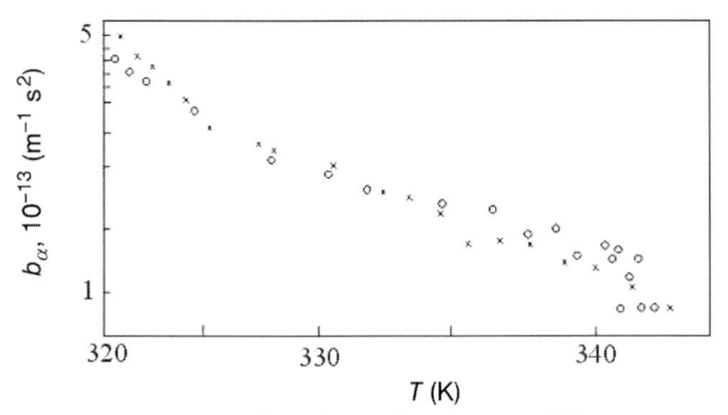

Figure 4.6 Temperature dependence of b_α in BBBA at different frequencies: (O) 0.63 MHz [10]; (×) 15 MHz [9].

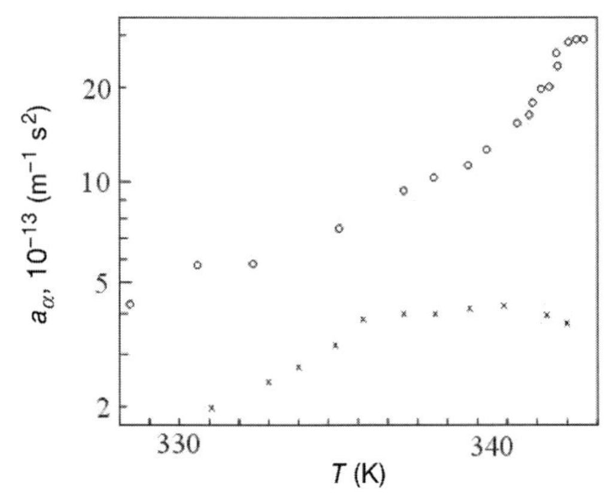

Figure 4.7 Parameter a_α in a nematic phase of BBBA as a function of temperature. (O) 0.63 MHz [10], × 15 MHz [9].

Table 4.1 Temperature dependence of ratio b_α/a_α in BBBA; $f=0.63$ MHz.

$-\Delta T_c$ (K)	22	21.4	19	15.3	13.1	11.4	8.4
b_α/a_α	9.0	5.0	1.9	0.83	0.58	0.47	0.31
$-\Delta T_c$ (K)	6.6	5.5	4.5	3.9	2.9	2.6	2.4
b_α/a_α	0.23	0.16	0.16	0.11	0.10	0.09	0.08

This dependence is more pronounced at approaching NI transition, which indicates the critical nature of relaxation contribution entering into a_α coefficient (see Equation 4.38). It is of practical importance that the accuracy of the determination of parameters b_α and a_α essentially depends on their ratio. The dependence of this ratio on temperature is presented in Table 4.1.

It follows from Table 4.1 that near NI transition ($\Delta T_c \leq 3$ K), coefficient a_α makes a major contribution to anisotropy of an absorption coefficient at least at low frequencies. It results in large errors in b_α parameter, extracted from angular dependences mentioned above. More reliable data in this case can be obtained from ultrasonic studies at higher frequencies. Results of such studies with additional information extracted from shear flow investigations were used in Ref. [38] for determining some Leslie coefficients of BBBA. The final results of such calculations are presented in Figure 4.8.

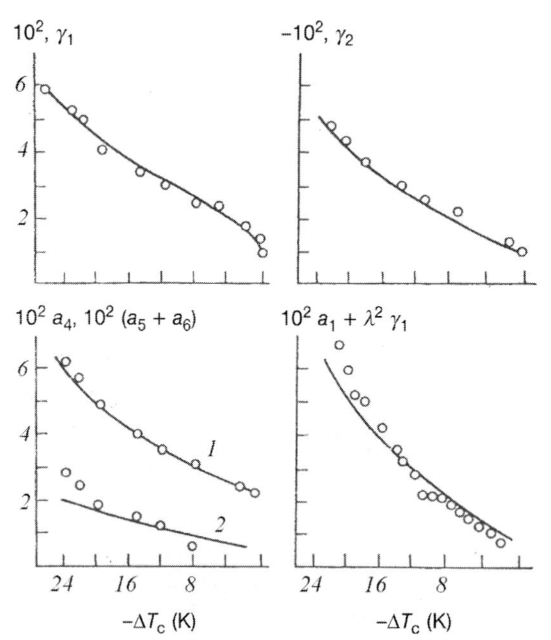

Figure 4.8 Temperature dependences of Leslie coefficients (in Pa s): (1) α_4 and (2) $\alpha_5+\alpha_6$. (After Pasechnik *et al.* [38].)

Analogous conclusions can be made from ultrasonic study of PAA performed by using a resonator method in the range 0.8–5 MHz [27]. It was found that there was only critical relaxation contribution to the parameter a_α, whereas the parameter b_α did not depend on frequency.

Of course, it would be interesting to estimate the Leslie coefficient α_1 from ultrasonic measurements. Such an attempt was made in the work [20] for a nematic mixture ZhK440 (2 : 1 mixture of *p*-*n*-butyl-*p*-methoxyazoxybenzene and *n*-butyl-*p*-heptanoyloxyazoxybenzene). The choice of this object was made because of a rather wide temperature interval in a nematic phase, which takes place in the mixture mentioned above. Additional information obtained by different methods for the mixture and separate components [39, 40] is important too. The calculations were performed using values of a twist viscosity coefficient γ_1, obtained from ultrasonic study of this mixture in a rotating magnetic field at variation of pressure and temperature.

To estimate the only unknown parameter λ^2, values of the flow orientation angle measured for the first component of the mixture [40] were used. The result of such calculation is presented in Table 4.2.

One can see that this coefficient has a negative sign and essentially smaller values compared to the rotational viscosity coefficient, at least far from the clearing point, which is in accordance with data obtained for MBBA [41]. Though the presented results demonstrate a possibility of determining the coefficient α_1 from ultrasonic measurement results, it is clear that in the case of small values of this coefficient such procedure demands rather precise measurements of parameters γ_1 and λ. Nevertheless, for practical purposes, even the estimate of α_1 seems to be very useful.

Table 4.2 Temperature dependence of α_1 (ZhK440).

T (K)	α_1 (Pa s)
297.0	−0.0055
301.0	−0.0085
309.0	−0.0058
315.0	−0.0073
319.0	−0.0100
321.0	−0.0111
323.0	−0.0111
327.0	−0.0142
328.0	−0.0137
333.0	−0.0172
335.0	−0.0166
336.0	−0.0164
337.0	−0.0167
339.0	−0.0178
339.8	−0.0194
340.5	−0.0194
341.5	−0.0189
342.0	−0.0185

4.2.2.2 **Dynamic Regime**

The first ultrasonic investigation of a nematic liquid crystal placed into a rotating magnetic field was performed in 1980 [42]. Simultaneously, the primary theoretical interpretation of experimental results in the framework of a hydrodynamic model was proposed. Later on, such technique was widely used for the study of nematic liquid crystals [6, 20–22]. Experimental results show good agreement with main predictions of the simple hydrodynamic model in a synchronic regime. In this case, the time dependence of an ultrasonic absorption anisotropy repeats the angular dependence of this parameter with the phase shift φ determined by expression (4.41). The form of the curve $\Delta\alpha(t)$ became unchangeable at a variation of an angular velocity of magnetic field rotation up to a critical value ω_H^c. An example of such dependence obtained from ultrasonic studies of MBBA under varying temperature and pressure [22] is shown in Figure 4.9a. The data were obtained at rotation of 3 kG magnetic field. The form of the curve $\Delta\alpha(t)$ became unchangeable at a variation of an angular velocity of magnetic field rotation up to a critical value ω_H^c defined in Equation 4.42. The dependence of the phase shift φ on the angular velocity ω_H shown in Figure 4.10 is also in accordance with the theoretical result (4.41). A slight difference between the theoretical curve and the experimental results occurs only in the vicinity of a critical velocity.

The time dependence $\Delta\alpha(t)$ is drastically changed with an increase in angular velocity above a critical value ω_H^c (Figure 4.9b). Both huge phase and amplitude distortions take place in this case. The simple hydrodynamic model [42] partly

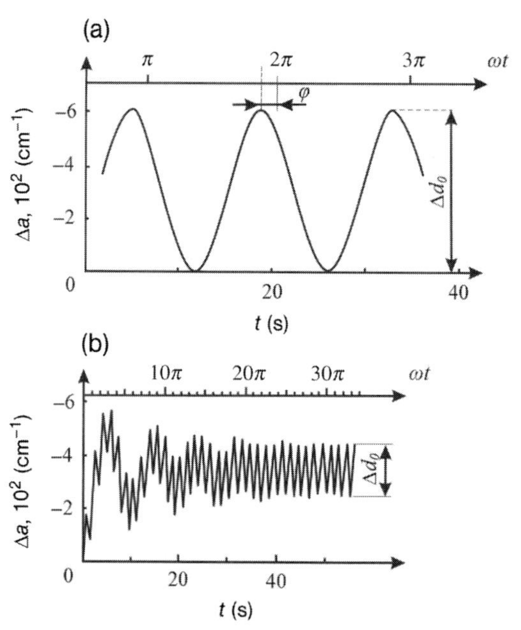

Figure 4.9 Variation with time of the ultrasound adsorption coefficient for MBBA at $p = 2 \times 10^7$ Pa, $T = 314.2$ K; (a) synchronic regime and (b) asynchronous regime (After Pasechnik *et al.* [22].)

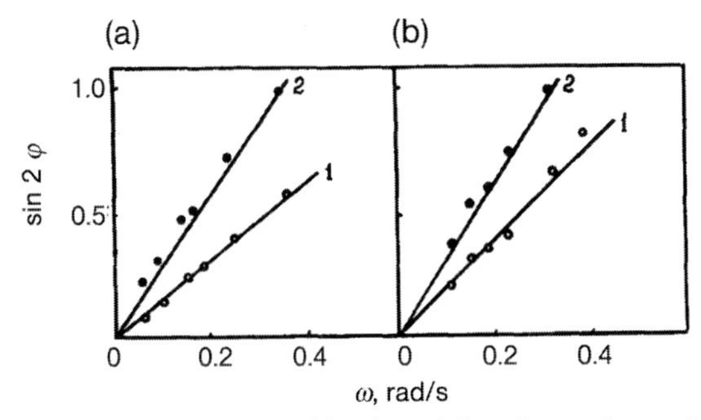

Figure 4.10 Dependence of the phase shift angle φ on the angular velocity ω_H of magnetic field rotation: (a) for MBBA at $T = 303.6$ K; $1 - P = 10^5$ Pa, $2 - P = 4 \times 10^7$ Pa; (b) for BBBA at $T = 303.6$ K; $1 - P = 10^5$ Pa, $2 - P = 6 \times 10^7$ Pa. (After Pasechnik *et al.* [22].)

predicts such distortions, but it is useless to explain significant decrease in the maximal amplitude of $\Delta\alpha(t)$ dependence with increasing speed of field rotation. An attempt to clarify this question was made [22] on the basis of a Fokker–Planck-like model of dynamic behavior of NLCs in a rotating magnetic field [43]. The latter predicts for an asynchronous regime some disorder arising due to uncorrelated rotation of compact regions of a liquid crystal with different orientations. The system becomes more chaotic with increasing angular velocity ω_H that results in a corresponding decrease in the amplitude of $\Delta\alpha(t)$ dependence, shown in Figure 4.9b. In spite of the roughness of the original model, the main idea, namely, the emergence of chaos due to quickly rotating magnetic field, seems to be correct. Additional experimental and theoretical work is needed for better understanding the propagation of ultrasound in an asynchronous regime. In particular, the role of near-wall defects considered in Chapter 5 has to be analyzed too.

In any case, two parameters, the shift angle φ measured in a synchronic regime and the critical angular velocity ω_H^c, can be used to obtain the ratio $\gamma_1/\Delta\chi$. A lot of ultrasonic investigations of liquid crystals in rotating magnetic fields conducted so far [6, 20–22] confirm this conclusion.

Such type of ultrasonic technique has turned out to be very useful for extracting the twist viscosity coefficients at both pressure and temperature variations. Indeed, there are only a very few studies on viscous properties of liquid crystals at variable pressure done by using alternative methods (see, for example, Ref. [44]).

Data obtained from such measurements are useful both for experimental check of microscopic theories' predictions on dependencies of a twist viscosity coefficient on P,T-state parameters and for direct practical applications as the value of γ_1 is a key parameter responsible for dynamic properties of LC devices.

As an example in Figure 4.11, we present the temperature dependence of the ratio $\gamma_1/\Delta\chi$ at different pressures obtained by an ultrasonic method [22] for BBBA.

The critical increase in this parameter observed at low temperatures can be referred to as the contribution from fluctuations of a smectic order parameter in the vicinity of phase transition nematic–smectic A [45]. So, the data presented

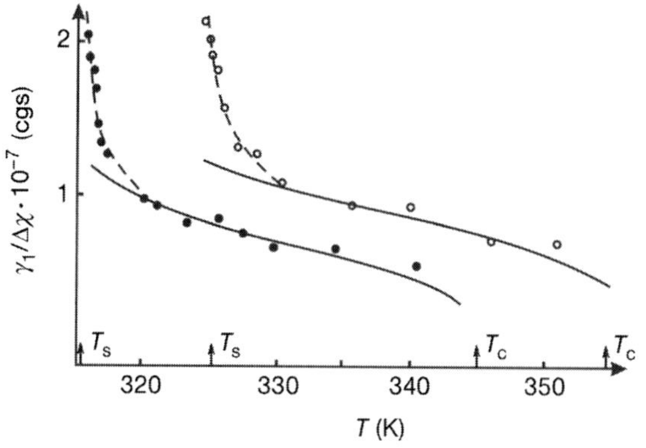

Figure 4.11 Temperature dependence of $(\gamma_1/\Delta\chi)$ for BBBA: (●) $p = 10^5$ Pa; (O) $p = 4 \times 10^7$ Pa. (After Pasechnik *et al.* [22].)

demonstrate the possibility of using the ultrasonic method for studying critical phenomena in liquid crystals, which will be discussed below.

This method was successfully applied also for a number of LC mixtures with wide nematic intervals [6, 20, 46].

Numerical values of the rotational viscosity coefficient γ_1 and other parameters (the combination b_α, the density of LC, and the ultrasonic velocity) extracted from ultrasonic measurements at different P,T-state parameters for nematic mixture ZhK440, are presented in Table 4.3.

The connection $\Delta\chi \sim S$ (where S is the order parameter, which is assumed to depend only on the difference between clearing temperature T_c and the given temperature T) was taken into account when calculating absolute values of the rotational viscosity coefficient. The pressure dependence of the clearing temperature T_c was obtained from the described ultrasonic measurements, as the total absorption anisotropy $\Delta\alpha$ goes to zero at the clearing point. This dependence is linear in the

Table 4.3 Material parameters of ZhK440 evaluated from ultrasonic measurements. (After Larionov *et al.* [20]).

		P (MPa)					
T (K)		**0.1**	**10**	**20**	**30**	**40**	**60**
297.0	γ_1 (Pa s)	0.1333	—	—	0.1425	0.1457	0.1560
	ρ (kg/m^3)	1106.8	—	—	1117.8	1120.3	1126.1
	c (m/s)	1553.9	—	—	1660.1	1685.5	1729.5
	b (10^{14} m^{-1} s^2)	53.4	—	—	76.0	81.0	95.0
337.0	γ_1 (Pa s)	0.0222	0.0252	0.0289	0.0319	0.0354	0.0418
	ρ (kg/m^3)	1088.7	1093.3	1097.9	1102.8	1107.5	1115.5
	c (m/s)	1392.3	1442.9	1496.2	1533.3	1571.8	1632.5
	b (10^{14} m^{-1} s^2)	−7.4	0.1	6.5	10.3	15.8	24.3

experimentally realized range of pressure (10^5–6×10^7 Pa):

$$T_c(P) = T_{oc} + k_c \cdot P, \tag{4.44}$$

where the temperature of nematic–isotropic phase transition at atmospheric pressure is $T_{oc} = 345.7$ K, and the coefficient k_c is equal to 3.28×10^{-7} Pa^{-1} K.

It was shown that experimental dependence of $\gamma_1(T)$ for ZhK440 is well approximated by the empirical dependence

$$\gamma_1 = A_1\left(1 - \frac{T}{T^*}\right)^{2\beta} \exp\left(\frac{B}{T - T_0}\right) \tag{4.45}$$

proposed in [44]. In this expression, T^* is the clearing point at pressure P defined as

$$T^* = T_0^*\left(\frac{P}{a} + 1\right)^C. \tag{4.46}$$

The best result was obtained using the next parameters of approximation: $T_0^* = 353.7$ K; $a = 4.757 \times 10^8$ Pa; $c = 0.6$; $A_1 = 0.0269$ Pa s; $\beta = 0.3536$; $B = 243.26$ K; $T_0 = 213.04$ K.

Using the data on ρ and c at varying pressure and temperature (see Table 4.3), one can get the dependence of γ_1 upon pressure and temperature at a constant volume, which is shown in Figure 4.12.

The presented dependence is close to the exponential one, so the activation energy at constant volume,

$$E_V = R\left(\frac{\partial(\ln\gamma_1)}{\partial(1/T)}\right)_V, \tag{4.47}$$

can be calculated. This parameter reflects the sole influence of temperature without the superimposed influence of density. The value of the activation energy E_V was found to be independent of the liquid crystal density and equal to 23 kJ/mol in the temperature interval $\Delta T_c = T_c - T = 0$–35 K. Far from the clearing point

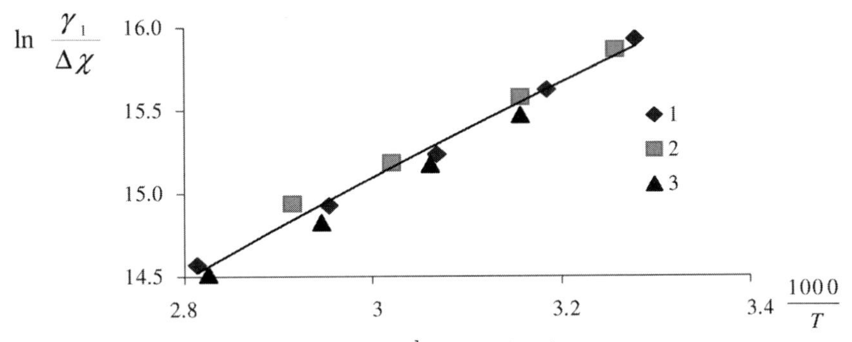

Figure 4.12 Dependence of $\ln(\gamma_1/\Delta\chi)$ upon T^{-1} at a specific volume $\upsilon \cdot 10^4$ (m^3/kg): (1) 9.03; (2) 8.99; (3) 9.08. (After Larionov *et al.* [20].)

Table 4.4 The activation energy E_p at different pressures. (After Larionov *et al.* [20]).

P (MPa)	0.1	10	20	30	40	50	60
E_p (kI/mol)	31.7	31.4	30.3	27.8	26.7	24.6	23.3

Table 4.5 Pressure dependence of the ratio (E_p/E_V), $\Delta T_c = 0$–35 K. (After Larionov *et al.* [20]).

P (MPa)	0.1	10	20	30	40	50	60
E_p/E_V	1.38	1.37	1.32	1.21	1.16	1.07	1.02

Table 4.6 Temperature dependence of the hole volume. (After Larionov *et al.* [20]).

T (K)	309	319	323	327	331	335	339	347
ΔV (10^5 m^3/mol)	0.64	0.82	1.05	1.30	1.60	1.84	2.08	2.42

($\Delta T_c = 35$–55 K), some increase in E_V (about 20%) took place. It may be connected with the decrease of the activation energy at constant pressure E_p when the pressure is increasing (Table 4.4).

The dependence of the activation energy E_p on pressure may be explained by means of the thermodynamic equation [20]

$$E_V = E_0 - RT^2 \left(\frac{\partial P}{\partial T}\right)_V \alpha^*_{\text{visc}}, \tag{4.48}$$

where α^*_{visc} is the slope of the $\ln \gamma_1 – P$ isotherms. While E_p should be proportional to the total molecule energy used to jump over a potential barrier and to form the "hole," E_V is connected only with the former part. So, the ratio (E_p/E_V) has to be more than 1, which is in accordance with experimental results. This parameter decreases in nematic phase with increasing pressure (Table 4.5).

Using the presented results, one can calculate the hole volume V (Table 4.6). This parameter increases with increasing temperature, which seems to be quite reasonable. Results presented above show that ultrasonic studies of liquid crystals in rotating magnetic fields provide a lot of useful information on rheological properties of liquid crystals.

4.3 Shear Waves in Liquid Crystals

4.3.1 Shear Waves in Isotropic Liquids

Shear waves differ from longitudinal ones by transverse direction of oscillations of particles with respect to the wave vector of an ultrasound wave.

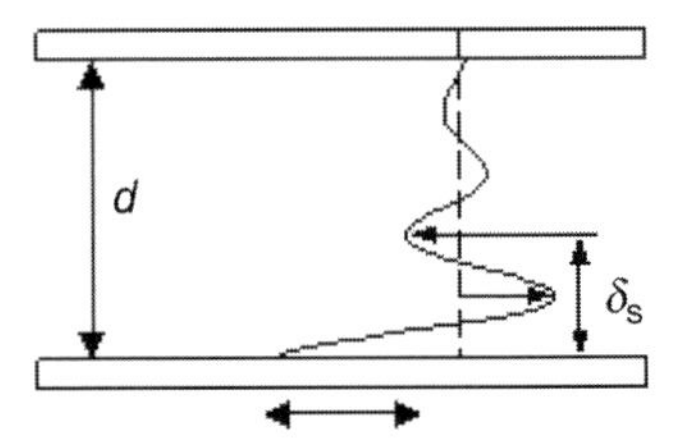

Figure 4.13 Shear waves in liquids.

In spite of the fact that these waves are generated by similar piezoelectric transducers as those described above, their propagation in liquid crystals is quite different from the longitudinal wave propagation.

It is well known that simple isotropic liquids can flow under extremely low shear stress, as their shear static modulus is equal to zero. That is why shear waves in such media are highly overdamped and can propagate only on a short distance comparable to a wavelength. Let us consider a liquid layer of thickness d confined between two plates, one of them oscillates in the plane of the layer with frequency ω (Figure 4.13).

The resulting motion of a liquid in this case depends on a ratio

$$r_S = \delta_S/d, \tag{4.49}$$

where

$$\delta_S = \sqrt{\frac{2\eta_S}{\rho\omega}} \tag{4.50}$$

is the depth of penetration of shear waves and η_S is a shear viscosity coefficient. Low-frequency regime ($r_S \gg 1$) corresponds to a simple oscillating shear flow with a linear velocity profile described in Chapter 3. At high-frequency regime ($r_S \ll 1$), oscillations of a liquid are concentrated in the boundary layer of a thickness δ_S. They can be described in terms of an overdamped shear wave:

$$\xi = \xi_0 \exp[i(\omega t - qz)], \tag{4.51}$$

with the wave vector expressed as

$$q = (1-i)\delta_S^{-1}. \tag{4.52}$$

So, according to (4.50), the thickness of the boundary layer is about 30 μm at ultrasonic frequency $f = 3$ MHz. It means that using shear waves at least in megahertz range of frequencies provides some information about viscosity properties referred to the near-boundary layer contrary to the case of bulk samples in experiments with longitudinal waves. This information is restricted by shear viscosities of liquids as shear waves are not connected with density variations and so can be considered by hydrodynamic models of incompressible liquids. That is why ultrasonic measurements at shear waves technique are insensitive to the above-mentioned relaxation processes contrary to the use of longitudinal ultrasound.

4.3.2
Peculiarities of Shear Waves in Liquid Crystals

Application of ultrasonic technique based on the use of shear waves in case of liquid crystals is possible with a proper account of difference between the structure and the rheological behavior of isotropic and anisotropic liquids.

First, a limited thickness of a boundary layer does not play an essential role in isotropic Newtonian liquids, although it is critical for the study of NLCs. Indeed, the most interesting results can be obtained for the well-oriented samples. It is difficult to prepare such samples by using only strong magnetic fields in a manner described above for the case of longitudinal ultrasound in bulk samples. Even in a rather strong field (about 1 T), the magnetic coherence length ξ_H (see Chapter 5) is about some microns, which is comparable to the shear wavelength in megahertz range of frequencies.

Second, as shown above, in the frame of hydrodynamics, shear waves described by expressions analogous to (4.27) can be considered as overdamped "fast" modes with the next dispersion relation between a wave vector $\boldsymbol{q}$ and a frequency ω [25]

$$q_m = (1-i)(\rho\omega/2\eta_m)^{1/2}, \qquad (4.53)$$

where $m = a, b, c$ are referred to as the three principal geometries at shear wave propagation (a: $\boldsymbol{q}\perp\boldsymbol{n}$, $\boldsymbol{n}||\boldsymbol{v}$; b: $\boldsymbol{q}||\boldsymbol{n}$, $\boldsymbol{n}\perp\boldsymbol{v}$; c: $\boldsymbol{q}\perp\boldsymbol{n}$, $\boldsymbol{n}\perp\boldsymbol{v}$) shown in Figure 4.14.

In the absence of magnetic field, the effective shear viscosity coefficients η_m are expressed as

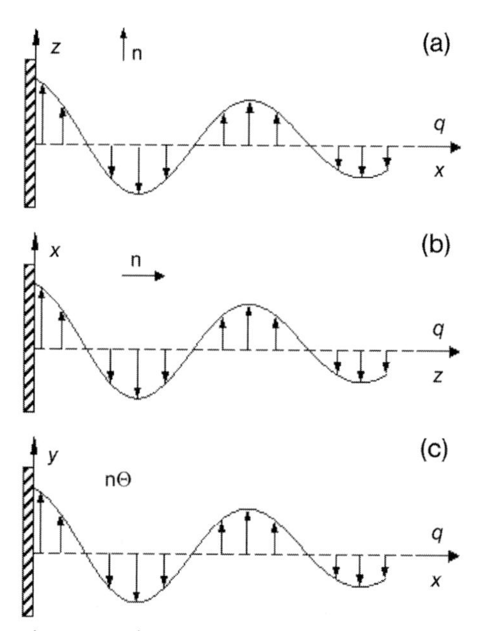

Figure 4.14 Three principal geometries for shear wave propagation.

$$\eta_a = \eta_b = \eta_1 - \frac{\alpha_3}{2[1+(\gamma_2/\gamma_1)]} = \eta_2 + \frac{\alpha_2}{2[1-(\gamma_2/\gamma_1)]} \tag{4.54}$$

$$\eta_c = \eta_3. \tag{4.55}$$

(Here and below we used notations from Ref. [25].)

Strong magnetic fields suppress the director motion. It allows, as in the case of steady shear flows, to consider NLCs as a Newtonian liquid with the shear viscosity depending on orientation. For three primary geometries shown in Figure 4.14, the shear viscosity coefficients become equal to the Miesowicz viscosities, namely,

$$\eta_a = \eta_1; \qquad \eta_b = \eta_2; \qquad \eta_c = \eta_3. \tag{4.56}$$

One can easily obtain the expressions for penetration lengths analogous to expression (4.50) and corresponding to different principal geometries taking into account the relation between this parameter and the imaginary part of the wave vector (q^{i})

$$\delta_m = q^{\mathrm{i}}_m = \sqrt{\frac{2\eta_m}{\rho\omega}}. \tag{4.57}$$

The difference between penetration lengths for different geometries can be essential (as the maximal ratio of Miesowicz viscosities often exceeds 5) and has to be taken into account when elaborating on a particular experimental setup.

It is obvious that shear wave methods can provide some useful information about shear viscosities and Leslie coefficients of nematic liquid crystals.

4.3.3 Experimental Methods for Shear Wave Studies

In fact, the same types of ultrasonic methods, pulse and resonator, were applied both for longitudinal and for shear waves. Nevertheless, there is a big difference in a particular realization and acoustic chambers.

First, the *impulse phase ultrasonic method* was applied to rheological study of liquid crystals by Martinoty and Candau [47]. As in the case of longitudinal ultrasound, the amplitude and the phase of the output impulse were the primary measurable parameters. In shear wave experiments, they are changed at reflection from a solid–liquid crystal boundary.

It is well known that the reflection of any type of waves (electromagnetic or acoustic) can be described in terms of impedances (Z_{i}) referred to different media under consideration. In the case of shear waves, mechanical impedance Z is defined as the ratio (with minus sign) of shear stress to the rate of shear displacement. In general, Z is a complex value that includes the real (Z_{r}) and imaginary (Z_{i}) parts. For isotropic Newtonian fluid in the absence of viscoelastic relaxation, $Z_{\mathrm{i}} = Z_{\mathrm{r}} = (\rho\omega\eta_s/2)^{1/2}$. It corresponds to the zero value of the real part of the complex shear elastic modulus G. This equality breaks for viscoelastic liquids where the difference between Z_{r} and Z_{i} increases with frequency. In the case of solid-like behavior with small losses, the real part Z_{r} essentially exceeds the imaginary part Z_{i}.

In the simplest case of reflection from one boundary, the coefficient R, which represents the ratio of the complex amplitudes of incident (A_1) and reflected (A_2) waves, is defined as

$$R = A_2/A_1 = (Z_1 - Z_2)/(Z_1 + Z_2), \tag{4.58}$$

where complex amplitudes are connected with the velocities of shear oscillations in incident (v_1) and reflected (v_2) waves by usual way

$$v_1(t,\vec{r}) = \mathrm{Re}[A_1 \exp i(\omega t - \vec{q}\vec{r})], \qquad v_2(t,\vec{r}) = \mathrm{Re}[A_2 \exp i(\omega t + \vec{q}\vec{r})]. \tag{4.59}$$

Expressions for acoustical impedance of LC Z_2 and for reflection coefficient R were obtained and analyzed in detail [25] for the case of normal incidence of shear waves. For overdamped waves under consideration, an acoustic impedance Z is a complex parameter expressed as

$$Z_2^m = (1-i)\sqrt{\frac{\rho\omega\eta_m}{2}}, \tag{4.60}$$

where η_m is referred to as the three principal shear viscosities η_a, η_b, η_c defined by (4.54–4.56).

It is simple to get expression for a complex reflection coefficient R using (4.58) and (4.60). Usually, the latter parameter is written in the form

$$R^m = IR^m I \exp(i\varphi_m), \tag{4.61}$$

where the modulus $|R^m|$ and the phase φ_m of the reflection coefficient define the changes of an amplitude and of a phase of the reflected impulse. The latter parameters are registered in experiments. It helps to determine the complex acoustic impedance by the next expression, easily obtained from (4.58), (4.60), and (4.61),

$$Z_2^m = Z_1 \frac{1 - IR^m I^2 - 2iIR^m I \sin\varphi_m}{1 + IR^m I^2 + 2IR^m I \cos\varphi_m}. \tag{4.62}$$

The unknown value of impedance Z_1 of the first media can be obtained by calibration using isotropic liquid, for example. It is worthwhile to note that according to (4.60) the absolute values for real and imaginary parts of the acoustic impedance Z_2^m are equal, so the next expression for the phase shift φ_m has to be made

$$\sin\varphi_m = \frac{1 - IR^m I^2}{2IR^m I}. \tag{4.63}$$

It means that only one parameter (e.g., amplitude of the reflected impulse) has to be measured to calculate the acoustic impedance of liquid crystals. So, the information about shear viscosities and Leslie coefficients can be extracted from shear wave measurements under a proper geometry. Indeed, the first experiments of such type [47] were made at oblique incidence of shear waves (see Figure 4.15).

In this case, the expressions presented above have to be modified to describe experimental results. The temperature dependences of the two principal viscosity coefficients of MBBA obtained by analyzing the data obtained in this case [47] are shown in Figure 4.16.

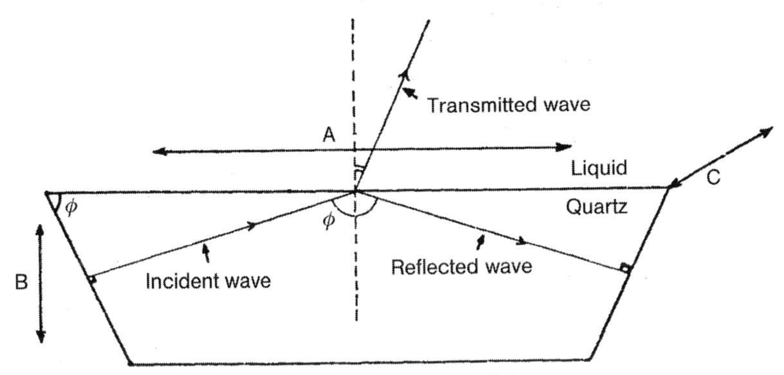

Figure 4.15 Scheme of experiments on the reflection of shear waves at oblique incidence relative to the boundary liquid crystal quartz. (After Martinoty and Candau [47].)

The correspondence between η_a and η_0 in the nematic phase is rather good except for the vicinity of the clearing point, which can be attributed to the increase in the flow alignment angle (see Chapter 5). The essential difference in the isotropic phase was referred to as the structural relaxation of the short-range order [48]. So, in the last case, the shear wave method is not suitable to correctly determine a static shear viscosity.

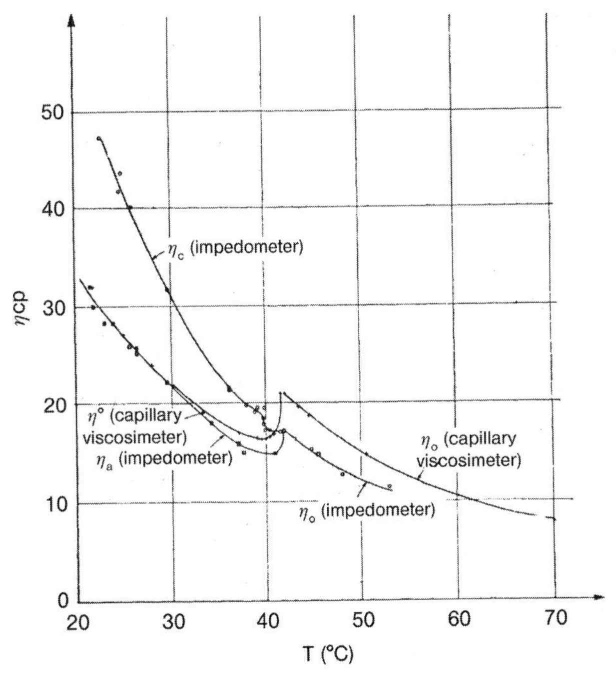

Figure 4.16 Principal shear viscosities η_a and η_c of MBBA determined via shear wave measurements. The data are compared with viscosity η_0 of flow-aligned sample obtained using capillary viscosimeter. (After Martinoty and Candau [47].)

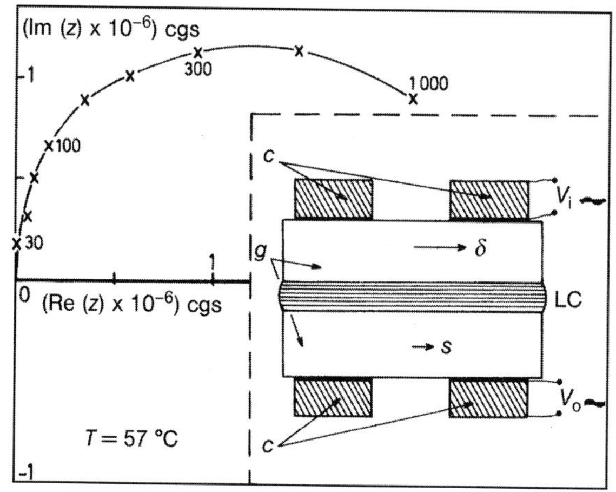

Figure 4.17 Complex shear modulus in polar coordinates for a smectic A phase of 40.8; the parameter on the curve is the frequency. Inset, schematic drawing of the sample geometry: c, piezoelectric ceramics; g, glass plates; LC, liquid crystal; ν_i, input voltage inducing the shear displacement δ; ν_0, output voltage due to the transmitted force *S*. (After Cagnon and Durand [49].)

In the literature, one can find a description of different experimental setups based on shear vibrations propagating in LC media and proposed for resolution of some particular problems in rheology of liquid crystals. For example, low-frequency measurements were realized in the work [49]. The scheme of such experiments is illustrated in the inset in Figure 4.17.

In this case, a 30–200 μm thick LC layer was placed between two glass plates connected to piezo ceramics. The input shear oscillations were induced by the periodic displacement of the upper glass plate via input AC voltage V_i (10–100 V) of frequency 1 Hz–10 kHz applied to the upper piezo ceramic. The lower ceramic was used to register the shear stress transmitted by LC layer and transformed into output voltage V_0 (in the range 100–1000 μV). The combination $(V_0/V_i)\sin(i\varphi)$ (where φ is the phase shift between input and output signals) is proportional to the complex shear modulus (*Z* in the author's notation). In the described experiments, it provided ultrasonic study of rheological behavior of smectic A and smectic B phases in butyloxy-benzylidene octylaniline (40.8). In particular, it was found that smectic A behaved as a viscous liquid, whereas smectic B showed an elastic response with a shear modulus $C_{44} \sim 10^6$ cgs units. It is worthwhile to note that the characteristic length of shear oscillation penetration, defined by (4.57), is higher or comparable to the LC layer thickness practically in the entire frequency range. It corresponds to the vibration regime and is different from experiments of Martinoty and Candau [47] with shear wave reflection.

The alternative *resonator method* is based on the use of continuous shear oscillations produced by resonators of different types (see, for example, Refs [50, 51]). As in the case of longitudinal ultrasound, the changes in elastoviscous properties of a liquid media connected with piezo elements result in variations of both resonance frequency f_r and Q-factor of a resonator. The latter parameters can be connected with an acoustic impedance Z of a liquid. Therefore, it is possible to determine the elastic modulus (G) and dynamic shear viscosity (η), which enters into complex shear modulus $G^* = G - i\omega\eta$ via expressions

$$G = \frac{Z_r^2 - Z_i^2}{\rho} \tag{4.64}$$

$$\eta = \frac{2Z_r Z_i}{\omega\rho}. \tag{4.65}$$

Such technique was mostly used for the study of smectic phases including freestanding films [50, 51]. In particular, the torsion pendulum was applied in the work [50] to measure the shear mechanical properties of *n*-hexyl-4′-*n*-pentyloxybiphenyl-4-carboxylate (65OBC) in its stacked hexatic smectic B phase. The experimental geometry is shown in Figure 4.18.

The freestanding film of LCs was formed between a torsion disk and a supported ring. A torsional fiber was attached to the disk that rotated it about an axis normal to the film leading to in-plane shear stress on the film. The small displacement (of order 100 Å) was registered to calculate the shear modulus of the film. Experiments performed at frequency 45 Hz showed that the hexatic B phase did not exhibit an in-plane shear modulus, which was in agreement with X-ray structural studies indicating only short-range in-plane correlations. It was also established that in freestanding 65OBC films, crystalline surface layers can coexist with hexatic B interior layers.

Summarizing literature data, we can conclude that investigation of shear waves and oscillations in liquid crystals provide additional information about rheological properties of liquid crystal phases especially of the smectic phases with nonzero shear

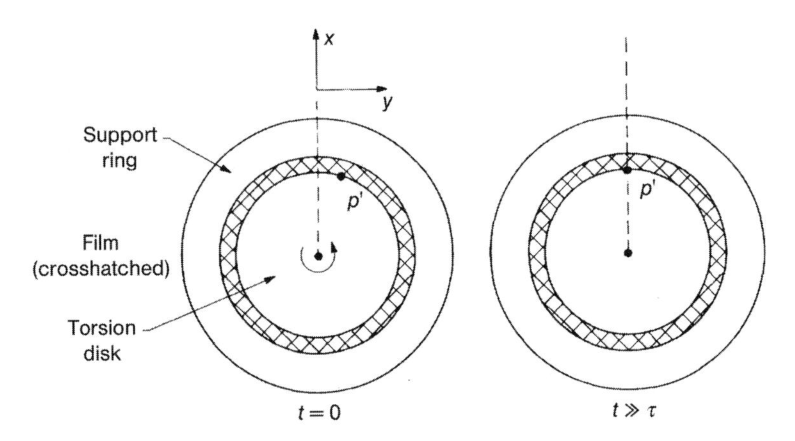

Figure 4.18 Experimental geometry for the strain relaxation measurements. T is the mechanical relaxation time of the film (After Pindak *et al.* [50].)

elastic modulus. In many cases, experiments were performed at low frequencies referred to as sound or even infrasound range when the regime of vibrations was realized. Of course, it essentially differs from experiments with longitudinal waves in the ultrasonic frequency range.

4.4 Ultrasonic Parameters and Viscoelastic Properties at Phase Transitions

4.4.1 Phase Transitions and Critical Phenomena in Liquid Crystals: General Aspects and Peculiarities of Ultrasonic Studies

Among the different classes of condensed matter, liquid crystals are of special interest from the point of view of variety of dynamical processes [48]. Such processes originate from both individual and collective motions of mesogenic molecules. Depending on the concrete liquid crystal phase (nematic, smectic A, and smectic C; Figure 4.19), there are different types of collective molecular motions. In the simplest case of nematic liquid crystals, these motions can be subdivided into very slow motion of a local optical axis (director **n**) and relatively fast motions of long molecular axes relative to the director that can be considered as changes of a degree of an orientation order parameter $S = 3/2\,\langle\cos^2\theta_i - 1/3\rangle$ (θ_i is the angle between a long axis of the individual molecule and the director **n**), where $\langle\cdots\rangle$ is a statistical average value. In a smectic A phase, where molecules are arranged in a system of equidistant layers, there are the additional degrees of collective motions connected with distortions of smectic layers. These distortions can be described in terms of the vector of a layer displacement $\boldsymbol{u}$, directed along the layer normal ν. So, in a smectic A phase, the complex order parameter $\psi(z) = |\psi|e^{i\varphi}$ (analogous to that for λ-transition in ^{4}He) is usually used [48] to take into account the modulation of density $\rho(z)$ in the layer structure

$$(\rho(z) = \rho_0[1 + 2^{-1/2}|\psi|\cos(k_0 z - \phi)], \qquad (4.66)$$

where $k_0 = 2\pi/l$, l is the mean distance between smectic layers and ϕ is the arbitrary phase. In a smectic C phase, molecules are inclined relative to the layer normal ν.

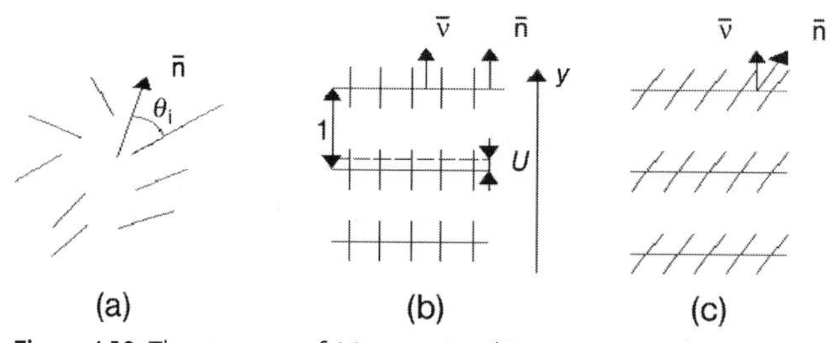

Figure 4.19 The structure of (a) nematic, (b) smectic A, and (c) smectic C phases.

Thus, to describe A–C transition, the vector order parameter $\psi = [\mathbf{n}\nu]$ can be introduced [52] (its modulus is identically zero in a smectic A phase and differs from zero in smectic C). The difference in the structure of nematic, smectic A, and smectic C phases as well as corresponding characteristics referred to as order parameters introduced above are presented in Figure 4.19.

Liquid crystals can be considered very interesting objects from the point of view of the general theory of phase transitions and critical phenomena in condensed matter. In most cases, phase transitions in liquid crystals mentioned above can be attributed to second-order transitions (such as nematic–smectic A, smectic A–smectic C) or to weak first-order transitions (nematic–isotropic liquid). The jumps of the first derivatives of the thermodynamic potentials (such as a specific volume or entropy) are negligible or small corresponding to the first and the second cases. It is accompanied by strong pretransitional anomalies in a number of equilibrium (such as compressibility and specific heat capacity) and nonequilibrium (such as viscosity or thermoconductivity) properties of liquid crystals [45, 53–55]. It originates from fluctuations of the order parameter corresponding to the given phase transition [45]. In general, these fluctuations are connected with fluctuations of other thermodynamic parameters, such as density. The intensity of fluctuations increases drastically upon approaching the second-order transitions. It results in the universal critical behavior for a number of physical parameters in the close vicinity of the transition. Such behavior is described by universal theoretical models operating with multicomponent order parameters in “isotopic” space [45, 52].

However, a universal description of phase transitions cannot be applied to polymorphic transformations in liquid crystals. It occurs due to the first-order character in the case of nematic–isotropic phase transition. The existence of a nonzero order parameter in a nematic phase results in suppressing critical fluctuations. At the same time, it is difficult to reach truly critical region upon approaching a clearing point from an isotropic phase due to essential difference (of order 1 K) between the transition temperature T_c and the critical temperature T^* [45].

In smectic phases, the strong initial anisotropy serves as a background for critical phenomena [52]. This anisotropy could not be observed in common liquids in which critical phenomena have been fully studied. This feature leads to a number of peculiarities of both an experimental and a theoretical character.

In particular, the totally universal behavior takes place over a very narrow temperature range near the transition at infralow frequencies. This temperature–frequency range is not achievable experimentally. Real experiments are conducted in the crossover regime. The fluctuations in this case are not strong enough to lead to the whole isotropization of the smectic phase. That is why to interpret the experimental data in this case (say, at phase transition smectic A–smectic C), an exact analysis of the critical behavior at the strong bare anisotropy background is needed [52].

From the experimental point of view, ultrasonic studies of critical phenomena in liquid crystals are accompanied by a number of restrictions.

First, they demand monodomain LC samples of a large volume, a precise mutual orientation of the initial wave vector and director, and an accurate account of the director surface distortions.

Second, liquid crystal phases usually exist in relatively narrow temperature ranges. So, the pretransition regions, which are associated with various critical points, can overlap each other. It makes it difficult, in general, to extract critical contributions in acoustical parameters corresponding to the given phase transition.

Third, noncritical relaxation processes in the high-frequency region arise due to the same mechanisms as in isotropic liquids (e.g., intramolecular rotations of the molecular fragments [56]). These processes also contribute to the frequency dependences of acoustic parameters.

Finally, low-frequency measurements are of key importance due to critical slowing down of dynamic processes upon approaching the phase transition points. At the same time, as it was shown for smectics A and C [52, 57, 58], strong fluctuations of smectic layers far from any transitions cause contributions to sound absorption and bulk viscosities. In particular, contributions to the bulk viscosity diverge as ω^{-1} in low-frequency limit.

Due to the reasons mentioned above, the analysis of acoustic data becomes a nontrivial problem, even in the case of nematics. Nevertheless, acoustic methods can be considered as unique practical tools for studying critical dynamics and viscoelastic properties at phase transitions in liquid crystals, as shown below.

4.4.2
Nematic–Isotropic Transition

Despite numerous experimental and theoretical studies of nematic–isotropic (N–I) phase transition, critical processes in the vicinity of clearing point have not yet been fully understood. Traditional phenomenological description in the framework of Landau–de Gennes theory fails to explain thermodynamic properties in proper way [45], as the given transition shows simultaneously the features of the first- and second-order transitions. It is also true for ultrasonic properties in the vicinity of nematic–isotropic liquid transition. One can find a number of original papers, some reviews, and books with description of the results of ultrasonic studies in both nonoriented and oriented samples of nematics (see, for example, Refs [2, 6, 59]). Here, we will briefly discuss the most important features of the behavior of ultrasonic parameters, especially at low frequencies in the vicinity of clearing point.

Typical examples of temperature dependences of ultrasonic attenuation and velocity obtained in the vicinity of N–I transition [60] are shown in Figure 4.20.

Measurements in the frequency range 0.15–1.2 MHz were performed by a resonator method via acoustic camera shown in Figure 4.2b. Liquid crystal samples were oriented by the magnetic field 0.29 T in the direction normal to the wave vector. The nonsymmetric (relative to T_c) type of $\alpha/f^2(T)$ and $c(T)$ dependencies is typical for all nematics [2, 6]. The critical peak-like increase in α/f^2 upon approaching T_c and the sharp minimum on $c(T)$ dependence are well registered in the low-frequency range indicated above. Ultrasonic parameters were found to be independent of frequency excluding the nearest vicinity of a transition (about 0.5 K), where only the lowest frequency 0.15 MHz was used for measurements due to a strong increase in ultrasonic absorption. It can be considered as an experimental confirmation of

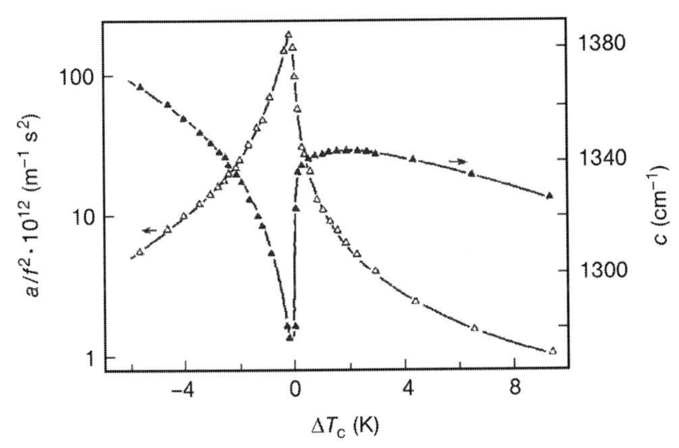

Figure 4.20 Temperature dependences of the low-frequency ultrasonic attenuation coefficient α and the velocity c in the vicinity of the clearing point T_c for BBBA. (After Pasechnik *et al.* [60].)

realization of the low-frequency limit $\omega\tau \ll 1$. In accordance with expression (4.30), the value of the attenuation coefficient $\alpha^{\perp}$ at the normal reciprocal orientation of a director **n** and a wave vector $\boldsymbol{q}$ is defined only by two viscosity coefficients μ_1 and α_4. The latter can be simply extracted from shear viscosity measurements (see Chapter 5). Its temperature dependence for BBBA shown in Figure 4.8 is not able to describe the strong critical increase in the attenuation upon approaching T_c. Moreover, the estimate made in accordance with (4.30) shows that the contribution to attenuation near T_c, defined by α_4 coefficient

$$\frac{\alpha^{\perp}}{f_{\text{incom}}^2} = \frac{2\pi^2}{\rho_0 C_0^3}\alpha_4 \tag{4.67}$$

is about $0.2 \times 10^{-12}\,\text{m}^{-1}\,\text{s}^2$. It means that the low-frequency attenuation in the vicinity of N–I transition shown in Figure 4.20 is almost completely defined by the bulk viscosity coefficient μ_1. The analogous conclusion is definitely valid for any other orientation of the wave vector as all Leslie coefficients entering into Equations 4.29–4.32 are of the same order of magnitude (or smaller), and attenuation anisotropy in a nematic phase is relatively small (not more than 20% at frequencies lower than 10 MHz). So, the bulk viscosities μ_1 can be determined from ultrasonic measurements of oriented samples. In the case of ultrasonic studies of nonoriented samples, widely represented in literature [2, 45, 59, 61], the expression for the average ultrasonic attenuation $(\alpha/f^2)_{\text{av}}$ is derived from (4.29) to (4.32) by substitutions

$$\cos^2\theta \rightarrow \cos^2\theta_{\text{av}} = 1/3; \qquad \cos^4\theta \rightarrow \cos^4\theta_{\text{av}} = 1/5 \tag{4.68}$$

taking into account a chaotic distribution of local orientation in bulk samples of liquid crystals. At low frequencies, the combination b_α in Equation 4.29 is essentially smaller than the a_α combination (see Table 4.1). In this case, by using the substitutions mentioned above, it is easy to derive the simple connection between ultrasonic attenuation coefficients measured at two principal orientations and in nonoriented samples:

$$\frac{\alpha^{\|}-\alpha_{av}}{\alpha_{av}-\alpha^{\perp}}=2, \tag{4.69}$$

where $\alpha^{\|}=\alpha\ (\theta=0°)$. It was found that this expression approximately holds even at frequencies higher than 1 MHz [2, 6]. It confirms the isotropic-like distribution of local directors in nonoriented samples and makes it possible to obtain some additional combinations of viscosity parameters.

Of course, the bulk viscosity coefficients strongly depend on frequency due to different relaxation mechanisms mentioned above. In the case of the unique relaxation process with a relaxation time τ, the frequency dependence of μ_i is expressed as

$$\mu_i=\frac{\mu_i^0}{1+(\omega\tau)^2}, \tag{4.70}$$

where μ_i^0 is the value of the viscosity in the low-frequency limit $\omega\tau\ll 1$. The temperature dependence of the latter parameter is defined by the nature of the relaxation process under consideration. For example, for the critical process attributed to the order parameter relaxation, the general expression for the complex elastic modulus is defined by (4.36). The comparison of this expression with (4.70) results in the next temperature dependence of the low-frequency parameter μ_i^0:

$$\mu_i^0\sim\tau\sim\left(\frac{\Delta T_c}{T_c}\right)^{-1}. \tag{4.71}$$

The increase in the relaxation time with temperature in accordance with the simple power law, and the critical index equal to -1 is quite a general result for both the mean field theory (the Landau–Halatnikov mechanism) and the dynamic scaling theory taking into account strong pretransitional fluctuations [45, 62]. The latter mechanism can also contribute to anomalies of acoustic parameters in a nematic phase and is totally responsible for analogous anomalies in an isotropic phase. In the low-frequency limit, dynamic scaling also predicts the similar temperature dependence for the fluctuation contribution $(\alpha/f^2)_{fl}$ to the attenuation coefficient [45]:

$$\left(\frac{\alpha}{f^2}\right)_{fl}\sim\tau^{y+1}\sim\left(\frac{\Delta T_c}{T_c}\right)^{-(y+1)}, \tag{4.72}$$

where y is the ratio of critical indexes for heat capacity and for relaxation time. In accordance with the scaling theory, the former index is equal to zero or to a small value (0.08). So, we conclude that both mechanisms, the relaxation of the order parameter and the relaxation of fluctuations of the order parameter, result in critical contributions to the ultrasonic attenuation with critical indexes equal (or close) to -1 for low-frequency limit.

Experimental dependences of low-frequency ultrasonic attenuation coefficient [60], shown in Figure 4.21, are in good agreement with the theory both for

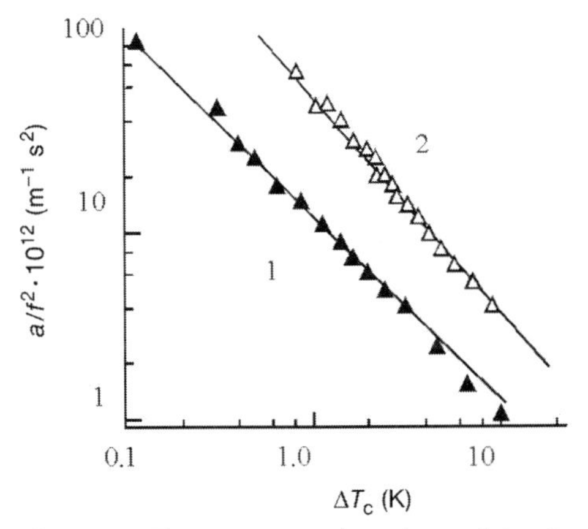

Figure 4.21 The temperature dependence of the ultrasonic attenuation α (the wave vector is normal to the director) in the isotropic (1) and nematic (2) phases of BBBA in the vicinity of N–I transition (T_c) at frequency $f = 0.15$ MHz. (After Pasechnik *et al.* [60].)

nematic and for isotropic phases. In the latter case, the critical index is slightly higher than −1, which can be partially attributed to the above-mentioned shift of the virtual critical temperature T^* with respect to the clearing point T_c [48]. We have to note that far from the clearing point, the low-frequency attenuation becomes small (about $1 \times 10^{-12}\,m^{-1}\,s^2$ at $\Delta T_c = 10$ K in isotropic phase and $2 \times 10^{-12}\,m^{-1}\,s^2$ at $\Delta T_c = -15$ K). It means that one can neglect all noncritical contributions in the analysis of experimental data. Of course, it does not hold for higher ultrasonic frequencies. The simple power laws are already broken at frequencies slightly higher than 1 MHz [62]. Critical anomalies in ultrasonic parameters essentially decrease with increasing frequency. It strongly prevents correct separation of critical and noncritical contributions, especially in compounds such as MBBA, BBBA, and so on, where intramolecular relaxation processes with comparable relaxation frequencies essentially modify the critical behavior of ultrasonic parameters [30, 61, 62]. The situation is essentially better for well-studied nematic PAA, with a rather simple molecular structure and relatively high clearing point (135 °C), where only critical processes contribute to ultrasonic parameters. The same critical index (−1) for low-frequency ultrasonic attenuation in a nematic phase was obtained during investigations into frequency range 0.8–5 MHz via a resonator method [27]. A slightly higher value of the critical index (about −1.5) was derived in isotropic phase when taking into account the shift (about 1 K) between T_c and second-order transition point T^*. The examples of analysis of experimental data obtained in frequency range 1–500 MHz for both individual nematic compounds and nematic mixtures can be found in [6, 61, 63].

Of course, the critical processes mentioned above also define pretransitional behavior of ultrasonic velocity c and corresponding real part of complex elastic modulus (see, for example, (4.36)). In isotropic liquids, the latter is inversely proportional to the adiabatic compressibility. Usually, critical contributions to these parameters do not exceed some percents. Nevertheless, the experimental error upon measuring the ultrasonic velocity is extremely low (of order 0.01% at measurements of relative changes, induced by temperature variations). So, some useful results can be extracted from such data too [60–62]. A lot of information on elastic bulk modulus and viscosities of nematics obtained via ultrasonic measurements can be found in [6, 63, 64].

4.4.3 Nematic–Smectic A Transition

Nematic–smectic A phase transition has been under intense attention for the last few decades in connection with the possible existence of the tricritical point, where the first-order N–S_A is transformed to the second-order transition. Indeed, the jumps of the order parameter, density, and entropy registered at N–A transition fall to zero with increasing temperature range (ΔT_N) of a nematic phase, characterized by a parameter $\Delta = 1 - \Delta T_N/T_{NA}$. Precise calorimetric measurements [45, 65–67] do not reveal any latent heat in the transition to values $\Delta \sim 10^{-2}$. The second-order character of the given transition leads to strong anomalies in pretransitional behavior of a number of physical parameters (a specific heat capacity, thermal expansion coefficient, correlation lengths, some shear viscosities, and so on [45]). These anomalies originate from strong fluctuations of a smectic order parameter describing the layer structure of a smectic A phase. In a nematic phase, such fluctuations were observed by de Vries [68] as a number of local regions with a layer-like structure, cybotactic clusters. As in the case of N–I transition the critical anomalies of ultrasonic parameters were registered near transition point. Results presented in Figure 4.22 were obtained at ultrasonic studies of terephthal-bis-*p*-*p*′-butylaniline (TBBA) [69].

The transition temperature in this compound is rather high. In general, it results in shifting relaxation spectra to a high-frequency region. That is why essential anomalies of acoustic parameters were registered at relatively high frequencies as it is shown in Figure 4.22.

At the same time, for a liquid crystal with relatively low transition temperature, such as BBBA, the above-mentioned anomalies become pronounced at frequencies lower than 1 MHz, as shown in Figure 4.23.

A similar behavior was also found in the case of CBOOA (4-cyanobenzylidene-4′-octyloxyaniline), where critical anomalies of ultrasonic parameters at N–A transition were registered at frequencies lower than 2 MHz at different orientations of a wave vector [71].

The mutual influence of different phase transitions can prevent the correct determination of critical contributions to acoustic parameters, connected with the given transition. It is clearly demonstrated by the results obtained for BBBA (Figure 4.23), where such procedure is applicable only in a wide-range nematic phase whereas it is impossible in the smectic A phase due to strong influence of the smectic A–smectic B phase transition. The real character of critical anomalies is much better seen for LC compounds such as 409 (*n*-butyloxyphenylether of *n*-nonyloxybenzoic acid) in the well-defined vicinity of N–A transition (Figure 4.24).

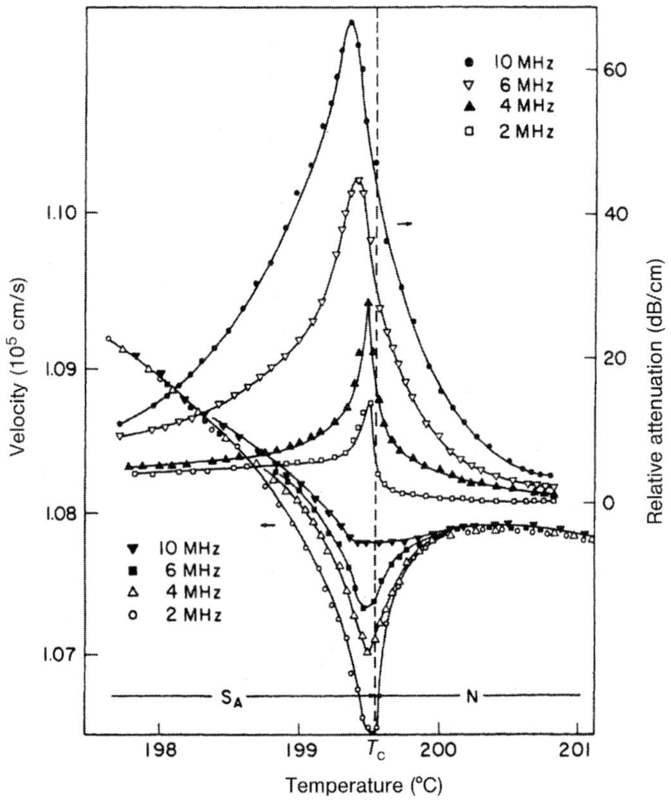

Figure 4.22 Temperature dependences of the velocity and the attenuation in TBBA in the vicinity N–A transition at the parallel reciprocal orientation of a wave vector and a director ($\theta = 0$). (After Bhattacharya *et al.* [69].)

The general difference of the critical behavior shown in Figures 4.23 and 4.24 from that registered for N–I transition in the same compounds is some shift of the relaxation spectra to the low-frequency region. It results in the frequency dependence of an ultrasonic attenuation coefficient observed at very low ultrasonic frequencies (lower than 0.5 MHz) in the near-transition region ($|\Delta T_{\mathrm{NA}}| = |T - T_{\mathrm{NA}}| \leq 1$ K). At the same time, for frequencies higher than 3 MHz often used in ultrasonic investigations of LCs [6, 59], critical anomalies become small (as for 409) or even comparable to experimental errors (as for BBBA). Partly, such behavior is explained by a possibility of approaching close to the critical point characterized by a strong slowdown of dynamic processes. Such possibility is absent for an isotropic phase in the vicinity of the clearing point due to the first-order nature of N–I phase transition (see above). That is why one has to take care when comparing experimental results with theoretical expressions valid for a low-frequency regime.

The wide temperature range of a nematic phase, such as in BBBA and 409, is important for two reasons. First, it provides correct separation of critical contributions,

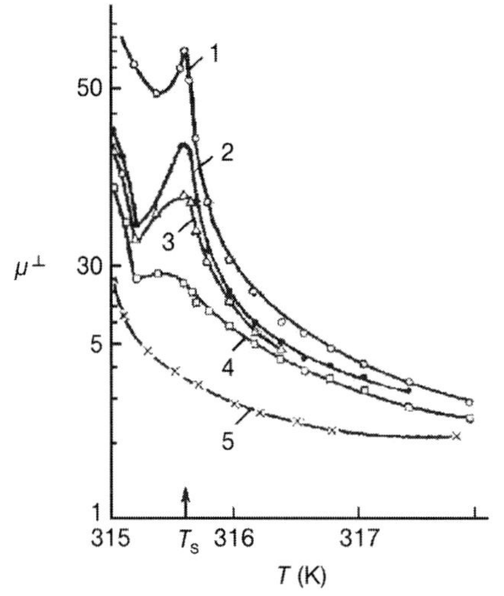

Figure 4.23 Temperature dependences of the attenuation coefficient $\mu^\perp = \alpha^\perp/f^2$, $10^{-13}\,\mathrm{m}^{-1}\,\mathrm{s}^2$ for BBBA in the vicinity N–A transition at the normal reciprocal orientation of a wave vector and a director ($\theta = 90°$): (1) 0.15 MHz; (2) 0.27 MHz; (3) 0.51 MHz; (4) 1.25 MHz; (5) 3.01 MHz. (After Balandin *et al.* [70].)

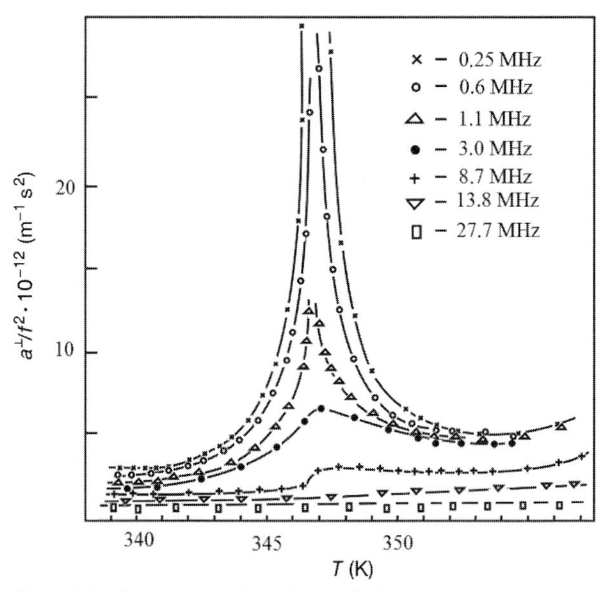

Figure 4.24 Temperature dependence of ultrasonic attenuation coefficient at phase transition – nematic–smectic A of 409 ($\theta = 90°$). (After Kashitsin [17] and Pasechnik [37].) The data were obtained by the resonator (0.25–1.1 MHz) and impulse (3.0–27.7 MHz) methods.

connected correspondingly with N–I and N–A phase transitions. Second, the latter transition is of the second order for compounds with a wide nematic range, as it was mentioned above. In this case, critical behavior can be mostly determined by strong fluctuations. The values of a ratio T_{NA}/T_c are equal to 0.918 and 0.958 for BBBA and 409, respectively. The scaling behavior of the critical contribution to the heat capacity was found in LCs with close values of this parameter [72]. So, one can wait to see that the same behavior will take place in a dynamic case too. The latter was considered in two different ways. In the first approach [73], the critical contribution $(\alpha/f^2)_c^{\perp}$ was attributed to the corresponding increase in the bulk viscosity coefficient ν_4 (in Forster's notation 4.31). It predicts a critical index for low-frequency contribution $(\alpha/f^2)_c^{\perp}$ equal to $-1/2$ or $-1/3$ in the mean field or scaling approximation. The interaction of fluctuations of the order parameter and the density is taken into account in the second approach [14, 74]. It results in higher values of critical indexes mentioned above ($-3/2$ in the mean field approximation and -1 in the dynamic scaling approximation).

Experimental temperature dependences of the critical contribution $(\delta\alpha^{\perp}/f^2)$ in the nematic phase of BBBA and 409 are shown in Figure 4.25.

These are obtained by deleting the critical contribution connected with N–I phase transition from the experimental data. One can see that in the closest vicinity to N–A transition ($\Delta T_{NA} < 1$ K), the simple power law

$$\frac{\delta\alpha^{\perp}}{f^2} \sim (\Delta T_{NA})^{-\gamma} \tag{4.73}$$

holds only for extremely low ultrasonic frequencies. In this case, the values of critical index γ are close to 1 for both compounds (BBBA and 409). It confirms the above-mentioned prediction of the dynamic scaling theory [14]. Moreover, the experimentally obtained variation of the asymptotic critical index with frequency (Figure 4.26) is also in accordance with this theory.

In particular, according to [14] the critical index of ultrasonic attenuation in high-frequency limit ($\omega\tau_s \gg 1$) has to be twice higher than the renormalized critical index of the thermal capacity (equal to 0.08 in scaling theory). It is close to experimental values, shown in Figure 4.26. It is of sense to comment on the results of analogous investigations of CBBOA [71], where the same value of the critical index (-1) was obtained at frequency 0.6 MHz only relatively far from the transition, whereas this index became smaller in the vicinity of N–A transition. Taking into account the results presented above, one can propose that the low-frequency limit was not achieved in the latter case.

A very interesting feature of the universal dynamic theories of phase transitions in liquid crystals is the connection between the temperature dependence of critical ultrasonic attenuation and the frequency dependence. For example, the same critical index has to describe the frequency dependence of α in the high-frequency limit and the temperature dependence of this parameter in the low-frequency limit. The high-frequency regime can be easily realized in the closest vicinity of N–A phase transition

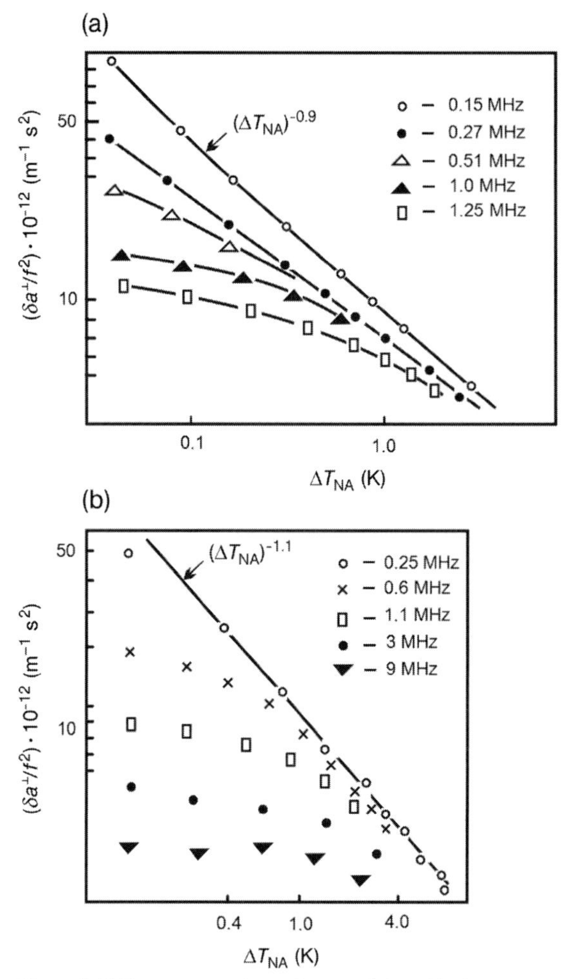

Figure 4.25 Temperature dependences of the critical contributions ($\delta\alpha^{\perp}/f^2$) to ultrasonic attenuation α (the wave vector is normal to the director) in nematic phases of BBBA (a) and 409 (b) in the vicinity of N–A transition (T_{NA}).

due to a critical increase in the relaxation time. Data presented in Figure 4.27 demonstrate such possibility – the simple power law

$$\alpha/f^2 \sim \omega^{-\gamma} \tag{4.74}$$

for the frequency dependence of ultrasonic attenuation really takes place at $\Delta T_{NA} = 0.05$ K. It can be considered a strong confirmation of the dynamic scaling theory.

It is worthwhile to note that the alternative dynamic process attributed to the relaxation of the modulus of the smectic order parameter (the Landau–Halatnikov mechanism) can also contribute to ultrasonic parameters in high-temperature nematic phase [15]. It is explained by the conservation of uniaxial symmetry at N–A

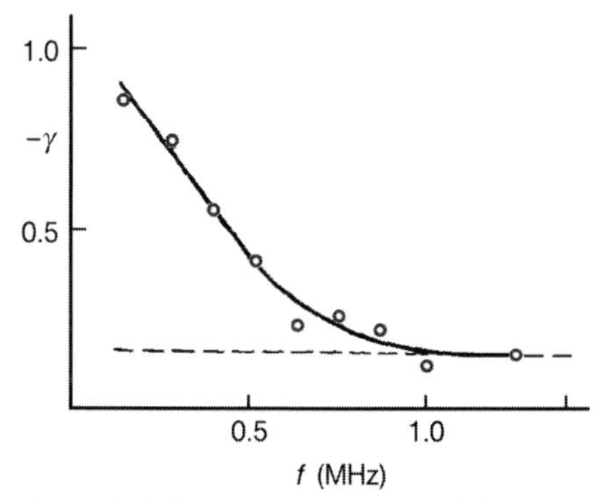

Figure 4.26 Frequency dependence of the asymptotic critical index γ in the nematic phase of BBBA. (After Balandin *et al.* [70].)

transition contrary to the case of N–I transition, when such symmetry is broken, and only fluctuation contribution exists in high-temperature isotropic phase. The relaxation contribution to ultrasonic attenuation $\delta\alpha^{\perp}/f_{\mathrm{r}}^{2}$ is expressed as

$$\left(\frac{\delta\alpha^{\perp}}{f^{2}}\right)_{\mathrm{r}}=\frac{2\pi^{2}\beta_{\perp}^{2}}{\rho c^{3}\eta}\frac{1}{1+\omega^{2}\tau_{\mathrm{s}}^{2}}, \tag{4.75}$$

where η is a kinetic coefficient with the reverse viscosity dimension and $\beta_{\perp}$ is a dimensionless parameter describing the dynamic connection between the smectic order parameter and the hydrodynamic parameters. As it was shown in [70], the presented expression failed to describe experimental results obtained in BBBA in a proper way. So, one concludes that the relaxation mechanism is of minor significance for describing ultrasonic attenuation in the nematic phase close to N–A transition.

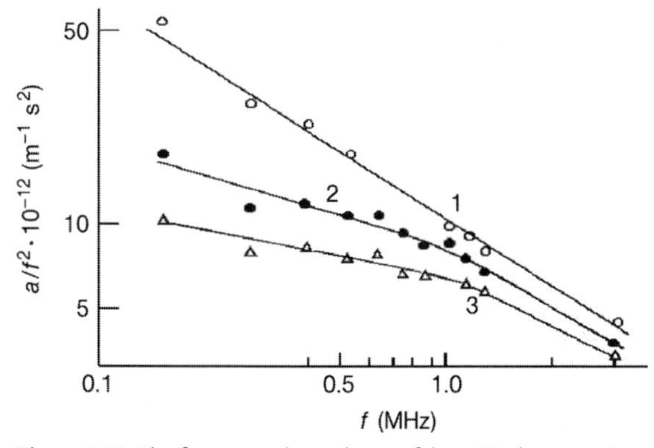

Figure 4.27 The frequency dependence of the critical attenuation α (the wave vector is normal to the director) in the nematic phase of BBBA in the vicinity of N–A transition (T_{s}) at different temperatures: (1) $\Delta T_{\mathrm{s}} - 0.05$ K; (2) $\Delta T_{\mathrm{s}} - 0.20$ K; (3) $\Delta T_{\mathrm{s}} - 0.37$ K. (After Balandin *et al.* [70].)

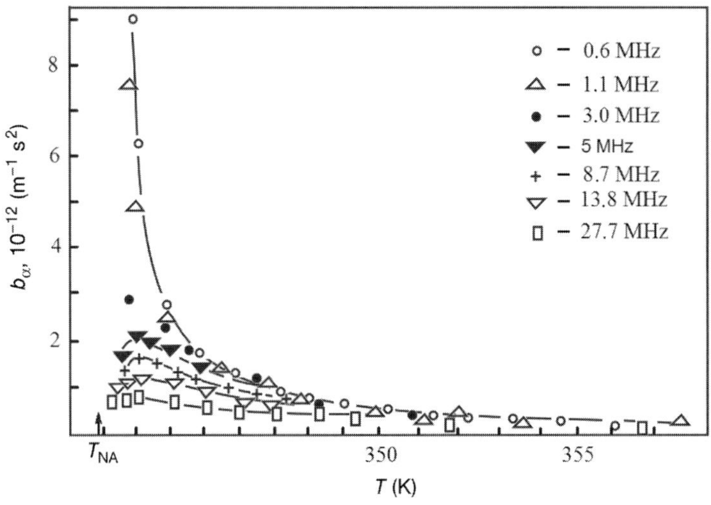

Figure 4.28 Temperature dependences of b_α in the region of N–A phase transition (409). (After Prokopjev *et al.* [13] and Pasechnik [37].)

Of course, the critical dynamic processes can contribute not only to the absolute values of acoustic parameters but also to the ultrasonic attenuation and velocity anisotropy. From practical point of view, the behavior of the coefficient b_α is interesting (see expression (4.32)), as it does not contain the bulk viscosity coefficients. That is why it does not show frequency dependence in nematic phase and can be used to get information about viscosity parameters assigned to incompressible liquid crystals (see above). The situation is drastically changed in the vicinity of a nematic–smectic A phase transition, where strong frequency dependence of b_α takes place (Figure 4.28).

It is seen that the b_α coefficient becomes frequency independent at frequencies lower than 1 MHz. In this case, according to (4.32) one can extract the viscosity coefficient combination $b = (\alpha_1 + \gamma_1 \lambda^2)$ (or $(\nu_1 + \nu_2 - 2\nu_3)$ in Forster notations). Close to T_{NA}, the low-frequency temperature dependence of such combination is described by the simple power law with critical index $-1/2$ as shown in Figure 4.29.

The observed behavior $b(T)$ in the vicinity of T_{AN} is in agreement with the conclusions of fluctuation theory [71], which predict the critical divergence of ν_1 coefficient with asymptotic index equal to $-1/2$ in mean field approximation. When $\Delta T_{AN} \geq 1$–5 K, the divergence index changes; this may be due to the influence of the relaxation process [15]. It is worth noting that in this temperature range, the noncritical contribution to the value of b does not exceed 10% and does not affect the conclusions.

The behavior of the coefficient $a = a_\alpha(\rho c^3/2\pi^2)$ is much more complicated as can be seen from its frequency dependence for different ΔT_{AN} shown in Figure 4.30. It may be attributed to the strong influence of the bulk viscosity coefficients entering into this combination. A detailed analysis of experimental results of velocity anisotropy in

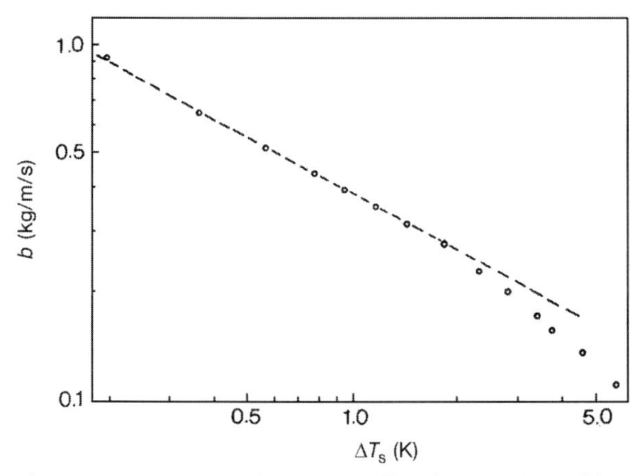

Figure 4.29 Temperature dependence of low-frequency limit of b in the vicinity of N–A transition for BBBA. (After Balandin *et al.* [19].)

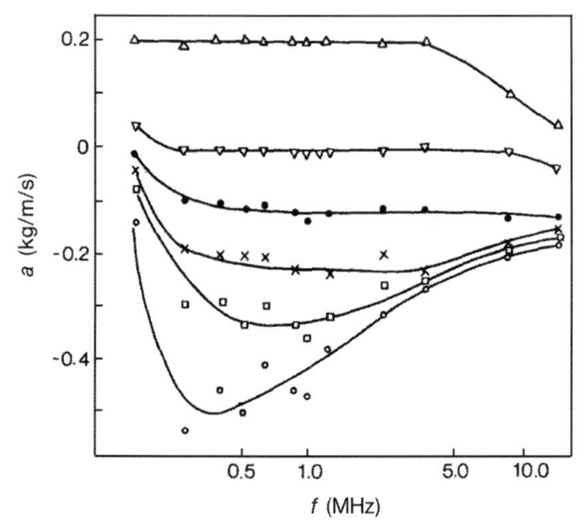

Figure 4.30 Frequency dependence of a at different $\triangle T_S$: (△) 25.0 K; (▽) 3.8 K; (●) 0.96 K; (×) 0.37 K; (□) 0.20 K; (O) 0.10 K. (After Balandin *et al.* [19].)

the vicinity of N–A phase transition can be found in Ref. [74]. The analysis of the pretransitional anomalies in smectic A phase is more complicated due to strong influence of bare anisotropy of smectic phase [17].

4.4.4 Critical Dynamics and Viscoelastic Properties at Smectic A–Smectic C Phase Transition

The experimental study of the critical dynamics at phase transitions between different smectic phases is of great interest. This is primarily due to the strong initial anisotropy that serves as a background for critical phenomena in liquid crystals as mentioned above. It is certainly true for the phase transition smectic A–C, which can be considered as a unique example of strongly anisotropic second-order phase transition.

Such anisotropy results in complicated elastic properties of smectic phases even far from phase transitions. They are described not by a single elastic modulus but by several constants. The longitudinal sound velocity possesses a very complicated dependence on the propagation direction. Owing to the strong anisotropy of smectics, universal models with multicomponent order parameters, which are traditional for the critical phenomena theory, considered in isotopic space, are not relevant for the description of the smectic C–smectic A phase transition [52]. The point is that a totally universal behavior takes place over a very narrow temperature range near the transition on infralow frequencies. This temperature–frequency range is not experimentally achievable. Real experiments are conducted in a crossover regime. Fluctuations in this case are not strong enough to lead to a complete isotropization of the smectic. That is why to interpret the experimental data in this case the exact analysis of the critical behavior at the strong bare anisotropy is needed.

The most detailed ultrasonic studies of A–C phase transitions were undertaken using the acoustic chambers shown in Figure 4.2. In these experiments, values of the velocity (c) and the absorption coefficient of ultrasound (α) were measured in the smectic A and C phases of *p*-(hexyloxy)phenylether-*p*-(decyloxy)benzoic acid. This compound possesses the following polymorphism [75]:

$$S_x\text{-}317\ \text{K}–S_C\text{-}350.8\ \text{K}–S_A\text{-}256.7\ \text{K}–\text{N}$$

Orientated smectic phases were obtained by cooling the sample from the nematic phase in a magnetic field with strength of 0.3 T for various angles between the wave vector and the magnetic field vector. Because of the large penetration depth of longitudinal sound, the orienting influence of the surface of the measuring devices was negligibly small. The measurements were carried out over a wide range of ultrasound frequencies $f = 0.15$–27 MHz (in the frequency interval 0.15–1.3 MHz by the resonator method, and in the frequency interval 3–27 MHz by the modified pulse phase method of a variable frequency).

Temperature dependences of an ultrasonic absorption in the vicinity of A–C transition at two principal geometries ($\boldsymbol{q}\|\nu$ and $\boldsymbol{q}\perp\nu$) are shown in Figures 4.31 and 4.32. The data presented show that the critical anomalies in ultrasonic attenuation connected with A–C phase transition are strongly anisotropic.

Indeed, for ultrasound propagating along the smectic layers ($\theta = 90°$), the small anomalies in the vicinity of the A–C transition are observed only at low frequencies and the frequency dependence of the absorption coefficient is essentially weaker than that in the case of normal propagation of the sound relative to the smectic layers ($\theta = 0°$). Such critical behavior is quite different from that of critical anomalies of ultrasonic attenuation registered at nematic–isotropic and nematic–smectic A phase transitions, characterized by relatively low anisotropy.

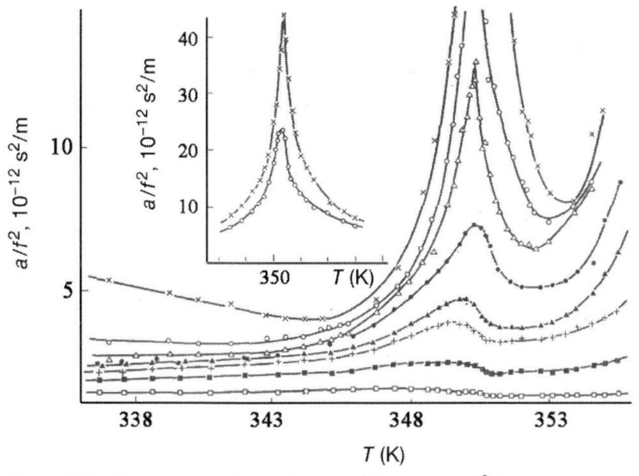

Figure 4.31 Temperature dependences of the ratio α/f^2 (α is the ultrasound absorption coefficient) at $\theta = 0°$: (×) 0.36 MHz; (O) 0.69 MHz; (△) 1.2 MHz; (●) 3 MHz; (▲) 5 MHz; (+) 8.7 MHz; (■) 15.8 MHz; (□) 27.7 MHz. (After Balandin *et al.* [76].)

The analogous conclusion is true for the critical behavior of ultrasonic velocity shown in Figure 4.33. In spite of very high resolution achieved at relative ultrasonic measurements of C, the latter parameter does not show any pronounced anomaly at $\theta = 90°$ even at low ultrasonic frequencies whereas the well-defined minima on $C(T)$ dependence was registered at $\theta = 0$. The depth of this minimum is of order 1 ms^{-1}, which is essentially smaller than the corresponding values registered at N–I and N–A phase transitions (see for comparison Figure 4.20). Moreover, it decreases with increasing frequency and vanishes at frequencies higher than 5 MHz. In the latter case, the phase transition is accompanied by a change in the velocity temperature coefficient (see Figure 4.34) at $\theta = 0°$ and $\theta = 30°$ whereas no changes in this parameter were registered at $\theta = 60°$ and $\theta = 90°$ over the complete frequency range.

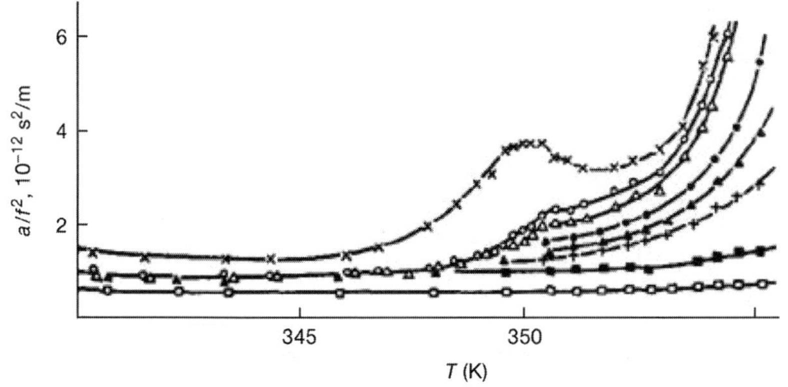

Figure 4.32 Temperature dependence of the ratio α/f^2 at $\theta = 90°$ (notations as the same as in Figure 4.31.

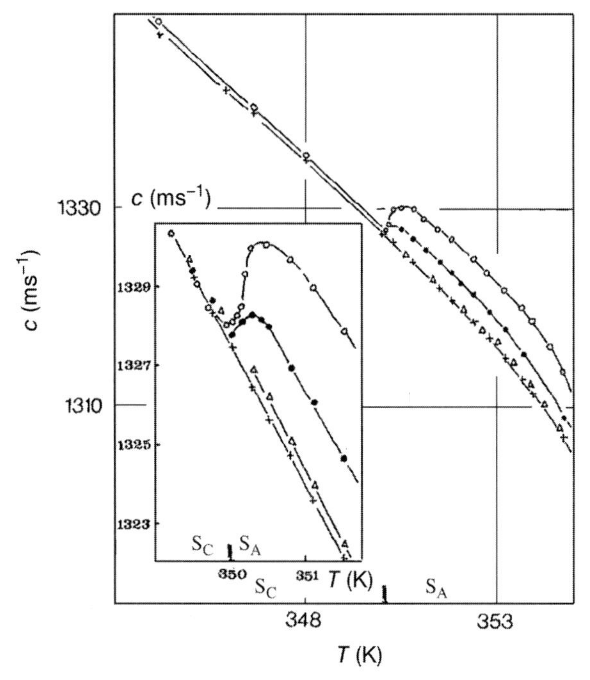

Figure 4.33 Temperature dependence of the longitudinal sound velocity in smectic C and A phases at 0.3 MHz for different values of the angle θ between the wave vector and the normal to the smectic layers: ○, θ = 0°; ●, θ = 30°; △, θ = 60°; +, θ = 90°. (After Balandin *et al.* [16].)

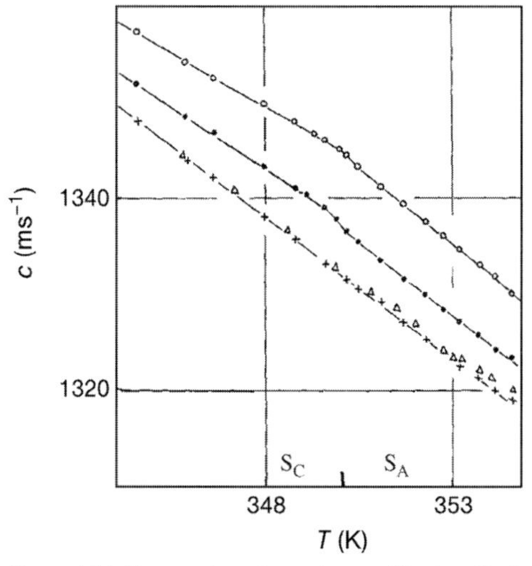

Figure 4.34 Temperature dependence of the longitudinal sound velocity in smectic C and A phases at frequency 27 MHz for different values of the angle θ (notations as the same as in Figure 4.33).

The exception is the very low-frequency regime, where only small changes in the temperature velocity coefficient are visible (Figure 4.33). A very high sensitivity to velocity at $\theta = 0°$ and very low sensitivity at $\theta = 90°$ to the vicinity of the transition demonstrate the monodomain structure of the sample.

Note that the character of the critical anomalies of the acoustic parameters in the vicinity of the A–C phase transition does not vary substantially when a sample of oriented A phase is cooled either in the presence or in the absence of the relatively weak magnetic field used in the experiment.

Experimental data on α and c obtained in experiments were used to calculate viscosity coefficients and elastic moduli of smectic phases. It is worthwhile to note that one can find different notations for viscoelastic parameters of smectic phases entering into acoustic parameters [18, 24, 52]. We will follow the notations used in the book by Kats and Lebedev [52] and in the paper [76].

Below, we briefly discuss the critical behavior of liquid crystals near the smectic C–smectic A phase transition following the paper [76]. Additional details of the calculation can be found in [52, 77, 78], which are especially dedicated to the theoretical investigation of S_A–S_C transition.

First, we have to note that the temperature region of the developed fluctuations near this transition is quite broad. It is comparable to the width of the existence region of the smectic phases themselves. By virtue of this, the conclusions obtained as a result of considering the critical fluctuations are to a significant extent of a general nature since the smectic phases can almost always be considered as being close to the investigated transition. As it was mentioned above, in equilibrium smectics are systems of equidistant layers, whose normal vector we take to be directed along the z-axis. To describe the deviation of smectic layers from equilibrium, one can introduce the smectic variable u, which plays the role of a displacement vector of the layers along the z-axis. The elastic energy associated with the curvature of smectic layers to first order has the standard form [77]

$$\frac{B(\nabla_z u)^2 + K(\nabla_\perp{}^2 u)^2}{2}.$$

Here, B is the compression modulus of smectic layers and K is a coefficient analogous to the Frank modulus in nematics.

The main term in the energy density associated with the inhomogeneous deformation of the director in space has essentially the same form as in nematics (see Equation 2.15).

In the energy expansion, there are two terms that include powers of the smectic C order parameter ψ:

$$\frac{1}{2}A\psi^2 + \frac{1}{4}U\psi^4.$$

The parameter A characterizes the phase transition: when A is greater than zero, the A phase is realized and when A is less than zero, the C phase; U is a constant at the four-point vertex.

The density ρ and the specific entropy σ also enter into the complete set of thermodynamic variables of the smectic in addition to the smectic variable and the order parameter. In the vicinity of the transition, only the order parameter ψ fluctuates strongly. All other variables fluctuate weakly. Their deviations from equilibrium can be conveniently described by the components of the dimensionless vector φ:

$$\varphi_\nu = \{\varphi_\rho, \varphi_\sigma, \varphi_u\} = \{\delta\rho/\rho, \delta\sigma/\sigma, \nabla_z u\}.$$

Here, $\delta\rho$ and $\delta\sigma$ are deviations from equilibrium values of the density and the specific entropy, respectively.

As usual, to investigate the peculiarities of the phase transition, it is sufficient to keep the terms of lowest order in ψ in the energy density. In the interaction terms of the order parameter with weakly fluctuating quantities φ_ν, it is sufficient to keep the terms that are linear in the latter quantities. In the part of the energy that does not contain ψ, it is sufficient to keep only the terms quadratic in weakly fluctuating variables. As a result, the energy density in the vicinity of the smectic A–smectic C transition acquires the following form:

$$E = \frac{1}{2}\left\{A\psi^2 + K_1(\nu(\nabla\psi))^2 + K_2(\nabla\psi)^2 + K_3((\nu\nabla)\psi)^2 + \frac{1}{2}U\psi^4 + \varphi g^- \varphi + D\varphi\psi^2\right\}. \quad (4.76)$$

In expression (4.76), the vector $D = \{D_\rho, D_\sigma, D_u\}$characterizes the magnitude of the interaction of the order parameter with the fluctuations of the density, the specific entropy, and the displacement of the smectic layers.

In this expression, the elastic modulus matrix g^- is also introduced as

$$g^-_{\mu\nu} = \frac{\partial^2 E}{\partial\varphi_\mu \partial\varphi_\nu}. \quad (4.77)$$

In particular,

$$\begin{aligned} g^-_{uu} &= \frac{\partial^2 E}{\partial(\nabla_z u)^2} = B \\ g^-_{u\rho} &= \frac{\partial^2 E}{\partial\rho\partial\nabla_z u} = -B\left(\frac{\partial \ln l}{\partial \ln \rho}\right)_\sigma \\ g^-_{\rho\rho} &= \frac{\partial^2 E}{\partial\rho^2} = \rho\left(\frac{\partial P}{\partial\rho}\right)_\sigma. \end{aligned} \quad (4.78)$$

This notation in contrast to the conventional notation [79, 80] $g^-_{\rho\rho} = A$, $g^-_{uu} = B$, and $g^{--}_{\rho u} = C$ makes it possible to use matrix formalism that simplifies the otherwise cumbersome formulas. In Equation 4.78, P is pressure, g^-_{uu} is a compression modulus of the smectic layers (B), and an elastic modulus, $g^-_{\rho\rho}$, is the inverse compressibility. The elastic moduli associated with the compression of the smectic layers satisfy the following relation:

$$\frac{g_{u\rho}}{g_{uu}} = \left(\frac{\partial \ln l}{\partial \ln \rho}\right)_\sigma \sim -1, \quad (4.79)$$

where l is the equilibrium spacing between smectic layers.

It is worthwhile to present typical values of the material parameters of smectic: the compression modulus of the smectic layers has the typical value

$$g_{uu}^{-} = B \sim 10^7\ \mathrm{J/m^3} \tag{4.80}$$

the compressibility of the smectic has a value of the same order of magnitude as in ordinary liquids:

$$g_{\rho\rho}^{-} \sim 10^9\ \mathrm{J/m^3}. \tag{4.81}$$

It follows from this that in smectic phases there exists a small parameter

$$g_{uu}/g_{\rho\rho} \sim g_{u\rho}/g_{\rho\rho} \sim 10^{-2}, \tag{4.82}$$

which shows that the density in the smectic phase is weakly modulated because the latter is close to a nematic. The components of the vector **D** that figure in the energy density (4.76) and that describe the contribution to the energy from the interaction of the order parameter with the weakly fluctuating quantities are equal in order of magnitude to the compression modulus of the smectic layers.

The quantity U has the same order of magnitude:

$$U, D_{\rho}, D_{\sigma}, D_{u} \sim 10^7\ \mathrm{J/m^3}. \tag{4.83}$$

Finally, the characteristic value of the Frank moduli in smectic phase coincides with their value in nematic phase:

$$K_{1,2,3} \sim 10^{-11}\ \mathrm{J/m}. \tag{4.84}$$

As the analysis of Kats and Lebedev [78] has shown neither in the mean field theory nor in the wide range of developed fluctuations do the fluctuations of the order parameter yield corrections to the gradient terms of the energy density (4.76). In the latter region, the Frank moduli are not renormalized. Renormalization of the Frank moduli corresponding completely to the universal behavior of the model with a two-component order parameter does not take place in the real situation. As the general result, it was shown that the region of developed fluctuations corresponds to the region of the nonuniversal critical behavior. In this region, the critical behavior is described by nonuniversal indices, which depend on the nonrenormalized ratios of the Frank models K_1/K_2 and K_1/K_3.

In contrast to the Frank moduli, the elastic moduli are very sensitive to how close the system is to the smectic A–smectic C transition. In the mean field theory, these moduli undergo a jump at the transition:

$$\delta g_{\nu\mu}^{-} = -D_{\nu} D_{\mu}/2U. \tag{4.85}$$

It follows from Equations 4.80 and 4.83 that the elastic moduli decrease by a jump at the transition from the A to the C phase of the order of magnitude of g_{uu}^{-} that is two order smaller than the elastic moduli $g_{\rho\rho}^{-1}$ responsible for an absolute value of ultrasonic velocity.

In the region of developed fluctuations, critical corrections to the elastic moduli arise. The renormalized elastic moduli have the following form [81]:

$$\tilde{g}_{\mu\nu}^{-} = g_{\mu\nu}^{-} - \frac{D_{\mu} D_{\nu} F}{1 + (DgD)F}, \tag{4.86}$$

where D and g^{-} are nonrenormalized quantities and the magnitude of F is determined by the correlator F of the order parameter:

$$F = \int \frac{\langle \psi^2(0) \psi^2(r) \rangle}{4T} d^3 r, \tag{4.87}$$

where T is the temperature and F depends on the dimensionless parameter Θ describing the nearness of the system to the phase transition:

$$\Theta = \frac{T - T_{AC}}{T_{AC}}. \tag{4.88}$$

Equations 4.85 and 4.86 were used to explain low-frequency anomalies in the elastic constants and therefore in the ultrasonic velocity registered in the vicinity A–C phase transitions. They can be described by critical indexes that are not universal. In particular, the correlator (4.87) has a critical singularity $\Theta^{-\alpha}$ near the transition point. The specific heat index α is a small quantity. In the standard model [82] with a two-component order parameter, it is close to zero. By virtue of the nonuniversality index, α lies within the limits 0.06–0.14.

The quantity A figuring in expression (4.76) varies as $\Theta^{-\gamma}$ in the region of developed fluctuations. The susceptibility index γ varies depending on the starting ratios of the Frank moduli in the range 1–1.25.

In the mean field theory, the order parameter behaves like $\Theta^{1/2}$ approaching the transition point and in the region of developed fluctuations, like $\Theta^{2\beta}$. The magnitude of the index β of the order parameter lies within the range $\sim$0.43–0.45 and the index of the correlation length $\nu \sim 0.58$–0.62.

A consistent description of the dynamic pretransitional effects registered by measurements of the ultrasonic velocity and attenuation at nonzero frequency is more complicated. The viscosity tensor of the smectics in the vicinity of A–C transition includes three bulk viscosities $\eta_{2,4,5}$ and two shear viscosities $\eta_{1,3}$. The dynamic viscoelastic properties in the vicinity of A–C transition are determined by the components of the complex elastic matrix $G_{\mu\nu}$ that is expressed by the equation similar to (4.86). The principal difference is that the correlator F becomes dependent on frequency.

The existence of the small parameter defined by Equation 4.82 makes it possible to find out the explicit expressions for the spectra of the first (1) and second (2) sounds. In the vicinity of the transition, they have the following form:

$$\frac{\rho\omega^2}{q^2} = \rho c_{(1,2)}^2 - i\omega\eta_{(1,2)}, \tag{4.89}$$

where

$$\rho c_{(1)}^2 = \tilde{g}_{\rho\rho}^- - 2\tilde{g}_{\rho u}^- \frac{q_z^2}{q^2} + \tilde{g}_{uu}^- \frac{q_z^4}{q^4} \tag{4.90}$$

$$\begin{aligned}\eta_{(1)} &= [\eta_1 + (\eta_2 + \delta\eta_2 + \delta'\eta_2)q_\perp^4/q^4 + (\eta_5 + \delta\eta_5 + \delta'\eta_5)q_\perp^4/q^4] \\ &\quad + 2[\eta_3 + (\eta_4 + \delta\eta_4 + \delta'\eta_4)]q_z^2 q_\perp^2/q^4\end{aligned} \tag{4.91}$$

$$\rho c_{(2)}^2 = \tilde{g}_{uu}^- \frac{q_z^2 q_\perp^2}{q^4} \tag{4.92}$$

$$\begin{aligned}\eta_{(2)} &= \eta_3 (q_z^2 - q_\perp^2)^2/q^4 + [\eta_1 + (\eta_2 + \delta\eta_2 + \delta'\eta_2) \\ &\quad + (\eta_5 + \delta\eta_5 + \delta'\eta_5)q_\perp^4/q^4] - 2(\eta_4 + \delta\eta_4 + \delta'\eta_4)]q_z^2 q_\perp^2/q^4.\end{aligned} \tag{4.93}$$

Here, $\boldsymbol{q}_z$ and $q_\perp$ are the components of the wave vector $\boldsymbol{q}$ along the normal ν and perpendicular to it, respectively, η_i are the nonrenormalized viscosity coefficients, $\delta\eta_i$ are the critical corrections to the bulk viscosity coefficients, $\delta'\eta_i$ are the corrections to the bulk viscosity coefficients associated with the fluctuations of the smectic layers, and $\tilde{g}^-$ are the renormalized dynamic elastic moduli of the smectic phase defined by (4.86).

The sound described by the spectrum (4.89)–(4.91) is analogous to the ordinary sound in liquids or the longitudinal sound in solids. The wave vector $\boldsymbol{q}$ coincides in this mode with the direction of the velocity vector. The mode (4.89), (4.92) and (4.93) describes the second sound that is analogous to the shear sound in solids. The second sound degenerates into the diffusion mode when the wave vector coincides with the normal to the layer or lies in it.

It is important to note that fluctuations contribute only to the bulk viscosity coefficients. Nevertheless, according to (4.91) and (4.93), they define the attenuation of both the first and the second sounds.

The simple power laws for critical contributions to the elastic moduli and viscosity coefficients were obtained for the case of relatively weak fluctuations. In the hydrodynamic and fluctuation regions, they have the following forms, respectively:

$$\begin{aligned}&\delta g_{\mu\nu}^- \sim \Theta^{-\alpha}, \qquad \delta\eta_{2,4,5} \sim \Theta^{-(z\nu+\alpha)}, \\ &\delta g_{\mu\nu}^- \sim \Theta^{-\alpha/z\nu}, \qquad \delta\eta_{2,4,5} \sim \Theta^{-(1+\alpha/z\nu)}.\end{aligned} \tag{4.94}$$

The critical corrections to the elastic moduli of a smectic C satisfy the following relation:

$$\delta g_{uu}^- : \delta g_{\rho u}^- : \delta g_{\rho\rho}^- = (D_u/D_\rho)^2 : (D_u/D_\rho) : 1. \tag{4.95}$$

These theoretical conclusions have found a strong experimental confirmation through the analysis of temperature and frequency dependences of the ultrasonic parameters presented above. Temperature dependences of elastic moduli are shown in Figures 4.35 and 4.36.

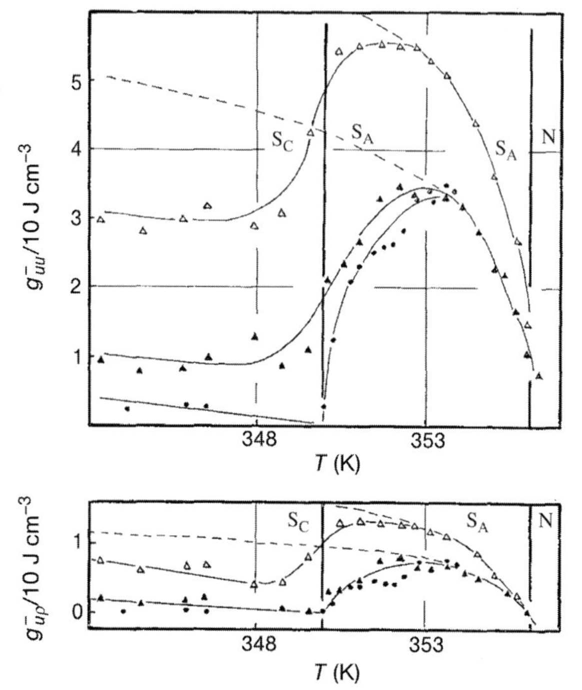

Figure 4.35 Temperature dependence of the elastic moduli g^-_{uu} and $g^-_{u\rho}$: (●) 0.36 MHz; (▲) 5 MHz; (△) 15.6 MHz. (After Balandin *et al.* [16].)

From the calculated values of the elastic moduli (Figures 4.35 and 4.36), it can be seen that over the entire investigated frequency–temperature region, the ratio (4.82) of the elastic moduli renormalized by the critical fluctuations remains small. It confirms the possibility of getting correct theoretical predictions based on this assumption.

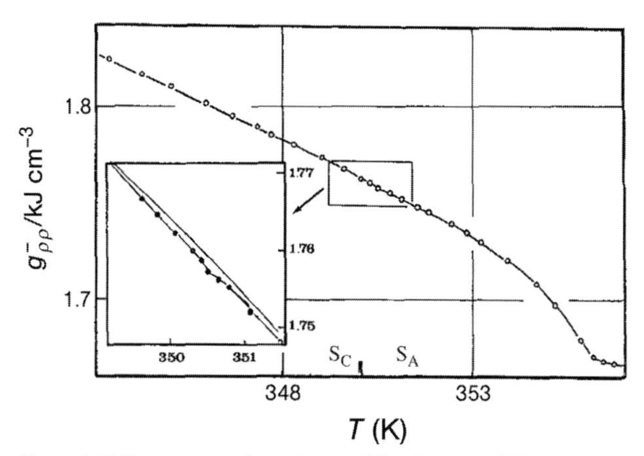

Figure 4.36 Temperature dependence of the elastic modulus $g^-_{\rho\rho}$ at the frequency of 15 MHz. (After Balandin *et al.* [16].)

The frequency dependence of the elastic moduli g^{-}_{uu} and $g^{-}_{u\rho}$ far from the transition is apparently related to the appearance at high frequencies of noncritical mechanisms of molecular dissipation. In addition, in the smectic A phase the behavior of the elastic moduli is complicated by the proximity of the nematic–smectic A phase transition. The regions of the developed fluctuations of these transitions overlap, hindering the analysis of the dispersion properties of the elastic moduli. Nevertheless, at low frequencies, as it can be seen from Figure 4.35, the frequency dependence disappears. This indicates the attainment of the hydrodynamic limit.

From Equation 4.79, it is possible to use the starting values of the elastic moduli to calculate the density (pressure) dependence of the interlayer distance l. The calculations show that, as the density increases, the interlayer distance l decreases. In this case, in the smectic A and C phases we obtain for the ratio (4.79)

$$\frac{g_{u\rho}}{g_{uu}} = \left(\frac{\partial \ln l}{\partial \ln \rho}\right)_{\sigma} \sim -0.2. \tag{4.96}$$

In general, it is quite possible that the interlayer distance also increases with increase in the density. The magnitude of the expression (4.96) does not depend on the temperature within the limits of the measurement error and is the same in both phases.

The analysis of the experimental data also provides the extraction and the comparison of critical contributions in different elastic moduli. In particular, one can calculate the ratio of critical contributions defined by the expression (4.95). Results of such calculations are presented in Figure 4.37.

It can be seen that in the vicinity of the transition, this quantity does not depend on either the frequency or the temperature, which is in accordance with theoretical predictions.

Critical contributions to the bulk viscosity coefficients also correspond to theoretical conclusions. Figures 4.38 and 4.39 present temperature and frequency dependences of the critical contribution to the bulk viscosity η_5. The solid lines in these figures correspond to the following value of the combination of critical indices (see (4.94)):

$$z\nu + \alpha = 1.1. \tag{4.97}$$

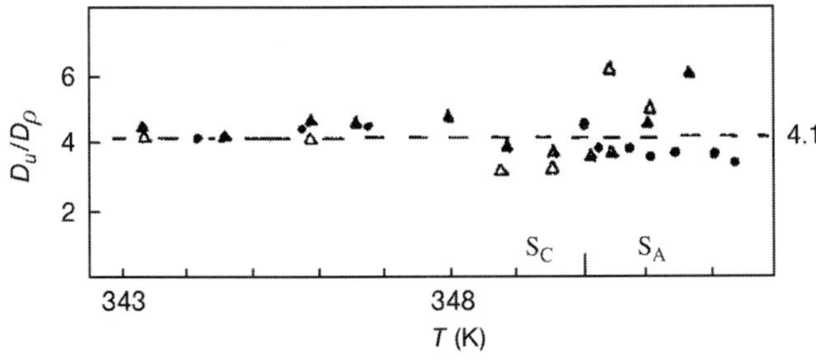

Figure 4.37 Temperature dependence of the ratio D_u/D_ρ: (●) 0.36 MHz; (▲) 5 MHz; (△) 15.6 MHz. (After Balandin *et al.* [16].)

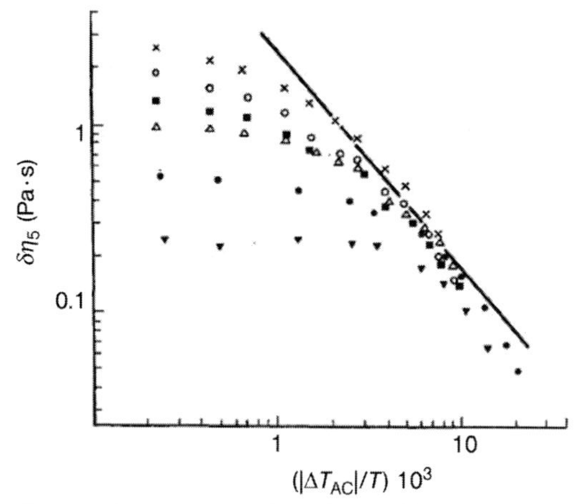

Figure 4.38 Temperature dependence of $\delta\eta_5$ in the S_c phase: (×) 0.36 MHz; (O) 0.69 MHz; (■) 1.2 MHz; (△) 3 MHz; (●) 5 MHz; (▼) 8.7 MHz. (After Balandin *et al.* [76].)

As it might have been expected, indices ν and α are the same in both smectic phases. Their numerical values agree with theoretical values obtained in Refs [48, 81, 83]. The deviation of the critical corrections from the power law dependences (4.94) determines the boundary of the hydrodynamic and fluctuation regions.

It is worthwhile to note that results of the ultrasonic study of smectic A–smectic C phase transition by using shear wave technique are also in accordance with the theoretical analysis [76]. Nevertheless, the second sound shows essentially lower sensitivity to critical fluctuations. Moreover, the agreement of the experimental orientational dependences with theoretical orientational dependences breaks down

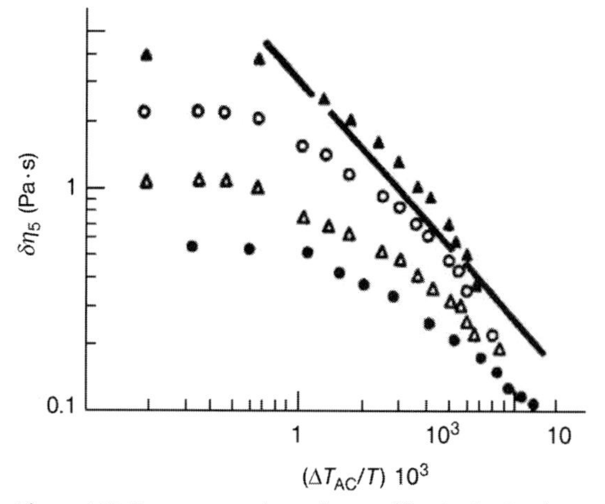

Figure 4.39 Temperature dependence of $\delta\eta_5$ in the S_A phase: (▲) 0.36 MHz; (O) 0.69 MHz; (△) 1.2 MHz; (●) 3 MHz. (After Balandin *et al.* [76].)

in the propagation angle ranges 0–25° and 75–90°. This has to do with the fact that at $\theta = 0°$ and 90° the second sound degenerates into a diffusion mode. The sound penetration depth in a sample of the investigated medium in this case is comparable to the depth of influence of the boundary conditions on the quartz disk. This may explain the difference between experimental and theoretical orientational dependences of the absorption of the second sound at these angles. The shear wave technique turned out to be more useful for studying standing smectic films, as it was mentioned above. For bulk samples, ultrasonic measurements at longitudinal waves are more useful from the point of view of the study of anisotropic critical dynamics at phase transitions in liquid crystals.

4.4.5 Ultrasonic Studies of Phase Transitions in Confined Liquid Crystal Systems

Expansion of ultrasonic studies to the case of liquid crystal systems with spatial confinement seems to be very interesting. First, correct measurements of ultrasonic parameters at frequencies 0.1–10 MHz in the entire temperature range of a liquid crystal state demand the use of a rather big amount of liquid crystals (typically more than 1 cm^3). It essentially restricts the practical application of ultrasonic methods for the study of newly synthesized LC materials. The situation is changed at studies of phase transitions that are often accompanied by a critical increase in an ultrasonic absorption coefficient, at least at low frequencies. It results in an essential decrease in diffraction losses responsible for experimental errors at the measurement of ultrasonic velocity and attenuation in weakly absorbed liquids. So, it makes it possible to elaborate on resonator-like ultrasonic chambers of a rather small volume (of order 0.1 cm^3) providing registration of phase transition [84, 85]. It seems to be important as additional means of control of phase diagrams in liquid crystals at the variation of PVT parameters (see above). Critical contributions to the viscoelastic parameters can also be extracted and analyzed at measurements of such type. In principle, layered systems (a number of liquid crystals and solid layers) can be considered an example of confined liquid crystals. Such systems are convenient for creating the given orientation of LC by the appropriate surface treatment of solid plates. Besides, in thin layers of liquid crystals, it is simple to change orientation with the help of electric fields. Generally, the combination of orienting action of fields and surfaces results in the spatially nonuniform distribution of the director within a liquid crystal layer. The theory of such deformed structures is rather well developed in connection with the use of layer systems in liquid crystal displays. Nevertheless, propagation of elastic waves in systems of LC layers has not been considered theoretically yet. A unique experimental study [32] was performed with ultrasonic waves propagating normally to a number of LC layers (1 mm of thickness) separated by thin brass films. The thickness of these films (0.1 mm) was essentially smaller than the ultrasonic wavelength to provide high acoustic transparence. Both electric and magnetic fields were used to control the orientation of liquid crystal and to suppress the influence of surfaces. Though the experiments confirmed the new possibilities for the study of viscous properties of bulk samples of LC, they did not involve the problem of phase

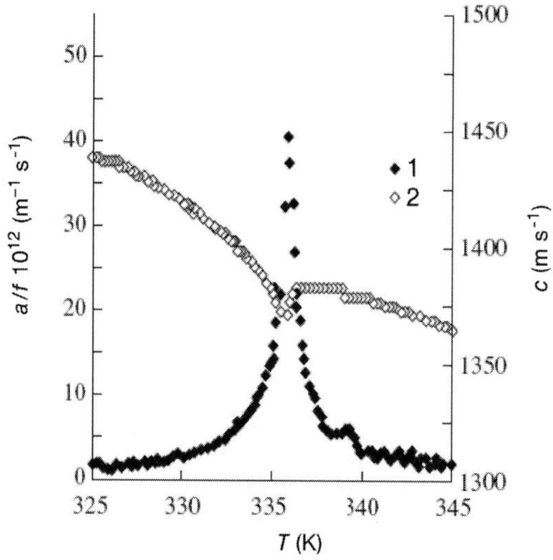

Figure 4.40 Typical plots of the acoustic parameters (absorption coefficient α/f^2(1) and sound velocity c(2)) versus temperature for an LC sample of small volume (0.15 cm^3); $f = 0.68$ MHz [84].

transitions. Moreover, the total amount of LC samples (about 10 cm^3) restricts the usage of such a camera construction in practical applications. This restriction was overcome in the simplest case miniresonator mentioned above, where only one LC layer about 2 mm thick was used without special surface treatment. Of course, in this case we can say only about average orientation of the sample and average critical anomalies of ultrasonic parameters at *very weak confinement*.

Results of measurements [84] of ultrasonic parameters in the vicinity of nematic–isotropic phase transitions of 4′-*n*-amylphenyl ester of 4′-*n*-hexyloxyphenylcyclohexane-2-carboxylic acid obtained by using a miniresonator of the inner volume 0.15 cm^3 are shown in Figure 4.40.

It was found that for both nematic and isotropic LC phases, the critical index for the ultrasonic absorption coefficient was about 1 for the temperatures differing by more than 1.5 K from the phase transition temperature decreased inside this interval. Such behavior is typical of bulk LC samples (see above). It confirms the possibility of studying viscoelastic properties of small samples of LC at phase transitions by ultrasonic methods. An example of temperature dependences of average value of a bulk viscosity coefficient in the above-mentioned compound extracted from ultrasonic measurements is shown in Figure 4.41.

The possibility of using mini-resonators for acoustic studies of phase diagrams and critical phenomena in newly synthesized liquid crystals with different types of polymorphism was also confirmed by the results presented in Ref. [85]. In the given work, some complex mesogene compounds of ethers of alkyloxyphenyl-cyclohexanecarboxylic acid synthesized in the Problem laboratory of liquid crystals of the Ivanovo State University were investigated. The temperature dependences of relative changes of acoustic parameters for such compounds are shown in Figure 4.42.

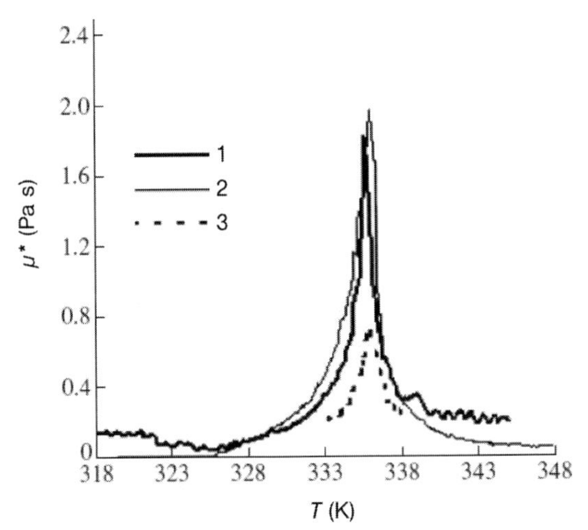

Figure 4.41 Plots of the effective viscosity $\mu^*(T)$ for various frequencies (MHz): (1) 0.68; (2) 1.37; (3) 1.63. (After Maksimochkin *et al.* [84].)

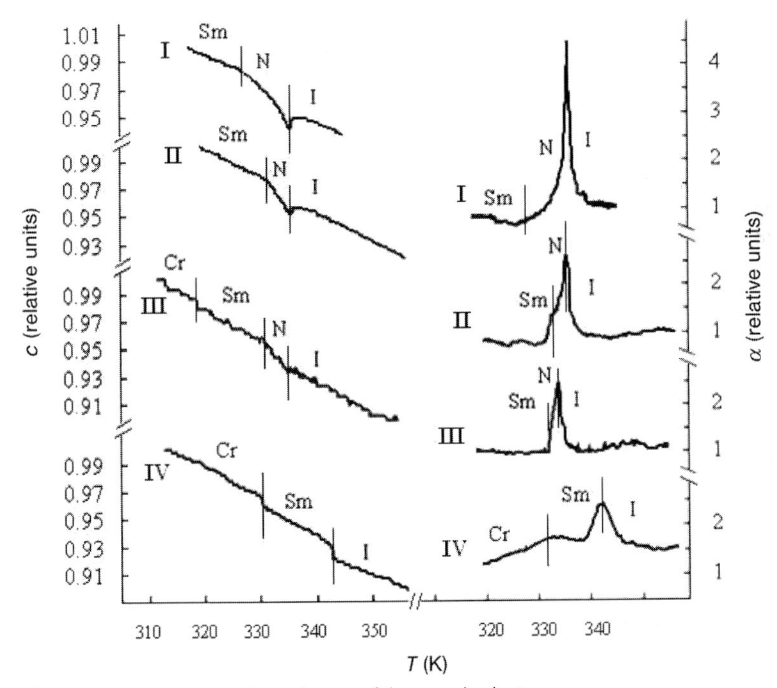

Figure 4.42 Temperature dependences of the sound velocity c and the absorption coefficient α at frequency 0.68 MHz for compound (I) of volume 0.15 cm^3 and (II–IV) of volume 0.06 cm^3.

According to the thermal polarizing microscopy, polymorphism of the following kind was registered: in compound I – Cr 313 K N 333 K Sm 328 K I; in compound II – Cr 309 K Sm 335 K N 339 K I; in compound III – Cr 318 K Sm 340 K N 341 K I; and in compound IV – Cr 342 K Sm 339 K I (Cr, Sm, N, I – crystal, smectic, nematic, and isotropic phases, respectively). It is obvious that in most cases acoustic parameters show some peculiarities at phase transition temperatures.

Moreover, not only the temperature of transition points but also the types of critical anomalies can be extracted from measurements in acoustic resonators of very small volume. It affords good prospects of using ultrasonic methods in routine laboratory studies.

For the basic science, the case of *strong confinement* is of much greater interest. For the last few decades, a lot of research has been devoted to the confined LC systems (see, for example, Ref. [86]). It was shown that the confinement of LCs in general depresses phase transition temperatures [87–89]. The effect of confinement strongly depends on the ratio R/ξ, where R stands for the characteristic linear size of the confinement and ξ is the relevant order parameter correlation length. For $R \approx \xi$, a change in phase transition character can take place [90–92]. Moreover, the spectrum of relaxation times in such systems was also strongly influenced by confinement [86]. Nevertheless, only a limited number of experimental methods focused on the dynamics of critical phenomena in confined LC systems. Although an attempt was made to use acoustic methods to study relaxation properties of suspensions of small solid particles (aerosil) immersed in LCs [93], it was not connected with phase transition problems and critical behavior of confined LC systems. Ultrasonic studies of phase transitions in such systems were experimentally realized [85, 94, 95] in the case of liquid crystal emulsions (LCEs) – micro- and submicro droplets of liquid crystals immersed in isotropic liquids.

Liquid crystal emulsions can be considered as the composite LC material, which is most suitable for ultrasonic studies. Indeed, the acoustic impedance of liquid crystals is very close to that of isotropic liquids (e.g., water) that provides a low-intensity acoustic scattering especially when the acoustic wavelength essentially exceeds the average droplet size. The latter is performed well at relatively low frequencies used in experiments (see below). Densities of LC and isotropic liquids can be chosen close to each other enabling the production of stable emulsions, which is important for experiments with phase transitions. It is also important for exclusion of a specific relaxation contribution to ultrasonic attenuation proportional to the difference of densities for nonhomogeneous media, considered previously [96]. The letter mechanism can be important for previously studied [93] dispersions of solid particles in liquid crystals. It is also worthwhile to note that the use of water as an isotropic matrix provides the registration of very slight changes in an acoustic absorption induced by liquid crystal droplets, as the basic absorption in water is very small. So, the measured ultrasonic absorption can be almost completely attributed to the dissipative processes taking place inside LC droplets.

Typical temperature dependences of an ultrasonic absorption coefficient in LC–water emulsion of different drop diameters are shown in Figure 4.43. The

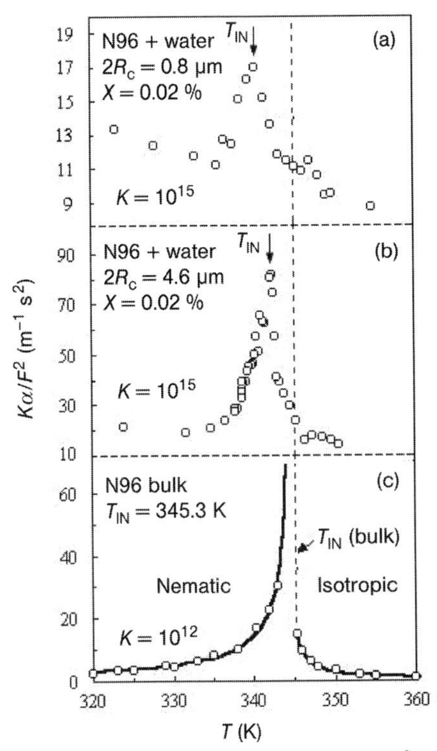

Figure 4.43 Temperature dependence of $K\alpha/f^2$ with $F = 2.7$ MHz (open circles) for LCE samples with $2R_c = 0.8$ μm (a) and $2R_c = 4.6$ μm (b) with $K = 10^{15}$. The data for the bulk N96 reference sample are also shown in (c) with $K = 10^{12}$. Solid lines represent the power law approximation.

measurements were performed at frequency $F = 2.7$ MHz. The LCE samples were prepared by ultrasound mixing (at frequency 45 kHz, temperature 303 K) the nematic liquid crystal and distilled water. The nematic mixture N96 ($T_{IN} = 345.3$ K, Charkov production) was chosen as an object. It obeyed a wide temperature of nematic phase and showed high chemical stability when contacted with water. The structure of emulsion and drop size distribution were controlled by microscopy and photon correlation spectroscopy.

Data presented in Figure 4.43 clearly demonstrate the temperature shift of the absorption peak (which has to be close to the clearing point T_c) relative to that obtained in the bulk sample. The value of this shift increases with decrease in drop diameter, which is typical for systems with strong confinement [88–91]. It is interesting to note that the critical indexes for ultrasonic absorption obtained in LC emulsions turned out to be close to the values presented above for the bulk samples of LCs in the vicinity of a clearing point. Presumably, this fact is connected with relatively large radius of droplets R compared to a correlation length ξ. It would be interesting to perform analogous measurements for a case $R \sim \xi$ where the space restrictions on the size of fluctuations can result in essential modification of critical behavior. It can be important from the point of view of the general theory of critical phenomena in liquid crystals. So, future experimental and theoretical work in this direction promises new and interesting results.

The results presented above can be summarized as follows:

1. Ultrasonic methods and techniques are well elaborated and widely used for studying different classes of condensed matter including liquid crystals.
2. The longitudinal ultrasonic waves are well suited for investigations of bulk samples of liquid crystals. In this case, one can neglect the orienting action of surfaces and obtain uniform orientation of liquid crystals with the use of moderate magnetic fields.
3. Theoretical description of orientational changes induced by fields in bulk samples of LCs is essentially simpler than that in the case of confined samples. It affords good prospects for elaboration of ultrasonic methods of determination of viscous and elastic parameters of liquid crystals.
4. At the same time, a number of relaxation processes contribute to the viscoelastic behavior of liquid crystals.
5. Such contributions can be formally described in terms of the so-called bulk viscosities that in most cases are essentially higher than the dissipative parameters of incompressible liquid crystals.
6. Nevertheless, such practically important parameter as a rotational viscosity coefficient can be easily extracted from an ultrasonic study of nematic liquid crystals under rotating magnetic field.
7. Ultrasonic technique shows a lot of promise at investigations of viscous elastic properties at variation of PVT-state parameters.
8. Ultrasonic methods can also be considered as unique tools for the study of critical dynamics of liquid crystals at phase transitions of different types.
9. Ultrasonic investigations of phase transitions in bulk samples open opportunity for an experimental check of some conclusions of the general dynamic theory of phase transitions. So, it stimulates the progress of the theory.
10. It is also possible to apply ultrasonic technique for studying phase transitions in strongly confined systems. The critical dynamics of such systems is practically unknown so far.
11. Shear wave methods can be effectively applied to the study of viscoelastic properties of incompressible liquid crystals.
12. Such techniques turned out to be useful in the case of smectic phases including freely suspended smectic films.

13. At the same time, the penetration length of ultrasonic shear waves is relatively small and the surface-induced orientation has to be accounted for when analyzing experimental results. As a result, shear wave methods usually deal with confined samples in contrast to the case of longitudinal ultrasound applicable for bulk samples.

References

1 Mason, W.P.(ed.) (1964) *Physical Acoustics*, Academic Press, New York.
2 Natale, G.G. (1978) The contribution of ultrasonic measurements to the study of liquid crystals. 1. Nematics. *J. Acoust. Soc. Am.*, **63**, 1265.
3 Truell, B., Elbaum, Ch., and Chick, B.B. (1969) *Ultrasonic Methods in Solid State Physics*, Academic Press, New York.
4 McSkimin, H.J. (1965) Ultrasonic methods for measuring the mechanical properties of liquids and solids, in *Physical Acoustics*, vol. 1A (ed. W.P. Meson), Academic Press, New York, p. 272.
5 Nozdrev, V.F. (1963) *Application of Ultrasonics in Molecular Physics*, Gordon and Breach, New York.
6 Khabibullaev, P.K., Lagunov, A.S. and Gevorkian, E.V. (1994) *Rheology of Liquid Crystals*, Allerton Press, New York.
7 Litovitz, T.A. and Davis, C.M. (1965) Structural and shear relaxation in liquids, in *Physical Acoustics*, Vol. 2A (ed. W.P. Meson), Academic Press, New York, p. 282.
8 Derjaguin, B.V., Bazaron, U.B., Lamazhapova, Kh.D. and Tsidypov, B.D. (1990) Shear elasticity of low-viscosity liquids at low frequencies. *Phys. Rev. A*, **42**, 2255.
9 Shmelioff, O.Ya. (1983) Anisotropic propagation of ultrasound in liquid crystals with the nematic–smectic A phase transition. PhD thesis (in Russian).
10 Prokopjev, V.I. (1988) Low frequency ultrasonic studies of liquid crystals in a rotating magnetic field. PhD thesis, Moscow Institute of Instrument Engineering (in Russian).
11 Kononenko, V.S. (1987) Precise method of measurement of ultrasonic absorption coefficient in liquids at frequencies 0.1–20 MHz. *Akust. Zh.*, **23**, 688 (in Russian).
12 Rjashikov, A.S. (1983) Relaxation phenomena in liquid crystals oriented by magnetic field. PhD thesis, Moscow Institute of Instrument Engineering (in Russian).
13 Prokopjev, V.I., Ryashchikov, A.S. and Kononenko, V.S. (1985) Ultrasound absorption factor measuring device, patent SU1142786 (in Russian).
14 Swift, J. and Mulvaney, B.J. (1979) Sound attenuation and dispersion near the nematic–smectic A phase transition of a liquid crystal. *J. Phys. Lett.*, **40** (13), 287.
15 Liu, M. (1979) Hydrodynamic theory near the nematic–smectic-A transition. *Phys. Rev. A*, **19**, 2090.
16 Balandin, V.A., Gurovich, E.V., Kashitsin, A.S. and Pasechnik, S.V. (1991) An ultrasonic investigation of the critical behavior of the elastic moduli near the smectic C–smectic A phase transition. *Liq. Cryst.*, **9**, 551.
17 Kashitsin, A.S. (1989) Acoustical studies of dynamical processes in oriented liquid crystals with phase transition smectic A–smectic C. PhD thesis, Moscow Institute of Instrument Engineering (in Russian).
18 Miyano, K. and Ketterson, J.B. (1975) Ultrasonic study of liquid crystals. *Phys. Rev. A*, **12**, 615.
19 Balandin, V.A., Pasechnik, S.V., Prokopjev, V.I. and Shmelyoff, O.Ya. (1988)

Ultrasound absorption anisotropy in the vicinity of smectic A–nematic transition. *Liq. Cryst.*, **3**, 1319.

20 Larionov, A.N., Larionova, N.N. and Pasechnik, S.V. (2004) Viscous properties of nematic mixture at variation of PVT-state parameters. *Mol. Cryst. Liq. Cryst.*, **409**, 459.

21 Bogdanov, D.L., Gevorkian, E.V. and Lagunov, A.S. (1980) Acoustic properties of liquid crystals in rotating magnetic field. *Akust. Zh.*, **26**, 284 (in Russian).

22 Pasechnik, S.V., Larionov, A.N. and Balandin, V.A. *et al.* (1984) Etude acoustique de cristaux liquides nématiques sous champ magnétique pour différentes temperatures et pressions. *J. Phys. France*, **45**, 441.

23 Forster, D., Lubensky, T.C., and Martin, P.C. *et al.* (1971) Hydrodynamics of liquid crystals. *Phys. Rev. Lett.*, **26**, 1016.

24 Martin, P.S., Parodi, O., and Pershan, P.S. (1972) Unified hydrodynamic theory for crystals, liquid crystals, and normal fluids. *Phys. Rev. A*, **6**, 2401.

25 Stephen, M.J. and Straley, J.P. (1974) Physics of liquid crystals. *Rev. Mod. Phys.*, **46**, 617.

26 Pasechnik, S.V., Prokop'ev, V.I. and Shmelev, O.Ya. *et al.* (1987) Relationship between the dissipative coefficients and the anisotropic acoustic parameters of a nematic liquid crystal. *Russ. J. Phys. Chem.*, **67**, 1675.

27 Thiriet, Y. and Martinoty, P. (1979) Ultrasonic study of the nematic-isotropic phase transition in PAA. *J. Phys. France*, **40**, 789.

28 Jahnig, F. (1973) Dispersion and absorption of sound in nematics. *Z. Physik*, **258**, 199.

29 Kogevnikov, E.N. (1994) Statistical theory of acoustical anisotropy of a nematic liquid crystal. *Akust. Zh.*, **40**, 613 (in Russian).

30 Nagai, S., Martinoty, P., Candau, S. and Zana, R. (1975) The intramolecular ultrasonic relaxation of nematic liquid crystals far below the transition temperature. *Mol. Cryst. Liq. Cryst.*, **31**, 243.

31 Lagunov, A.S., Nozdrev, V.F. and Pasechnik, S.V. (1981) Anisotropy of ultrasonic absorption in flowing nematic liquid crystals. *Sov. Phys. Acoust.*, **27**, 304.

32 Pasechnik, S.V. and Neronov, N.A. (2001) Relaxation of nematic oriented by magnetic and electric fields: acoustical investigation. *Mol. Cryst. Liq. Cryst.*, **367**, 19.

33 Zwetkoff, V. (1939) Bewegung anisotroper flussigkeiten im rotierenden magnetfeld. *Acta Physicochim. URSS*, **10**, 555.

34 Gasparoux, H. and Prost, J. (1971) Détermination directe de l'anisotropie magnétique de cristaux liquides nématiques. *J. Phys. France*, **32**, 963.

35 Lord, A.E. Jr. and Labes, M.M. (1970) Anisotropic ultrasonic properties of a nematic liquid crystal. *Phys. Rev. Lett.*, **25**, 570.

36 Kemp, K.A. and Letcher, S.V. (1971) Ultrasonic determination of anisotropic shear and bulk viscosities in nematic liquid crystals. *Phys. Rev. Lett.*, **27**, 1634.

37 Pasechnik, S.V. (1998) Ultrasound at phase transitions and structural transformations in liquid crystals. DSc thesis, Moscow State Academy of Instrument Engineering and Computer Science, Moscow.

38 Shmelioff, O.Ya., Pasechnik, S.V., Balandin, V.A. and Tsvetkov, V.A. (1985) Temperature dependences of Leslie coefficients of butoxybenzilidene-aniline. *Z. Fizicheskoi Khimii*, **LIX**, 2036 (in Russian).

39 Barnik, M.I., Belyaev, V.V. and Grebenkin, M.F. (1978) Electrical, optical, and viscous-elastic properties of LC mixture of azoxy-compounds. *Krisstallografiya*, **23**, **4**, 805 (in Russian).

40 Beens, W.W. and de Jeu, W.H. (1983) Flow-measurements of the viscosity coefficients of two nematic liquid crystalline azoxybenzenes. *J. Phys. France*, **44**, 129.

41 Kneppe, H. and Shneider, F. (1981) Determination of the viscosity coefficients of the liquid crystal MBBA. *Mol. Cryst. Liq. Cryst.*, **65**, 753.

42 Bogdanov, D.L., Lagunov, A.S. and Pasechnik, S.V. (1980) Acoustical properties of liquid crystals in varying magnetic fields. *Primenenie Ultra-akoustiki k Issledovaniyou Vechtchestva*, **30**, 52 (in Russian).

43 Kuznetsov, A.N. and Kulagina, T.P. (1975) K teorii magnitogidrodinamicheskogo effekta v nematicheskih zhidkih kristallov. *Zh. Eks. Teor. Fiz.*, **68**, 1501 (in Russian).

44 Dörer, H., Kneppe, H., Kuss, E. and Schneider, F. (1986) Measurement of the rotational viscosity γ_1 of nematic liquid crystals under high pressure. *Liq. Cryst.*, **1**, 573.

45 Anisimov, M.A. (1991) *Critical Phenomena in Liquids and Liquid Crystals*, Gordon and Breach, Philadelphia.

46 Lagunov, A.S. and Larionov, A.N. (1986) Orientational relaxation in a solution of nematic liquid crystals. *Z. Fizicheskoi Khimii*, **LX**, 2206 (in Russian).

47 Martinoty, P. and Candau, S. (1971) Determination of Viscosity Coefficients of a Nematic Liquid Crystal Using a Shear Waves Reflectance Technique. *Mol. Cryst. Liq. Cryst.*, **14**, 243.

48 de Gennes, P. (1974) *The Physics of Liquid Crystals*, Clarendon Press, Oxford.

49 Cagnon, M. and Durand, G. (1980) Mechanical shear of layers in smectic-A and smectic-B liquid crystal. *Phys. Rev. Lett.*, **45**, 1418.

50 Pindak, R., Sprenger, W.O. and Bishop, D.J., *et al.* (1982) Mechanical measurements on free-standing films of smectic liquid-crystal phases. *Phys. Rev. Lett.*, **48**, 173.

51 Tarczon, J.C. and Miyano, K. (1981) Shear mechanical properties of freely suspended liquid-crystal films in the smectic-A and smectic-B phases. *Phys. Rev. Lett.*, **46**, 119.

52 Kats, E.I. and Lebedev, V.V. (1994) *Fluctuational Effects in the Dynamics of Liquid Crystals*, Springer, Berlin.

53 Zywociñski, A., Wieczorek, S.A. and Stecki, J. (1987) High-resolution volumetric study of the smectic-A-to-nematic transition in 4-(*n*-pentyl) phenylthiol-4′-(*n*-octyloxy)benzoate (8-S5) and octyloxycyanobiphenyl (8OCB). *Phys. Rev. A*, **36**, 1901.

54 Sohl, C.H., Miyano, K., Ketterson, J.B. and Wong, G. (1980) Viscosity and surface-tension measurements on cyanobenzylidene octyloxyaniline using propagating capillary waves: critical behavior. *Phys. Rev. A*, **22**, 1256.

55 Marinelli, M., Mercuri, F. and Foglietta, S., *et al.* (1996) Anisotropic heat transport in the octylcyanobiphenyl (8CB) liquid crystal. *Phys. Rev. E*, **54**, 1604.

56 Lamb, J. (1965) Thermal relaxation in liquids, in *Physical Acoustics*, Vol. 2A (ed. W.P. Meson), Academic Press, New York, p. 203.

57 Mazenko, G.F., Ramaswamy, S. and Toner, J. (1982) Viscosities diverge as $1/\omega$ in smectic-A liquid crystals. *Phys. Rev. Lett.*, **49**, 51.

58 Mazenko, G.F., Ramaswamy, S. and Toner, J. (1983) Breakdown of conventional hydrodynamics for smectic-A, hexatic-B, and cholesteric liquid crystals. *Phys. Rev. A*, **28**, 1618.

59 Kapustin, A.P. and Kapustina, O.A. (1986) *Akustika zhidkih kristallov*, Nauka, Moscow (in Russian).

60 Pasechnik, S.V., Balandin, V.A., Prokopjev, V.I. and Shmelyoff, O.Ya. (1989) Critical dynamics and acoustical parameters of a nematic in the vicinity of the clearing point. *Z. Fizicheskoi Khimii*, **LXIII**, 471 (in Russian).

61 Eden, D., Garland, C.W. and Williamson, R.C. (1973) Ultrasonic investigation of the nematic-isotropic phase transition in MBBA. *J. Chem. Phys.*, **58**, 1861.

62 Anisimov, M.A. (1987) Universality of the critical dynamics and the nature of the nematic-isotropic phase transition. *Mol. Cryst. Liq. Cryst.*, **146**, 435.

63 Alekhin, Yu.S., Lagunov, A.S. and Pasechnik, S.V. *et al.* (1976) Temperature dependence of the relaxation parameters in nematic liquid crystals. *Sov. Phys. Acoust.*, **23**, 193.

64 Pasechnik, S.V., Larionov, A.N. and Balandin, V.A. (1982) Acoustic viscosimetry of nematic liquid crystals at varying pressure and temperature. *Sov. Phys. JETP*, **56** (6), 1230.

65 Thoen, J., Marynissen, H. and Van Dael, W. (1984) Nematic–smectic A tricritical point in alkylcyanobiphenyl liquid crystals. *Phys. Rev. Lett.*, **52**, 204.

66 Litster, J.D., Garland, C.W. and Lushington, K.J.,*et al.* (1981) Experimental studies of liquid crystal phase transition. *Mol. Cryst. Liq. Cryst.*, **63**, 145.

67 Ocko, B.M., Birgeneau, R.J. and Litster, J.D.,*et al.* (1984) Critical and tricritical behavior at the nematic to smectic A transition. *Phys. Rev. Lett.*, **52**, 208.

68 de Vries, A. (1970) Evidence for the existence of more than one type of nematic phase. *Mol. Cryst. Liq. Cryst.*, **10**, 31.

69 Bhattacharya, S., Sarma, B.M. and Ketterson, J.B. (1981) Critical attenuation and dispersion of longitudinal ultrasound near a nematic–smectic-A phase transition. *Phys. Rev. B*, **23**, 2397.

70 Balandin, V.A., Pasechnik, S.V. and Prokop'ev, V.I. *et al.* (1987) Low-frequency acoustical parameters of a nematic liquid crystal near the nematic-smectic A phase transition. *Sov. Phys. Acoust.*, **33** (4), 342.

71 Kiry, F. and Martinoty, P. (1978) Ultrasonic attenuation in CBOOA near the nematic–smectic A transition. *J. Phys. France*, **39** (9), 1019.

72 Holmurodov, F. (1983) Adiabatic calorimetry near phase transitions in liquid crystal mixtures. PhD thesis, Moscow Institute of Instrument Engineering.

73 Jähnig, F. (1975) Critical damping of first and second sound at a smectic A-nematic phase transition. *J. Phys. France*, **36** (4), 315.

74 Pasechnik, S.V. and Balandin, V.A. (1982) Anisotropic nature of the velocity of ultrasound in the vicinity of the nematic-smectic-A phase transition. *Sov. Phys. JETP*, **56** (1), 106.

75 Demus, D. and Demus, H. (1973) *Fllissige Kristalle in Tabellen*, VER Deutscher Verlag für Grundstoffindustrie, Leipzig.

76 Balandin, V.A., Gurovich, E.V. and Kashitsin, A.S. *et al.* (1990) Experimental study of the critical dynamics in the vicinity of the smectic A–smectic C phase transition. *Sov. Phys. JETP*, **71** (2), 270.

77 Kats, E.I. and Lebedev, V.V. (1983) Nonlinear dynamics of smectic liquid crystals. *Sov. Phys. JETP*, **58** (6), 1172.

78 Kats, E.I. and Lebedev, V.V. (1985) Nonlinear dynamics of smectic liquid crystals with orientational ordering in the layer. *Sov. Phys. JETP*, **61** (3), 484.

79 Bhattacharya, S. and Ketterson, J.B. (1982) Anomalous damping of sound in smectic-A liquid crystals: breakdown of conventional hydrodynamics? *Phys. Rev. Lett.*, **49**, 997.

80 Landau, L.D. and Lifshitz, E.M. (1986) *Theory of Elasticity*, 3rd edn, Butterworth-Heinmann, Oxford.

81 Gurovich, E.V., Kats, E.I. and Lebedev, V.V. (1988) Critical dynamics at the smectic-A–smectic-C phase transition. *Sov. Phys. JETP*, **67** (4), 741.

82 Wilson, K.G. (1971) Renormalization group and critical phenomena. I. Renormalization group and the kadanoff scaling picture. *Phys. Rev. B*, **4**, 3174.

83 Hohenberg, P.C. and Halperin, B.I. (1977) Theory of dynamic critical phenomena. *Rev. Mod. Phys.*, **49**, 435.

84 Maksimochkin, G.I. and Pasechnik, S.V. *et al.* (2007) Studying viscoelastic properties of liquid crystals during nematic-isotropic phase transition by low-frequency ultrasound probing of small-volume samples. *Tech. Phys. Lett.*, **33** (6), 505.

85 Usol'tseva, N.V., Maksimochkin, G.I., Pasechnik, S.V. and Bykova, V.V. (2009) Ultrasonic and viscoelastic properties of small-volume mesogen samples at the phase transition. *Int. J. Thermophys.*, in press.

86 Grawford, G.P. and Žumer, S. (1996) *Liquid Crystals in Complex Geometries*, Taylor and Francis, London.

87 Zidanšek, A., Kralj, S., Lahajnar, G. and Blinc, R. (1995) Deuteron NMR study of liquid crystals confined in aerogel matrices. *Phys. Rev. E*, **51**, 3332.

88 Kralj, S., Zidanšek, A., Lahajnar, G., Žumer, S. and Blinc, R. (1998) Phase behavior of liquid crystals confined to controlled porous glass studied by deuteron NMR. *Phys. Rev. E*, **57**, 3021.

89 Kutnjak, Z., Kralj, S., Lahajnar, C. and Žumer, S. (2003) Calorimetric study of octylcyanobiphenyl liquid crystal confined to a controlled-pore glass. *Phys. Rev. E.*, **68**, 021705.

90 Golemme, A., Žumer, S., Doane, J.W. and Neubert, M.E. (1988) Deuterium NMR of polymer dispersed liquid crystals. *Phys. Rev. A*, **37**, 559.

91 Kralj, S., Žumer, S. and Allender, D.W. (1991) Nematic-isotropic phase transition in a liquid-crystal droplet. *Phys. Rev. A*, **43**, 2943.

92 Kralj, S. and Žumer, S. (1992) Fréedericksz transitions in supra-µm nematic droplets. *Phys. Rev. A*, **45**, 2461–2470.

93 Sperkach, V.S., Glushenko, A.V. and Yaroshchuk, O.V. (2001) Structure of filled liquid crystals studied by acoustic methods. *Mol. Cryst. Liq. Cryst.*, **367**, 463.

94 Pasechnik, S.V., Maksimochkin, G.I. and Tsvetkov, V.A. (2003) Structure and properties of liquid crystal emulsions: acoustic and optical investigation. Proceedings of 7th ECLC, Jaca, Spain, p. 25.

95 Maksimochkin, G.I. and Pasechnik, S.V. (2003) Acoustical properties and critical dynamics of liquid crystal emulsions. Proceedings of XIII Session of the Russian Acoustical Society, Moscow, p. 178.

96 Boguslavskii, Yu.Ya. (1978) On absorption and dispersion of sound waves in two-phase medium (O pogloshchenii i dispersii zvukovih voln v dvuhfaznoi srede). *Akust. Zh. (Russia)*, **24**, 46.

5
Experimental Determination of Elastic and Viscous Parameters of Liquid Crystals

In this chapter, experimental methods elaborated for the determination of visco-elastic parameters of liquid crystals will be considered. First, we describe a number of optical methods and corresponding techniques used for measuring Frank's elastic moduli K_{11}, K_{22}, K_{33}. Special attention will be paid to the analysis of advantages and disadvantages of proposed methods from the viewpoint of routine laboratory measurements. We will show that an accurate determination of the modulus K_{22} is more complicated compared to the measurement of K_{11} and K_{33} constants. After that, we will describe in detail experimental methods for measuring the rotational viscosity coefficient that is responsible, in the first approximation, for inertial properties of liquid crystal devices. We will consider this problem for both nematic and ferroelectric liquid crystals taking into account the practical importance of theses materials for display industry. This will be followed by the description of methods for measuring anisotropic shear viscosities in different types of flows. It will be shown that traditional technique involving strong magnetic fields is of limited use for studying newly synthesized liquid crystals. So, the alternative optical method based on stabilization action of surfaces instead of fields will be considered too. After that, we will go to experimental methods applied for determining different Leslie coefficients as the latter parameters can contribute essentially to the operating times of some modes used in modern displays. Finally, we will consider the methods of determination of such parameters as anchoring strength and viscosity of gliding that describe very complicated phenomena in near-surface layers of liquid crystals.

5.1
Methods for Measurements of Frank's Elastic Constants of Liquid Crystals

The dependence of the free energy F of nematics on gradients of the director field (see Equation 2.15) is the unique property of liquid crystals. That is why experimental methods for the determination of elastic constants K_{11}, K_{22}, and K_{33} entering in this equation have no analogues among the viscosimetric methods elaborated for studying isotropic liquids and polymers.

Liquid Crystals: Viscous and Elastic Properties
Sergey V. Pasechnik, Vladimir G. Chigrinov, and Dina V. Shmeliova

ISBN: 978-3-527-40720-0

The existence of elastic energy of curvature leads to a number of physical effects usually studied by optical methods. Moreover, some of these effects are widely used in modern applications of LC (e.g., in LCD industry). Such effects essentially depend on the values of Frank's elastic moduli, so the experimental determination of these parameters described in a number of reviews (see, for example, [1–3]) has great importance.

The main idea of experimental methods for K_{ii} measurements is related to the registration of spatial distortions in LC structure induced by different factors (magnetic/electric) fields, surfaces, and thermal fluctuations. It is possible to apply measurements of different anisotropic properties of LC for this purpose. For example, Pieranski *et al.* [4, 5] used both thermal conductivity measurements and polarizing microscopy to study static and dynamic variations of orientational distortions induced by magnetic field in homeotropic and planar monodomain samples of MBBA. Martins *et al.* [6, 7] applied well-elaborated NMR technique to determine the ratio K_{33}/K_{11} in polymer LCs from the analysis of time-dependent NMR spectra. Nevertheless, most of numerous data on Frank's elastic moduli were obtained using optical methods including polarizing microscopy and light scattering spectroscopy. The particular technique is well elaborated compared to viscosity measurements.

Below we describe some typical experiments to reveal advantages and disadvantages of a particular experimental technique from the point of view of LC applications.

5.1.1
Optical Methods Based on Fréedericksz Transition

It is well known that Fréedericksz transitions are induced by an external electric (**E**) or magnetic (**B**) field applied in the direction normal to the optical axis of monodomain samples stabilized by surfaces (for positive sign of dielectric ($\Delta\varepsilon$) or diamagnetic ($\Delta\chi$) permittivity anisotropy) or in the parallel direction (when $\Delta\varepsilon$ or $\Delta\chi$ are negative). In both cases, monodomain orientation becomes unstable at the threshold field E_c (or B_c). Values of the threshold fields can be determined by minimizing the total free energy F including contributions from the elastic energy of curvature and external fields, which are defined in Chapter 2 by Equations (2.15), (2.40) and (2.41). For the problem under consideration, only anisotropic parts g_h^a and g_e^a of the free energy density g_h and g_e of magnetic and electric fields, respectively, are important. So, the general expression for F can be written as

$$\begin{aligned} F &= \int_V (g_k + g_e^a + g_h^a)\,dV \\ &= \frac{1}{2}\int_v \{[K_{11}(\vec{\nabla}\cdot\vec{n})^2 + K_{22}(\vec{n}\cdot\vec{\nabla}\times\vec{n})^2 + K_{33}(\vec{n}\times\vec{\nabla}\times\vec{n})^2] \\ &\quad -\mu_0^{-1}\,\Delta\chi(B\cdot n)^2 - \varepsilon_0\,\Delta\varepsilon(E\cdot n)^2\}\,dV. \end{aligned} \tag{5.1}$$

Expressions for E_c (or B_c) for different geometries (i) corresponding to different basic types of the director deformation (splay – $i=1$, twist – $i=2$, bend – $i=3$) under action of electric or magnetic fields can be written in a similar way [3]:

$$B_c^{(i)} = \frac{\pi}{d}\left(\frac{\mu_0 K_{ii}}{\Delta\chi}\right)^{1/2},$$
$$E_c^{(i)} = \frac{\pi}{d}\left(\frac{K_{ii}}{\varepsilon_0\,\Delta\varepsilon}\right)^{1/2}, \tag{5.2}$$

where d is the layer thickness.

So, measurements of critical fields E_c or B_c corresponding to the principal geometries described in Chapter 2 (Figure 2.13) with defined anchoring conditions can be used to determine Frank's three moduli K_{11}, K_{22}, K_{33}. Although such possibility was first declared in the 1930s by Fréedericksz *et al.* [8, 9], the detailed theoretical background and practical application of magnetically induced Fréedericksz transition for measurements of K_{ii} was realized by Saupe only in 1960 [10]. In this experiment, splay geometry was used to determine the splay (K_{11}) and bend (K_{33}) constants of *p*-azoxyanisol (PAA). To obtain K_{33}, Saupe proposed to use small distortions of the initial planar orientation induced by magnetic field of induction B slightly higher than the threshold induction B_c. A maximal deflection angle θ_m with respect to the boundary orientation in the midplane of the cell depends on the magnetic field induction in accordance with

$$\sin^2\theta_m = 4(K_{33}/K_{11})(B/B_c-1). \tag{5.3}$$

Optical detection of this deflection is essentially the same as in the case of flow-induced distortions (Chapter 3). Namely, field-induced distortions are visualized in the transmitted light by placing the sample cell between crossed polarizers oriented at 45° relative to the director tilt plane (as shown in Chapter 3). The field-induced changes in the intensity of transmitted light $\Delta I = I(\mathrm{B}) - I(B=0)$ arise due to a decrease in the phase delay δ between an extraordinary and ordinary rays. The latter parameter can be connected with field-induced changes of a polar angle and expressed in terms of θ_m defined above (5.3). So, registration and analysis of $I(B)$ dependence provide an estimation of the splay constant K_{11}(from threshold field), the ratio (K_{33}/K_{11}), and so the bend constant K_{33}. More accurate determination of the latter parameter can be made using the initial homeotropic orientation of the sample [11] stabilized by boundaries. It can be shown that for homeotropic boundary orientation, the changes in light intensity are more sensitive to the deflection angle θ_m than in the case of planar orientation. It results in sharp clearing of initially dark LC cell at magnetic fields slightly stronger than the threshold field, so the value of B_c (and the bend constant K_{33}) can be determined from visual observation. Further increase in B leads to a number of interference extrema (Figure 5.1), so more care (compared to a planar cell) is needed when operating with $I(B)$ dependences. In particular, to obtain uniform deformation in the liquid crystal in the vicinity of the critical field, it was necessary to limit the scanning rate by13 Oe/min.

A thorough experimental procedure can help determine and compare both splay (K_{11}) and bend (K_{33}) moduli. In particular, Haller [11] established that the ratio (K_{33}/K_{11}) for MBBA was about 1.25 and independent of temperature within

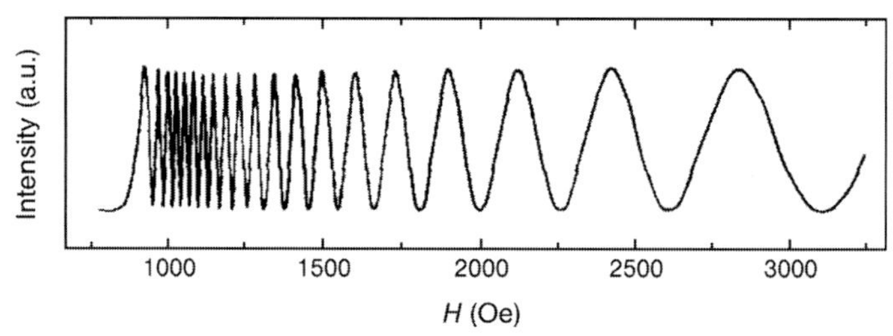

Figure 5.1 Light intensity ($\lambda = 0.589\,\mu m$) versus magnetic field; MBBA, B is the deformation of the layer thickness $-76\,\mu m$. (After Haller [11].)

experimental accuracy. This behavior is similar to the case of PAA ($K_{33}/K_{11} = 2.4$) [10], though this ratio is much smaller in MBBA. The interpretation of this difference in terms of molecular properties is not obvious.

Magnetic field acting in the plane of LC layer produces the simplest type of distortions – twist deformation described by only one elastic modulus K_{22} both at linear and at nonlinear deformations. It is simple to obtain a normal orientation of the field relative to the initial sample orientation that corresponds to this Fréedericksz transition geometry. Nevertheless, the optical detection of this transition and determination of field-induced deflection in azimuthal angle are not so straightforward as in the case of the splay and bend distortions described above. An observation of the parallel beam at experimentally available thickness shows that there are no changes in the intensity of transmitted polarized light induced by twist deformation as polarization plane follows adiabatically for the azimuthal rotation of the local optic axes (Mauguin effect).

Cladis [12] proposed to overcome this problem by conoscopic observation of the cell. It is well known that for the case of planar boundary conditions conoscopic image of the undistorted cell looks like two sets of hyperbolas sensitive to twist deformation (more detailed description can be found in the book by Oswald and Pieranski [3]). In particular, for the geometry considered, the angle ψ of the hyperbola rotation induced by twist-like distortions φ can be expressed as [12]

$$\tan 2\Psi = \langle \sin 2\varphi \rangle / \langle \cos 2\varphi \rangle, \tag{5.4}$$

where $\sin 2\varphi$ and $\cos 2\varphi$ are averaged through the layer. It means that a deformation of the conoscopic image will be visible for any nonzero twist angle in the center of the layer φ_m even for small distortions induced by the field. It allows to register the Fréedericksz transition in described geometry. Dependence of ψ and φ_m on the field strength obtained by Cladis [12] is shown in Figure 5.2.

One can observe a very sharp change in these parameters near the critical point (H_c) corresponding to the Fréedericksz transition. It helps determine the ratio ($K_{22}/\Delta\chi$) rather precisely. For example, Cladis [12] reported that for MBBA ($K_{22}/\Delta\chi$) 2.88 ± 0.03 (cgs), which corresponds to errors about 1%.

Later on, Leenhouts and Dekker [13] slightly improved this method. They pointed to the particular problem arising in the conoscopic study of twist deformation. Namely, if $\mathbf{H}$ is oriented exactly perpendicular to $\mathbf{n}_0$, the conoscopic pattern can rotate

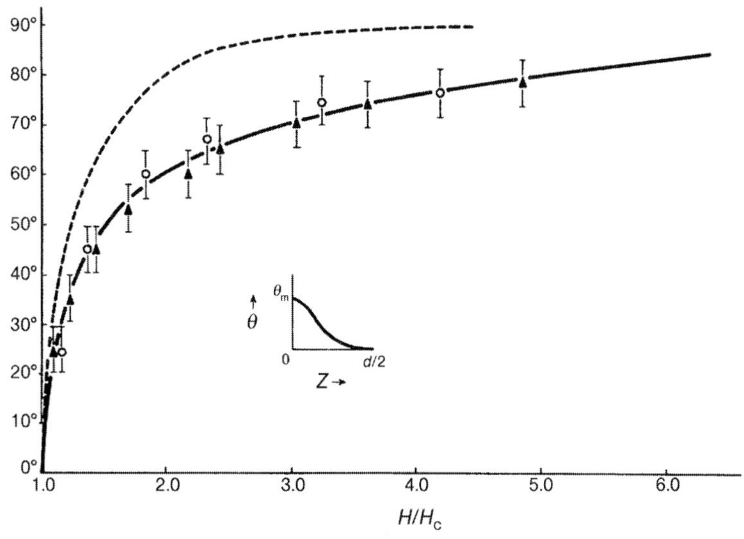

Figure 5.2 The maximal twist angle θ_m (dashed curve) and the rotation angle ψ of the interference figure (solid curve) as functions of reduced field H/H_c: MBBA, room temperature. (After Cladis [12].)

either clockwise or counterclockwise, since these two senses of rotation are equally probable. Often, therefore, a pattern originating from two domains rotating in opposite senses is observed. In that case, an accurate determination of the rotation angle is practically impossible. To avoid this difficulty, they proposed to use the initial orientation of the sample ($\mathbf{n}_0$) not exactly normal to $\mathbf{H}$ but at some angle ($90° - \varphi_0$). At some complication of expression (5.4), it results in more reliable results for K_{22} modulus. Regarding determination of K_{11} and K_{33} constants, the measurements were close to those described above. Experiments in splay and bend geometries were performed by recording the intensity of a laser beam (He–Ne; $\lambda = 633$ nm) traversing the nematic layer that was subjected to a magnetic field. The field was swept very slowly (−7 Oe/min) through the Fréedericksz transition. Values of K_{11} and K_{33} were obtained by fitting the experimental data by nonlinear theoretical expressions derived by Saupe [10].

Measurements of elastic constants of Schiff bases presented in this paper have shown that temperature dependences of these parameters can be approximated by the unique empirical expression

$$K_{ii}(T/T_c) = K_{ii}^0\left(1-\beta\frac{T}{T_c}\right)^{\gamma}, \tag{5.5}$$

where β and γ are the approximation parameters.

It was found that for compounds with a rigid molecular structure the ratio K_{33}/K_{11} increases with increasing molecular length to width ratio (Figure 5.3). At the same time, K_{22} was found to be independent of the length of the alkyl chain.

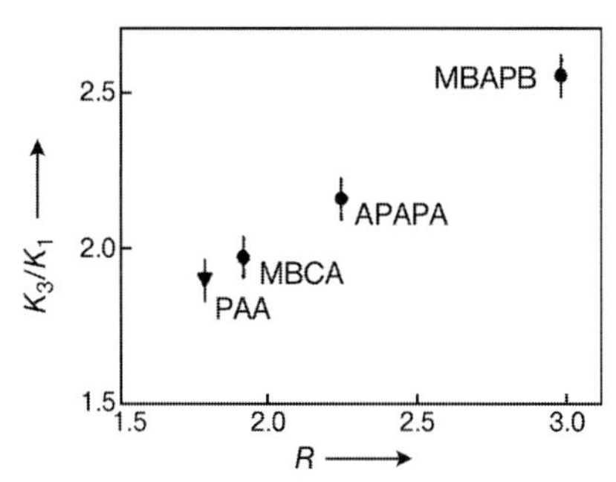

Figure 5.3 $K_{33}/K_{11}(T/T_c = 0.95)$ versus R, where $R + 1$ is the spherocylindrical length to width ratio. (After Leenhouts and Dekker [13].)

Such results are of practical importance as both parameters (K_{33}/K_{11} and K_{22}) determine technical characteristics of LC displays. So, the proposed technique can be useful for the optimization of LC materials (see Chapter 6). It is worthwhile to note that the accuracy of the determination of relative values of Frank's moduli is higher than that for the absolute value of these parameters (about 5% for K_{11} and K_{33}). In principle, such accuracy can be considered satisfactory for testing measurements at chemical synthesis and for simulation of some electrooptical modes used in LC displays (see Chapter 6).

That is why magnetic fields have been used until now with this aim. In particular, magnetically induced Fréedericksz measurements were performed in the nematic liquid crystal phase to extract elastic constants of a terminal–lateral–lateral–terminal trimer [14]. In this compound, connections to the first and third mesogens are at the end of the mesogen and both attachments to the central mesogen are lateral.

Polymeric liquid crystals based on this unit have negative Poisson ratios, which is interesting for practical applications. In this work, the optical method was combined with electric capacity measurements to extract reliable data on elastic constants. Temperature dependences of these parameters are shown in Figure 5.4. They are similar to those for monomer nematics. Deviations of the ratios K_{33}/K_{11} and K_{11}/K_{22} from typical monomer values were referred to as the influence of the central mesogen that made the molecule somewhat bulky in the center.

As all materials, excluding ferromagnetics, show magnetic permittivity close to that of a vacuum (about 1), the magnetic field can be considered as the most "pure" tool to provide the given geometry for experiments described above. Nevertheless, there are obvious disadvantages that prevent the use of magnetic fields for routine measurements of Frank's moduli.

First, according to the expression (5.2), values of diamagnetic anisotropy and the gap of the cell have to be measured rather precisely, which is not easy especially with respect to $\Delta\chi$. For example, Kang *et al.* [14] performed additional measurements of the optical birefringence Δn to minimize errors originating from the rather big scatter (more than 10%) in measured values of $\Delta\chi$. The normal orientation between

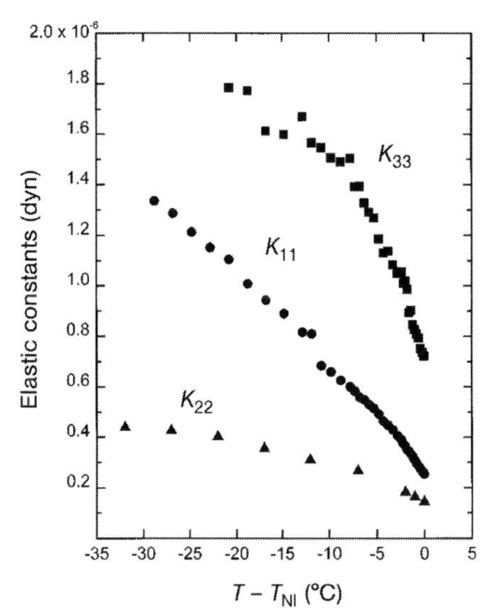

Figure 5.4 Splay (K_{11}, ●), twist (K_{22}, ▲), and bend (K_{33}, ■) elastic constants versus $T - T_{NI}$. (After Kang *et al.* [14].)

the vector of magnetic induction and the director in an undistorted sample is of primary importance too. At least, possible deflections have to be under control and taken into account during data processing [13]. Only practically homogeneous fields with well-defined orientation are applicable for this purpose.

That is why electric fields instead of magnetic ones are often applied for K_{ii} measurements. For small orientational distortions, the action of radiofrequency electric field on the initially monodomain sample of LCs with a positive value of $\Delta\varepsilon$ is equivalent to the action of magnetic field (see Section 5.1.1). A detailed analysis of the use of electric fields at different geometries of Fréedericksz transitions can be found in Ref. [3]. It is quite clear that electric measurements are mostly convenient for the realization of splay-like small distortions. In this case, a homogeneous electric field with strictly orthogonal orientation with respect to the initial planar orientation is formed in a nondistorted sample. In accordance with expression (5.2), only one additional parameter $\Delta\varepsilon$ is needed to calculate the modulus K_{11} from the experimental value of threshold voltage $U_f = E_c d$ [15, 16]). The latter is determined using highly précise methods with typical errors of about 1% [17].

Bend-like distortions needed to measure K_{33} modulus are induced at the application of high-frequency electric field to the initially homeotropic sample of LCs with a negative dielectric permittivity anisotropy. The main peculiarity of this case is the generation of field-induced distortions relative to the azimuthal angle. It results in nonhomogeneous picture slightly above threshold voltage. So, the changes in the intensity of transmitted light can depend on a local distribution of optical axes inside the scanned area. It introduces some additional errors in threshold voltage and K_{33}

values. Some combinations of elastic constants can be found from distortions of mixed type induced by electric field. For example, the threshold voltage U_f for twist-nematic planar cell depends on three Frank's modulus and is described by expression (see, for example [1])

$$U_f = \pi\sqrt{\frac{1}{\varepsilon_0\,\Delta\varepsilon}\left[K_{11} + \frac{(K_{33}-2K_{22})}{4}\right]}. \tag{5.6}$$

Recently, "in-plane" electric field was realized to achieve a twist-like deformation needed to extract K_{22} modules [18]. Authors used conoscopic illumination and a rotating input polarizer to study optic axis orientation of a sample as a function of angle of incidence. Conoscopic images of the cell were compared with theoretical predictions based on a combination of a 1D nematic and an extended Jones optical method. The comparison allowed to determine values for the twist elastic constant K_{22} and the azimuthal surface anchoring strength W_φ. The latter is important as this parameter can vary for different surface treatments (e.g., rubbing and photoalignment treatment). It is obvious that the obtained value of $W_\varphi = 5\times10^{-6}\,\mathrm{J/m^2}$ at $K_{22} = 5.0$ pN for nematic mixture E7 oriented by a thin rubbed layer of PVA (polyvinyl alcohol) does not correspond to strong anchoring (see Section 5.5). So, the influence of weak anchoring on the threshold voltage has to be taken into account at a layer thickness of order 25 μm used in experiments.

As a result, one can say that optical measurements of elastic moduli via electrically induced Fréedericksz transition are rather simple and can be done using relatively cheap, standard laboratory equipment. At the same time, they are not as universal as magnetooptical studies. The combination of both magnetic and electric fields in such experiments was also considered very useful [19].

5.1.2
Light Scattering Method

Contrary to isotropic liquids, nematic liquid crystals show a strong scattering of visible light. This optical property can be used as the most simple way to identify nematic phase and is known since the time liquid crystal state was discovered (see review by Gray [20]), but the description of this phenomena in the framework of a continuum theory of NLC was done only in the 1960s [21, 22]. One can find it in many books and reviews on physics and applications of liquid crystals (see, for example, Ref. [23]). So, we will only summarize some theoretical conclusions important for the measurement of Frank's moduli.

It is known from the general theory of light scattering (see, for example, Ref. [24]) that this phenomenon originates from the presence of inhomogeneities in the dielectric constant of the medium. For time-dependent inhomogeneities $\delta\varepsilon(\mathbf{r}, t)$, the local dielectric tensor $\varepsilon(\mathbf{r}, t)$ of the media is expressed as

$$\varepsilon(\vec{r}, t) = \bar{\varepsilon} + \delta\varepsilon(\vec{r}, t), \tag{5.7}$$

where $\bar{\varepsilon}$ is an average part. The intensity of the scattered light depends on the magnitude of dielectric fluctuations while its frequency is affected by their temporal behavior. The magnitude of the scattered electric field E_s at a large distance R from the scattering volume V with a wave vector $\mathbf{k_f}$ and polarization along the unit vector $\mathbf{f}$ is given by [24])

$$E_s(t) = \frac{E_0}{4\pi R}\exp\left[i(\vec{k}_f\vec{R}-\omega t)\right]\int_V \exp(i\vec{q}\vec{r})\{\vec{f}\cdot[\vec{k}_f\times\vec{k}_f\times(\delta\varepsilon(\vec{r},t)]\}d\vec{r}, \qquad (5.8)$$

where $\boldsymbol{q}$ is the scattering wave vector defined by

$$\vec{q} = \vec{k}_i - \vec{k}_f \qquad (5.9)$$

and $\mathbf{k}_i$ is the wave vector of the incident light with polarization along the unit vector $\mathbf{i}$. The phase difference Δ between two rays scattered from points separated by $\mathbf{r}$ is written as

$$\Delta = \vec{r}\vec{k}_i - \vec{r}\vec{k}_f = \vec{r}\vec{q}. \qquad (5.10)$$

So, Equation (5.8) represents the sum of phase-shifted waves emitted from different parts across the volume V of the scattering medium.

For a nematic phase, the dielectric tensor can be written as

$$\varepsilon_{\alpha\beta}(\vec{r}) = \varepsilon_\perp + \Delta\varepsilon\cdot n_\alpha(\vec{r})n_\beta(\vec{r}). \qquad (5.11)$$

So, the fluctuations of dielectric tensor $\delta\varepsilon(\mathbf{r}, t)$ entering into expression (5.8) for scattered light field E_s may be expressed in terms of small thermal fluctuations of the director $\delta\mathbf{n}(\mathbf{r}, t)$ relative to its mean value $\mathbf{n}_0$ by the next expression

$$\delta\varepsilon_{\alpha\beta} = \Delta\varepsilon(\delta n_\alpha n_{0\beta} + n_{0\alpha}\delta n_\beta). \qquad (5.12)$$

Equations presented show the physical background for calculation of light scattering in nematic liquid crystals induced by thermal fluctuations of the director. The general expression for an average intensity I_s of scattered light is written by (see, for example, Ref. [24])

$$I_s(\vec{q}) = \frac{k_f^4 I_0(\Delta\varepsilon)^2 k_B T}{16\pi^2 R^2}\sum_{\alpha=1,2}\frac{(i_\alpha f_3 + i_3 f_\alpha)^2}{K_{\alpha\alpha}q_\perp^2 + K_{33}q_{II}^2 + \Delta\chi H^2}. \qquad (5.13)$$

It corresponds to the vector $\mathbf{n}_0$ oriented along X_3 axis in geometries shown in Figure 5.5. Unit vectors $\boldsymbol{i}$ and $\boldsymbol{f}$ define the polarization state in the incident and scattered light.

In general, expression (5.13) for the scattered light intensity includes two contributions arising from the normal mode fluctuations of the director in $\boldsymbol{q}$-space that correspond to splay–bend and twist–bend deformations (see Figure 5.5). Strong magnetic field $\mathbf{H}$ acting in the direction of $\mathbf{n}_0$ (or electric field in the case $\Delta\varepsilon > 0$) stabilizes the initial orientation via suppression of thermal director fluctuations. It is possible to select the special geometry at which only one type of principal deformations

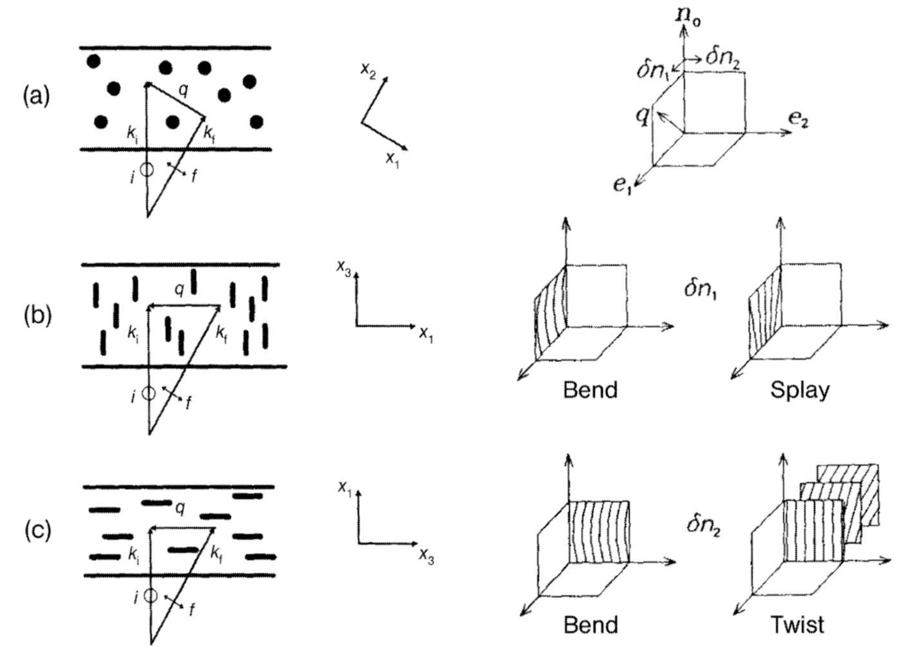

Figure 5.5 Light scattering geometries for the splay (a), twist (b), and bend (c) modes.

(splay, bend, or twist) will be responsible for light scattering. For example, at geometry *a* shown in Figure 5.5 at $i_1 = i_2 = f_3 = f_3 = q_{\|} = 0$, $i_3 = f_1 = 1$, only splay (K_{11} deformation) determines light scattering. Twist deformation will play the dominant role at $i_1 = i_3 = f_2 = q_{\|} = 0$, $i_2 = 1$ (Figure 5.5b) and bend deformation at $i_1 = i_3 = f_2 = q_{\perp} = 0$, $i_2 = 1$ (Figure 5.5c). So, it is possible to measure each of the Frank moduli by measuring the intensity of scattered light at a proper geometry.

Usually in experiments, the dependences of this parameter on the scattering angle or the applied field are studied. In the first case, only ratios of Frank moduli are determined, whereas using fields provides estimation of absolute values of K_{ii} too.

Although the given optical technique seems to be rather simple and effective for practical use, results obtained from such experiments often essentially differ from the independent ones. As an example, Figure 5.6 shows the ratio K_{11}/K_{22} obtained by different techniques for OHMBBA (*o*-hydroxy-*p*-methoxybenzilidene-*p*′-butylaniline) [25].

Detailed analysis and experimental determination of possible errors at such measurements were done by Chen *et al.* [26]. They showed that the experimental annoyances prevented accurate determination of K_{ii} values even for optimal scattering geometries. In particular, the method is inherently susceptible to stray light: it was impossible to distinguish the signal from the stray light, both of which depended on the scattering angle and the polarization condition. Moreover, for precise intensity measurement of scattering angle dependence, very delicate adjustment of optical system was required. As a result, there is a large (about 10%) scattering of experimental data obtained in an independent series of measurements (Figure 5.7).

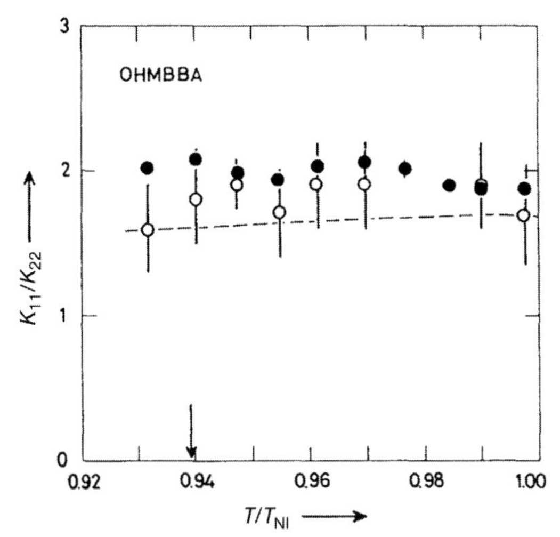

Figure 5.6 The elastic ratio K_{11}/K_{22} versus the reduced temperature T/T_{NI}. The dots are from the intensity measurements. The circles are from the spectral density measurements. The broken curve is from Fréedericksz transition data [13]. (After van der Meulen and Zijlstra [25].)

It was also shown that photon correlation spectroscopy is more useful for determining both elastic and viscous properties of liquid crystals. It is based on the study of the frequency spectrum of the scattered light in terms of the time correlation function $C_I(q, \tau)$ of the scattered intensity [24]:

$$C_I(\vec{q}, \tau) = \langle I_s(t) I_s(t+\tau) \rangle = \langle E_s^*(t) E_s(t) E_s^*(t+\tau) E_s(t+\tau) \rangle. \quad (5.14)$$

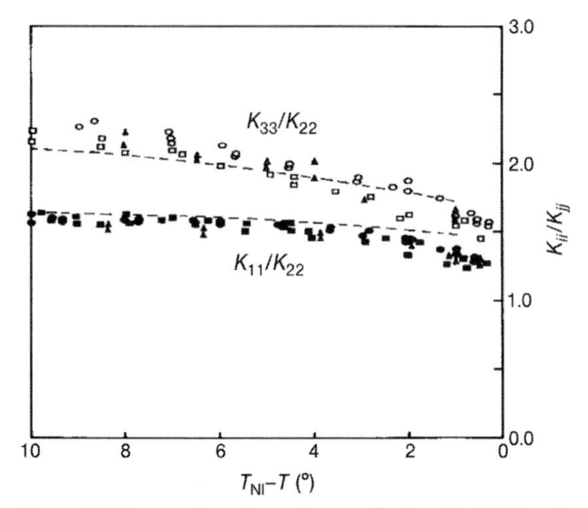

Figure 5.7 Temperature dependences of ratios (K_{33}/K_{22}) and (K_{11}/K_{33}) in independent series of measurements for 5CB. (After Chen *et al.* [26].)

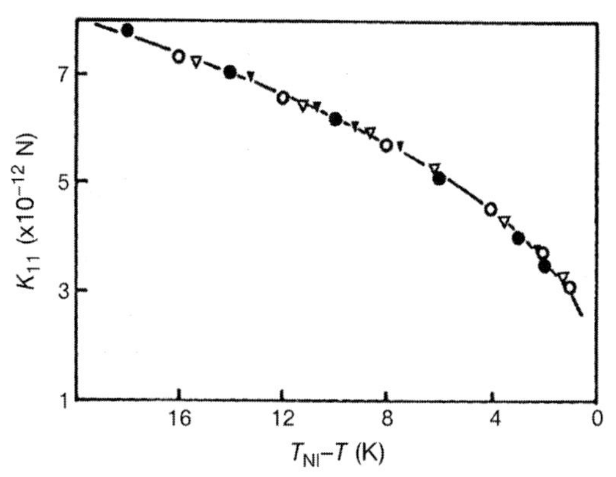

Figure 5.8 Variation of K_{11} with reduced temperature for 5CB. (▽) Results of Coles and Sefton [27]; (▼, ●, ○) independent data.

In this approach, the time correlation function $C_l(\boldsymbol{q}, \tau)$ is computed via electronic manipulation with the scattered light registered by the photocathode of a photomultiplier.

For a diffusive mode with a decay rate Γ, the simple expression for $C_l(\boldsymbol{q}, \tau)$ is valid:

$$C_l(\vec{q}, \tau) = I_s^2(\vec{q})[1 + \exp(-2\Gamma\tau)], \tag{5.15}$$

where the average intensity of the scattered light $I_s^2(\mathbf{q})$ is defined by (5.13).

The spectrum of the scattered light reflects the dynamics of director fluctuations that are related to viscosity coefficients. Photon correlation technique is able to accurately determine the linewidth of the scattered light. A variety of different scattering geometries make it possible to extract different combinations of elastic and viscoelastic parameters of liquid crystals. Electric field was found to be useful for precise determination of Frank modulus [27]. As an example, we present data on temperature dependence of K_{11} modulus obtained in this work (Figure 5.8). This method will be described in Section 5.4 in more detail.

We can summarize experimental methods for Frank moduli measurements as follows:

1. Elastic moduli could be measured from electro (and magneto)-optical characteristics of Fréedericksz transition such as threshold voltage U_F or threshold magnetic field $H_{c.}$ To find the value of the elastic modulus ratio K_{33}/K_{11}, it is also possible to measure either the birefringence $\delta = \Delta nd/\lambda$ (Δnd is the optical path difference and λ the wavelength of light) or the capacitance C versus voltage U curve in electrically induced Fréedericksz transition.

   We can either

   (i) measure the steepness of the curve near the threshold field (or at very high voltages) as a function of $1/U$, or
   (ii) fit the total experimental curve by means of a computer procedure [28].

The accuracy of the above-mentioned techniques depends on the following factors:

(i) Accuracy of the determination of $n_{\|}$, $n_{\perp}$, $\varepsilon_{\|}$, $\varepsilon_{\perp}$, and the layer thickness d.
(ii) The value of conductivity σ of a liquid crystal, that is, the following inequality must be satisfied:

$$\sigma^* \ll \varepsilon^* f,$$

where σ^* and ε^* are the average conductivity and dielectric constants of the liquid crystal, respectively, f is the frequency of the applied field.

The elastic coefficient K_{22} could be measured according to either from the threshold of twist distortion of the homogeneous alignment induced by magnetic field or from the threshold of initially twisted director alignment.

It is also possible to measure the unwinding voltage U_{unw} of the cholesteric to nematic transition [23]:

$$U_{unw} = \pi^2 d / P_0 (K_{22}/\varepsilon_0\,\Delta\varepsilon)^{1/2}, \tag{5.16}$$

where P_0 is the equilibrium value of the cholesteric pitch.

It is worthwhile to point out that due to the reasons mentioned above measurements of K_{22} are performed with essentially higher errors (more than 10%) than those of K_{11} and K_{33}. That is why there are very few materials for which all three elastic constants have been determined with sufficient accuracy, and attempts to improve experimental methods can be found in recent publications [18, 29]. The combination of different techniques (e.g., optical and capacitance [29]) makes results more reliable. Nevertheless, the scattering of experimental data for K_{22} constant even in the case of well-studied LCs such as 5CB is rather high (compare Figure 5.9b and Figure 5.10b).

2. The elastic ratio K_{33}/K_{11} could be determined by measuring the relative phase difference in homeoplanar (hybrid) liquid crystal cell [30]. To provide a high degree of accuracy in the measurements, we should verify whether the homogeneous and homeotropic boundary angles really take place on the substrates of the cell.

3. The elastic moduli ratios K_{33}/K_{11} and K_{22}/K_{11} could be evaluated by measuring differential sections of light scattering induced by thermal fluctuations of the director as a function of the angle between the scattering vector and the director for different scattering geometries. However, experimental data on elastic moduli obtained from scattering are not as reliable as those obtained in a much more direct way using Fréedericksz transition.

Experimental data on Frank elastic moduli can be found in original papers, including those mentioned above, and in a number of fine reviews (see, for example, Refs. [1, 31, 32]).

A lot of available experimental information is presented and summarized in Ref. [33]. As an example, we present in Figures 5.9 and 5.10 the temperature

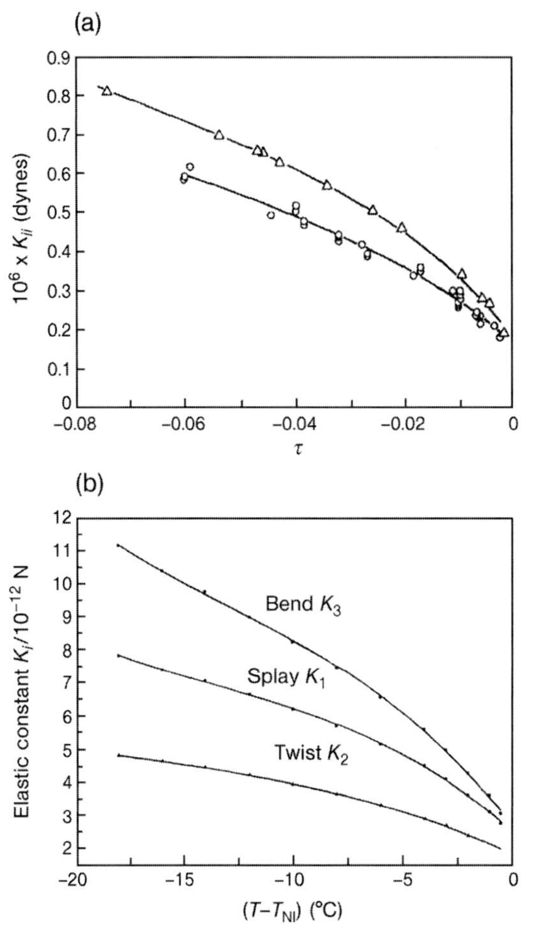

Figure 5.9 Elastic constants of MBBA as functions of reduced temperatures, $\tau = (t - t_{NI})/(t_{NI} + 273.2)$ (a). Splay is shown as circles, bend as triangles. The solid lines correspond to least squares fits of K_{11} (μdyn) $= 1.50|\tau^{0.5}| + 2.1|\tau| + 0.102$, and K_{33} (μdyn) $= 2.43|\tau^{0.5}| + 0.851|\tau| + 0.089$, respectively. Data obtained by [11] via optical study of magnetically induced Fréedericksz transitions. Estimated errors are about 8%. Splay, twist, and bend elastic constants (averaged values) of 5CB as a function of shifted temperature (b). Original data are obtained via Fréedericksz transitions and inelastic light scattering. (After Dunmur [2].)

dependences of Frank elastic moduli for some nematic liquid crystals including some particular results for two standard nematics – 5CB and MBBA. Values of elastic moduli of two of them (MBBA and 5CB) are often used for computer modeling of a number of nonlinear physical phenomena taking place in liquid crystals (see, for example, Chapter 3). Numerical values of Frank moduli (at room temperature) for typical nematic materials of practical use with different values and signs of dielectric permittivity anisotropy can be found in Ref. [33].

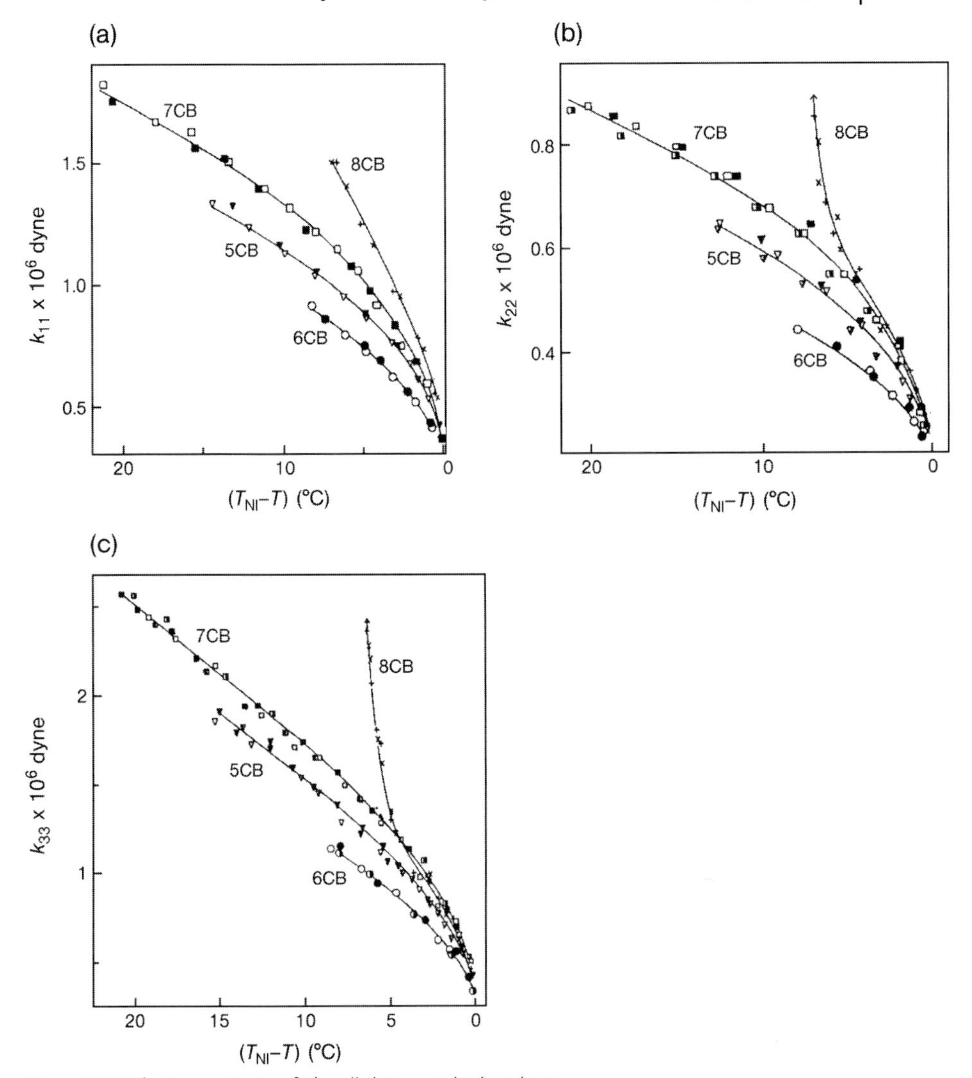

Figure 5.10 Elastic constants of 4′-*n*-alkyl-4-cyano-biphenyls as functions of the relative temperature: (a) K_{11}; (b) K_{22}; and (c) K_{33}. Results of different experiments are marked separately. (After Stannarius [1].)

Let us highlight the main information on liquid crystal elastic modules.

1. As a consequence of the complexity of molecular interactions in the nematic phase, there is no general quantitative relationship between the elastic constants and the molecular structure. For example, the ratio K_{33}/K_{11} playing an essential role in display applications was found to be more or less than 1 for different classes of LCs (in most cases $K_{33}/K_{11} > 1$).

Usual relationships between K_{33}/K_{11} and K_{22}/K_{11} are as follows [34]:

$$\begin{aligned} 0.5 &< K_{33}/K_{11} < 3.0 \\ 0.5 &< K_{22}/K_{11} < 0.8. \end{aligned} \tag{5.17}$$

2. Several general rules for structure/elasticity relationship have been extracted empirically from a large number of experiments. In particular, in homologous series, there is a pronounced odd/even alternation of the K_{33}/K_{11} ratio with the number of chain segments (see, for example, Ref. [32]) and this ratio gradually decreases with increasing alkyl chain length.

3. Temperature dependence of the elastic moduli correlates with that of the square of the order parameter

$$K_{ii} \sim S^2(T), \quad i = 1, 2, 3. \tag{5.18}$$

4. Near the nematic–smectic A phase transition, the K_{22} and K_{33} divergence takes place (see, for example, the data on 8CB in Figure 5.10). It means that (5.18) is no longer valid. Moreover, presmectic ordering in the nematic phase is observed even in the case where the smectic phase itself is absent. The physical origin of the presmectic ordering could be explained by the difference in the polarizability and steric factors for the central core of the molecule and its flexible alkyl chains, which results in the preferred orientation of the molecular cores closer to each other. The presmectic ordering increases with the number of the carbon atoms that is the length of the alkyl chain in the homologous series.

5. The dimer and trimer formation considerably affects the elastic ratio K_{33}/K_{11}. This may take place due to a different degree of overlapping for molecules from different chemical classes. As a result, the molecular unit, which defines the elastic properties of a liquid crystal, is formed not only of a single molecule but also of a molecular aggregate.

6. Remarkable elastic properties are observed in mixtures of weak and strong polar compounds. The situation is simple for the concentration dependences of K_{11} and K_{22} that obeyed a simple additivity law:

$$K_{\text{mix}}^{1/2} = x\, K_{\text{w}}^{1/2} + (1-x) K_{\text{s}}^{1/2}, \tag{5.19}$$

where K_{mix}, K_{w}, and K_{s} are elastic moduli of the mixture, weak, and strong compounds, respectively, and x is the molar fraction of a weak compound. However, both K_{33} and K_{33}/K_{11} concentration dependences possess a well-pronounced minimum (depression) in the region of $x \sim 0.4$–0.6 [35]. Physical interpretation of the phenomenon is based on the fact that high concentration of dimers is formed both in the mixture of pure polar compounds ($x = 1$) and weak ones ($x = 0$). In the latter case, dimers are formed either due to the dispersion interaction between the molecules or due to the interaction dipole–induced dipole.

There are a number of experimental data on elastic moduli of liquid crystals that are explained only qualitatively. Existing molecular approaches do not directly correspond to the real situation because molecules are considered to be spherocylinders or hard rods that is far from reality. For instance, the ratio K_{22}/K_{11} according to present approaches is about 1/3, which is two times lower than the corresponding experimental range (5.17). However, the above-mentioned data are quite sufficient for developing liquid crystal mixtures with required elastic properties. For instance, to get a minimum value of the elastic ratio of K_{33}/K_{11}, special liquid crystal molecules were constructed [36].

Here, in one molecule the authors use three structural elements that are known to diminish the K_{33}/K_{11} ratio an alkyl end group in the form of the pyridine ring and strong polar NCS group. The resulting experimental mixture possesses the ratio $K_{33}/K_{11} = 0.5$–0.6 obtained in a broad temperature range.

5.2 Rotational Viscosity of Nematic and Smectic C Liquid Crystals: Experimental Methods and Techniques

Rotational viscosity of liquid crystals can be considered as the specific dissipative property most important for practical applications. Indeed, liquid crystal devices are based on reorientation of an optical axis under the action of electromagnetic fields or shear flows. A coefficient of rotational (twist) viscosity γ_1 defines in first approximation dissipative losses connected with such motion and viscous forces which slow director motion. So, it is responsible for operating times of liquid crystal devices.

That is why values of the rotational viscosity coefficient are included in the description of all commercially produced nematics, and a number of experimental methods and particular techniques were proposed. They can be classified by different schemes [37, 38]. We will consider two main categories depending on the influence of boundary conditions on obtained results.

5.2.1 Measurements in Bulk Samples of Nematic Liquid Crystals

In general, the theoretical description of the director motion based on nonlinear hydrodynamic equations (see Chapter 3) is rather complicated and demands the use of numerical methods. It can be essentially simplified for homogeneous director fields with zero derivatives with respect to the space coordinates. Usually, such approximation holds for inner parts of LC samples oriented by strong magnetic (electric) fields or intensive shear flows. Corresponding inequality is written as

$$d \gg \xi,$$

where the coherence ξ [3, 23] is defined for magnetic (**B**) or electric (**E**) fields by expressions

$$\xi(B) = \sqrt{\frac{\mu_0 K_{ii}}{\Delta\chi} \frac{1}{B}} \qquad (5.20)$$

$$\xi(E) = \sqrt{\frac{K_{ii}}{\varepsilon_0\,\Delta\varepsilon}}\frac{1}{E}. \tag{5.21}$$

The coherence length is referred to as the boundary layer with nonhomogeneous director distribution imposed by the orienting action of boundaries. Dynamic behavior of such “bulk samples” is described by simplified hydrodynamic equations that provide straightforward ways for calculation of the rotational viscosity coefficient.

5.2.1.1 Method of Rotating Magnetic Field

This method is based on the study of behavior of bulk samples of liquid crystals under the action of rotating magnetic field. Two main variants of the method were used in experimental setups depending on the values of a rotation angle and a rotation rate.

In the first type of experiments, a liquid crystal sample was subjected to the action of magnetic field rotating around LC samples with a constant angular velocity making at least some revolutions until some stationary regime was established. Theoretical description of the dynamic behavior of LC at these conditions has to be based on nonlinear hydrodynamic equations of liquid crystals presented above that can be solved at some approximations.

In experiments of the second type, the magnetic field was rotated quickly to a relatively small rotation angle and after that a slow relaxation of director to a new position was studied. This makes it possible to use linearized hydrodynamic equations to describe the time evolution of an orientation of LC samples usually expressed by simple exponential laws. A number of experimental techniques suitable for bulk samples were proposed to determine the rotational viscosity coefficient from such type of experiments.

5.2.1.1.1 Permanent Rotation of Magnetic Field

Let us consider strong magnetic field **H** rotating with an angular velocity ω and acting on a homogeneous sample defined by the director **n** (Figure 5.11).

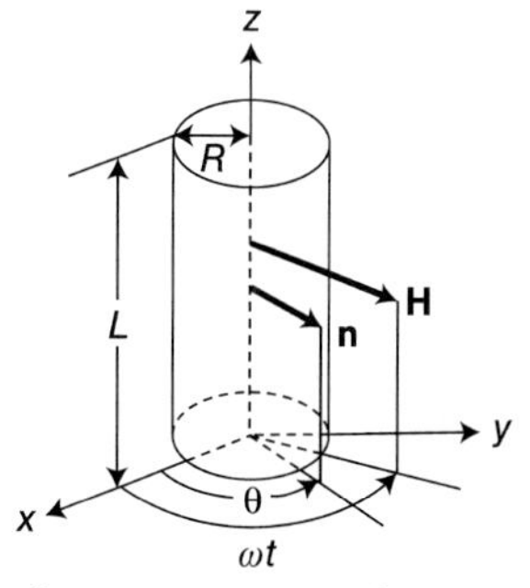

Figure 5.11 Geometry for experiments in rotating magnetic fields.

The time evolution of the angle of the director rotation θ is described by equation [39]

$$\gamma_1 \frac{d\theta}{dt} = \frac{1}{2}\mu_0\,\Delta\chi H^2 \sin 2(\omega t - \theta). \tag{5.22}$$

A solution of Equation 5.22 for the initial condition $\theta(t=0)=0$ can be expressed in terms of the angle $\varphi(t)$ describing a phase lag between magnetic field **H** and the director **n** as

$$tg\varphi(t) = \frac{1-\exp(-2\omega t\sqrt{\varepsilon^2-1})}{\varepsilon+\sqrt{\varepsilon^2-1}-(\varepsilon-\sqrt{\varepsilon^2-1})\exp(-2\omega t\sqrt{\varepsilon^2-1})}, \quad \varepsilon > 1, \tag{5.23}$$

$$tg\varphi(t) = \frac{tg(\omega t\sqrt{1-\varepsilon^2})}{\sqrt{1-\varepsilon^2}+\varepsilon tg(-2\omega t\sqrt{1-\varepsilon^2})}, \quad \varepsilon < 1, \tag{5.24}$$

where

$$\varepsilon = \frac{\mu_0\,\Delta\chi H^2}{2\gamma_1\omega} \tag{5.25}$$

is a dimensionless parameter, defining "synchronic" (5.23) or "asynchronous " (5.24) regimes of the director motion.

In the synchronic regime ($\varepsilon < 1$) after a transition time,

$$\tau = \frac{1}{2\omega\sqrt{\varepsilon^2-1}} \tag{5.26}$$

the director rotates synchronically with magnetic field, and the stationary phase lag φ_s is expressed as

$$\sin 2\varphi_s = 2\frac{\gamma_1\omega}{\mu_0\,\Delta\chi H^2}. \tag{5.27}$$

In the asynchronous regime ($\varepsilon > 1$), the director oscillates around the field direction. The critical frequency ω^c for the change of regimes is determined by expression

$$\omega^c = \frac{\mu_0\,\Delta\chi H^2}{2\gamma_1}. \tag{5.28}$$

The motion of the director in rotating magnetic field can be registered by using different physical properties of LC sample. In his pioneering work, Tsvetkov [43] considered the mechanical viscous torque M_v transmitted from bulk sample to the container walls. The main ideas realized in his experiments and in many later modifications turned out to be very fruitful – accurate measurements of viscous torque in rotating field provide the most direct and precise data on the rotational viscosity coefficient.

In the original Tsvetkov's setup, the magnetic field of a variable strength (0–4.3 kG) was achieved using a powerful electromagnet rotating around the chamber containing a glass cylindrical tube (with an inner diameter $d = 5$ mm) filled with a nematic sample. In synchronic regime, the viscous torque transmitted to the inner wall of the tube can be expressed as

$$M_v = \gamma_1 \omega V, \tag{5.29}$$

where V is the sample volume. To measure this moment, the chamber was suspended by a tungsten filament with a diameter and length of 0.1 mm and 300 mm, respectively. Under the action of the rotating magnetic field, the filament became twisted until the viscous torque was compensated for by the elastic one. Measuring the twisted angle α provided the estimation of the rotational viscosity coefficient by expression

$$\gamma_1 = \alpha D / \omega V, \tag{5.30}$$

where D is the torsion constant of the filament that can be easily determined by independent methods with high accuracy ($D = 20$ dyn cm in Tsvetkov's setup). So, described technique provides direct determination of the rotational viscosity coefficient.

Detailed analysis of possible errors of such type of measurements [37] has shown that they are mostly connected with orientational distortions arising in surface layers under the action of a rotating magnetic field. In particular, the possible generation of inversion walls results in an increase in the thickness of the surface layer with distorted orientation approximately exceeding by order the coherence length defined by expression (5.20). It means that strong magnetic fields are required to neglect the possible influence of the boundary layers. This condition can be expressed by the corresponding inequality:

$$d \gg 2k\xi. \tag{5.31}$$

For typical nematics, ξ is about 10 μm for magnetic field of induction $B = 0.3$ T. So, at $k \approx 10$, the diameter of LC sample has to exceed essentially 0.2 mm. According to expression (5.25), a synchronic regime in this case corresponds to the range $0 < \omega < 1$ rad/s for a typical value of $\gamma_1 \approx 0.05$ Pa s. Estimates obtained show that a good precision of measurements is possible using strong and homogeneous magnetic fields provided by powerful and massive magnets rotating around the sample with a variable frequency. As a result, experimental setups of such type became cumbersome and included a lot of mechanical parts.

Modifications of such experimental technique included the use of permanent magnets (see, for example, Refs [41, 42, 44]) or rotation of the container with LC sample [45] placed between poles of a powerful electromagnet (200 mm pole shoe diameter, 50 mm pole gap, $B = 0$–1.3 T). In the latter case, experimental errors were minimized (down to 0.3%) and the most precise measurements of the rotational viscosity coefficient in nematics were obtained. A usual amount of liquid crystals needed for precise measurements is about some cubic centimeters, which prevents the use of such technique for newly synthesized liquid crystal materials. Nevertheless, the precise temperature (and pressure [44]) dependence of the rotational viscosity coefficient is important for checking the molecular theories (see, for example, Ref. [38]) of viscous properties of liquid crystals.

Nonmechanical methods of registration of the director motion under permanently rotating magnetic field can be based on different physical effects sensitive to an orientational structure of liquid crystals. For example, NMR and EPR [46–48] techniques turned out to be useful for this purpose. The use of very strong (up to 10 T) magnetic fields needed for high-resolution NMR (EPR) spectrometry provides very perfect homogeneous orientation of LCs. So, the above-mentioned approximate expressions derived for bulk samples are applicable for liquid crystal samples rotating in magnetic field. Values of the angle φ between director and field direction extracted from NMR spectra allow one to calculate the rotational viscosity coefficient in accordance with expression (5.27). In this case, the anisotropy of magnetic susceptibility $\Delta\chi$ has to be determined by independent experiments.

The use of ultrasonic waves passing through anisotropic bulk samples for the study of liquid crystals under rotating magnetic field was discussed in detail in Chapter 4.

5.2.1.1.2 **Step-Like Rotation of Magnetic Field**

Measurements of the viscous torque arising due to the rotation of magnetic field and transmitted to the container walls can be obtained not only at permanent rotation of field (as described above) but also at restricted step-like rotation (short review of investigations can be found in [37]). Such types of experiments were performed by Bock *et al.* [49]. As in the work [45], a rotation of field was replaced by a step-like rotation of a container with LC sample in strong magnetic field. After that, relaxation of the container suspended from the torsion filament to the new position was observed. The time variation of the rotation angle for small initial distortions is well-described by a simple exponential law with the relaxation time τ defined by expression

$$\tau = \gamma_1 V/D. \tag{5.32}$$

Such modification of Tsvetkov's method turned out to be useful for liquid crystals with high viscosity as in this case the critical angular frequency of rotation expressed by (5.28) is too low for precise measurements in a permanently rotating magnetic field.

The step-like rapid rotation of the sample in strong magnetic field was also applied in [46–48]. The rotational viscosity coefficient was extracted by analyzing time variations of NMR spectra imposed by the director relaxation with the relaxation time

$$\tau = \frac{\gamma_1}{\mu_0 \Delta\chi H^2}. \tag{5.33}$$

As in the case of a permanent rotation, only a ratio $\gamma_1/\Delta\chi$ can be obtained using formula (5.33). So, knowledge of independent measurements of $\Delta\chi$ is necessary for the calculation of the rotational viscosity coefficient. It is worthwhile to note that the same is true for any experimental technique based on the registration of time variations of the angle φ between the field and the director. The latter can be achieved by alternative experimental techniques useful for routine measurements of the rotational viscosity coefficient.

For example, in the work [50] the rotational viscosity as a function of temperature for several commercial mixtures was studied via measurements of the dielectric constant. After rapid step-like rotation of magnetic field by a small angle, the director relaxed to the new position with the relaxation time described by expression (5.33) that resulted in time changes of the capacity of the measuring cell filled with LC. The proposed method is relatively simple and can be easily automatized. The thickness of the LC layer (1 mm) was smaller than typical sizes of the samples in experiments described above. So, boundary layers can in principle contribute to the parameter under determination at least for fields of moderate (of order 0.3 T) strength. It is worthwhile to note that the above-mentioned problem of the generation of the inverse walls does not exist for small angles of rotation, and influence of backflow effects is restricted to the boundary layers of thickness ξ that has to decrease experimental errors.

5.2.2
Measurements in Thin Layers of Nematics

Determination of the rotational viscosity coefficient via experiments with thin layers of liquid crystals is very attractive for practical applications as it demands small sample volumes. In all experiments of such type, the initial orientational structure of the layer defined by surface anchoring is disturbed with the help of magnetic (electric) field and nonequilibrium reorientation of the director induced by fields or boundaries is investigated. In general, an orientational motion of the director is quite complicated due to backflow effects, and its description is possible by numerical solving of nonlinear hydrodynamic equations including a set of Leslie coefficients. Nevertheless, in some geometry, the influence of backflow is negligible and reliable values of the rotational viscosity coefficient can be obtained.

The pure twist deformation of a liquid crystal is of special interest, as the rotation of the director in this case is decoupled with a translational motion. Usually, magnetic field $\mathbf{B}$ applied in the plane of LC layer in the direction normal (or approximately normal) to the initial planar orientation $\mathbf{n}_0$ is used for experimental arrangement of such deformation. Most reliable data can be obtained when the azimuthal angle $\varphi(z)$ describing deformed state of the layer is small enough to use linearized equations. The case of normal orientation of $\mathbf{B}$ with respect to $\mathbf{n}_0$ corresponds to well-studied Fréedericksz transition. With a strong surface anchoring, the reorienting field must be sufficiently strong, $H > H_c$, to overcome the elastic restoring force, and experimental determination of the critical field provides an estimation of elastic constant K_{22} (see Section 5.1) by observation of changes in conoscopic pattern. After turning the magnetic field off, the deformed structure relaxes to the initial state with the characteristic time defined by the rotational viscosity coefficient.

Relaxation processes at turning the magnetic field on and off are described by the next nonlinear equation with corresponding initial conditions:

$$K_{22}\frac{\partial^2\varphi}{\partial z^2} + (\mu_0\,\Delta\chi H^2)\sin\varphi\cos\varphi = \gamma_1\frac{\partial\varphi}{\partial t}, \tag{5.34}$$

where $H = 0$ at the relaxation of director to the initial planar state. The solution of this equation problem can be found via Fourier transformation as [51]

$$\varphi(z,t) = \sum_n C_n(t)\cos[(2n+1)\pi z/d]. \quad (5.35)$$

For switching the field off, the eigenfunctions of $\varphi(z, t)$ decay with characteristic times τ_n which are expressed as

$$\tau_n^{-1} = (K_{22}/\gamma_1)[(2n+1)\pi/d]^2. \quad (5.36)$$

The time constant τ_0 corresponds to the slowest decay. For relatively small values of $\varphi(z, 0)$, a simple exponential law describes the final stage of relaxation:

$$\varphi(z,t) = \varphi(z,0)\exp(-t/\tau_0) \quad (5.37)$$

with

$$\varphi(z,0) = \varphi_{\mathrm{m}}\cos(\pi z/d) \quad (5.38)$$

and

$$\tau_0 = \frac{\gamma_1 d^2}{\pi^2 K_{22}} = \frac{\gamma_1}{\mu_0\,\Delta\chi H_{\mathrm{c}}^2}. \quad (5.39)$$

The authors had shown that the full stationary solution of (5.34) is comparable to the approximate one (5.38) at $\varphi_{\mathrm{m}} < 40°$. Experimentally, the rotation angle $\delta(t)$ of the conoscopic figure instead of $\varphi(z, t)$ was studied. For low values of deformation angles, $\delta(t) \sim \varphi_{\mathrm{m}}$. So, a simple relaxation law is sufficient for the description of the director relaxation (Figure 5.12). It paves the way for straight calculation of γ_1 using (5.39) by extracting the relaxation time from time dependence of the rotation angle $\delta(t)$ and using the value K_{22} obtained by studying Fréedericksz transition (see Section 5.1.1). The temperature dependences of the rotational viscosity coefficient for a homologous series *p,p′*-dialkyloxybenzenes are presented in Figure 5.12b.

In practice, determination of γ_1 is difficult if $\mathbf{H}$ is perpendicular to $\mathbf{n}_0$ as the director can rotate either clockwise or counterclockwise [13]. It results in a multidomain structure of a conoscopic image useless for analysis.

van Dijk *et al.* [51] overcame this problem by orienting $\mathbf{H}$ at an angle of $(90° - \varphi_0)$ to $\mathbf{n}$. In this case, the deformation already starts at $H = 0$, and measurements can be more precise than those first performed by Cladis in 1972 [12]. The method described above can be considered reliable and very attractive as a small amount of LCs is needed. Nevertheless, it includes the processing of a number of digital images, which demands some additional efforts. Unfortunately, as it was pointed by Cladis [12], pure twist deformation may not be observed by usual microscopy in traditional geometry. Nor can it be observed by monitoring the dielectric constant (or thermal conductivity or any other anisotropic property of a uniaxial nematic) since the director remains in the same plane as glass plates in both the twisted and the untwisted configurations.

This problem can be solved by using alternative experimental geometries. In particular, in Refs [52, 53] a new geometry for an optical detection of pure twist deformation was proposed (see Figure 5.13). The general idea of experiments is to

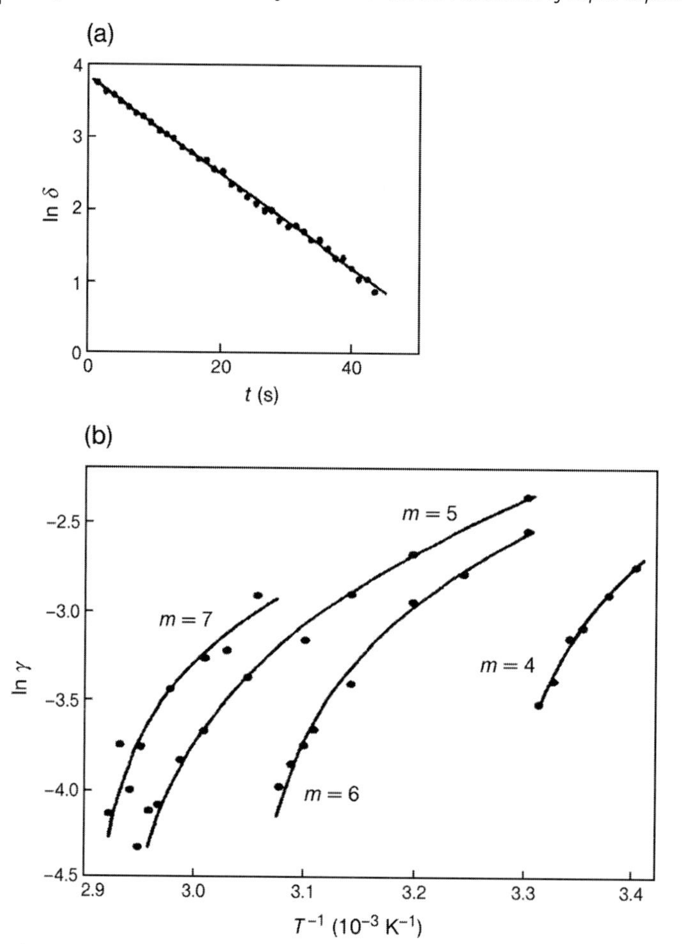

Figure 5.12 Typical decay of the rotation angle δ of the conoscopic figure (a) and temperature dependences of the rotational viscosity coefficient for a homologous series *p,p′*-dialkyloxybenzenes (b). (After Van Dijk [51].)

make possible microscopic observations of LC layer confined by a rectangular capillary from two orthogonal (x and z) directions. The general scheme of the cell used in the experiments is shown in Figure 5.14.

The channel of a rectangular cross section is formed by two pairs of glass plates. The upper and the bottom plates, 1.1 mm thick, are coated with ITO transparent electrodes to provide electric field inside the channel. Two other plates (thickness $d = 0.27$ mm or 1.1 mm) with polished edge surfaces are pressed between the first pair of the plates. So, the channel of the constant width b, shown in Figure 5.14, or the wedge-like channel of a rectangular cross section with width b linearly dependent on Y-coordinate [52] can be obtained in such manner. In the latter case, the small angle of the wedge (about 0.5°) makes it possible to consider the opposite polished edge

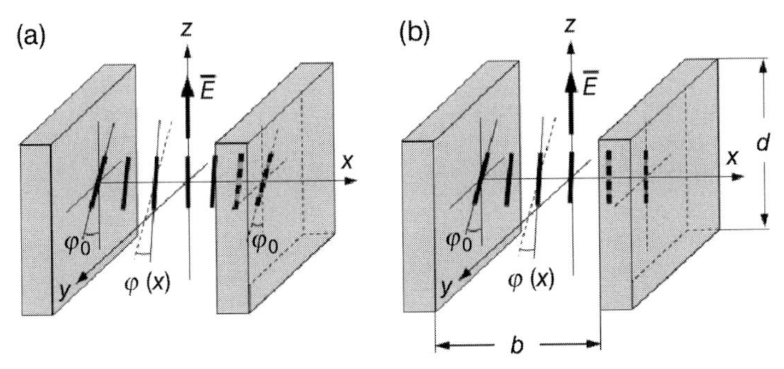

Figure 5.13 New geometry for the study of pure twist deformation; symmetric (a) and nonsymmetric (b) boundary conditions.

surfaces as locally parallel to simplify the hydrodynamic description. The main aim of such construction is to provide a slow variation of the aspect ratio $r = b/d$ of the channel, which plays a key role in the geometry under consideration. The values of b and d were small enough to make liquid crystal media inside the channel to be transparent both in Z- and in X-directions.

The case $r \gg 1$ corresponds to the traditional geometry mostly used for studying electrooptical effects in LCs via observation in Z-direction. The opposite case $r \ll 1$ corresponds to the nontrivial geometry. First, it was shown that electric field inside the gap can be considered a homogeneous one (contrary to the inhomogeneous field in the traditional geometry). The same conclusion is valid for thin LC cell placed into initially homogeneous field [3]. This makes it possible to use electric field instead of the magnetic one both for linear and nonlinear orientational deformations without

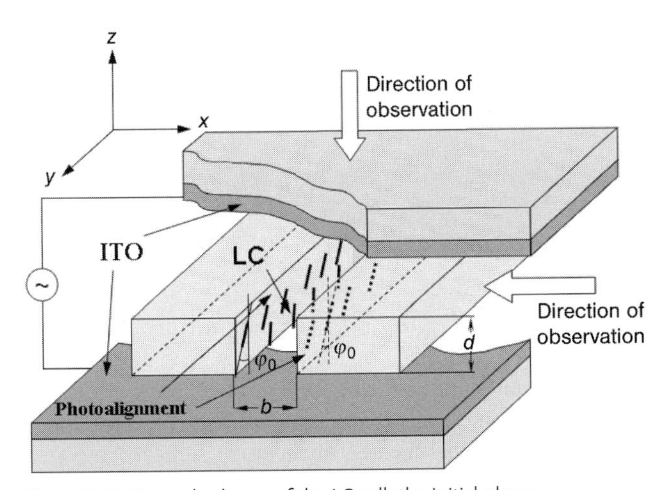

Figure 5.14 General scheme of the LC cell; the initial planar orientation is achieved by UV treatment of surfaces to obtain symmetric boundary conditions.

changing the theoretical description. Second, nonconoscopic observation of the layer in Z-direction provides registration of small angles of pure twist deformation that makes it possible to use linear approximation for the description of the orientational deformations of the layer.

The inner glass plates were treated by photoalignment (PA) technique [54] to get homogeneous planar orientation at a small angle φ_0 with respect to the Z-axes (symmetric boundary conditions) or the twist-like initial state (asymmetric boundary conditions). The upper and bottom plates were coated by a film of homeotropic surfactant to avoid possible disclination lines. So, maximal variations of an azimuthal angle φ did not exceed φ_0 in experiments under consideration. The channel was filled with a nematic liquid crystal (5CB or mixtures ZhK654 and ZhK616, NIOPiK production) with high positive values (11.5, 10.7, and 3.5) of $\Delta\varepsilon$. In case of symmetrical boundary conditions (Figure 5.13a), application of AC electric voltage produced twist-like deformation in LC layer, registered via microscopic observations in Z-direction (the channel was oriented at 45° with respect to the crossed polarizers). The latter was possible due to field-induced changes of the angle between the optical axes (**n**) and the direction of observation Z, analogous to those in traditional B-effect. Upon application of voltage, the angle φ became dependent on coordinate-X. It resulted in a number of interference stripes observed in the gap as the size of the layer in Z-direction was big enough even at a small initial angle φ_0. So, the resulting pattern showed very high sensitivity to small twist deformations induced by electric field. In the case of asymmetrical boundary conditions (Figure 5.13b), the field application resulted in changes in the initially twisted structure and in interference stripes as shown in Figure 5.15a and b. At high values of electric voltages, the image became dark almost everywhere. It corresponded to a homeotropic orientation except for the near boundary region of the twist-like deformation. In both cases, no essential changes were registered in X-direction, which is a typical behavior for a twist-like deformation in usual LC cells under Mauguin regime [3].

Processing of digital photos of the gap obtained at different electric voltages (Figure 5.15a and b) made it possible to determine the electric coherence length $\xi(E)$ defined by Equation 5.21 and shown in Figure 5.15b. This parameter characterizes the thickness of the boundary layer where the orientation is controlled by a surface. It is worthwhile to note that the new optical geometry provided the first visualization of such layer that was impossible for the traditional geometry. Values of ξ_E can be used to determine the Frank constant K_{22}. The value of the parameter $K_{22} = (3.5 \pm 0.4) \times 10^{-12}$ N obtained for the standard liquid crystal 5CB is in an accordance with the independent data (see Section (5.1). It made possible to propose the new method for measurement of this parameter [53].

Digital photos of the gap obtained after turning the electric field off (Figure 5.15) were processed to extract information about variations of local light intensity I with time. The latter can be connected with the time dependence $\varphi(x, t)$ by using a well-known expression:

$$I = I_0 \sin^2(\delta/2), \tag{5.40}$$

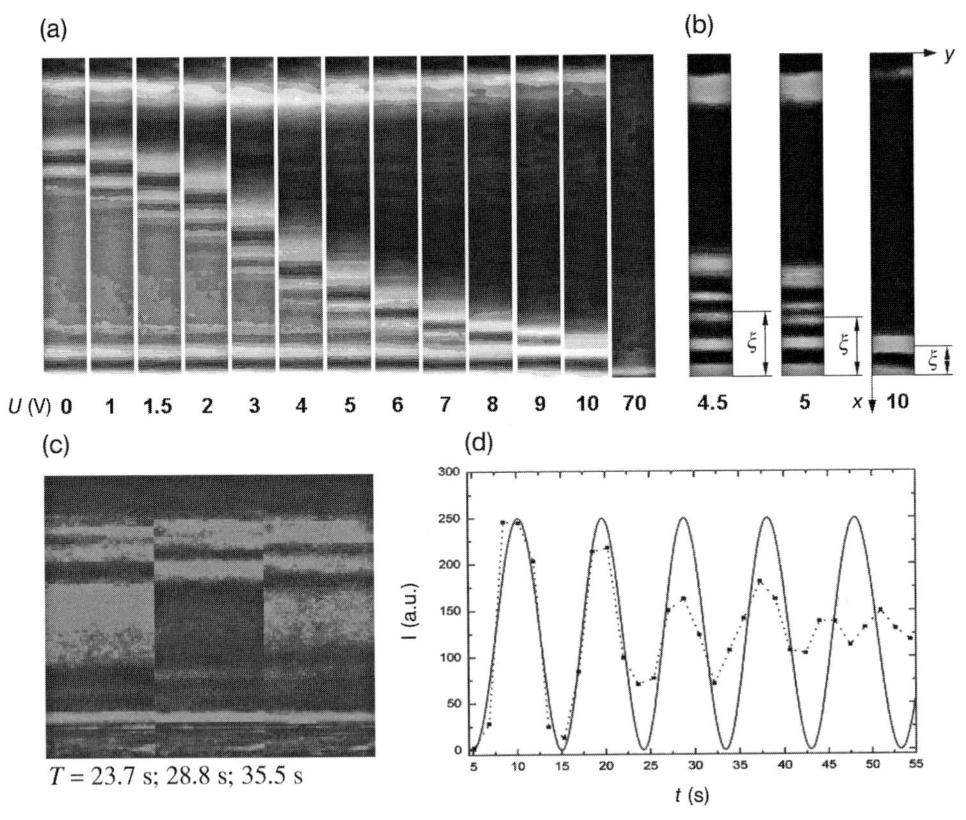

Figure 5.15 Microscopic images of the gap in Z-direction; (a) and (b) 5CB at asymmetric boundary conditions $\varphi_0 = 25^0$, $b = 62\,\mu m$, $d = 270\,\mu m$ and different voltages; (a) microscopic images in natural light, (b) results of image processing using red (R) component; (c) and (d) ZhK616 at symmetric boundary conditions $\varphi_0 = 25^0$, $b = 130\,\mu m$, $d = 1$ mm; (c) images are taken at different moments after turning the electric field off, $U = 400$ V; (d) time dependence of light intensity after turning the electric field off; points: results of image processing (in the center of the layer), line: theoretical curve.

where I_0 is the input light intensity and δ is the phase delay between the extraordinary ray and the ordinary one. As the relaxation process corresponds to very small values of φ, linearized equations analogous to (5.37)–(5.39) can be easily derived by using substitutions: $d \rightarrow b$, $\mu_0\, \Delta\chi H^2 \rightarrow \varepsilon_0\, \Delta\varepsilon E^2$.

For the geometry shown in Figure 5.13a, it results in the next expression for the time dependence of the light intensity in the middle ($x = 0$) of the channel:

$$I = I_0 \sin^2\{[\delta(x,t)]/2\} \tag{5.41}$$

with

$$\delta(x,t) = (2\pi\, \Delta n/\lambda)[\varphi(x,t)]^2. \tag{5.42}$$

The final stage of the light intensity changes is determined by the relaxation of the slowest harmonic, which results in the next expression for the angle $\varphi(x, t)$:

$$\varphi(x,t) = \varphi_0 - \varphi(0)\exp(-t/\tau_0)\cos(\pi x/b), \tag{5.43}$$

where

$$\tau_0 = \lambda_1 b^2/(K_{22}\pi^2) \tag{5.44}$$

and $\varphi(0)$ is the amplitude of the first harmonic in the Fourier transformation of the initial state of the layer $\varphi(x, t=0)$ before turning the electric field off. The maximal value of the latter function in the center of the layer ($x=0$) in the presence of electric field is expressed as

$$\varphi(x=0) = \varphi_E[1-1/\mathrm{ch}(b/2\xi)], \tag{5.45}$$

where ξ is electric coherence length (5.21).

A comparison between the experimental dependence of the light intensity on time obtained by processing digital images and the theoretical one is shown in Figure 5.15d). Positions of the interference extremes are well described by simple expressions presented above. So, such measurements can be used for determining the rotational viscosity coefficient γ_1. In particular, the value of this parameter for ZhK616 was found to be 0.23 Pa s.

Contrary to the case of pure twist deformation, described above, backflow effects determine the integral dynamic response of LC structure on step-like application of electric (magnetic) fields in general. It was first found and described by Pieranski *et al.* [5]. Detailed theoretical treatment and analysis of the possible errors induced by backflow effect in measured values of effective rotational viscosity can be found in Refs [40, 38]. The control of a birefringence sensitive to variations of polar angle θ provides an optical view of the director reorientation, and determination of the effective rotational viscosity γ_1^* is possible. The minimal difference (about 10%) between effective values γ_1^* of the rotational viscosity coefficient and the real ones corresponds to a small inclination of a director from the initial planar state. In the case of homeotropic initial orientation, this difference is essentially higher and a fitting procedure includes variation of a number of Leslie coefficients, which will be considered below. A complete numerical study of a two-dimensional nematic backflow problem [55] has shown that backflow effects depend critically on the geometry of the cell. It is important for the operation of LC displays. This question will be considered in Chapter 6.

5.2.3
Rotational Viscosity of Smectic C Phase

In general, the dynamic behavior of smectic C and smectic C* phases is quite complicated compared to that of the nematic phase. In accordance with low symmetry of these phases, hydrodynamic equations include a lot of elastic and viscous-like parameters responsible for the collective rotation of long molecular axes described by a director (**n**), the deformation of smectic layers (vector **a**, normal to the layer plane), the translation motion of a liquid (velocity **v**) and interactions between such types of motion (see, for example, Ref. [40]). Contrary to the case of nematics, the entire set of viscoelastic parameters is not available yet. Fortunately, in most cases, essential simplifications important for applications are possible.

First, it is reasonable to consider the orientational motion of an S_c director induced by fields or surfaces relative to the fixed layer structure. It is approved by a large difference in energies needed to distort the layer structure described by **a** vector and the mean molecular orientation (**n**). Such situation is excluded for smectic A phase, where **n** coincides with **a** and orientational changes are strongly connected with layer distortions. It means that relatively weak magnetic or electric fields are capable of changing the orientation that is similar to the case of nematics. The main difference is that such orientational motion is restricted in space by the condition

$$|\vec{\Psi}| = |\vec{n} \times \vec{a}| = \sin\theta = \text{const}, \tag{5.46}$$

which means that a director can move only along the conic surface with an axis of a cone defined by **a**. The value Ψ can be considered a vector order parameter of the smectic C phase (it is equal to zero in smectic A phase). The modulus of this vector coincides with the tilt angle θ of smectic C with respect to the layer normal at small values of θ. Such restriction of the director motion makes the origin of the orientational bistability to be controlled by electric (magnetic) field. This property is of primary interest for the practical application of smectic C. It was found both in thin layers and in bulk samples of smectic C, in particular via ultrasonic methods described above [56].

One can imagine the ideal layer structure as that shown in Figure 5.16a with $\mathbf{a}(\mathbf{r}) = \mathbf{a}_0 = \text{const}$.

Experimentally, such situation can be realized when cooling a bulk nematic sample oriented by strong magnetic field through nematic–smectic A and smectic A–smectic C phase transitions. Such method is usually used in NMR or ultrasonic studies [7, 56] of liquid crystals. In this case, a method of rotating magnetic field (if field rotates in the plane of layers) can be applied. Indeed, for this geometry, smectic C behaves like a two-dimensional nematic with a rotational viscosity coefficient γ_φ describing the phase delay of a director rotating along the smectic C cone with respect to the magnetic field.

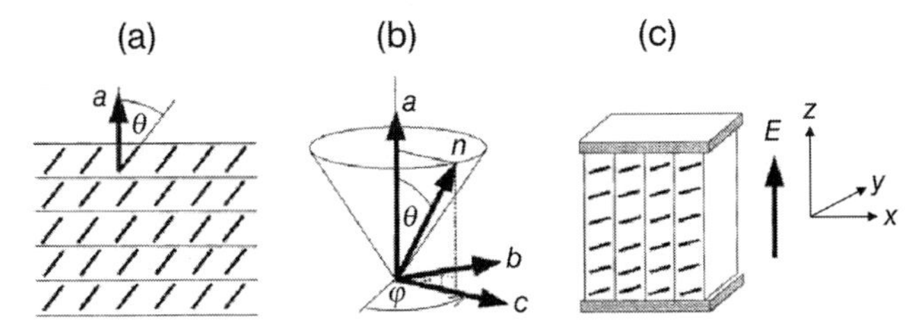

Figure 5.16 Possible regular layer structures of smectics C: (a) "planar homeotropic alignment," (b) description of the rotation of director **n** around the smectic cone. The **c** director is a projection of **n** onto the smectic planes and is described by the azimuth angle φ. The vector $\boldsymbol{b} = [\boldsymbol{a} \times \boldsymbol{c}]$ is introduced for convenience; (c) the "bookshelf" geometry. (After Stewart [40].)

Another possibility is to use a definite layer structure stabilized by surfaces. There are two possible structures of such type shown in Figure 5.16.

For the first structure (Figure 5.16a), smectic layers are arranged everywhere parallel to the plates. Practical realization of such "planar homeotropic alignment" geometry has been discussed by Beresnev *et al.* [57]. Obviously, it is the same case as considered above for bulk samples of smectic C.

In the "bookshelf" geometry shown in Figure 5.16c, smectic C layers are normally oriented at a normal direction relative to the plates. This geometry is commonly used for practical applications as for the realization of electrically controlled bistability. Recently, it was shown that backflow phenomenon (as in the case of nematics) can arise after the electric field is turned off in the "bookshelf" geometry [58]. It results in decrease in effective rotational viscosity coefficient. Usually, such possibility is omitted at the analysis of **c** director motion under the action of strong fields.

The reader can find only a very few experimental data (see, for example, Ref. [59]) on rotational viscosity coefficients in nonchiral smectic C. Some particular techniques, such as quasielastic light scattering [59], pirroeffect technique [60], NMR [7], are applicable for both nematic and smectic C liquid crystals. This concerns bulk samples too where NMR method can be applied. Extracted values of a rotational viscosity coefficient γ_φ turned out to be of the same order as in nematics.

At the same time, experimental determination of the rotational viscosity coefficients of chiral smectic C phase (ferroelectric smectics) has been of great interest for the past three decades for prospective applications of these materials. Ferroelectric smectic C (FLC) are known as the fastest liquid crystals controlled by low voltage. That is why they are of potential use not only for displays but also for a number of optical devices (see Chapter 6).

It is well known that switching times in electrooptical effects in the absence of flow FLCs are defined by rotational viscosities γ_φ *and* γ_θ that characterize the energy dissipation in the director reorientation process described by azimuthal (φ) and polar (θ) angles. In some FLC materials characterization, this viscosity is called the **n** and **c** director viscosity where **n** is a real FLC director and **c** director is its projection onto the plane perpendicular to the FLC layers normal (Figure 5.16). The simplest dynamic equations for *homogeneous* FLC samples take the form [61–63]

$$\gamma_\theta d\theta/dt + A\theta = 0, \quad \tau_\theta = \gamma_\theta/A, \tag{5.47}$$

$$\gamma_\varphi d\varphi/dt - P_S E \sin\varphi = 0, \quad \tau_\varphi = \gamma_\varphi/P_S E, \tag{5.48}$$

where $A = 2a(T_c - T)$ and $a > 0$ is an effective elastic coefficient for the tilt, τ_θ and τ_φ are characteristic response times for the θ and φ angles, respectively, of the FLC director, and P_s is polarization. The viscosity coefficient γ_φ can be rewritten as

$$\gamma_\varphi = \gamma_\varphi^0 \sin^2\theta, \tag{5.49}$$

where γ_φ^0 is independent of the angle θ. According to (5.49), $\gamma_\varphi \Rightarrow 0$ for $\theta \Rightarrow 0$, that is, γ_φ is very low for small tilt angles θ.

Far from the phase transition point T_{AC}, it is reasonable to consider only the azimuthal director angle φ because the angle θ is frozen. (Variations of the angle θ result in a density change that is energetically unfavorable.) However, near T_{AC} we can change the θ angle, for example, by applying electric field E; this effect is known as electroclinic effect [63].

The rigorous hydrodynamic theory of FLC is similar in most respects to the hydrodynamics of nonchiral smectic C. In particular, the total expressions for viscous stress including 20 viscosity-like parameters are identical for both SmC and SmC* liquid crystals [40]. As a result, unique methods (such as those based on rotating magnetic fields or quasielastic light scattering) are applicable for determining a rotational viscosity coefficient γ_φ. Nevertheless, the existence of a ferroelectric polarization provided an elaboration of a number of specific methods such as the electrooptic method [64, 65], the measurement of the polarization reversal current [66–68], and the dielectric method [69].

Clark and Lagerwall [64] were the first to establish an extremely fast dynamics for the optical study of two surface-stabilized FLC monodomains of opposite ferroelectric polarization separated by a domain wall. They used short electric pulses of opposite polarity to induce a sharp change in the bistable structure of FLC registered by optical responses in crossed polarizers. The corresponding geometry of the experiment for a "bookshelf" layer structure is shown in Figure 5.17a. This method is widely used in modern studies of FLC (see, for example, Refs [70–74]). In all cases, surface-stabilized ferroelectric liquid crystals (SSFLCs) were investigated. Experiments were conducted both for "bookshelf" and chevron structures (Figure 5.17a and b).

They differ from typical geometries for nonchiral smectic C by the polarization vector **P** defined as [40]

$$\vec{P} = P_0\vec{b} \quad \text{or} \quad \vec{P} = -P_0\vec{b}. \tag{5.50}$$

Rigorous dynamic equations for FLC can be found in Ref. [40]. In particular, the governing dynamic equation for a director, moving along the cone under the action of strong electric field **E** in the "bookshelf" geometry (Figure 5.17a), is written as

$$2\lambda_5\frac{\partial\varphi}{\partial t} = (B_1\sin^2\varphi + B_2\cos^2\varphi)\frac{\partial^2\varphi}{\partial z^2} + (B_1 - B_2)\left(\frac{\partial\varphi}{\partial z}\right)^2\sin\varphi\cos\varphi$$
$$+ P_0E\sin\varphi + \varepsilon_0\,\Delta\varepsilon E^2\sin\varphi\cos\varphi, \tag{5.51}$$

where λ_5 is a dissipative parameter referred to as the director rotation, B_1 and B_2 are elastic constants analogous to Frank elastic moduli. It is simple to show that in one constant approximation ($B_1 = B_2 = K_\varphi^0\sin^2\theta$), expression (5.51) coincides with the equation for C – a director motion derived earlier from simplified assumptions and usually used for the problem under consideration (see, for example, Refs [61, 69]):

$$\gamma_\varphi^0\sin^2\theta\frac{\partial\varphi}{\partial t} = K_\varphi^0\sin^2\theta\frac{\partial^2\varphi}{\partial z^2} + P_0E\sin\varphi + \varepsilon_0\,\Delta\varepsilon E^2\sin\varphi\cos\varphi \tag{5.52}$$

with

$$\gamma_\varphi^0 = \frac{2\lambda_5}{\sin^2\theta} = \frac{\gamma_\varphi}{\sin^2\theta}. \tag{5.53}$$

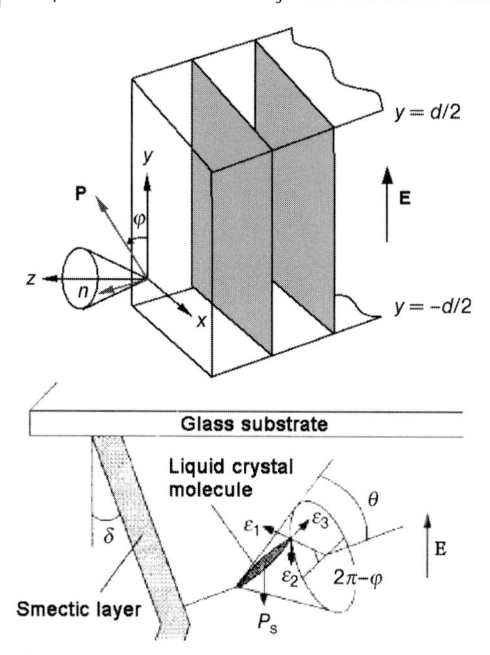

Figure 5.17 Geometries of experiments with a bookshelf (a) [72, 73] and chevron structures (b) [71].

As it was pointed out by Stewart [40], in the case of spatial dependence of the director influential due to elastic effects, the equation for a director motion should incorporate flow. It results in backflow effects and renormalization of the effective rotational viscosity that is similar to analogous effects in nematics and nonchiral smectic C [59].

Strictly speaking, expression (5.52) becomes correct only for a space-independent director rotation ($\varphi = \varphi(t)$). At low fields, a ferroelectric term (P_0) dominates and expression (5.52) will be applicable for a director motion everywhere excluding near-surface layers of thickness ξ defined as [40, 70]

$$\xi = \sqrt{\frac{B}{P_0 E}}. \tag{5.54}$$

For the optical detection of a director motion as uniform reorientation of the sample, this thickness has to be essentially smaller than the light wavelength λ [40]. It holds very well at typical values $B = 5 \times 10^{-12}$ N, and $P_0 = 100\,\mu\text{C/m}^2$, $\lambda = 0.4$–$0.7\,\mu$m for common values of $E = 1$–10 V/μm used in optical studies of smectic C* dynamics. In this case, the solution of (5.52) becomes rather simple [40]:

$$\varphi(t) = 2\tan^{-1}[\tan(\varphi_0/2)e^{t/\tau}], \tag{5.55}$$

where φ_0 is the initial angle and the relaxation time τ is defined by (5.48).

It is important to note that at moderate fields (1–10 V/μm), this time is essentially "faster" than the relaxation time τ_d in a dielectric regime ($P_0 = 0$):

$$\tau_d = \frac{\gamma_\varphi^0}{\varepsilon_0\,\Delta\varepsilon E^2}. \tag{5.56}$$

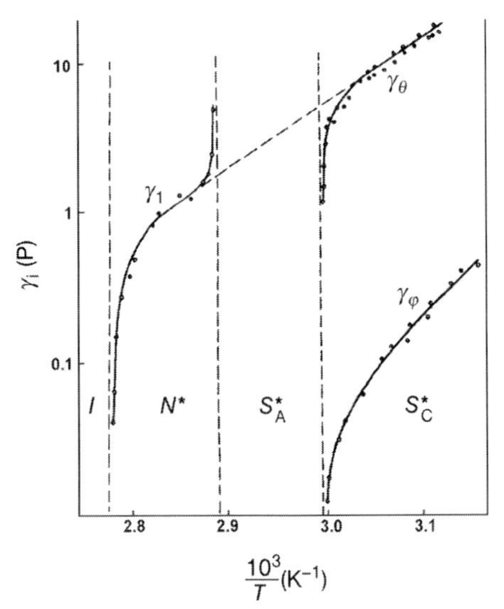

Figure 5.18 The temperature dependences of the rotational viscosity coefficients γ_1 in N* phase, γ_θ and γ_φ in smectic C* phase. (After Pozhidaev *et al.* [61].)

It is explained by the linear dependence of τ on the field strength E contrary to the quadratic dependence in dielectric regime (see expression (5.56)).

This property together with a high spontaneous polarization makes switching in SmC devices unique – they can switch equally fast in both directions as the field is reversed.

The comparison of different rotational viscosities in smectic C and nematic phase was first made in Ref. [61]. Experimental data obtained via optical and pirroelectric techniques are presented in Figure 5.18.

One can see that values of γ_θ rotational viscosity exceed the values γ_φ approximately by an order at comparable temperatures. At the same time, the latter parameter far from $S_A^*-S_C^*$ phase transition is close to the rotational viscosity coefficient γ_1 in a nematic phase.

Analogous electrooptical study was performed by Sako *et al.* [71] for more complicated chevron structure shown in Figure 5.17b. In an experimental part of this work, a pulse signal of amplitude E and duration t was applied to the cell, and the response time τ was defined as the shortest pulse duration for the director to switch. For the switching performance, the azimuthal angle of the director had to be changed at least from φ_0 to $\pi/2$ by applying electric field. In SSFLC cells studied in this work, the value of φ_0 was close to $\pi/2$, which made it possible to search the solution as a series in φ in some near neighborhood of $\pi/2$. By neglecting the second and higher order terms in φ, the response time τ as a function of E was calculated as

$$\tau = \frac{1}{kE^2+l}\log\left\{1+\left(\frac{\pi}{2}-\varphi_0\right)\frac{kE^2+l}{mE+n}\right\}, \tag{5.57}$$

where

$$k = \frac{\varepsilon_0}{\gamma_\varphi}\left[(\Delta\varepsilon \sin^2\theta - \partial\varepsilon)\cos^2\delta - \frac{\Delta\varepsilon}{4}\sin 2\theta \sin 2\delta\right] \tag{5.58}$$

$$l = 2n \cot 2\varphi_0 = -\frac{W}{\gamma_\varphi d}\cos 2\varphi_0 \tag{5.59}$$

$$m = \frac{P_z \cos\delta}{\gamma_\varphi} \tag{5.60}$$

In these expressions, two dielectric anisotropies $\Delta\varepsilon = \varepsilon_3 - \varepsilon_1$ and $\partial\varepsilon = \varepsilon_2 - \varepsilon_1$ are defined by three main values of the dielectric permittivity tensor (ε_3 is the permittivity parallel to the director, see Figure 5.17b). Finite value of the surface anchoring strength W is accounted via parameter l(5.59) that also depends on the layer thickness d. The spontaneous polarization P_s, which originates from the existence of molecular chirality, is in the direction of ε_2.

It is simple to show that at relatively low field strength and a strong surface anchoring, expression (5.57) becomes similar to the simplified analogue (5.48). The latter was used to estimate the rotational viscosity γ_φ from the experimental dependence of the electrooptical response of a "bookshelf structure" as follows [61, 65]:

$$\gamma_\varphi = P_S E \tau_\varphi, \qquad \tau_\varphi = (t_{90} - t_{50})/\ln\sqrt{5},$$

where t_{90} and t_{50} are the corresponding times for 90 and 50% transmission from the maximum level of light transmittance.

In the experiments, the pulse voltage composed of alternate positive and negative pulses at regular intervals of 10 ms was applied to the cell (thickness $d = 1.4\,\mu$m) filled with the mixture SCE8 ($\Delta\varepsilon < 0$) supplied by Merck Ltd. It was possible to measure the response time τ by changing the pulse duration. The dependence $\tau(E)$ is shown in Figure 5.19.

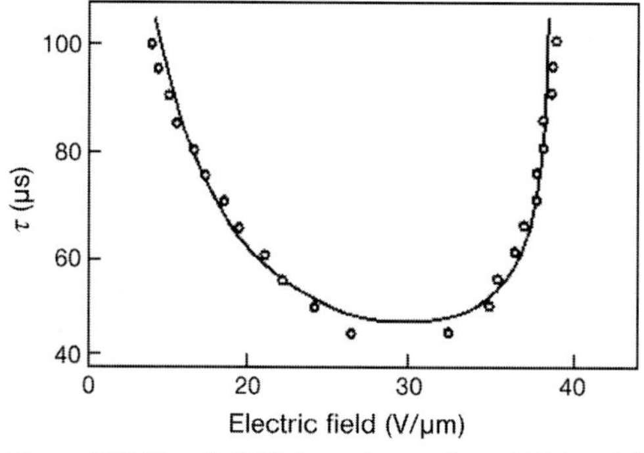

Figure 5.19 Electric field dependence of τ at 25 °C. Solid line: approximation in accordance with (5.57). (After Sako *et al.* [71].)

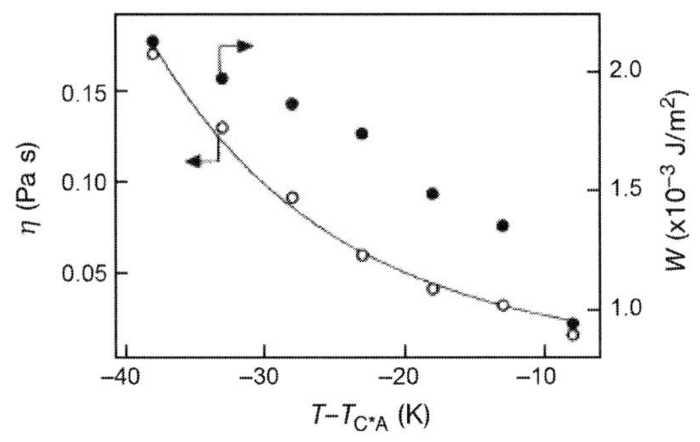

Figure 5.20 Temperature dependences of the effective rotational viscosity $\eta = \gamma_\varphi$ and anchoring strength W. (After Sako [71].)

It was fitted by (5.57) to define parameters of the problem. So, both the rotational viscosity and the anchoring strength could be extracted as functions of temperature (Figure 5.20).

Besides the study of dynamic optical response, electrical measurements are also possible to provide information about rotational viscosity of FLC. Dahl *et al.* [68] proposed to use measurements of polarization reversal current to yield the rotational viscosity as well as the spontaneous polarization and the effective elastic constant K_φ entering into the governing dynamic equation (5.52). The electric current I due to the polarization reversal is equal to the time derivative of the polarization charge times the electrode area A and can be connected with the dynamics of C director via the next expression:

$$I = A\frac{\mathrm{d}(P\cos\varphi)}{\mathrm{d}t} = -A\sin\varphi\frac{\mathrm{d}\varphi}{\mathrm{d}t}. \tag{5.61}$$

Authors considered the dynamic response of a "bookshelf" sample (Figure 5.17a) on the applied square wave (or sinusoidal) voltage. Time dependences of the current were calculated and compared with the experimental ones. In particular, the width of the observed hysteresis loop was derived in terms of physical parameters under determination. The proposed model was applied in evaluating data from the measurements on the substance MBRA-8 [2-hydroxy-4-(2-methyl-butyloxy)-benzilidene-4-*p*-*n*-octyl-aniline]. The reported value of effective rotational viscosity $\gamma_\varphi = 0.0166$ (N s/m^2) at $t = 33\,^\circ$C was determined with errors of about 10%.

The described method is widely used in modern research of FLCs as it allows to determine several parameters of the liquid crystal in a single cell via routine electrical measurements. A simple scheme for such type of measurements is shown in Figure 5.21.

A square wave voltage is applied to the LC cell. It provides more correct measurements compared to those made by using a triangular wave [77]. The current I flowing through the cell is integrated by capacitor resulting in time changes of output voltage U_{out}. The information on a rotational viscosity coefficient and other parameters was extracted by processing time dependence of output voltage. An example of

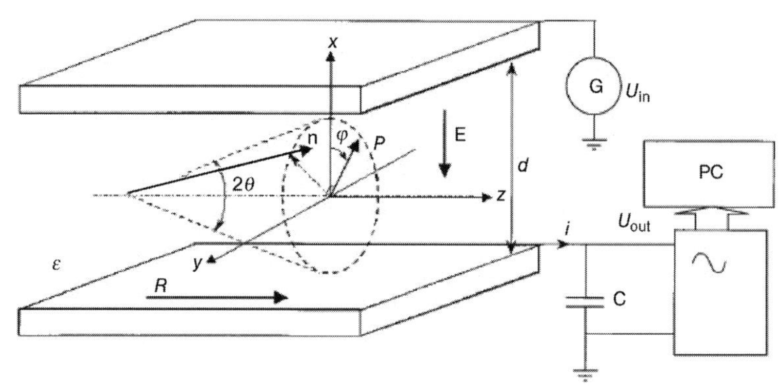

Figure 5.21 FLC cell in the "bookshelf" geometry coordinate system and schematic diagram of the setup. **R** – rubbing direction, **n** – molecular director, **E** – electric field vector, **P** – polarization vector, φ – azimuthal angle, and θ – molecular tilt angle. The smectic layers are parallel to the *XY* plane. (After Panov *et al.* [76].)

temperature dependences of the effective rotational viscosity and electric conductivity obtained by this technique is shown in Figure 5.22.

The LC used was a two-component commercial mixture, SCE8, supplied by Merck Ltd with the next transition temperatures: SmC*–59 °C–SmA–79 °C–N*–100 °C–I. This mixture has a relatively small spontaneous polarization P_s (about 60 μC/m^2 at 30 °C) and a large birefringence Δn (~0.16). All measurements were performed in LC cell with an area A of the electrodes equal to 29.3 mm^2 and the cell thickness

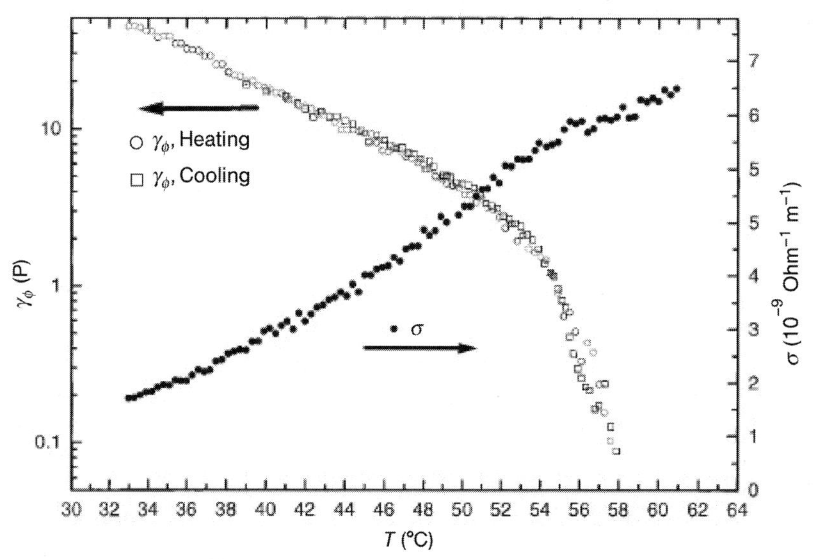

Figure 5.22 Temperature dependences of the rotational viscosity γ_φ and electric conductivity σ. (After Panov *et al.* [76].)

$d = 11\,\mu$m. It is worthwhile to note the abnormally high values of the effective rotational viscosity obtained far from the smectic A–smectic C* phase transition.

Another type of electric measurements of the rotational viscosity coefficient, first proposed by Gouda *et al.* [69], is based on the study of the spectra of dielectric relaxation. Dielectric properties of FLCs strongly depend on the temperature (especially near phase transition points) and frequency of the field. The dielectric susceptibility of an FLC could be defined as

$$\chi = (\varepsilon - 1)/4\pi = \lim_{E \to 0} \frac{\langle P \rangle}{E}. \tag{5.62}$$

Two modes contribute to the value of the averaged value of polarization $\langle P \rangle$. The first of them, *soft mode*, is induced due to the amplitude change of the polarization, that is, variation of the tilt angle θ. This mode is the most important near the phase transition point of the FLC phase and results in the electroclinic effect [78]. The second mode, called the *Goldstone mode*, is responsible for the variation of the phase of polarization, that is, the azimuthal director angle φ.

The dielectric response of the smectic A near the phase transition into smectic C* phase is approximately one order of magnitude weaker than the corresponding response of the ferroelectric phase. The only contribution to the smectic A response is made by the soft mode with characteristic relaxation frequency:

$$f_s^{(A)} \propto \frac{1}{\tau_\theta^A} \propto a\,(T - T_{CA}), \quad \text{for } T > T_{CA}, \tag{5.63}$$

where $a \approx 10$–15 kHz/K. In the smectic C* phase, the soft mode effect on polarization is fairly small and sharply decreases with decreasing temperature. The characteristic relaxation frequency obeys the well-known law for the second-order phase transitions [62]:

$$f_s^{(C)} \propto \frac{1}{\tau_\theta^C} \propto 2a\,(T - T_{CA}), \quad \text{for } T < T_{CA}. \tag{5.64a}$$

At the A ⇔ C phase transition point, the characteristic frequency of the dielectric relaxation does not tend to be zero and is defined by the Goldstone mode [62]. The characteristic frequency of the Goldstone mode

$$f_s^{(C^*)} \propto \frac{1}{\tau_\varphi} \propto \frac{K_{22} q_0^2}{\gamma_\varphi}, \quad \text{for } T < T_{CA}, \tag{5.64b}$$

where τ_φ is the response time of the FLC director azimuthal distortion φ, K_{22} is the elastic coefficient related to the deformation of the helix, and $q_0 = 2\pi/R_0$ is the helix wave vector. Typical values for the Goldstone mode relaxation frequency are 100 Hz–1 kHz [63].

This frequency can be determined from the measurements of the real (ε') and imaginary (ε'') parts of complex dielectric constants that are usually presented as a Cole–Cole diagram (Figure 5.23), and the temperature dependence of rotational viscosity coefficient γ_φ^0 calculated from such data is shown in Figure 5.24.

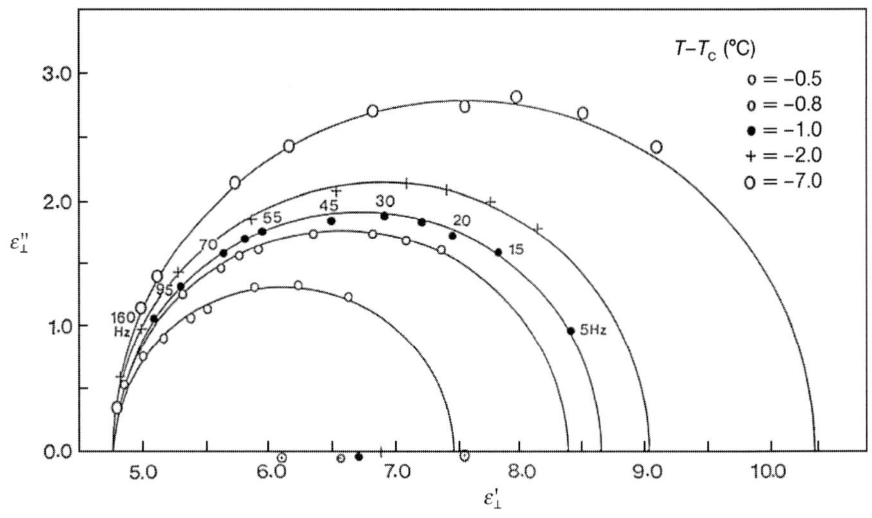

Figure 5.23 The Cole–Cole diagram for FLC 4-(2-methylbutyloxy) phenyl-4-*n*-decyloxybenzoate; $T_c = T_{CA} = 48\,^\circ$C. (After Gouda *et al.* [69].)

One can see that γ_φ^0 viscosity does not show the critical behavior and follows the Arrhenius law everywhere excluding in the vicinity of the phase transition. Such behavior is similar to that of the rotational viscosity in a nematic phase. Moreover, values of γ_φ^0 are of the same order as those of γ_1 in nematics.

The method described above is very useful due to well-elaborated technique for dielectric spectroscopy. In modern investigations of FLCs, it is often applied in combination with an electrooptical study to obtain reliable information about parameters of these materials [72, 74]. In particular, Piecek *et al.* [72] have studied a number of FLC mixtures and found that in some cases dielectric and optic methods give different results (see Figure 5.25).

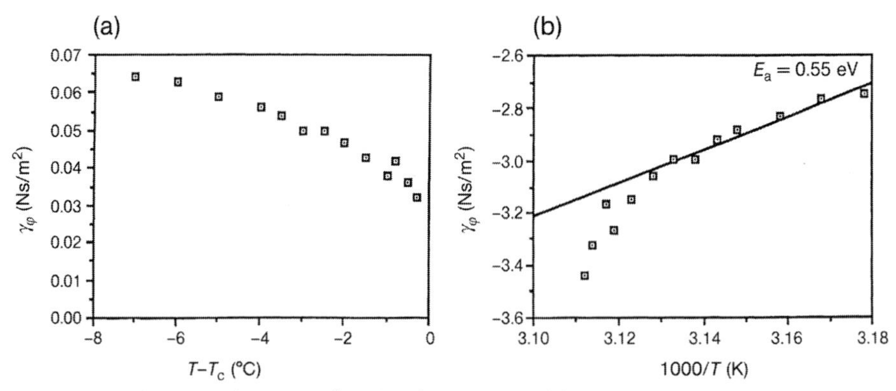

Figure 5.24 Rotational viscosity γ_φ^0 in the C* phase (a) and the corresponding Arrhenius plot (b). (After Gouda *et al.* [69].)

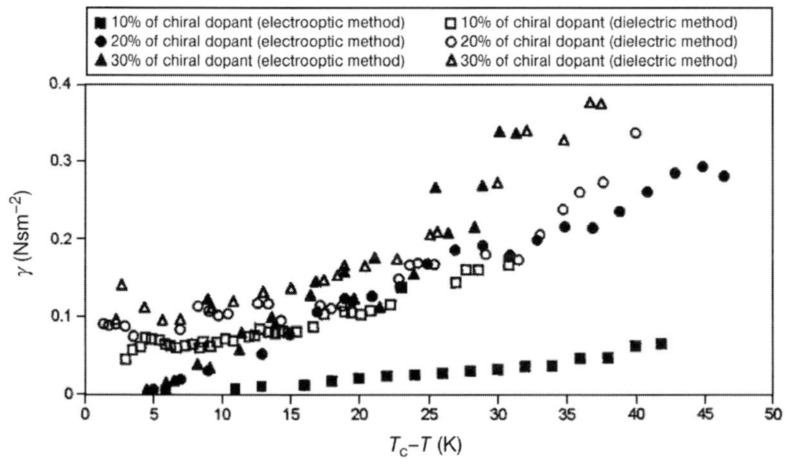

Figure 5.25 Rotational viscosity γ_φ in the smectic C* phase of some mixtures obtained from electrooptic and dielectric methods of measurement. (After Piecek *et al.* [72].)

They tried to explain this fact by essentially different amplitudes of motion of a director at electrooptic and dielectric studies. A similar difference was obtained in the recent work of Hemine *et al.* [74] where both methods were used. Data on two pure compounds of homologous biphenyl benzoate series showing a wide temperature diapason for the smectic C* phase were reported. The author found that the Goldstone viscosity obtained from the dielectric measurements is higher than that determined by the electrooptical method. The explanation of this fact was essentially the same as in Ref. [72]. Namely, the authors connected the difference mentioned above with the magnitude of the molecular movements: the higher the amplitudes of molecular rotations, the smaller the rotational viscosity is. It seems not to be very reasonable because the rotational viscosity was introduced in nonlinear hydrodynamics as a parameter independent of amplitude of a rotational motion. So, some methodical errors are very much possible in these cases. Alternative explanation of this difference was proposed in work [75]. The authors proposed that disagreement between results obtained by different experimental techniques could be referred to as the difference in the motion of the rigid central core of the molecules and the terminal aliphatic chains. It seems to be rather probable, so some corrections in theoretical backgrounds of described methods may have to be done.

We can summarize the description of experimental methods for rotational viscosity measurements in liquid crystals:

1. There are a number of well-elaborated methods for measuring rotational viscosity coefficients both in nematic and ferroelectric smectic C phases.

2. In nematic phase, both bulk samples and thin layers were investigated, while in case of FLCs, the measurements were done mostly in thin layers.

3. Different experimental techniques can be used to measure the rotational viscosity coefficient in nematics depending on the desirable accuracy. Obviously, direct measurements of a mechanical moment induced in bulk samples by rotating magnetic fields provide the best precision (errors lower than 1%). Such accuracy is needed for basic research in liquid crystals, for example, for experimental checking of microscopic theories of liquid crystals. Nevertheless, the use of strong magnetic fields, finest mechanical assumption, and a large amount of liquid crystals restricts application of such experimental setups for routine laboratory measurements.

4. For most practical applications and computer simulations, measurements with errors 1–5% can be considered satisfactory. Such accuracy can be achieved using both bulk samples and thin layers of liquid crystals.

   In the case of bulk samples, different nonmechanical methods of registration of a director motion (NMR, EPR, and dielectric measurements), for example, are applicable. The use of ultrasonic method for this purpose is considered in Chapter 4. The accuracy of such measurements is mostly restricted by errors from diamagnetic anisotropy determination, as only the ratio $(\gamma_1/\Delta\chi)$ is obtained in this case.

   In the case of thin layers, the most reliable results were obtained from optical study of twist-like deformation induced by magnetic field. This method can be considered very attractive, as a small amount of LCs is needed. The disadvantage of such technique is connected with the use of magnetic fields and processing of conoscopic images.

5. It is very desirable for practical application: first, to replace magnetic fields by electric fields and, second, to analyze optical birefringence instead of conoscopic pictures. For the twist-like deformation, it can be achieved using the new experimental geometry described above.

6. The rotational viscosity coefficient can be determined by optical studies of splay, bend, or combined deformations. In this case, backflow effects have to be taken into account. The latter essentially depend on the type of the geometry and the degree of orientational distortions. It introduces additional errors in measured values of γ_1.

7. It is possible to yield a rotational viscosity coefficient by the photon correlation spectroscopy or by measurements of anisotropic shear viscosities of nematics. Such possibilities will be considered below (Sections 5.3.1 and 5.3.2).

8. Experimental methods for measuring of a rotational viscosity of ferroelectric smectic C are mostly based on specific phenomena (such as relaxation of a Goldstone mode) and essentially differ from those applied in nematics.

9. There are different types of experimental techniques such as optical (electrical) studies of dynamic response of a smectic C director to applied voltage or dielectric spectroscopy measurements sensitive to a Goldstone mode. Usually, a number of

related parameters (such as elastic modules, spontaneous polarization) are determined from experiments. The simultaneous use of different techniques makes results more reliable. Nevertheless, some disagreement between such results was found in several special cases.

There are a number of fine reviews of experimental data on the rotational viscosity coefficient in nematics summarized and analyzed both on phenomenological and on microscopic levels [37, 38, 79]. So, we will make only short comments concerning this topic.

From the practical point of view, there are three aspects of great interest:

1. **The temperature dependence of the rotational viscosity:** Characteristic temperature dependences of this parameter are shown in Figure 5.13.

   The first general feature for all nematics is the critical decrease in rotational viscosity when approaching the clearing point (γ_1 is equal to zero in an isotropic phase). The simplest dependence on the order parameter S comparable to the symmetry of nematic phase is $\gamma_1 \sim S^2$. At the same time, it was found that far from the clearing point, the temperature dependences for most nematics could be approximately described by the Arrhenius law:

   $$\gamma_1 \propto \exp\left(\frac{\Delta E}{RT}\right), \tag{5.65}$$

   where ΔE is the activation energy that varies in the range 30–50 kJ/mol for different liquid crystals. A number of particular dependences originating from macroscopic and microscopic approaches were proposed for the description of $\gamma_1(T)$ dependence. A detailed discussion of this question can be found in review [37, 38, 79]. Indeed, experimental confirmation of the given theoretical dependence is not simple. It is restricted by the narrow temperature range of a nematic phase (which is typical for individual compounds) and there is not enough high accuracy of measurements. Moreover, rotational viscosity in some compounds shows critical divergence upon approaching nematic–smectic phase transition. Some examples of dependences of the rotational viscosity coefficients on PVT-state parameters obtained via ultrasonic methods are presented in Chapter 4.

   Temperature dependences of the rotational viscosity coefficient γ_φ of smectic C* (corresponding to the motion of C director) to some extent are similar to those in nematics: this parameter is also proportional to the squared value of the order parameter. As the latter approaches zero in the vicinity of the transition point smectic A–smectic C*, values of γ_φ become very small compared to those of nematics. At the same time, the ratio rotational viscosity/order parameter squared is of the same order of magnitude in both phases and approximately follows the Arrhenius law. A detailed analysis of the temperature dependences of γ_φ and γ_θ rotational viscosities can be found in Ref. [61].

   The temperature dependence of the rotational viscosity is highly important as it is directly connected with operating times of LC devices (see Chapters 6 and 8).

2. **The connection of the rotational viscosity with the molecular structure of mesogens:** Inertial properties of LC displays and other devices can be improved via optimization of the chemical structure of mesogens. That is why it is important for chemicals to find links between molecular structure and rotational viscosity. A lot of information obtained for different homologous series is presented and analyzed in the review by Beljaev [79]. Here, we point out only some general features [37, 79]:
   (a) the existence of odd–even effects (which can be opposite for different chemical classes);
   (b) the increase in γ_1 with increasing length axis/short axis ratio;
   (c) a strong influence of polar substituents on the rotational viscosity due to possible aggregation and antiparallel ordering; and
   (d) complicated types of concentration dependences in mixtures of nematics – only in some cases [80] the rotational viscosity of the resulting nematic mixture is connected with the concentrations of components (x_1) and (x_2) by the simple expression:

$$\ln \gamma = x_1 \ln \gamma_1 + x_2 \ln \gamma_2. \tag{5.66}$$

We direct the reader to refer to the above-mentioned reviews for more information. It is worthwhile to point out a good possibility of optimizing the rotational viscosity coefficient of multicomponent LC mixtures showing smectic C* phase [81]. It is of key importance for fast liquid crystal devices based on ferroelectric smectic C. This question will be discussed in more detail in Chapter 6.

5.3
Viscosimetry of Liquid Crystals in Shear Flows

The rheological study of liquid crystals in shear flows of different types [82–105] provides the most reliable data on anisotropic viscosities important for applications. Moreover, precise measurements of such type can be used to determine the complete set of Leslie coefficients describing dynamic characteristics of any particular mode realized in the given device (as it will be shown in Chapter 6). Below, we will consider the most typical and reliable experimental setups to answer the question "why such technique is not widely used in laboratories?" We will also show principal ways of improving this technique.

The first systematic viscosity measurements of viscosity of nematic liquid crystals (see, for example, Ref. [82]) were performed with traditional viscosimeters used for rheological studies of isotropic liquids.

It was found that in general liquid crystals showed non-Newtonian behavior. In particular, the apparent shear viscosity depended on a number of factors such as the shear rate, the diameter of a capillary, and the initial boundary orientation [83]. The description of these results in the framework of modern nematodynamics can be found in Ref. [83]. It was shown [83] that at high enough shear rates, the apparent viscosity becomes independent of the shear for different types of flows and equal to the value η_0 corresponding to the flow-induced orientation of liquid crystals (at least

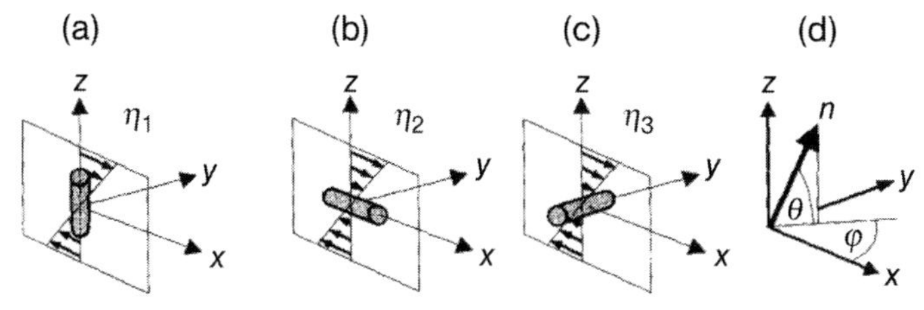

Figure 5.26 Three principal geometries for shear viscosity measurements.

for LC oriented by flow in one preferred direction close to the flow direction). So, such measurements provide some information on anisotropic shear viscosities of nematics and traditional viscometers are still in use (see review [38]).

Nevertheless, the most essential results were obtained after modification of older instruments and elaboration of new ones to provide viscosity measurements at different orientations of liquid crystals fixed by strong external fields. Miesowicz [84,85] was the first to introduce three basic geometries for viscosimetric measurements used until now (see Figure 5.26):

(i) Director parallel to the flow velocity gradient, $n||\nabla v$ (the maximal value η_1 of a shear viscosity coefficient).
(ii) Director parallel to the flow velocity, $n||v$ (the minimal value η_2 of a shear viscosity).
(iii) Director orthogonal to both the flow and the velocity gradient, $n||(v\times\nabla v)$ (the intermediate value η_3 of a shear viscosity).

It is worthwhile to note that alternative notations of the principal viscosity coefficients η_i can be found in the literature. In particular, the definitions of η_1 and η_2 are frequently interchanged and coincide with those proposed by Miesowicz [85]. In a general case of arbitrary orientation of director to v and ∇v defined by the angles θ and Φ, respectively, the expression for the shear viscosity coefficient can be written as

$$\eta(\theta,\varphi)=\eta_2\cos^2\theta+(\eta_1+\eta_{12}\cos^2\theta)\sin^2\theta\cos^2\varphi+\eta_3\sin^2\theta\sin^2\varphi. \qquad (5.67)$$

The additional parameter η_{12} entering into (5.67) cannot be considered as an independent viscosity coefficient as there is no restriction on its sign resulting from positive value of an entropy production. The maximal influence of this parameter on the viscous losses takes place at $\theta=\pi/4$, $\varphi=0$, which corresponds to the intermediate orientation of LC in the shear plane and the value of the shear viscosity coefficient expressed as

$$\eta\left(\frac{\pi}{4},0\right)=\eta_2+\frac{1}{2}\left(\eta_1+\frac{\eta_{12}}{2}\right). \qquad (5.68)$$

Expression (5.68) provides a simple way of calculating parameter η_{12} by measurements of the shear viscosity coefficients at three different orientations. Nevertheless, such procedure demands a very high precision of measurements as in typical nematics η_{12} is essentially less than the maximal shear viscosity coefficient η_1 [87].

The principal shear viscosity η_i and the parameter η_{12} can be easily connected with Leslie coefficients α_i via well-known expressions [23, 34]

$$\eta_1 = \frac{1}{2}(-\alpha_2 + \alpha_4 + \alpha_5) \tag{5.69}$$

$$\eta_2 = \frac{1}{2}(\alpha_2 + 2\alpha_3 + \alpha_4 + \alpha_5) \tag{5.70}$$

$$\eta_3 = \frac{1}{2}\alpha_4 \tag{5.71}$$

$$\eta_{12} = \frac{1}{2}\alpha_1 \tag{5.72}$$

For LCs aligned by intensive flow, an additional expression for the shear viscosity coefficient η_0 can be derived from (5.67)

$$\eta_0 = \eta(\theta_0, 0) = \eta_2 \cos^2\theta_0 + (\eta_1 + \eta_{12}\cos^2\theta_0)\sin^2\theta_0, \tag{5.73}$$

where the flow alignment angle θ_0 is expressed as

$$\cos 2\theta_0 = -\gamma_1/\gamma_2 = \frac{\alpha_2 - \alpha_3}{\alpha_2 + \alpha_3}. \tag{5.74}$$

In principle, expressions (5.69)–(5.74) can be used to determine all Leslie coefficients via anisotropic shear viscosity measurements. Nevertheless, for most of the studied liquid crystals, the flow alignment angle far from the clearing point is on the order of some degrees, which makes such calculations not reliable. So, independent methods (such as the optical one [88]) are often used to determine the flow alignment angle. In spite of the observations made above, measurements of anisotropic shear viscosities provide the most direct and universal way to get full information about Leslie coefficients describing dynamic behavior of nematics under fields and flows. In the following section, we consider in more detail the experimental technique used in such type of measurements.

5.3.1 Measurements of Anisotropic Shear Viscosities in Flows of Liquid Crystals Stabilized by Fields

The main idea of measurements is to make the flow and surface-induced orientational distortions to be negligible by using a strong magnetic field. In particular, such a field effectively suppresses possible hydrodynamic instabilities described in Chapter 3 and provides bulk monodomain samples of LCs excluding near-surface region defined by magnetic coherence length ξ_B(5.20).

In this case, LCs have to show Newtonian-like behavior with a constant viscosity.

5.3.1.1 Poiseuille Flow in Flat Capillary

The shear flow produced by a pressure gradient inside a capillary is known as a Poiseuille flow. Viscosimetric measurements in isotropic liquids are usually performed in capillaries with a circular cross section of a radius R and of a length L. For laminar flows usually used in viscometry, the instant volumetric flow of isotropic fluid $Q=(\mathrm{d}V/\mathrm{d}t)$ is determined by the well-known Poiseuille formula:

$$Q=\frac{\pi R^4}{8\eta}G, \qquad (5.75)$$

where

$$G=|\Delta P/L|=(P_1-P_2)/L, \quad P_1>P_2 \qquad (5.76)$$

is the value of a pressure gradient, L is the capillary length, $\Delta P=P_1-P_2$ is the pressure difference applied to the capillary, η is an effective shear viscosity. Taking into account the analogy between hydrodynamic flows and an electric current, the parameter

$$Y=\frac{\pi R^4}{8\eta L} \qquad (5.77)$$

can be considered as hydrodynamic conductivity of a capillary, and the expression

$$Q=Y\,\Delta P \qquad (5.78)$$

is applicable for capillaries with different shapes of a cross section. In particular, for a capillary with a rectangular cross section of large aspect ratio ($r=b/h$) shown in Figure 5.27,

$$Y=\frac{h^3 b}{12\eta L}. \qquad (5.79)$$

Although the shape of the cross section is not important for the study of isotropic liquids, it plays a key role in a viscometric measurement in liquid crystals. In particular, a capillary with a rectangular cross section provides well-defined orientation of LCs with respect to the directions of flow and flow gradients controlled by the magnetic field. So, three principal viscosities η_i and the parameter η_{12} defined above can be determined from measurements of the instant volumetric flow through a flat capillary.

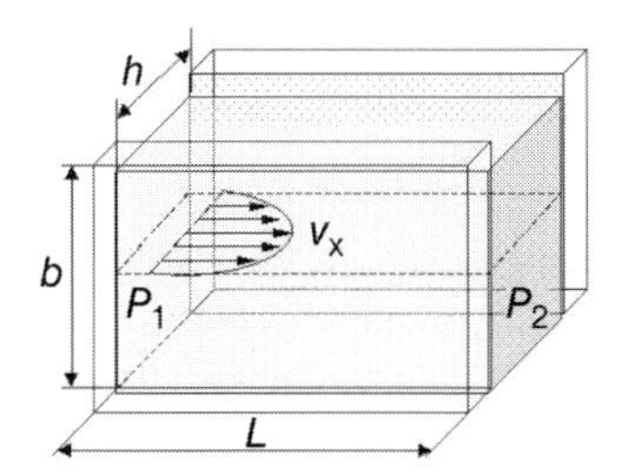

Figure 5.27 Poiseuille flow in flat capillary of a rectangular cross section; h and b are the gap and the width of a capillary.

Gähwiller [88] was the first to realize such measurements.

The rectangular capillary in his experiments was formed by two glass plates (typically 5 cm long and 4 mm wide with the thickness varying between 0.2 and 0.4 mm). It provided one-dimensional velocity distribution and the possibility not only for viscometric measurements but also for the determination of the flow alignment angle via an optical control of birefringence of LC layer. So, a complete set of Leslie coefficients could be calculated from the data on anisotropic viscosity coefficients in accordance with (5.69)–(5.74).

The flow inside a capillary was induced by an air pressure gradient (in the range 4–20 Pa/cm) applied to free surfaces of LC in two round tubes connected with capillary menisci to appear in two round tubes. The average flow speed was determined by timing the motion of one meniscus. Special care was taken to ensure that two menisci were at the same level in order to eliminate the effect of gravity. The pressure gradient was maintained constantly for a long time by controlling pressure in two buffer containers with needle valves. The author estimated the overall errors of measurements at about 7%, which originated mainly from the finite width of a capillary and friction forces induced by the motion of menisci in open tubes.

The method and construction proposed by Gähwiller seem to be rather attractive for applications, as they require small amount (less than 1 cm^3) of LCs and provide an optical control of orientation inside the channel. The data reported for two principal viscosity coefficients (η_2 and η_3) are in good agreement with those obtained via more precise measurements (see below). It does not hold for the maximal principal viscosity η_1, which shows a systematically lower (about 20%) value.

Detailed estimates of possible experimental errors arising from such type of experiments can be found in Refs [89,90] and in the review by Kneppe and Schneider [37].

Two main sources of errors related to specific properties of liquid crystal media were revealed.

The first one can be referred to as the influence of near surface layers with the orientation (and the effective viscosity) different from those provided by magnetic field in the central regions of a capillary. Estimates made for typical values of material parameters of nematics [37] have shown that errors in η_i imposed by this factor are of about 2% even for a very strong field ($B = 1$ T, $\xi_B \sim 3\,\mu$m) and a big thickness of the capillary ($d \sim 500\,\mu$m). Extrapolation of experimental dependences $\eta_i(B)$ from the region of extremely strong fields can be used to correct obtained values of viscosities. Another possibility, namely, the use of special surface treatment for internal surfaces of a capillary was considered in Ref. [90].

The second source of errors originates from orienting action of the flow that is most essential for the first geometry shown in Figure 5.26, so it can be responsible for the incorrect determination of η_1.

Estimates of influence of this factor on the measured value of η_1 under the experimental conditions reported by Gähwiller [88] ($B = 0.6$ T, shear rates u up to 20 s^{-1}) were first given by Summerford *et al.* [89]. They used the expression for the angle θ between a director and a flow velocity previously obtained by Helfrich [91] for a strong magnetic field H, stabilizing the geometry shown in Figure 5.26:

$$\tan\theta = (\Delta\chi H^2)/(-\alpha_2 u). \tag{5.80}$$

The maximal deviation $\Delta\theta$ of the orientation from that stabilized by field $\theta = 90^\circ$ was found to be about 14°. As it can be shown from Equation 5.67, such angle corresponds to a systematic error of about 10% that provides values of η_1 lower than the real one. It explains to a great extent a discrepancy in numerical values obtained by Gähwiller [88] and those reported in more recent papers [87, 89].

In addition, one should expect a strong influence of the surface alignment and of the unfavorable cross section of the rectangular capillaries in Gähwiller's experiments.

As it was shown by Kneppe and Schneider [37], such errors can be avoided by using small capillary thickness, small pressure gradients, and strong magnetic fields. For example, at $B = 1$ T and $h = 500\,\mu$m, the pressure gradient has to be lower than 10 Pa/cm to provide 1% of errors at η_1 measurements.

Some additional errors are common to measurements of isotropic liquids and liquid crystals. For example, the influence of the velocity gradient in the direction orthogonal to the flow plane can result in errors at measurements of η_3 coefficient that are about some percentage when aspect ratio b/h does not exceed 10 [37]. In summary, one can conclude that the precise measurements (with errors about 5% or smaller) of the anisotropic shear viscosity coefficients are possible only by a thorough account of possible sources of errors mentioned above and by optimization of experimental setups and measuring procedures.

At least, two experimental setups for precise measurements of such type are described in literature [90, 87].

In the first case [90], the experimental setup and the measuring procedure in many respects are similar to those proposed by Gähwiller. In particular, optically transparent rectangular capillary ($h = 205\,\mu$m, $b = 9$ mm, $L = 27.8$ mm) was used to provide measurements of both the shear viscosity coefficients and the flow alignment angle θ_0. In the latter case, the best results were obtained with a relatively thick ($500\,\mu$m) capillary. A rather powerful electromagnet with induction up to 1.1 T was used to achieve a perfect orientation of LC samples. The system providing pressure difference control needed to calculate shear viscosity is similar to that used in experiments of Kneppe and Schneider [87].

At the beginning of each experiment, an initial pressure difference ΔP_0 of about 100 Pa was created manually. After it, the pressure difference was slowly decreased with time due to the motion of the LCs inside a capillary. The obtained $\Delta P(t)$ dependence could be related to the instant volumetric flow, so only one control parameter was needed to calculate the shear viscosity coefficient. Measurements of the flow alignment angle θ_0 were performed in the same way as in Gähwiller's experiments [88]. For this purpose, a strong magnetic field was applied along the flow velocity. Turning the field off resulted in the rotation of an optical axis by angle θ_0 and in the changes $\Delta\delta$ in a phase lag δ between ordinary and extraordinary rays. The analysis of changes in the intensity of polarized light passed through the LC cell

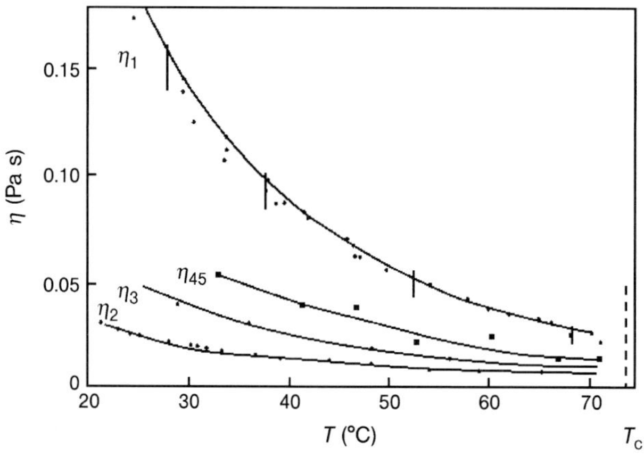

Figure 5.28 Shear viscosities of N4; solid line for η_1 is calculated in accordance with (5.82). (After Beens and de Jeu [90].)

allows, in accordance with (5.40), to determine the value of $\Delta\delta$. For small flow alignment angles, the next expression is valid:

$$\Delta\delta = -\frac{1}{2} n_e h \left[\left(\frac{n_e}{n_0} \right)^2 - 1 \right] \sin^2\theta_0. \tag{5.81}$$

It provides an estimation of the flow alignment angle θ_0. The temperature dependences of the shear viscosity coefficients for N4 are presented in Figure 5.28.

Numerical values of these viscosities together with the data on the flow-alignment angle and the rotational viscosity coefficient are provided in Table 5.1. The analysis of these data confirmed the expression

$$\eta_1 = \eta_2 + \gamma_1 \left(\frac{1 + tg^2\theta_0}{1 - tg^2\theta_0} \right) \tag{5.82}$$

derived from hydrodynamics of NLCs. In particular, the shear viscosity η_1 presented in Table 5.1 was calculated by Equation 5.28 and compared with the results of independent direct measurements (see Figure 5.28). The agreement can be considered as satisfactory (less than 5% difference). It means that the rotational viscosity coefficient γ_1 can be determined from the measurements of anisotropic shear viscosities. At small values of the flow alignment angle θ_0 that usually takes place far from transition point, the approximate expression

$$\gamma_1 \approx \eta_1 - \eta_2 \tag{5.83}$$

is useful for the estimation of the rotational viscosity of liquid crystals. One can easily check it by using data from Table 5.1.

Indicated errors of the measurements mentioned above are about some percentages. Partly, they can originate from additional sources, such as the underpressure caused by the surface tension of meniscus on both sides of a capillary, different wetting properties of surfaces, and so on [37, 87]. In the alternative variant of such setups useful for studying a small amount of LCs [92], similar precision was achieved only at the measurements of η_3 and η_0 shear viscosity coefficients. The magnetic field

Table 5.1 Viscous parameters of N4. (After Beens and de Jeu [90]).

T (°C)	η_2 (Pa s)	η_3 (Pa s)	η_{45} (Pa s)	θ_0	γ_1 (Pa s)	η_1 (Pa s)
26	0.024	0.047		4.2^0	0.152	0.178
30	0.021	0.040		4.3^0	0.122	0.144
34	0.017	0.034	0.051	4.5^0	0.099	0.117
38	0.015	0.028	0.044	4.7^0	0.081	0.097
42	0.013	0.025	0.039	5.1^0	0.067	0.081
46	0.012	0.021	0.034	5.5^0	0.056	0.069
50	0.011	0.018	0.030	6.1^0	0.046	0.058
54	0.010	0.015	0.027	6.8^0	0.038	0.049
58	0.009	0.013	0.023	7.7^0	0.032	0.042
62	0.008	0.012	0.019	8.8^0	0.026	0.035
66	0.007	0.011	0.015	10.4^0	0.022	0.031
70	0.007	0.010	0.013	13.2^0	0.017	0.026

induction (about 0.5 T) was obviously insufficient to minimize errors in η_1 for a capillary thickness equal to 300 μm due to arguments considered above.

The most precise measurements of anisotropic shear viscosities of nematic liquid crystals were performed by Kneppe and Schneider [87]. They used folded capillary of total length of about 1 m to achieve a rather high pressure difference at relatively big thickness of the channel that was needed to minimize possible errors mentioned above.

The capillary consists of 20 horizontal brass plates that are assembled together as shown schematically in Figure 5.29. The cross section of the capillary is 0.3 mm

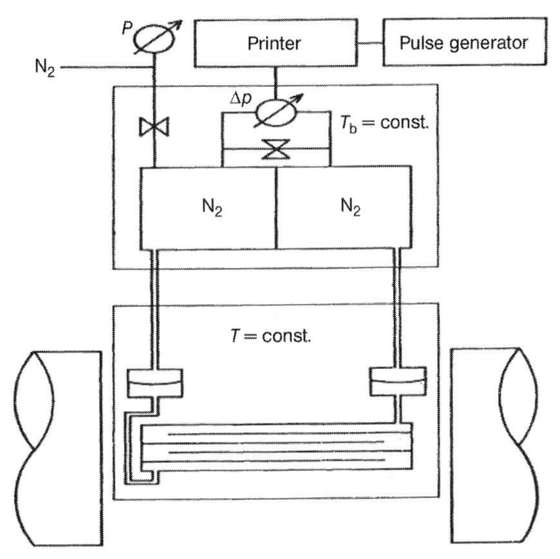

Figure 5.29 The schema of the experimental setup for anisotropic shear viscosity measurements. (After Kneppe and Schneider [87].)

16 mm and the total length amounts to 85.5 cm. The inlet and outlet of the capillary block are connected to two glass cylinders of 15 mm inner diameter for the storage of the liquid crystal. The capillary block and one half of the glass cylinders are filled with the degassed liquid crystals. Complete wetting is achieved by coating the inner surface of the glass cylinders with SnO_2. The glass cylinders are coupled by brass tubes to two buffer volumes of 85 cm^3. The remaining part of the glass cylinders, the brass tubes, and the buffer volumes are filled with dry nitrogen. As in experiments of Beens and de Jeu [90], only the time-dependent pressure difference $\Delta P(t)$ between two buffer volumes varying in the range 300–600 Pa was determined to calculate the effective shear viscosity coefficients. The alignment of the liquid crystals was achieved via magnetic field ($B' = 0$–1.1 T). The rotation of the magnetic field and utilization of two capillary blocks with different orientations allowed the determination of three principal shear viscosity coefficients η_1, η_2, η_3 and the shear viscosity coefficient $\eta(\pi/4, 0)$ needed for the calculation of η_{12} in accordance with (5.68). Experimental errors reported by the authors are very small (about 0.3%), which makes the precise determination of Leslie coefficients possible. The dependences of η_1, η_2, η_3, and η_{12} on temperature for MBBA are shown in Figure 5.30. These data are usually used for precise calculation of various dynamic effects taking place in liquid crystals. At the same time, a relatively big amount of LCs (about 50 cm^3) needed to fill a measuring cell restricts the use of such setup.

The authors found a complicated nonlinear characteristic for $\ln \eta_i(T^{-1})$ dependences that usually occurs in nematics. In principle, a decrease in LC amount can be achieved by using electric field instead of the magnetic one for the proper orientation of LC samples. Indeed, the ratio of the electric coherence length defined by expression (5.21) to the channel thickness is inversely proportional to the voltage

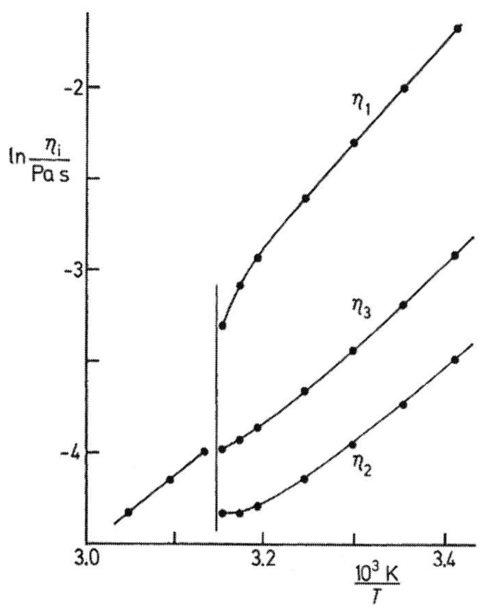

Figure 5.30 Viscosity coefficients η_i as a function of temperature. (After Kneppe and Schneider [87].)

and does not depend on the thickness at a fixed voltage. It means that the influence of boundary layers on the effective viscosity can be minimized even in the case of relatively thin (10–100 μm) layers of LCs with a positive value of dielectric permittivity anisotropy. Some preliminary results of such kind with the use of electric fields were obtained in Ref. [93]. At the same time, a well-defined orientation is hardly achieved for liquid crystals with negative sign of $\Delta\varepsilon$.

5.3.1.2 Direct Determination of Shear Viscosity Coefficients in a Simple Shear Flow

One can consider a simple shear flow as a more simple type of flow compared to a Poiseuille one at least for theoretical analysis (see Chapter 3). So, it seems to be reasonable to extract information about anisotropic shear viscosities from direct measurements of mechanical friction forces acting on the moving plates. Indeed, in pioneering experiments performed by Miesowicz [84, 85], a damping oscillating motion of a thin glass plate immersed into a liquid crystal was used to determine three principal shear viscosities defined by (5.68)–(5.70). Nevertheless, such types of viscometric measurements have been rarely used. To answer the question "why?", we will consider in detail the simplest variant of experimental setup based on direct measurements of a friction force at a steady simple shear flow [89].

5.3.1.2.1 Viscosity Measurements at a Steady Simple Shear Flow

The main features of the experimental setup are illustrated by Figure 5.31.

A steady shear flow in this experiment was realized in the gap between two surfaces of a thin copper vane (0.025 cm thick, 2.50 cm wide, and 5.33 cm long) and the inner walls of a rectangular container (2 cm × 5 cm × 15 cm) filled with LC (MBBA). With the vane remaining stationary, this container was pulled vertically downward at a constant rate that provided extremely low shear rate ($u = 0.079\,\mathrm{s}^{-1}$) in a large gap (about 1 cm). For measurements of friction force (F_η), the vane was attached to the precise automatic electrobalance with the help of thin (0.004 cm in diameter) gold wires in a manner as to exclude its rotation around a vertical axis. The vane was completely immersed into LCs, and the buoyant effect of a gold wire when it left the

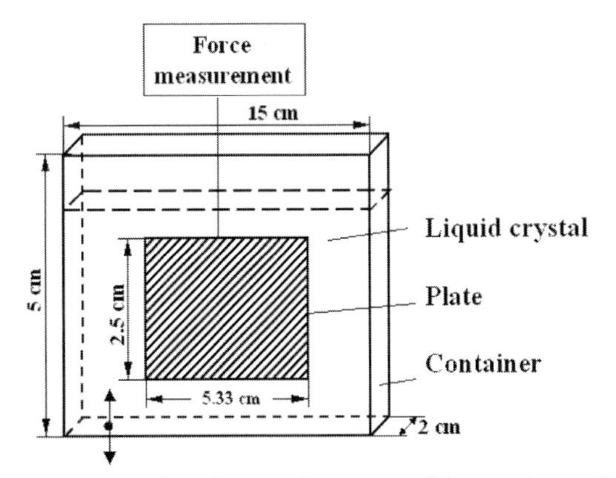

Figure 5.31 The schema and geometry of the experiment (in accordance with [89]).

liquid crystal was negligible. The relatively moderate magnetic field ($B = 3.5$ kG) was used to stabilize and to control the orientation of LCs by rotating a magnet around a vertical axis. It corresponds to the variation of the angle φ at $\theta \approx 90°$ in expression (5.67) for the geometry shown in Figure 5.31. So, one can measure two principal viscosities η_1 and η_3 corresponding to $\varphi = 0°$ and $\varphi = 90°$. Estimates made on the basis of expression (5.80) have shown that under the experimental conditions mentioned above the angle θ was equal to 87.3° that resulted in small errors (smaller than 1%) in the viscosity coefficient caused by replacing $\cos^2(87.3°)$ with $\cos^2(90°)$. Errors introduced by the boundary layers are also negligible as the ratio ξ/d is very small (about 10^{-3}). It is interesting to estimate a friction force that is a unique measurable parameter in the experiment. The friction force acting on the moving plate is determined by the expression valid for Newtonian fluids:

$$f_\eta = 2\eta S u, \tag{5.84}$$

where S is a square of the plate (about 13 cm^2 for the vane described above).

For values of η_1 ($\sim$0.15 Pa s) measured for MBBA at room temperature, it results in a friction force $F_\eta \sim 3 \times 10^{-5}$ N, which corresponds to a gravity force acting on a body with a mass of 3 mg. It is not easy to provide precise (better than 1%) measurements of such small forces even by using precise analytic balances. So, routine application of such technique is restricted by large amount of LCs originating from big size of the moving plate and the container needed to minimize possible errors. The latter also increase with a decrease in the viscosity. It may explain to some degree why the numerical value for η_3 coefficient of MBBA (0.79 P at 23 °C) obtained by Summerfold *et al.* [89] was essentially higher than that in experiments of Kneppe and Schneider (0.45 P at the same temperature), whereas values of η_1 (1.54 P – Summerfold *et al.*; 1.52 P – Kneppe and Schneider) agree quite well.

The steady regime of a shear flow close to that described above can also be achieved in the gap between two cylinders rotating with different angular velocities (Couette flow). The flow profile is practically the same as in the case of a simple shear flow. Using homeotropic surface orientation, the measured viscosity for low shear rates turned out to be close to the principal viscosity coefficient η_1 [94].

5.3.1.2.2 **Measurements in Low-Frequency Oscillating Flows**

The scheme of the experimental setup for such type of measurements proposed by Miesowicz [84, 85] many years ago is in many respects similar to that shown in Figure 5.31. It also includes the thin (glass) plate (32 mm × 22 mm) attached to the balances and moving inside the container filled with LCs. The main difference is in the type of motion. In the Miesowicz [84] experiments, the plate slowly oscillates for a period of about 5 s and with an initial amplitude of about 3 mm. Such choice of parameters of oscillations is important from two points of view. First, at such low frequency, oscillations are spread all over bulk layers of LCs separating the plate from the inner walls of the container. Indeed, estimate of the penetration length ξ_u for shear oscillations (see also Chapter 4) in accordance with expression $\xi_u = \sqrt{\eta/\rho\omega}$ gives the value ξ_u of about 3 mm at such frequency and $\eta \sim 0.1$ P, which is comparable to the gap (about 3 mm). It corresponds to the vibration mode for a shear motion of LCs (contrary to the case of higher frequencies where overdamped shear waves take place). Second, the maximal shear rate for such oscillations is rather low (about 1 s^{-1}). It means, in accordance with (5.80), that experimentally reasonable fields ($B \sim 1$ T) can provide the same value of the angle θ as that estimated above for a steady flow case. That is why one can wait to see the measured values of viscosity coefficients at $\varphi = 0°$ and $\varphi = 90°$ to be close to the two principal viscosities η_1 and η_3. The principal viscosity η_2 in Miesowicz experiments was extracted from LC samples aligned by flow that is correct only for small values of flow-induced angle θ_0. The main parameter under direct measurement was the damping time of plate oscillations.

To estimate the influence of different characteristics of the experimental setup, we will consider as the simplest model the decay oscillations of a pendulum of effective mass M damped by a friction force (5.84). As usual, the restoring force F_r is proportional to the displacement z of a pendulum.

$$F_r = -kz. \tag{5.85}$$

The equation of motion for such system is defined by Newton's law:

$$M\frac{d^2z}{dt^2} = F_\eta + F_r. \tag{5.86}$$

After dividing by M it is transformed to a standard form of differential equation for a plate motion:

$$\frac{d^2z}{dt^2} + 2\beta\frac{dz}{dt} + \omega_0^2 z = 0, \tag{5.87}$$

where

$$\beta = \frac{\eta S}{Mh} \tag{5.88}$$

is a coefficient, proportional to the shear viscosity,

$$\omega_0 = \sqrt{k/M} \tag{5.89}$$

is the frequency of nondecaying oscillations ($\beta = 0$).

The solution of the equation depends on the relative values of the damping coefficient β and frequency ω_0. For low-viscous LCs (like for PAA in Miesowicz experiments) and big value of a gap $\beta^2 < \omega_0^2$, this solution corresponds to decay oscillations:

$$z(t) = A\exp(-\beta t)\cos(\omega t + \Phi). \tag{5.90}$$

Increasing viscosity and (or) decreasing gap (which is desirable) result in an increase in the damping coefficient β. In the case $\beta^2 > \omega_0^2$, the solution is described by the expression:

$$z(t) = A_1 \exp[-(\beta - \delta)t] + A_2 \exp[-(\beta + \delta)t], \tag{5.91}$$

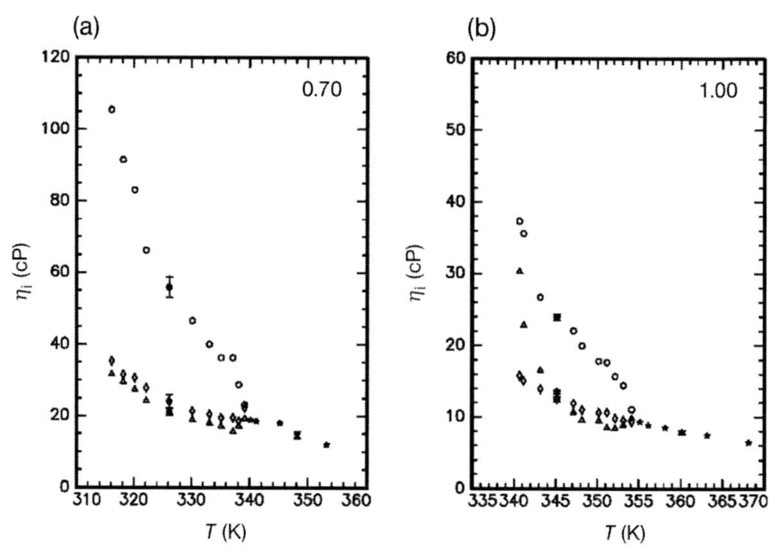

Figure 5.32 Typical Miesowicz viscosity coefficients (η_1 – ○, η_2 – △, η_3 – ◇) dependences on temperature; mixtures with the weight percentage of 8OCB 0.70 (a) and 1.00 (b). (After Janik *et al.* [96].)

where $\delta = (\beta^2 - \omega_0^2)^{1/2}$. It corresponds to nonperiodic motion that was observed in modern experiments of Hennel *et al.* [95].

A precise control of a small (less than 2 mm) displacement of the plate in these experiments was achieved via laser detection. It provided the viscosity measurements for both low and high viscous LCs. The same construction was applied [96] for the determination of principal viscosities in nematic mixtures 8OCB (4-cyan-4′-octyloxybiphenyl) and 4TPB (4-butyl benzoate 4′-isothiocyanate phenyl). At the external magnetic field of 0.5 T, errors in viscosity coefficients were estimated at 5–10%. Results presented in Figure 5.32 show that such accuracy is not enough to separate η_2 and η_3 values, although the difference between η_1 and η_2 is well determined.

Viscosity measurements using high-frequency shear oscillations were described in Chapter 4.

5.3.2
Measurements of Anisotropic Shear Viscosities in Flows of Liquid Crystals Stabilized by Surfaces

Though highly precise measurements of anisotropic viscosities are possible using magnetic fields, some physical restrictions considered above prevent such methods from being widely used. To create magnetic fields needed to overcome orientational action of flow and surface, one has to use cumbrous and expensive equipment such as heavy powerful magnets. It is out of use for other routine experiments in laboratories

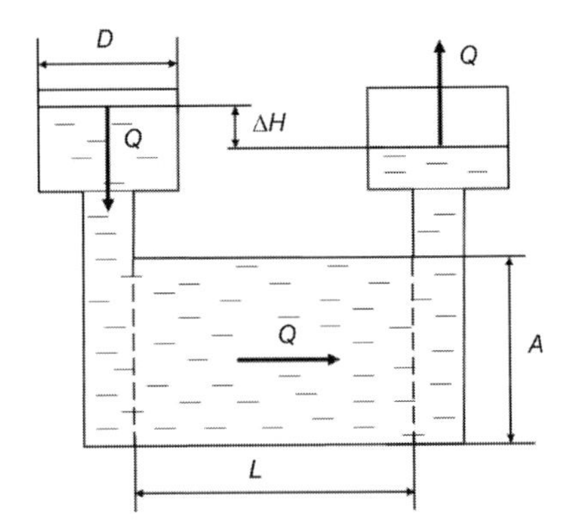

Figure 5.33 Scheme of a decay flow in the channel of the length L, width A, and gap d. The flow is produced by the instant pressure difference $\Delta P(t) = \rho g \Delta H(t)$ slowly decreasing with time. The fluid volumetric flow rate $Q = dV/dt$ is a constant in different cross sections of the hydrodynamic circuit and can be expressed as $Q = (S/2)(d\,\Delta H/dt)$, $S = \pi D^2/4$ – the cross section of the open tube. (After Pasechnik *et al.* [93].)

dealing with elaboration and use of liquid crystal materials. So, alternative decisions more suitable for practice are of great interest.

As an example, we describe here the new experimental method applicable for any NLC [97]. The main idea of the method is to use properly treated surfaces, instead of fields, to provide well-defined orientation of a liquid crystal. Stabilization of orientation by surfaces was considered previously [90, 94, 98]. In particular, it was found [90] that additional homeotropic treatment of the surface provided better results for the measurement of the viscosity coefficient η_1. Nevertheless, measurements of all principal shear viscosities without using any fields were first performed via the method under consideration. The idea of the method was proposed and experimentally checked first in Ref. [93]. A specific type of pressure-induced shear flow was observed in the cell with two open edges. Namely, decay flow was realized from measurements of anisotropic shear viscosities in liquid crystal samples. Hydrodynamic instabilities arising under the nonlinear decay flow in the cell of variable gap were described in Chapter 3. Below, we will focus on the linear regime of a director motion in the flow plane. The scheme illustrating a decay flow is shown in Figure 5.33.

A decay shear flow arises in the channel via the difference of levels ΔH in the open tubes of diameter D connected by the rectangular channel ($L \times A \times d$), d is the gap of the channel.

There is no flow through this cell in the equilibrium state (the levels of LCs in the tubes are the same). But when the initial position of the levels is changed by putting some amount of LCs into one tube or by inclining the cell, shear flow will be induced by the hydrostatic pressure difference ΔP. This difference will decrease with time, and, obviously, a certain time is required for complete cessation of the LC motion that depends on the shear viscosity. Actually, such a decay flow takes place not only in

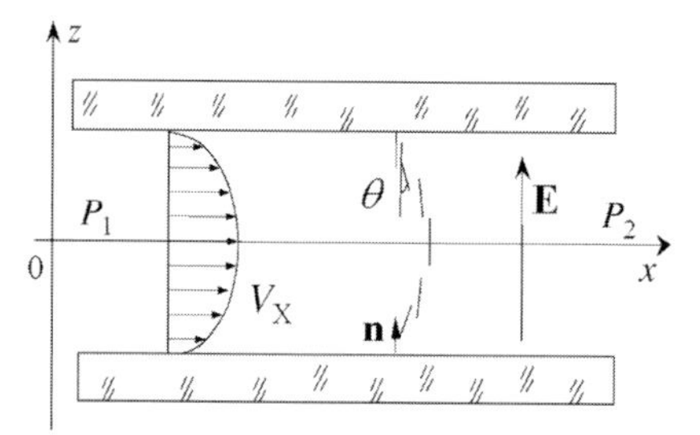

Figure 5.34 The linear flow of LC induced by a pressure difference $\Delta P = P_2 - P_1$. A parabolic velocity profile $v(z)$ occurs in the case of small values of the angle $\theta(z)$ describing orientational changes in the homeotropic LC sample; E is the electric field usually used for additional control of LC orientation. (After Pasechnik *et al.* [97].)

liquid crystals but also in isotropic fluids. In the latter case, a Newtonian-like behavior with a constant value of shear viscosity is realized. By contrast, there is a connection between the velocity gradients and the LC alignment in the liquid crystalline media. Thus, the shear flow can change the initial LC orientation (e.g., homeotropic as shown in Figure 5.34) and thus a non-Newtonian flow is observed (the shear viscosity depends on time).

Electric or magnetic fields can be used to stabilize the initial homeotropic LC alignment in this case [90, 93]. However, for a weak flow the linear approximation is valid and the flow can be considered as quasi-Newtonian (the shear viscosity is approximately constant). At the same time, slight flow-induced variations of the initial homeotropic structure described by the angle θ induce the phase difference δ that is quite sufficient to be detected by measuring the intensity I of polarized light in accordance with (5.40). This parameter correlates with the time-dependent pressure variations determined by the shear viscosity coefficient.

The theory of a linear decay flow is based on the system of coupled hydrodynamic equations (3.32) and (3.33) describing a time evolution of the velocity $V_x(z, t)$ and orientation $\theta(z, t)$ fields for a plane flow. These equations become uncoupled if the quasistationary approximation is valid [100]. It takes place when the characteristic time of pressure variations is much longer than the time of an orientational relaxation ($\tau \gg \tau_o = \gamma_1 d^2/K_{33}$, d is the thickness of the layer). It means that the LC director $n = \{\sin\theta(z, t), 0, \cos\theta(z, t)\}$ will vary in phase with pressure. So, two independent equations for the velocity V_x and the polar angle θ are written as

$$\eta_1 \frac{\partial^2 V_X}{\partial z^2} = \frac{\partial P}{\partial x} \tag{5.92}$$

$$k_{33} \frac{\partial^2 \theta}{\partial z^2} = \alpha_2 \frac{\partial V_X}{\partial z} + \varepsilon_0 \Delta\varepsilon E^2 \theta, \tag{5.93}$$

where η_1 corresponds to the maximal shear viscosity.

The equation for velocity V_X is quite the same as in the case of isotropic liquids. So, the common parabolic profile of velocity field for a laminar plane flow of an isotropic liquid takes place in this case too:

$$V_x(z,t) = -\frac{G(t)}{2\eta_1}\left(z^2 - \frac{d^2}{4}\right), \tag{5.94}$$

where the pressure gradient $G(t) = \Delta P(t)/L$ (L is the length of the channel) can be considered as a parameter slowly varying with time. The obtained expression for V_x (z, t) and Equation 5.93 can be used to derive the time-dependent field of LC orientation $\theta(z, t)$ that becomes rather simple in the absence of external fields:

$$\theta(z,t) = -\frac{\alpha_2}{6K_{33}\eta_1} z\left(z^2 - \frac{d^2}{4}\right) G. \tag{5.95}$$

This result is valid only for small deformations of the initial homeotropic orientation (in the opposite case, "out-of-plane" motion of a director described in Chapter 3 can arise for the flow under consideration [99]).

As LC director deviation $\theta(z, t)$ is proportional to the time-dependent pressure difference, the latter can be extracted by analyzing polarized light intensity passing through the cell. Indeed, the expression for a phase delay δ between ordinary and extraordinary rays is written as

$$\delta(t) \cong \frac{2\pi d}{\lambda} \frac{n_o(n_e^2 - n_o^2)}{2n_e^2} \langle\theta^2\rangle = \frac{1}{15120} \frac{\pi d}{\lambda} \frac{n_o(n_e^2 - n_o^2)}{2n_e^2} \left(\frac{\alpha_2 G d^3}{K_{33}\eta_1}\right)^2. \tag{5.96}$$

Expression (5.96) states that the phase difference δ is proportional to the square of the pressure gradient and therefore to the square of the hydrostatic pressure difference ΔP applied to the open edges of the cell:

$$\delta(t) \sim G^2(t) \sim [\Delta P(t)]^2. \tag{5.97}$$

The expression obtained provides control of the time-dependent pressure difference $\Delta P(t)$ via the time-dependent phase difference $\delta(t)$ that can be found through the analysis of the flow-induced changes of polarized light intensity $I(t)$.

It is easy to show for the given quasistationary decay flow that in every moment of fluid motion the hydrostatic pressure difference

$$\Delta P(t) = \rho g\, \Delta H(t) \tag{5.98}$$

is compensated for by the viscous pressure losses (ΔP_η) arising in the channel, which connects the open edges of the cell. For the rectangular channel having a length L, a width A, and a constant thickness d such losses are proportional to the fluid volumetric flow rate $Q = dV/dt$ (V – the volume of a liquid flowing through the cross section of the channel). Using (5.78) and (5.79), it is easy to get

$$\Delta P_{\eta_1} = -\frac{12\eta_1 L}{d^3 A} Q = -RQ, \tag{5.99}$$

where R is the hydrodynamic resistance of the capillary,

$$R = Y^{-1} = \eta_1 / K_c \tag{5.100}$$

$$K_c = \frac{d^3 A}{12L}, \tag{5.101}$$

the constant that depends only on the geometrical size of the capillary, $\eta_1 = 1/2(\alpha_4 + \alpha_5 - \alpha_2)$– the viscosity that corresponds to the homeotropic orientation of LC layer.

It is simple to derive from geometry of the problem shown in Figure 5.33 (the connection between the volumetric flow rate Q and the rate of change of the level difference ΔH):

$$Q = \frac{S}{2}\frac{d(\Delta H)}{dt}, \tag{5.102}$$

where $S = \pi D^2/4$ is a cross section of the open tubes. Taking into account the obvious connection $\Delta P = \rho g \, \Delta H$, one can easily obtain from (5.99) to (5.102) the differential equations for the level and pressure differences ΔH and ΔP:

$$\frac{d(\Delta H)}{dt} = -\frac{(\Delta H)}{\tau} \tag{5.103a}$$

$$\frac{d(\Delta P)}{dt} = -\frac{(\Delta P)}{\tau}, \tag{5.103b}$$

where

$$\tau = \frac{\eta_1}{2\rho g}\frac{S}{K_c} = 2\frac{\eta_1}{K} \tag{5.104}$$

$$K = \frac{4K_c}{S}\rho g = \frac{4}{3}\left(\frac{d^3 A}{L \pi D^2}\right)\rho g, \tag{5.105}$$

the parameter that includes the size of both the capillary and the open tubes.

By solving Equations (5.103a) and (5.103b), we come to the following conclusions:

1. The hydrostatic pressure difference ΔP (proportional to the difference of levels ΔH in the open tubes) that induces the shear flow decays exponentially with time:

$$\Delta P(t) = \Delta P_0 \exp(-t/\tau). \tag{5.106}$$

2. The characteristic decay time τ is proportional to the shear viscosity coefficient η_1:

$$\tau \sim \eta_1. \tag{5.107}$$

3. According to Equation 5.97, the phase difference also exponentially decays with time:

$$\delta(t) = \delta_0 \exp(-t/\tau_\delta), \tag{5.108}$$

where the relaxation time of the phase difference

$$\tau_\delta = \tau/2 = \eta_1/K. \tag{5.109}$$

Let us note that similar conclusions are also valid for the channel of the variable thickness shown in Figure 3.7. It is easy to show that in this case (for small enough value of the angle of the wedge-like cell), the parameter K can be expressed as

$$K = \frac{A(d_{max} + d_0)(d_{max}^2 + d_0^2)}{3\pi D^2 L}, \tag{5.110}$$

where d_{max} and d_{min} are the maximal and minimal values of the gap in the wedge-like cell.

So, the shear viscosity coefficient η_1 can be extracted from the time dependence of a light intensity $I(t)$ obtained under a decay flow. Using a wedge-like cell is desirable for decreasing experimental errors. In the case of homeotropic boundary orientation according to (5.96), the sensitivity of LC layer to the pressure gradient increases with layer thickness. So, small deviations of LC orientation from the homeotropic one registered by the optical method first occurs in the thicker part of the cell. At the same time, thinner parts of the layer became unchanged that make the overall effective viscosity to be closer to the principal viscosity coefficient η_1 than in the case of a cell of a constant gap.

A similar construction of the cell can be used to obtain other principal shear viscosities η_2 and η_3. In this case, a narrow stripe with homeotropic orientation was formed in the part of the LC cell with lowest thickness, to register the time variation of the light intensity $I(t)$ (Figure 5.35).

The overall effective viscosity is determined by a decay flow in the planar part of the cell with a total area essentially bigger than the area of the homeotropic stripe. So, expressions (5.108) and (5.109) hold when η_1 is replaced with a proper shear viscosity (η_2 or η_3) for a planar orientation in the direction of a flow or in the direction normal to the flow plane.

Examples of time dependences of light intensity $I(t)$ and phase difference $\delta(t)$ obtained by using cells with different types of preferred orientation are given in Figure 5.36.

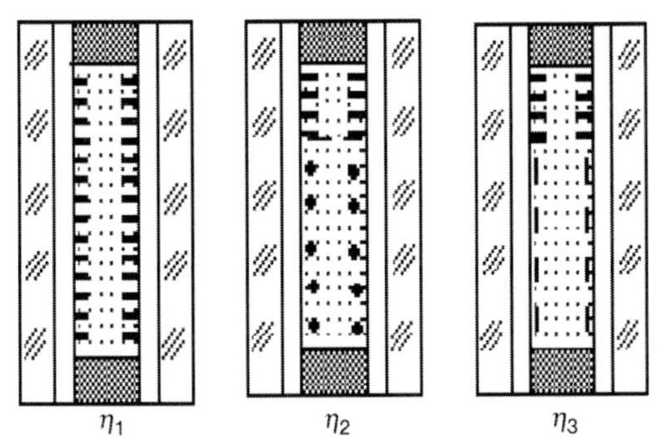

Figure 5.35 Construction of LC cells with combined orientation of LC for measurement of three principal viscosity coefficients. (After Pasechnik *et al.* [97].)

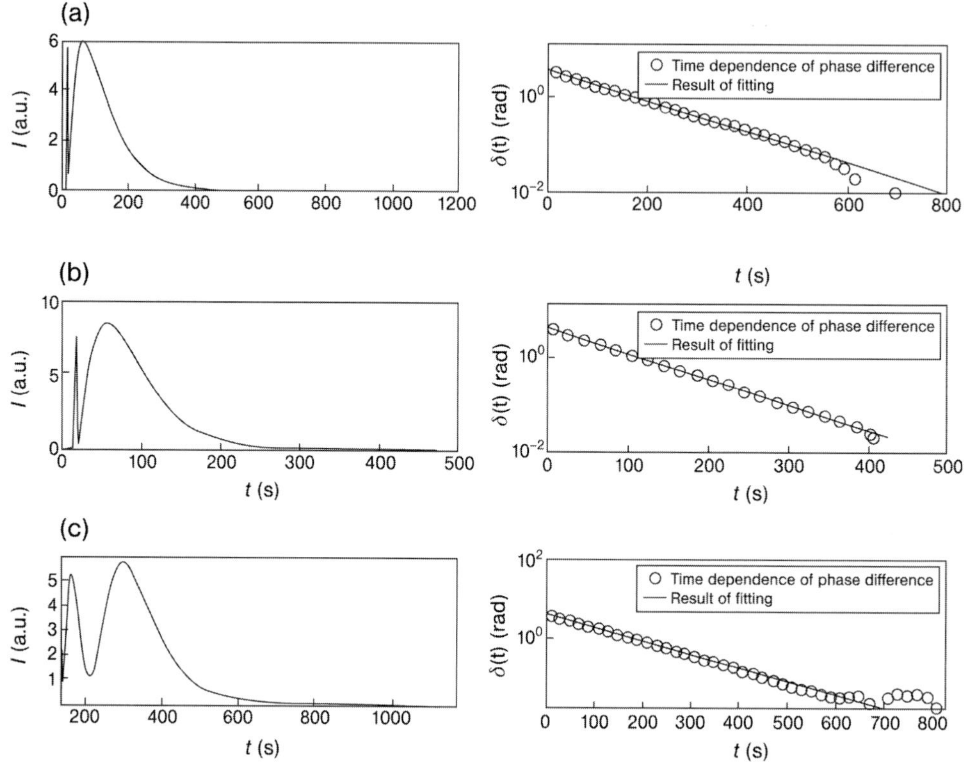

Figure 5.36 Optical response and time dependence of the measured phase difference in MLC-6609 for three principal geometries of a shear flow. The time-dependent phase retardations $\delta(t)$ define the corresponding decay times τ_δ and anisotropic shear viscosities η_i of the LC. (a) $\tau_\delta = 132$ s, $\eta_1 = 0.156$ Pa s; (b) $\tau_\delta = 81.8$ s, $\eta_2 = 0.022$ Pa s; (c) $\tau_\delta = 134.8$ s, $\eta_3 = 0.0371$ Pa s. (After Pasechnik *et al.* [97].)

Dependences $\delta(t)$ are in accordance with a simple decay law (5.108) that determines three principal shear viscosity coefficients. Estimates of possible errors show that they are about 5%, which seems to be satisfactory. This method is promising for rheological studies of newly synthesized liquid crystals as a relatively small amount of LCs (about 1 g) is needed for measurements.

In conclusion, we have to note that direct measurements of anisotropic shear viscosities in flows of nematic liquid crystals provide the most reliable information about dissipative material parameters of these materials important for applications. In principle, all Leslie coefficients can be extracted via such measurements. The main problem is to reach high enough accuracy of measurements. Though this problem can be solved by careful account and minimization of possible errors (see Ref. [87]), some factors such as the use of strong magnetic fields and large amount of liquid crystals make such setups unsuitable for routine measurements. Some excellent investigations (as [88]) were done with a relatively small amount of LCs. Unfortunately, errors arising upon the determination of the practically important viscosity coefficient η_1 are too large (more than 10%). In principle, these can be decreased by proper surface treatment of the inner surfaces of the channel. Nevertheless, even in this case, strong magnetic fields have to be used to overcome the flow-induced orientation. So, the elaboration of new measurement methods seems to be important. The most promising way to solve the problems mentioned above is the use of surface-induced orientation to stabilize monodomain structure in thin layers of LC (about 100 μm and smaller). For the realization of this idea, a new optical method for anisotropic shear viscosity measurements was proposed [97]. This method can be adopted for studying a small amount (less than 1 g) of LCs and does not require powerful magnets. The standard optical technique and simple measuring procedure make this method rather attractive for control measurements in the synthesis of new liquid crystal materials. Moreover, it can be applied for studying lyotropic nematics hardly oriented by magnetic fields.

The reader can easily find a number of shear viscosities of LC oriented by flow (see, for example, Ref. [38]) obtained via commercial viscosimeters. Though they are useful for some estimation, the knowledge of anisotropic shear viscosities is of primary interest for the optimization of new dynamic mode used in liquid crystal displays (see Chapter 6). Such results are known only for a restricted number of nematic materials. As a rule, they are obtained with different accuracy making their mutual comparison and analysis difficult. Some conclusions about temperature dependences of η_i can be found in reviews [37, 38].

In particular, the authors of the review [37] analyzed the temperature behavior of anisotropic shear viscosities on the basis of precise results (see Figure 5.37) obtained for a nematic mixture (Nematic Phase V, Merck) with a wide temperature range (268–347 K) of nematic phase.

The authors concluded that at low temperatures far from the clearing point curves of shear and rotational viscosity coefficients are parallel to each other, that is, the ratios η_i/η_j and γ_1/η_i remain more or less constant. For a given temperature, the activation energies of the coefficients are the same, but they change with temperature in the range 55–30 kJ/mol. Such changes far from the clearing point can be alternatively described (as for isotropic liquids) in terms of a free volume theory that predicts the next type of $\eta_i(T)$:

$$\eta_i = \eta_{0i}\exp\left(\frac{B}{T-T_0}\right) \tag{5.111}$$

with $T_0 = 180$ K. One can point out that the coefficients γ_1 and η_2 show the critical increase upon approaching the nematic–smectic A phase transitions that makes the temperature behavior of the shear viscosity coefficients more complicated for LC materials with such type of polymorphism [37]. In particular, the usual inequalities

$$\eta_1 > \gamma_1 > \eta_3 > \eta_2 \tag{5.112}$$

cannot be valid near transition point T_{NA}.

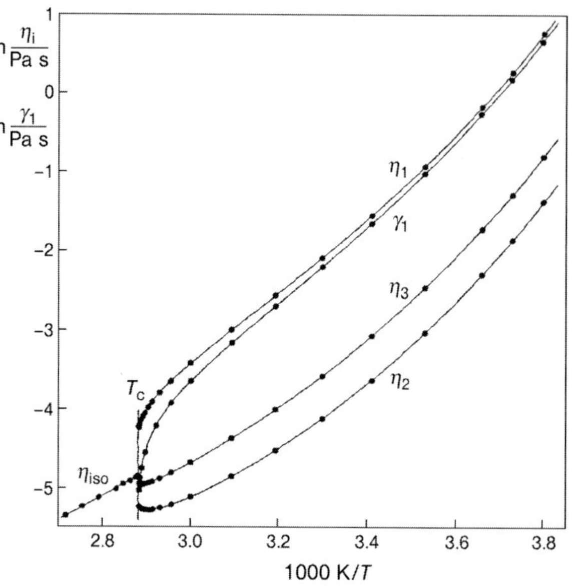

Figure 5.37 Principal shear viscosities (η_i), rotational viscosity coefficient (γ_1), and isotropic shear viscosity coefficient (η_{iso}) as a function of temperature for the liquid crystal Nematic Phase V. T_c – clearing point temperature. (After Kneppe and Schneider [37].)

5.4 Optical Methods for the Measurement of Leslie Coefficients

Below (see Chapter 6), we will show that inertial properties of some LC devices (e.g., new types of LCD) are controlled not only by a rotational viscosity coefficient but also by other combinations of Leslie coefficients. So, the knowledge of these parameters is important.

In principle, the complete set of Leslie coefficients can be extracted from shear viscosity measurements in flows of different types added by measurements of flow-induced angle (like in experiments of Gähwiller [88] and Beens and de Jeu [90].) or by a rotational viscosity coefficient. In practice, however, only precise measurements, such as those made by Kneppe and Schneider ([37,87]), provide reliable estimates of all Leslie coefficients. An example of such calculations is presented in Figure 5.38.

Above, we have shown that there are some physical restrictions on geometrical sizes of measuring cells used in rheological studies of LC oriented by fields. So, such precise viscosimetric setups are of minor importance for routine laboratory measurements of newly synthesized LCs, where only a small amount of LCs is available. Optical methods operating with thin layers are very attractive from this point of view. A number of optical techniques were proposed for this purpose. Below we will consider some examples most suitable for applications.

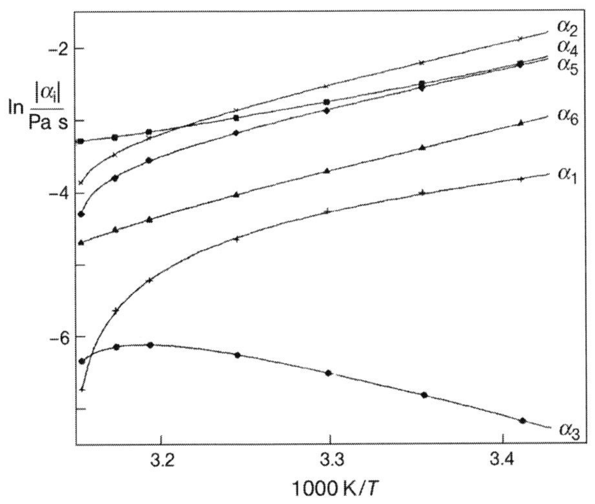

Figure 5.38 Leslie coefficients of MBBA at different temperatures (After Kneppe and Schneider [37].)

5.4.1 Flow Alignment Measurements

The alignment by flow can be considered as fundamental physical property of liquid crystals. For most liquid crystals, intensive laminar flows result in the mean orientation described by a flow alignment angle that is connected with Leslie coefficients α_2 and α_3 in accordance with (5.74). So, the measurements of θ_0 can be used to calculate a ratio α_3/α_2. Optical technique is most suitable for this purpose due to high sensitivity to director rotation. It was applied in rheological studies by Gähwiller [88] and Beens and de Jeu [90]) as an additional method.

5.4.1.1 Measurement in a Steady Simple Shear Flow

In their experiments, Wahl and Fisher [101] and Skarp *et al.* [102] conducted an optical study on orientational changes induced by a simple shear flow. The main idea of these experiments is illustrated in Figure 5.39.

In both experiments, a nematic liquid crystal layer is contained by capillarity between two glass disks (about 6 cm in diameter) treated in order to obtain homeotropic boundary conditions. One of the disks is fixed, while the other one is rotating with a constant angular frequency ω. In Wahl and Fisher experiments, the thickness d of the nematic layer was varied between 40 and 500 μm. It could be measured with an accuracy of ± 2 μm. The angular velocity ω ranged from 4.19×10^{-5} to $4.19 \times 10^{-2}\,\text{s}^{-1}$. This way, a wide range of shear velocities $v = \omega r$ and shear rates $u = (\omega r/d)$ were covered, r being the distance from the rotating axis. The disks were placed between crossed polarizers and illuminated with monochromatic light ($\lambda = 546$ nm) normal to the layer. The example of images obtained by Wahl and Fisher for MBBA is shown in Figure 5.39b. The black and dark rings originate due to the interference between the

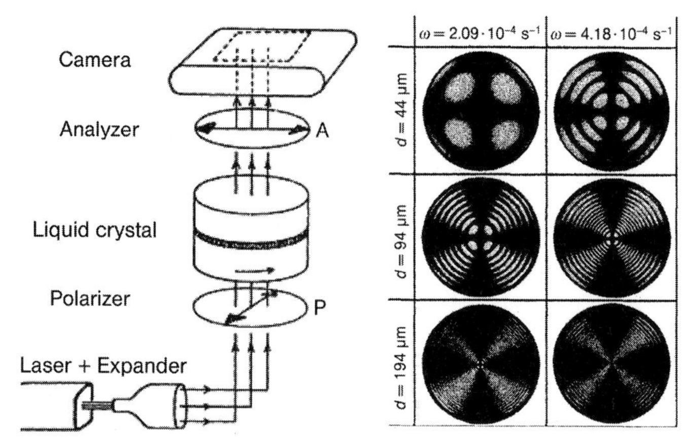

Figure 5.39 (a) Schematic picture of the experimental setup (after Skarp *et al.* [102] and (b) interference patterns seen in the flowing liquid crystal are recorded on a photographic film. (After Wahl and Fisher [101].)

ordinary and extraordinary rays propagating through the sample. It demanded a few minutes after beginning of the rotation to stabilize the images. After this period, the number and the positions of the rings became stable.

The description of this experiment is rather simple. The specific "out-off-plane" instability induced by Poiseuille flow at homeotropic boundary orientation does not exist for a steady simple shear flow (see Chapter 3). So, well-defined rings correspond to the tilt of the director in the orthoradial plane that depends on a radius. For small distortions, the angle θ of inclination of a director from the initial orientation can be found from the torque balance equation (expression (5.93)) that becomes very simple for this case:

$$K_{33}\frac{d^2\theta}{dz^2} = \alpha_2\frac{\omega r}{d}. \tag{5.113}$$

The solution of this equation can be expressed as

$$\theta(z) = \theta_0\left(\frac{1}{4}-\frac{z^2}{d^2}\right), \quad \text{with} \quad \theta_0 = \frac{-\alpha_2 d\omega r}{2K_{33}}. \tag{5.114}$$

It is simple to show that (5.114) results in quadratic dependence of the phase delay δ on d, ω, and r that are in accordance with images obtained in a regime of linear distortions. In a nonlinear regime, the angle θ saturates when approaching the flow alignment angle θ_0, defined by (5.74) throughout the cell except for thin boundary layers. Measurements of the phase delay δ can be done by taking into account a number of interference rings. It helps calculate the flow alignment angle and the ratio $(\alpha_2-\alpha_3)/(\alpha_2+\alpha_3)$. Values of these parameters reported by Wahl and Fisher for MBBA are equal to 81.8 ± 0.5^0 and 0.959 ± 0.005. It means that the ratio α_3/α_2 is positive and very small (about 0.02), which is the usual situation for flow aligning LCs. It was found by different techniques that in some liquid crystals – HBAB (hexylo-amino-benzo-nitrile) and CBOOA (cyano-benzylidine-octyloxy-aniline), for example – the steady orientation induced by flow breaks down [103–105]. This behavior was related to the influence of cybotactic smectic clusters arising at approaching nematic–smectic A (N–A) phase transition. In terms of Leslie coefficients, it corresponds to positive value of α_3. To study such crystals, Skarp *et al.* [102] slightly modified the experimental technique of Wahl and Fisher by including an additional AC electric field (of amplitude E_0 and frequency $f = 500$ Hz). The alignment angle θ induced by both flow and electric field can be expressed as

$$\tan\theta = \frac{\varepsilon_a E_0^2}{2u\alpha_3} \pm \left(\frac{\varepsilon_a^2 E_0^4}{4u^2\alpha_3^4} - \frac{|\alpha_2|}{\alpha_3}\right)^{1/2}, \tag{5.115}$$

where $\varepsilon_a = \varepsilon_0\,\Delta\varepsilon$ (it is worthwhile to mention that there was a technical mistake in this formula presented in the original paper).

The minus sign in Equation 5.115 has to be chosen by stability conditions in the case $\alpha_3 > 0.9$. Without the electric field, the argument in the square root will be negative implying a nonexisting equilibrium angle θ. With a nonzero field, the negative term in the square root can be compensated resulting in a steady solution for the angle θ. Such behavior was confirmed experimentally for 8CB ($\Delta\varepsilon > 0$). So, this made it possible to extract values of both α_3 and α_2 coefficients. The temperature dependence of α_3 in the vicinity (N–A) is shown in Figure 5.40.

So, the described technique turned out to be rather universal. Relatively, α_3 measurements can be considered as rather precise (errors of order 10%) taking into account the extremely small value of this parameter. Values of α_2 and the rotational viscosity coefficient $\gamma_1 = \alpha_3 - \alpha_2$ also may be determined by such technique.

5.4.1.2 **Measurement in Oscillating Simple Shear Flows**

The shear flow in the gap between two rotating disks described above provides a steady shear flow, close to one-dimensional flow at least at a distance far from the

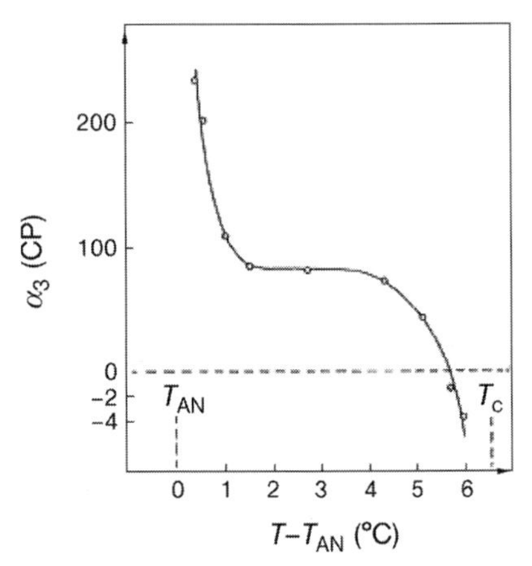

Figure 5.40 Viscosity α_3 as a function of temperature in 8CB. (After Skarp *et al.* [102].)

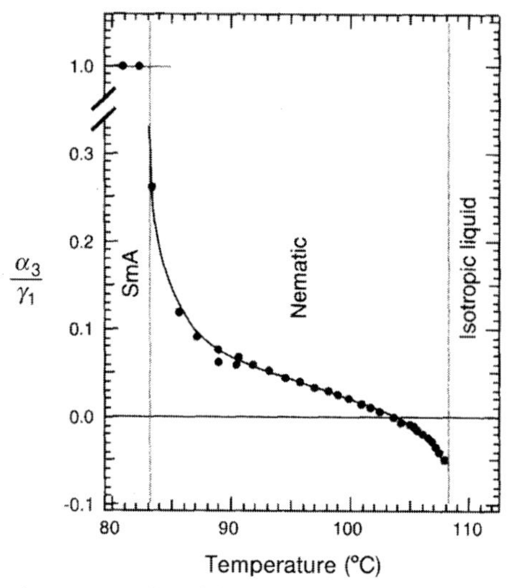

Figure 5.41 Value of the α_3/γ_1 ratio as a function of temperature in CBOOA. (After Oswald and Pieranski [3].)

centers of the disks. The real one-dimensional flow can be realized at a relative motion of two plates in one direction. It is difficult to maintain such steady motion for a long time; so, in experiments, it was replaced with low-frequency (a few s^{-1}) oscillating motion. In accordance with arguments presented in Section 5.4.1.1, it provides a regime of vibrations that is rather simple for a theoretical description. The flow-induced orientational changes were registered via changes in conoscopic figures. The final information on Leslie coefficients is essentially the same as presented in Section 5.4.1.1. A detailed description of such experiments and their results can be found in the book by Oswald and Pieranski [3]. Figure 5.41 demonstrates the temperature dependence of α_3/γ_1 ratio in CBOOA.

There is a change of the sign of α_3 upon approaching the nematic–smectic A transition, similar to the one shown in Figure 5.40. In spite of simplicity of the basic idea, experimental realization demands a fine mechanical system and a rather complicated analysis of conoscopic images needed to extract quantitative information about Leslie coefficients.

5.4.2
Measurements by Using Quasielastic Light Scattering

Such technique was applied to liquid crystals for the past four decades, beginning from the background publication [22, 106]. The theory of the method and the examples of its application are described in most of the books on the physics of liquid crystals (see, for example, Ref. [23]). So, in this section, we will present only a short review of this technique and the results obtained to estimate its application for routine measurements. In Section 5.5, we will also discuss the application of this method for studying rheological properties of LC at strong confinement.

It is well known (see, for example, Ref. [24]) that quasielastic light scattering can be considered a special class of optical spectroscopy that is preferred when the processes to be examined have relatively slow relaxation times of the order 10^{-6} s or longer. Such timescale is a characteristic of thermal orientational fluctuations in liquid crystals. As it was shown in Section 5.1, the analysis of the intensity of scattered light at different geometries helps determine Frank elastic constants K_{ii}. The information about viscous properties can be extracted from the line width of scattered light.

The frequency broadening $\Delta\omega_\alpha$ of the line in the optical spectrum of scattered light corresponding to the given (α) eigenmode of director fluctuations (see Section 5.1) can be expressed as

$$\Delta\omega_\alpha = \frac{K_{\alpha\alpha}q_\perp^2 + K_{33}q_\parallel^2}{\eta_\alpha(\vec{q}),}, \tag{5.116}$$

where the effective viscosity coefficients $\eta_\alpha(q)$ are expressed via combinations of Leslie coefficients [107]:

$$\eta_1(\vec{q}) = \gamma_1 - \frac{(q_\perp^2\alpha_3 - q_\parallel^2\alpha_2)^2}{q_\perp^4\eta_2 + q_\parallel^2 q_\perp^2(\alpha_1+\alpha_3+\alpha_4+\alpha_5) + q_\parallel^4\eta_1} \tag{5.117}$$

$$\eta_2(\vec{q}) = \gamma_1 - \frac{q_\parallel^2\alpha_2}{q_\perp^2\eta_3 + q_\parallel^2\eta_1}. \tag{5.118}$$

So, in principle, all Leslie coefficients are included in expressions (5.117) and (5.118) and one can try to extract them by analyzing the scattered light for different experimental geometries.

In particular, for the principal geometries ($1 - q_\parallel = 0$ and $2 - q_\perp = 0$), expressions for effective viscosity coefficients become rather simple:

$$\begin{aligned} \eta_1(q) &= \eta_{\text{splay}} = \gamma_1 - \alpha_3^2/\eta_2, \\ \eta_2(q) &= \eta_{\text{twist}} = \gamma_1 \qquad [\text{for geometry } 1(q_\parallel = 0)]: \end{aligned} \tag{5.119}$$

$$\eta_1(q) = \eta_2(q) = \eta_{\text{bend}} = \gamma_1 - \alpha_2^2/\eta_1 \qquad [\text{for geometry } 2(q_\perp = 0)]. \tag{5.120}$$

So, for the first geometry, two fluctuation modes, the splay mode and the twist corresponding to two Lorentzian-shaped modes of different widths in the spectra of scattered light, are observed.

For the second geometry, only one – the bend mode – determines one Lorentzian band in this spectra. Values of frequency shifts induced by fluctuations (about 10^2–10^3 Hz) are too small in comparison with initial light frequency (about 5×10^{14} Hz for red laser light). It cannot be resolved by usual spectroscopic technique and some alternative technique such as (heterodyne and homodyne) was used for this purpose. The heterodyne technique (light beating spectroscopy) is based on the interference of scattered light of the frequency $\omega = \omega_0 + \delta\omega$ changed by thermal fluctuations and the initial monochromatic light of frequency ω_0 that takes place at a photodetector. Due to the small value of ratio $\delta\omega/\omega_0$, such interference results in low-frequency beatings for the output signal from the photodetector of the frequency $\delta\omega$ convenient for further processing.

As a rule in modern experimental setups, homodyne or "self-beating" mode is used. In this mode, only the scattered light is registered by a photomultiplier, and the measured intensity is recorded and manipulated electronically to compute its time correlation function $C_I(q,)$ that is defined as [24]

$$C_I(q,\tau) = \langle I_s(t)^* I_s(t+\tau)\rangle, \tag{5.121}$$

where $I_s(t)$ is the instantaneous value of the scattered intensity at time t. This correlation function can be expressed in terms of fluctuation modes mentioned above. In general, the spectrum of scattered light consists of the sum of three Lorentzian functions. Fortunately, only one of them can be extracted by a proper choice of geometry. For example, in the geometries studied by van der Meulen and Zijlstra [25], only the twist term remains for small scattering angles ($<5^\circ$), while for angles exceeding $>20^\circ$ the splay term dominates. So, the effective viscosities defined by (5.119) and (5.120) can be determined by this technique.

Since the pioneering work by the Orsay Liquid Crystal Group [106] that tried to obtain all Leslie coefficients for PAA, several attempts [25, 27, 108] have been made to extract such type of information from Rayleigh light scattering. Although results of these first experiments were in accordance with the main theoretical predictions, estimated experimental errors were too high (about 20%). Later on, a lot of efforts were made to improve experimental setups and to optimize the measuring procedure. In particular, van der Meulen and Zijlstra [25] studied the scattering geometries rather systematically, obtaining the viscosity constants of a series of nematic compounds APAPA ($m = 1$–5, 9) and OHMBBA. In their experiments, they used the homodyne self-beat technique to detect noise intensity spectra of laser light scattered by a thin layer (50 μm) of the nematic liquid crystal. The schema of the experimental setup is shown in Figure 5.42.

Their studies were, however, still limited to the geometries where the incident beam was perpendicular or parallel to the director.

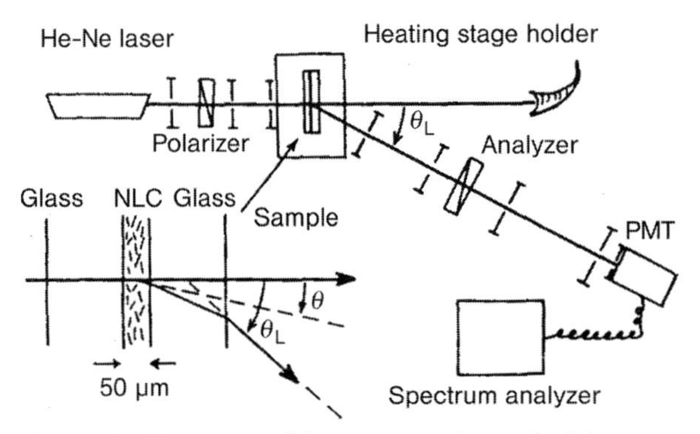

Figure 5.42 The schema of the experimental setup for light scattering measurements. (After van der Meulen and Zijlstra [25].)

Essential improvement of experimental accuracy was obtained by choosing alternative optimal geometries, with the incident beam oblique to the director, and by using precise computer-based photon correlation spectroscopy [109–111]. It turned out to be very useful for experimental investigations of the correlation between chemical structure and viscoelastic properties of liquid crystals [109]. Experimental data obtained for 4′-alkyl-4-cyanobiphenyls (nCB, $n=5$–8) and presented in Figure 5.43 confirm this possibility. In this experiment, optical cells with both homogeneous (parallel) alignment and homeotropic one were used. Cells with a thickness of 25 μm constructed from two glass plates (19 mm × 13 mm × 1 mm)

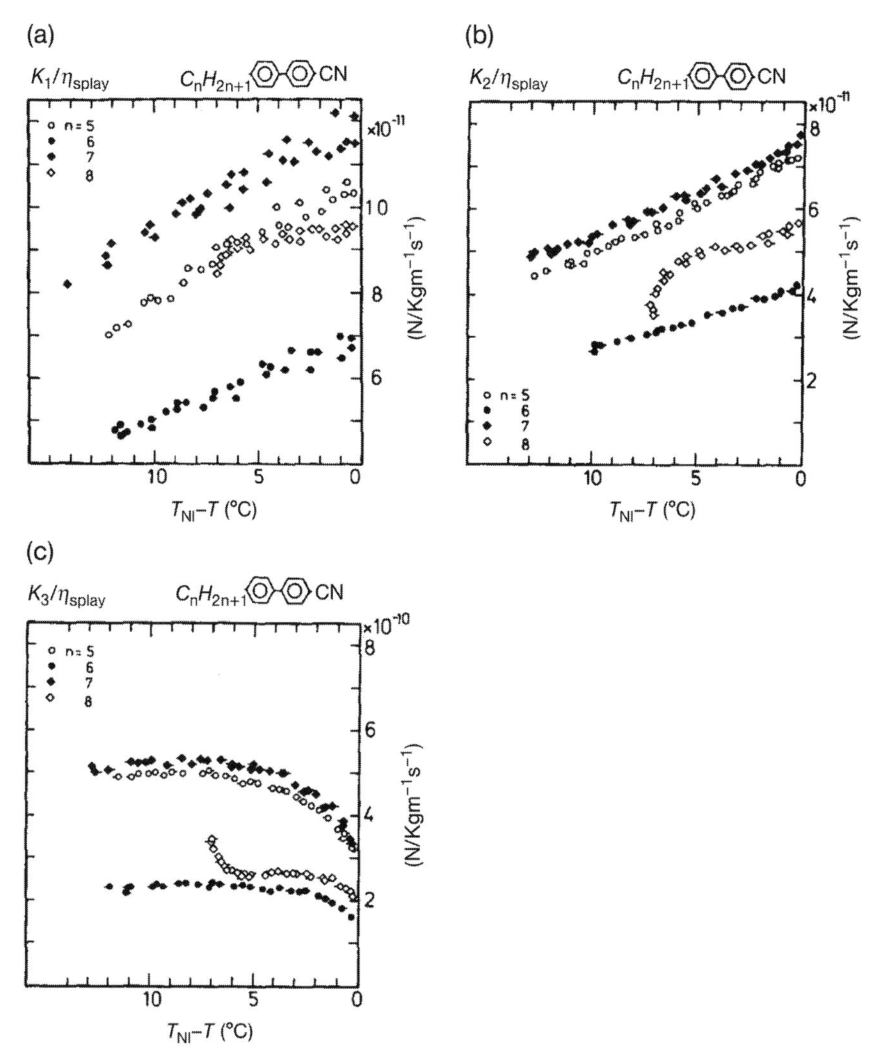

Figure 5.43 Obtained K_1/η_{splay}, K_{22}/η_{twist}, and K_{33}/η_{bend} ratio as a function of temperature $T_{NI}-T$ for nCB ($n=5$–8). (After Hirakata *et al.* [109].)

were leaned to obtain rather large (15–20°) scattering angles. In this case, the homodyne detection used the scattered photons most efficiently, resulting in high speed and high accuracy. Experimental errors in K_{22}/η_{twist} and K_{33}/η_{bend} ratios were estimated at about ±1.5% (for K_{33}/η_{splay} such error was slightly larger).

Such precision is enough to detect the so-called odd–even effect, clearly seen in Figure 5.43a–c; *n*CB with odd ($n = 2m - 1$) has the larger ratios than *n*CB with even ($n = 2m$). It means that viscoelastic properties can be effectively changed by slight manipulation of the chemical structure of liquid crystals. The strong (and nontrivial) influence of chemical structure on anisotropic viscosities was also confirmed by quasielastic light scattering measurements performed in nematic phases of a dialkoxyphenylbenzoate monomer and its dimer [111]. In particular, close values of splay viscosity (consistent with typical literature values) were found for the monomer and the dimer. At the same time, the bend viscosity of a dimer was essentially higher than that of a monomer (Figure 5.44).

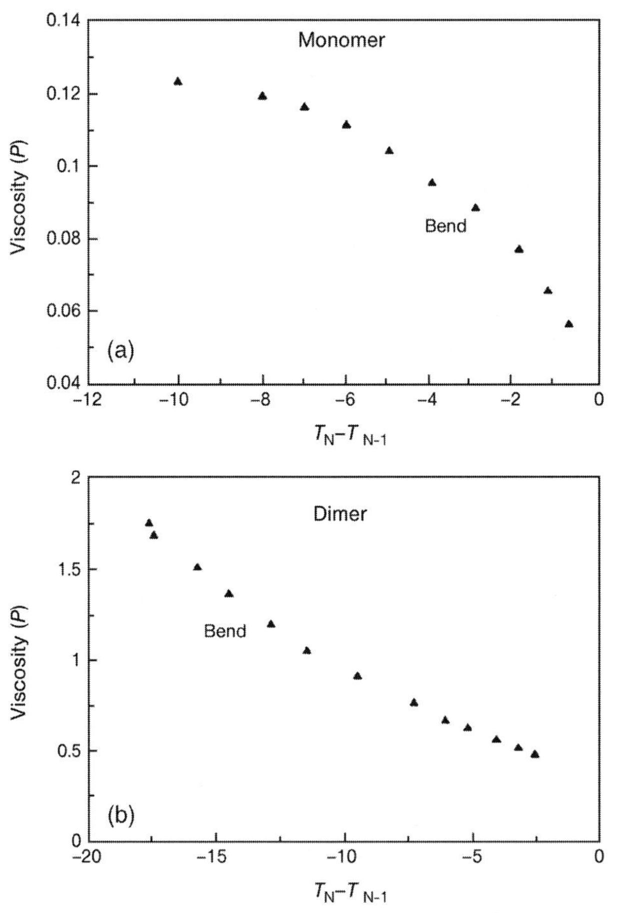

Figure 5.44 Bend viscosity as a function of temperature for a monomer (a) and dimer (b). (After DiLisi and Rosenblatt [111].)

The examples presented show that viscoelastic properties can be effectively changed via manipulation of chemical structure of liquid crystals. It is of primary importance for elaboration of new LC materials, as it will be shown below (see Chapter 6). The use of correlation photon spectroscopy operating with a small amount of LC can be considered as rather attractive for this purpose. Nevertheless, some reasons are possible to prevent the application of such technique for viscosity measurements in basic and applied research.

First, it is important to note that only the ratios (K_{ii}/η_i) can be extracted from the spectra of scattered light. It means that the primary source of error for the viscosities is uncertainty in the elastic constants since they are determined with errors of about 5%. So, the resulting errors cannot be reduced to values achieved at precise viscometric measurements described above.

Second, it is not possible to extract information about such Leslie coefficients as α_3 and α_1 and usually they are equaled to zero.

Finally, the preparation of the samples and the measuring procedure are not very simple. In particular, the cell preparation requires great care because light scattering is very sensitive to defects, such as dusts, scratches on glass plates, and static director distortions of various kinds. To provide fast measurements at different geometries modern (and rather expensive) complexes for correlation photon spectroscopy are demanded. So, the elaboration of more simple optical technique for routine laboratory measurements is still of interest.

5.4.3
Determination of Leslie Coefficients from the Dynamics of Fréedericksz Transitions

Beginning from the pioneering work of Pieranski and coworkers [5], there were some experimental works devoted to the determination of Leslie coefficients by optical study of dynamics of thin LC layers distorted by magnetic (or electric) fields. Rapid switching on (off) field results in the transient behavior of the director determined by the viscoelastic properties of LCs. For different types of initial boundary conditions and directions of applied field, the characteristic times of a relaxation process are defined by different combinations of Leslie coefficients. In the simplest case of a twist deformation, only the rotational viscosity coefficient determines the relaxation times and this parameter can be found from optical registration of such rotation (see Section 5.2).

At small distortions of initially planar (homeotropic) monodomain samples, the backward relaxation process is governed by splay (bend) viscosities defined by expressions (5.119) and (5.120). This process is described in the same manner as in the case of pure twist deformation (see expressions (5.34)–(5.39) by replacing values K_{22} with $K_{33}(K_{11})$ for the initial homeotropic-H (planar-P) orientation and by using effective viscosities η_{bend} (for H) and η_{splay} (for P) instead of rotational viscosity coefficient γ_1. So, the combination of Leslie coefficients $(\gamma_1 - \alpha_2^2/\eta_1)$ and $(\gamma_1 - \alpha_3^2/\eta_2)$ entering into the expressions for relaxation times can be extracted at properly performed optical experiments. A traditional scheme of such experiments, first proposed by Pieranski *et al.* [5], is shown in Figure 5.45. LC cell is placed between

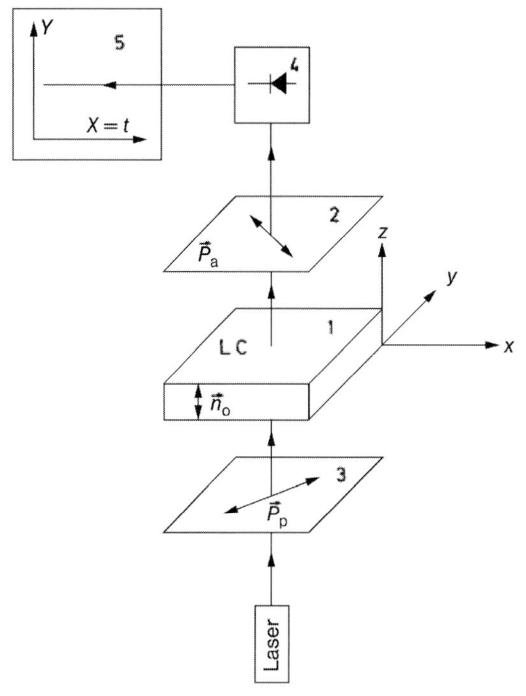

Figure 5.45 Optical setup used to measure the birefringence of LC film. (After Pieranski *et al.* [5].)

two crossed polarizers. Changes in polar angle θ induced by field (after turning the field on) or by boundaries (after turning the field off) result in time variations of the phase delay δ and the intensity *I* of output polarized light connected to δ by expression (5.40).

It is obvious that both the phase δ and the light intensity *I* are nonlinear functions of the polar angle θ. So, equal variations of this angle will result in different changes in *I*. Moreover, the light intensity shows nonmonotonic dependence on δ in accordance with (5.40). That is why computer fitting of experimental data by theoretical dependences is useful for such types of experiments. In principle, this procedure can help determine almost all Leslie coefficients at strong deformation of the initial structure under electric field [112]. Nevertheless, precision of such method is not very high. As for measurements via a photon correlation spectroscopy, errors in Frank constants contribute to the effective viscosity η_{splay} and η_{bend} under consideration. It provides a comparable precision of two optic methods for a rotational viscosity coefficient γ_1 (see Figure 5.46). Attempts to extract other Leslie coefficients [112] provided numerical values of low accuracy that can be used only for estimations. It is obvious that more precise initial data are needed for the four-parameter fitting used in this work taking into account a big difference in values of Leslie parameters under determination.

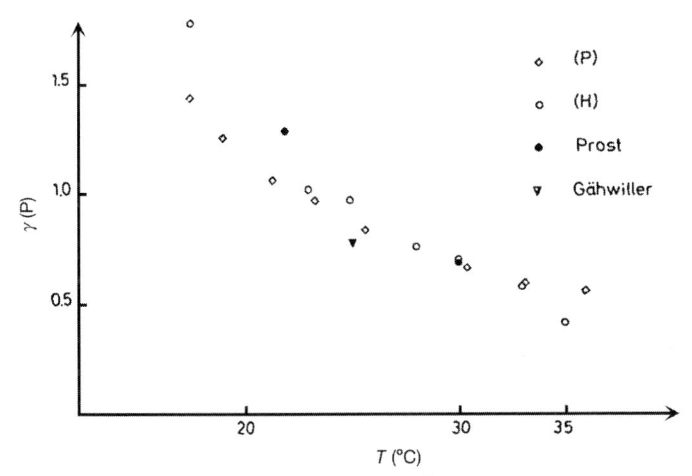

Figure 5.46 Temperature dependence of $\gamma_1(T)$ for MBBA obtained from the experiments in geometry with initial planar (P) and homeotropic (H) orientation. The results are compared with independent data. (After Schmiedel *et al.* [112].)

It is possible to draw some conclusions concerning the methods for determination of Leslie coefficients and obtained results.

1. Usually, different combinations, but not Leslie coefficients themselves, can be derived from experiments. The possibility to extract individual values of α_i depends on a number of factors, such as an accuracy of measurements of the whole combinations, relative values of entering Leslie coefficients, and an accuracy of measurements of additional parameters (e.g., Frank moduli).

2. There are well-defined optical methods for measuring α_3/α_2 ratio from flow alignment angle. As the rotational viscosity coefficient $\gamma_1 = \alpha_3 - \alpha_2$ is usually measured with satisfactory accuracy, both coefficients α_3 and α_2 can be determined separately.

3. The precise anisotropic shear viscosity measurements that provide the most reliable values of Leslie coefficients including α_1. Nevertheless, they are of limited use due to a large amount of LCs needed for measurements.

4. Some combinations, such as the bend viscosity, important for applications can be made with the study of inelastic light scattering. To calculate viscous parameters, one has to know Frank moduli as only viscoelastic ratios are determined. It is not simple to conduct precise measurements of elastic moduli, as it was shown in Section 5.1.

5. As a result, the entire set of Leslie coefficients is known only for a very restricted number of nematics. Meanwhile, relative values and temperature dependences of Leslie coefficients are under consideration in many phenomenological and microscopic theories describing dissipative properties of liquid crystals (see, for

example, Refs [37, 38, 79]). So, the lack of enough number of reliable data on this topic prevents the experimental check of theoretical models. The knowledge of the entire set of Leslie coefficients is also very desirable for computer simulation of different modes applied in modern LC displays (see Chapter 6).

5.5
Methods for Studying Surface Anchoring and Surface Dynamics of Liquid Crystals

One of the main ways of optimizing liquid crystal devices is to use solid-like or polymer surfaces with well-defined anchoring properties [113]. Besides traditional methods (e.g., rubbing), some new technologies such as photoalignment technology have been derived and tested in the past two decades [53, 114]. The study of static and dynamic properties of surface anchoring is of growing interest in modern basic and applied LC science [54, 114, 115]. One can find a number of reviews and books (see, for example, Refs [63, 115, 118–120]), where both the experimental methods and the results of determination of static anchoring properties (surface angle, anchoring strength, and angular dependence of surface anchoring energy) are considered in detail. So, below we will pay special attention to some experimental methods that provide reliable results on anchoring strength. We will also consider some methods and results concerning specific surface dynamics of liquid crystals that are of interest at present.

5.5.1
Surface Anchoring Parameters

It is well known that the local physical properties of both isotropic and anisotropic liquids in thin layers in the vicinity of liquid–solid (or gas) interface can be essentially different from the properties of bulk samples [121]. On a phenomenological level, it can be attributed to the additional contributions to the free energy density arising via molecular liquid–solid interactions [118]. The breaking of translation symmetry near the interface results in dependence of local physical parameters (such as density) on the distance z from the interface ($z = 0$). In particular, for nematic liquid crystals, the tensor order parameter Q_{ij} has to be considered as a function of z and the traditional description of nematic liquid crystals via a director $\mathbf{n}$ with fixed boundary orientation (strong anchoring) fails. The simplest way to take into account this situation is to introduce additionally to the bulk director $\mathbf{n}_b$ the surface director $\mathbf{n}_s$ describing the overall molecular orientation in near-surface layer (Figure 5.47).

In the equilibrium and in the absence of external torques, the orientation of vector $\mathbf{n}_s$ is the same as that of an easy axes $\mathbf{n}_e$ that corresponds to the preferable boundary orientation defined by surface treatment. Below, we will show that extremely slow changes (gliding) of an easy axis are possible in general. Nevertheless, in most cases $\mathbf{n}_e$ can be considered a time-independent parameter. Then, only the variations of the surface director $\mathbf{n}_s$ under the action of external factors (e.g., fields) have to be considered. Two additional parameters azimuthal W_φ and polar W_θ strengths are introduced to quantitatively characterize the surface anchoring (see Chapter 2). They

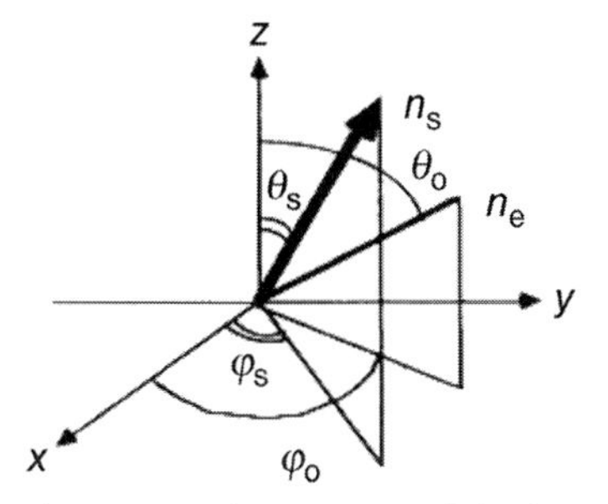

Figure 5.47 Schematic view of the director orientation at the interface. (After Faetti [118].)

can be determined as [118]

$$W_{\varphi} = \frac{1}{2}\frac{\partial^2 W(\varphi_s, \theta_s)}{\partial\varphi_s^2}\bigg|_{\varphi_s=\varphi_e;\theta_s=\theta_e} \tag{5.122}$$

$$W_{\theta} = \frac{1}{2}\frac{\partial^2 W(\varphi_s, \theta_s)}{\partial\theta_s^2}\bigg|_{\varphi_s=\varphi_e;\theta_s=\theta_e} \tag{5.123}$$

where $W(\varphi_s, \theta_s)$ vanishes for $\varphi_s = \varphi_e$ and $\theta_s = \theta_e$ and describes the anisotropic contribution to the surface tension function.

The values W_φ and W_θ correspond also to the parameters introduced in Rapini–Popoular model of interfacial interactions with the next type of the anchoring energy functions $W(\varphi_s)$ and $W(\theta_s)$ for azimuthal or polar variations of the surface director

$$W(\varphi_s) = W_\varphi \sin^2(\varphi_s - \varphi_e) \tag{5.124}$$

$$W(\theta_s) = W_\theta \sin^2(\theta_s - \theta_e). \tag{5.125}$$

It is also useful to introduce the extrapolation lengths (L_φ and L_θ) defined as

$$L_s^{\varphi(\theta)} = K_{ii}/W_{\varphi(\theta)}, \tag{5.126}$$

where K_{ii} corresponds to the given type of deformation. In particular, K_{22} is applicable for twist deformation and K_{33} (K_{11}) for bend (splay) deformations corresponding to the initial homeotropic (planar) orientation.

The ratio

$$w = d/L_S, \tag{5.127}$$

where d the thickness of LC layer can be considered as a dimensionless parameter used to define the anchoring strength. The value $w \gg 1$ corresponds to strong anchoring. The opposite case – weak anchoring – takes place when $w \ll 1$. As a rule for surfaces with well-defined orientation, the term "weak anchoring" means the essential difference from strong anchoring that is realized at $w \leq 1$ too.

Though Rapini–Popoular potential (5.124) and (5.125) successfully describes many effects, it fails in some cases (e.g., at describing the distortions of the director in strong external fields [122]). So, a number of alternative dependences such as Legendre polynomial functions, Fourier expansion, or unified functions depending on both azimuthal and polar angles were proposed for a surface anchoring potential [63, 115, 123, 124].

It is worthwhile to point out that the term "anchoring energy" is often used instead of anchoring strength, though these two parameters have a different physical meaning and do not coincide in general [118].

There are no principal problems in precise experimental determination of the surface angles [118]. It is not true for surface anchoring strengths that will be demonstrated below.

5.5.2 Methods of Measuring Surface Anchoring Strength

There are various experimental techniques for measuring anchoring strength that may be divided into two different groups according to whether an external perturbing field is applied (field-on technique) or not (field-off technique) [119]. In any case, the kind of distortion of the director field in a surface layer is of great importance. In different experiments for the determination of W, the parameters registered and varied were (i) phase delay, Bruster angle etc. in optical technique, (ii) the strength of the external field (electric, magnetic, and acoustic), (iii) the parameters of surface treatment of substrates, and (iv) chemical composition of liquid crystalline compounds (for details, see the reviews [63, 119, 123, 125, 126]).

Here, we shall only list the most important approaches and show some interesting examples of experimental techniques used in modern studies of LCs.

5.5.2.1 Field-Off Techniques

Perhaps the simplest way to evaluate both the polar and the azimuthal anchoring strength is the measurement of the thickness of domain walls separating regions of nematic liquid crystals with different director orientations. The energy of the wall depends on its thickness r, cell thickness and elastic moduli, and anchoring energy at the interface [127]. The azimuthal (W_φ) and polar (W_θ) anchoring strength can be determined from the corresponding thicknesses r of the wall. This method is mostly useful for studying weak anchoring surfaces. Usually, values of r are on the order of a few microns and can be determined from microscopic measurements. For example, in Refs [128, 129], the Neel wall method was applied to determine the azimuthal anchoring strength for 5CB and MLC-2051 (Merck, Japan) oriented by photopolymer film of poly(vinyl cinnamate) (PVCi).

The azimuthal anchoring strength can be extracted by measuring the thickness of the wall r in accordance with expression

$$W_\varphi = \frac{[-\ln(3-2\sqrt{2})]^2 d K_{11}}{2r^2}. \tag{5.128}$$

A thorough comparison of obtained results with the experimental data obtained by independent technique [129] has shown that the surface adsorption may prevent accurate measurement; therefore, the Neel wall method may be unsuitable for the accurate measurement of azimuthal anchoring energy.

5.5.2.1.1 Light Scattering Methods

Anchoring strength can also be calculated from the measurements of the intensity of small-angle light scattering by director fluctuations [119]. The wave vector $\boldsymbol{q}$ of director fluctuations depends on anchoring energy. For strong anchoring, $W \to \infty$ the lowest energy curvature mode has $q_\infty = \pi/d$. For finite W, the wave vector is smaller:

$$q_r = \frac{\pi}{d + 2L_s} \tag{5.129}$$

and the scattered intensity is concentrated at smaller scattering angles. This technique is especially useful for dimensionless anchoring strength of the order $0.1 < w < 10$.

The dependence of the wave vector on the anchoring strength modifies not only the angular dependence of the scattering but also the frequency broadening $\Delta\omega_\alpha$ of the line in the optical spectrum of scattered light corresponding to the given (α) eigenmode of director fluctuations (5.116). So, in modern investigations of liquid crystals, photon correlation spectroscopy described above was also applied. This method (see Ref. [130]) is based on the measurement of the relaxation time of thermal fluctuations in very thin nematic layers with a well-defined orientation. It has been applied successfully to determine temperature dependence of the zenithal (polar) and azimuthal anchoring on polymeric orienting layers that induce planar anchoring [131–133] and on DMOAP-silanated glass surfaces inducing homeotropic alignment [134].

One of the experimental geometries is shown in Figure 5.48.

This variant of light scattering studies seems to be rather attractive for routine measurements. Nevertheless, the cell quality in such technique is critical. The thickness of the layer has to be small enough to observe scattering induced only by the fundamental mode of the orientational fluctuations. It holds for the layer thickness less than 2 μm even for small scattering angles (about 3°). The precise control of the gap is important to determine anchoring strength with a reasonable accuracy. For example, in Ref. [134] it was provided by two independent methods:

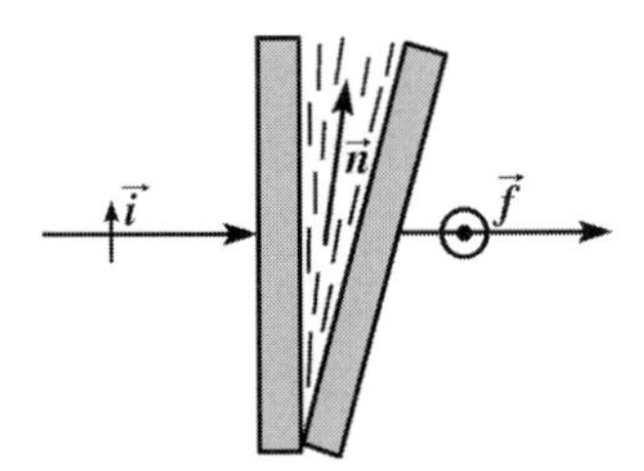

Figure 5.48 Schematic illustration of the light scattering experiment. Substrate's easy axes are parallel to the director **n** and to the polarization of the incoming light ***i***, while the polarization of the outgoing beam ***f*** is orthogonal. (After Vilfan *et al.* [131].)

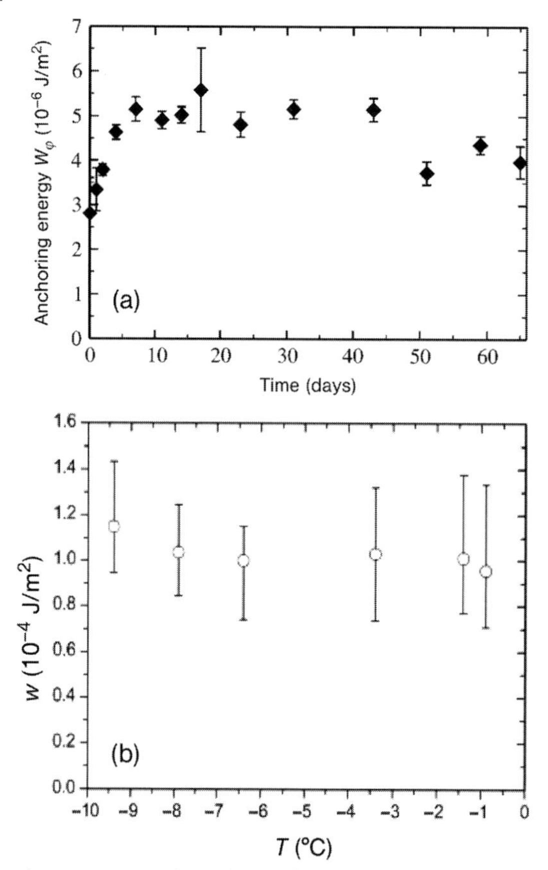

Figure 5.49 Time dependence of the azimuthal anchoring strength (energy) on the time for the interface 5CB–PVC film (a) (weak anchoring) (after Vilfan *et al.* [131]). Temperature dependence of the polar anchoring strength for 8OCB connected with DMOAP-silanated glass surfaces at homeotropic alignment (b) (strong anchoring) (after Škarabot [134]).

(i) spectrophotometry on empty cells and (ii) birefringence measurements on cells filled with a liquid crystal. Examples of experimental data on the azimuthal and polar anchoring strength obtained by such technique are presented in Figure 5.49. In the first case (Figure 5.49a), the azimuthal anchoring strength was measured for 5CB oriented by photoactive PVC film that changes configuration if illuminated with UV light. The obtained values of W_φ correspond to the weak anchoring ($w \leq 1$). At the same time, the strong polar anchoring ($w \geq 1$) took place for 8OCB connected with DMOAP(*N*,*N*-dimethyl-*n*-octadecyl-3-aminopropyltrimethoxysilyl)-silanated glass surfaces at homeotropic alignment (Figure 5.49b).

Data presented confirm the possibility of measurements of both polar and azimuthal anchoring strengths varying in a wide range with a reasonable accuracy.

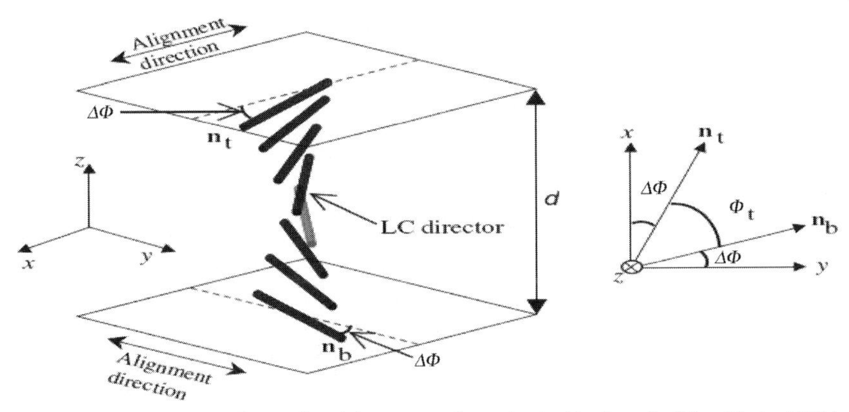

Figure 5.50 Schematics of TN cell and director configuration inside the cell. (After Okubo [129].)

So, a dynamic correlation spectroscopy can be considered as a universal technique useful for laboratory applications.

It is worthwhile to mention the additional possibility to use the angular dependence of Rayleigh scattering intensity for the determination within the accuracy of about $\pm 0.1°$ the surface angles φ_s and θ_s [135].

Nevertheless, light scattering measurements of anchoring strengths are of limited use due to the reasons mentioned above (see also Sections 5.1.2 and 5.4.2). So, more simple and convenient methods for routine measurements of azimuthal and polar anchoring strength are usually used.

5.5.2.1.2 **Torque Balance Method**

One of the most popular measurement methods for the azimuthal anchoring strength is the torque balance method, which is based on rather simple principles demonstrated in Figure 5.50.

In the configuration shown in Figure 5.50, the surface directors $\mathbf{n}_t$ and $\mathbf{n}_b$ are oriented at the angle $\Delta\Phi$ to the surface alignment direction due to elastic torque at twist deformation. This angle determined by an optical method increases with a decrease in azimuthal anchoring force W_φ. The latter can be calculated in accordance with the well-known expression [63]

$$W_\varphi = \frac{2K_{22}\Phi_t}{d\sin(2\,\Delta\Phi)}, \tag{5.130}$$

where Φ_t is the actual twist angle described by

$$\Phi_t = \Phi_t^0 - 2\,\Delta\Phi \tag{5.131}$$

Φ_t^0 is the angle between the directions of alignment treatment for the alignment films on two substrate surfaces, for example, rubbing direction or optical axis of the polarized UV exposure light.

For weak anchoring surfaces (e.g., after PA treatment), an easy axis formed by the absorbed layer of LC molecules can differ from alignment direction [136]. It results

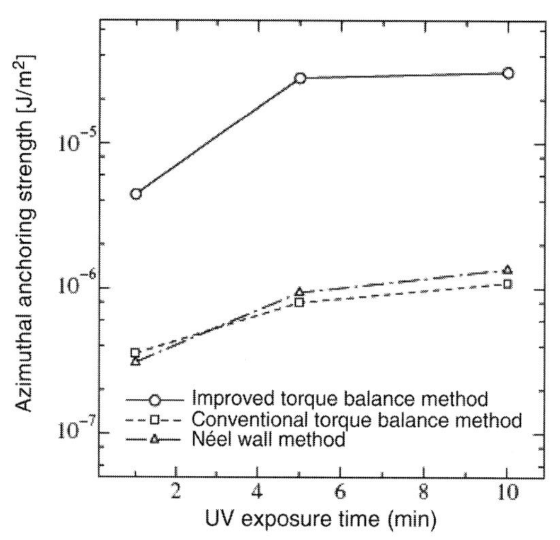

Figure 5.51 Azimuthal anchoring strength as a function of the exposure time for different techniques. (After Okubo [129].)

in incorrect values of the anchoring strength calculated via (5.129). In this case, preparation of the twist structure as shown in Figure 5.50 demands some efforts. In Ref. [129], it was done by a rotation of one glass plate with respect to the opposite one *after* forming initial planar structure via cooling LC from isotropic to a nematic phase. Experiments were conducted with the MLC-2051 (Merck, Japan) oriented by photopolymer film of polyvinyl cinnamate at UV exposure time. The dependences of an azimuthal anchoring strength on the latter parameter obtained by the improved torque balance method and the conventional one are shown in Figure 5.51. It is worth noting that the azimuthal anchoring strength measured by the improved torque balance method (and by on-field technique) is one order of magnitude greater than that measured by the conventional torque balance method and by the Neel wall method. This difference was referred to as molecular adsorption mechanism by the authors.

Some modification of the torque balance method was proposed in Ref. [137]. It is based on the modulation in the polarization state of an incident probing laser beam. The transmitted light intensity was analyzed through 2×2 Jones matrix method and compared with experimental results. It provided reliable determination of a rather high azimuthal anchoring strength. For example, it was determined as $1.5 \times 10^{-5}\,\mathrm{J/m^2}$ for 5CB contacted with a rubbed PVA film. This value corresponds to a strong anchoring ($w \sim 30$) for the thickness of the layer $10\,\mu\mathrm{m}$ used in experiments.

5.5.2.2 **Field-On Techniques**

These methods are based on the study of field-induced deformations of the orientational structure initially stabilized by surfaces. Such approach is applied for measurements of an anchoring strength (mostly for polar anchoring) and well-described in literature [63, 115, 118, 119]. Experimental setups can be assigned to the

Fréedericksz transition technique, high electric field technique, and saturation voltage method depending on the field strength [115].

The classical method for determining the anchoring strength is the use of the Fréedericksz transitions that were discussed in detail in Chapter 2. The orienting influence of the surface results in deformation of the director profile in a previously homogeneous or homeotropic liquid crystal cell providing its free rotation parallel or perpendicular to the external field due the dielectric interaction. The value of the threshold field, the shape of electric or optical liquid crystal response above the threshold, and the dynamics of the Fréedericksz transition make it possible to determine the corresponding anchoring strength $W_{\varphi(\theta)}$. As (5.129), $q_r < q_\infty$, the critical elastic energy Kq of the liquid crystal cell for a weak boundary anchoring is smaller and the threshold energy of the disturbing magnetic (H_c^w) or electric field (E_c^w) can be reduced. For example, the threshold strength H_c^w of magnetic field inducing distortions of the initial planar orientation is expressed as [40]

$$H_c^w = H_c h_c, \tag{5.132}$$

where H_c is the critical value for strong anchoring defined above and the parameter h_c can be found from a nonlinear equation

$$\pi w^{-1} h_c = \cot[(\pi/2)h_c], \tag{5.133}$$

where w is the dimensionless anchoring strength defined by (5.127).

For relatively strong anchoring ($w \gg 1$), one can obtain from (5.133)

$$h_c \approx 1 - 2w^{-1}. \tag{5.134}$$

It describes a slight decrease in the critical field strength due to the finite anchoring strength.

Similar expressions are referred to as the reduction of the threshold electric field of Fréedericksz transition.

It is clear from expression (5.134) that errors in determining an anchoring strength via Fréedericksz transition technique have to increase with an increase in parameter w. Usually, errors in measurements of threshold fields for Fréedericksz transitions are of about some percentage via a number of physical factors (see Chapter 2). It restricts the range of a polar anchoring strength ($W_\theta \leq 10^{-5}$ J/m^2) where such technique provides a suitable accuracy. It is also true for azimuthal anchoring strength due to peculiarities of optical detection of a Fréedericksz transition for pure twist deformation (see above). That is why many different particular techniques based on the use of strong fields have been proposed.

For example, Faetti *et al.* [138] applied relatively strong magnetic fields (0–0.9 T)) to the glass plate (diameter 15 mm, thickness 0.7 mm) treated with an oblique evaporation of SiO (planar orientation) and dipped in a nematic (5CB). The elastic torque arose due to a pure twist deformation induced by field in near surface layers of LC. As in the experiments of Zwetkoff [43] (see Section 5.2), a thin wire (quartz, diameter 30 μm) was used as a suspension to compensate for the very small (0.3×10^{-10} N m) magnetically induced torque. The azimuthal anchoring strength was determined by

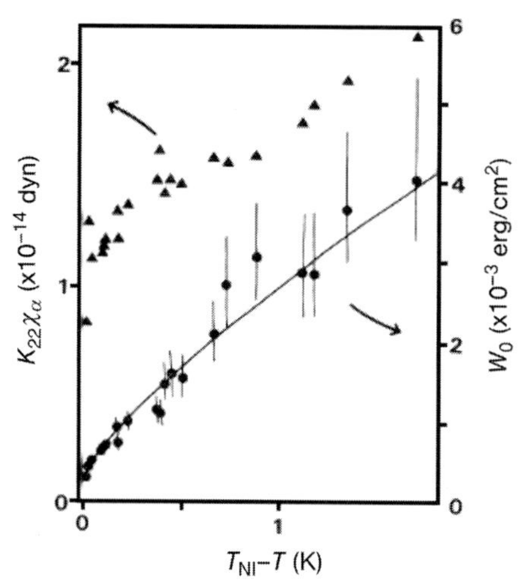

Figure 5.52 Temperature dependence of the azimuthal anchoring strength (W_0) and the Frank's modulus (K_{22}) multiplied by the diamagnetic permittivity anisotropy (χ_a) for 5CB oriented by SiO-treated glass surface. (After Faetti *et al.* [138], the authors' notations are used.)

measuring the rotation angle of the torsion pendulum as a function of the field strength. The temperature dependence of W_φ obtained in this experiment is presented in Figure 5.52.

It is evident that the used range of magnetic field induction is effective for the study of relatively weak surface anchoring as the errors grow essentially at an increase in the anchoring strength up to $5 \times 10^{-6}\,\mathrm{N/m^2}$. So, in modern investigations, strong electric fields instead of magnetic ones are usually applied to induce near surface deformation of orientation [115].

A common method to induce the twist-like deformation is the application of "in-plane" electric field. This regime (IPS mode) is used in some types of LC devices [113] including bistable ones (see, for example, Ref. [139]), where the exact control of azimuthal anchoring is of key importance. Different particular schemas of realization of "in-plane switching" for determination of the azimuthal anchoring strength were proposed (see, for example, Refs [140, 141]). Two of them are shown in Figure 5.53.

There are some peculiarities in optical measurements of high values of the azimuthal anchoring strength by using this field configuration. Indeed, the application of strong "in-plane" electric field induces not only the surface director rotation but also an interfacial director twist with a small characteristic length that affects the polarization state of the transmitted and the reflected light [141]. In the case of strong anchoring, this noisy effect especially relevant for the transmission can stimulate an increase (even by some orders of magnitude) in the apparent surface rotations compared to the true surface director rotation. Of course, it produces additional errors in the measured value of the anchoring strength (a decrease in this parameter).

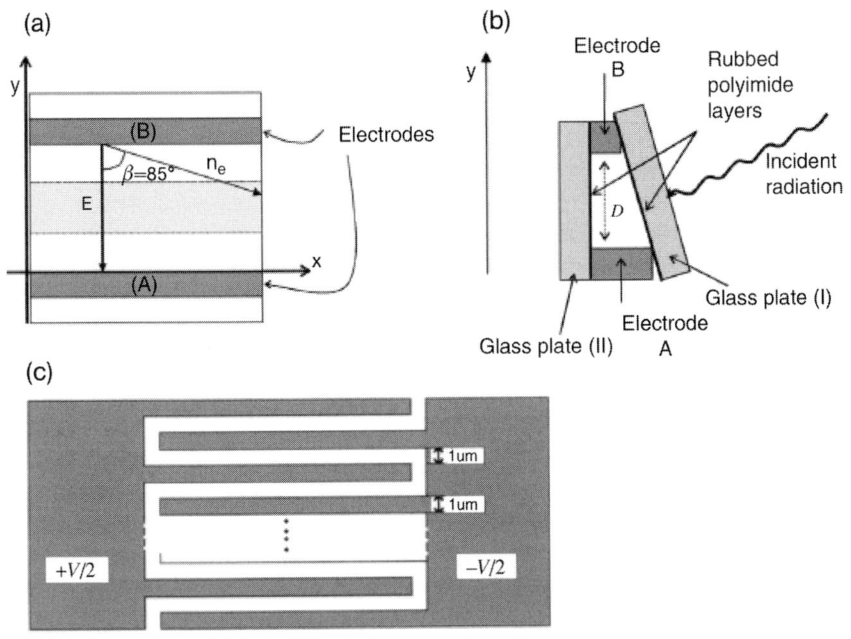

Figure 5.53 Possible schemas for the application of "in-plane" electric field; (a and b) typical geometry and wedge-like cell used for the study of strong surface anchoring via rubbed polyimide layers (after Faetti and Marianelli [141]); (c) a schematic diagram of the comb electrodes on an ITO glass substrate (after Zhang *et al.* [140]) used for a study of azimuthal surface anchoring.

That is why Faetti and Marianelli [141] applied for the measurement of W_φ the improved reflectometric technique insensitive to the presence of the bulk director twist in a high external fields regime. The value of anchoring strength for the interface 5CB-rubbed polyimide ($W_\varphi = 3.3 \times 10^{-3}\,\mathrm{J/m^2}$) turned out to be very high and comparable to polar anchoring strength for such type of surface treatment [115].

An accurate determination of the polar anchoring strength at strong and moderate anchoring is also of great importance as this parameter defines a number of technical characteristics of LC displays (Chapter 6). As optical methods meet certain problems in polar anchoring energy determination (e.g., for vertically aligned LC cells or for cells with high pretilt angle [142]), electric capacitance measurements are widely used in the study of polar anchoring [115]. It is based on changes in electric capacitance of LC cell due to the field-induced distortions of the near-surface layers of LCs.

A rather universal and simple technique of such type was proposed and checked recently [142]. The main idea of this technique is to use an additional LC cell with a perpendicular director orientation (planar or homeotropic). It helps determine the saturation level of the capacitance and exclusion of certain volume effects, such as the variation of the LC order parameter under a high-field application (see, for example, Ref. [140]).

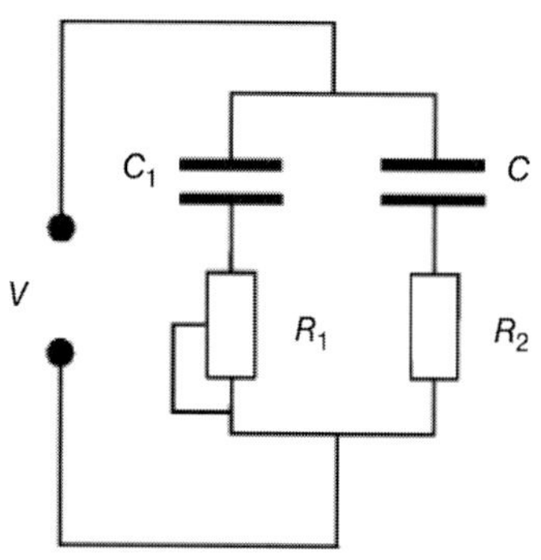

Figure 5.54 Principal scheme for two-channel capacity measurements. (After Murauski *et al.* [142].)

A rather simple electric schema for two-channel capacity measurements is shown in Figure 5.54. The voltage V from a function generator is applied simultaneously to two cells. Cell 1 has a planar LC orientation and cell 2 homeotropic LC alignment. Using an LC with a positive dielectric anisotropy, cell 1 was switched in an electric field. Cell 2 was used as a reference channel. Resistors R_1 and R_2 were used to measure the current at every channel. The resistance of LC cells ($\geq$0.10 MΩ m) was much higher than the values of R_1 and R_2 (10 kΩ m), so it was possible to consider only the capacitance of LC cells in the equivalent electrical circuit. This variable resistor R_1 was used for the equalization of two channels for empty cells.

The polar surface anchoring strength W_θ was determined by the next basic expression:

$$W_\theta = \frac{2}{\pi^2}\frac{V_{\text{th}}^2}{S}\frac{(C_{\parallel}-C_{\perp})^2}{C_{\perp}}\frac{B}{1-C_{\parallel}/C_{\text{inf}}}, \tag{5.135}$$

where S is the surface of a capacitor, B is the parameter dependent on the pretilt angle, $C_{\parallel}$ and $C_{\perp}$ are the capacity of an LC cell for uniform planar ($\theta = 0$) and homeotropic orientation of LC cells, respectively. The value $C_{\perp}$ can be obtained for a planar LC cell at the voltage $V < V_{\text{th}}$. The value C_{inf} can be measured in the limit of high voltages. At the same time, the value of $C_{\parallel}$ can be defined by applying a very high voltage that reorients LC molecules in LC bulk including molecules on the surface aligning layers. The basic problem of the capacity method is the correct determination of the ratio $C_{\parallel}/C_{\text{inf}}$. In traditional approach [115], it is solved by additional measurements of *optical* retardation at high voltage. In the described method, only simultaneous *electric* measurements of the relative capacity for planar and homeotropic cells were needed to solve the problem. In this case, the polar anchoring strength was determined from the ratio U_1/U_2 as a function of applied voltage V (U_1 and U_2 – the voltages measured on R_1 and R_2 resistors). Experimental data shown in Figure 5.55 demonstrate a possibility of the described method for measuring a polar anchoring strength at different pretilt angles. It makes the proposed technique useful for studying surfaces with high pretilt angles used in modern display technology (Chapter 6). Some problems arise for LC with relatively small negative dielectric anisotropy (high

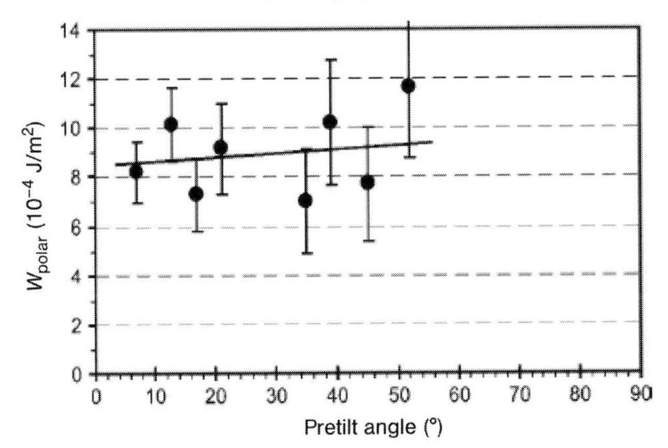

Figure 5.55 Polar anchoring energy dependence on pretilt angle for high-pretilt substrates, prepared with planar and homeotropic aligning layers. (After Murauski *et al.* [142].)

threshold voltage) as the correct electric capacity measurements are possible only at high voltages $V > 6V_{th}$ [115].

It is worthwhile to note that previously optical investigations in the saturation regime were realized when the applied field was much higher than the threshold of the Fréedericksz transition and the complete reorientation of a liquid crystal, including surface layers, took place. Using the field of such reorientation (the second threshold field) allowed to calculate the value of the anchoring strength W [143, 144].

The director deformation in an electric field may also take place due to the flexoelectric effect, which usually occurs near the cell boundaries. The flexoelectric deformation, as well as the deformation in Fréedericksz transition, is also modified by the value of the anchoring energy W [126]. However, the accuracy of the method is poor due to the great uncertainty in the values of the flexoelectric coefficients and the comparable contribution from the surface polarization [63].

There are dozens of studies where anchoring strength were measured for various substances. It should be noted that polar anchoring energy for homeotropically oriented samples is generally of the order 10^{-6}–10^{-5} J/m^2. For homogeneously oriented samples, the polar energy is generally higher, $W_\theta \sim 10^{-5}$–10^{-3} J/m^2. The anchoring energy for the nematic liquid crystals measured by various methods is collected in Refs [63, 119].

Contrary to the case of nematics, the anchoring strength of ferroelectric smectic C can be determined by specific methods based on measurements of the polarization reversal current [68, 145]. In particular, it was shown [145] that the shift of the hysteresis loop center at zero voltage was proportional to the difference in the polar anchoring strength of the FLC aligning surfaces. So, such shift can be used for estimating anchoring strength. Usually, obtained values of the anchoring strengths are similar to those measured in a nematic phase at the same surface treatment (compare, for example, with Ref. [146]).

5.5.3
Surface Dynamics of Liquid Crystals

A number of newly developed LC displays, for example, bistable nematic displays [139, 147], are based on the breaking of weak anchoring under the action of strong electric fields. The operation of these devices fully depends on the dynamic behavior of near-surface layers of liquid crystals. It is well known that a number of physical properties of both isotropic and anisotropic liquids can be modified by the surface, so they are different from properties of a bulk phase. In particular, transport processes such as diffusion can be slowed down in thin nanometer layers of polar liquids and liquid crystals [148–150].

In particular, direct NMR studies [148, 149] show the essential slowing of a translational and orientational diffusion in the surface-absorbed layers both for isotropic phase of LCs and isotropic phase of polar liquids. It is in accordance with the earlier results on the shear viscosity of polar liquids [150]. Recently, abnormally high shear viscosities of water and liquid crystals under nanoscale confinement were reported [151]. To some extent, such effects can be referred to as translational and orientational ordering induced by a surface. Such results make reasonable models that include near-surface intermediate layers with viscous properties essentially different from those of bulk samples. The principal question for a nematic phase: what is the nature and spatial structure of the intermediate surface layer formally described by $\mathbf{n}_s$ director? It was found that for an isotropic phase of liquid crystals a realistic order parameter profile is constant over a molecular distance from the wall and then decreases exponentially with the decay constant corresponding to the nematic coherence length [148]. The latter parameter increases at approaching the clearing point from above T_c [23]. It results in a corresponding increase in the thickness of a near-surface layer. The character of molecular ordering within the surface layer depends also on the boundary orientation. For example, for homeotropic boundary orientation, a thin surface layer with a constant order parameter has not only orientational but also positional, that is, presmectic order. In this case, an increase in the spin-lattice relaxation rate, larger in smaller pores, can be attributed to molecular reorientations induced by translational diffusion [152].

The absorption phenomena, the motion of polymeric chains, related to LC and other processes [116, 143, 153, 154], can also contribute to the surface dynamics of liquid crystal layers adjusted to the solid surface. This makes the surface dynamics of liquid crystal to be more complicated compared to well-studied bulk dynamics of LC layers. A strict theoretical description of viscoelastic behavior of LC layers adjusted to the surface is extremely complicated and under elaboration even on phenomenological level [155]. In general, mesoscopic approach operating with a tensor order parameter has to be used to describe relatively slow orientational motions in near-surface layers of liquid crystals.

It essentially differs from the case of a bulk nematic phase where the changes in the overall direction of long molecular axes and the mean deflection of individual molecules from this direction can be described separately in terms of a bulk director and a scalar order parameter.

Nevertheless, simple phenomenological models based on reasonable assumptions are rather useful for the explanation of specific surface dynamics of liquid crystals. Of course, such models have to take into account experimentally observed peculiarities in molecular ordering and motion in near-surface layers.

Some phenomenological description of surface dynamics was proposed by a number of authors [143, 153, 156–161]. In particular, they explain some features of extremely slow field-induced motion (gliding) of an easy axis $\mathbf{n_e}$ defining the preferable surface direction of a liquid crystal. Such slow surface dynamics, with a timescale from seconds to weeks, was registered and studied at weak anchoring surfaces of a different treatment [141, 143, 154, 157, 159–161]. Contrary to it, the fast surface dynamics has been studied essentially worthier.

In proposed theoretical models [156, 162, 163], it is referred to as the existence of near-surface layers (of thickness h_s) described by surface director $\mathbf{n_s}$ that moves slower compared to the bulk one. It produces additional viscous moments acting on the bulk director from the boundary layers, which modifies the dynamic response of LCs on the external force, applied to the layer. Usually, the value of this moment is described in terms of the so-called surface viscosity $\zeta = \gamma_1 h_s$. It denotes the phase shift between bulk and boundary layers and can be accounted for in a usual hydrodynamic approach via dynamic boundary conditions in a manner similar to that described in Chapter 3. The fulfilled estimates [162] have shown that the above-mentioned shift lies in microsecond range and can be hardly detected directly. So, some experimental technique [131, 164–167] was proposed to get information about fast surface dynamics and values of surface viscosity coefficients ζ (or corresponding thickness of a surface layer h_s). So far, the most reliable estimates of the latter parameter have been obtained by studying dynamic light scattering in liquid crystals under strong confinement [131, 165].

In particular, Mertelj and Cŏpič [165] investigated nematic liquid crystal (5CB) confined by cylindrical pores of polycarbonate (Nuclepore) membranes with radii 25–400 nm. It was found that the fundamental mode of the orientational fluctuations showed a crossover from bulk behavior dominated by bulk elastic constant K to a surface-dominated one, in which the relaxation rate was determined by the ratio of surface-anchoring strength W and viscosity η. In the smallest pores, the contribution of surface viscosity ζ was found to be also significant. The temperature dependence of the ratio η/ζ, corresponding to the thickness h_s of a near-surface layer, is shown in Figure 5.56.

The measured value for the surface layer thickness (20 nm at $T - T_{NI} = -4$ K) was found to be of the range of molecular size and essentially smaller than the extrapolation length L_s (about 100 nm at the same temperature) that defines the anchoring strength. Such small value has not to modify the dynamic behavior of the micron layers used in LC devices.

Contrary to it, in the case of a contact LC–photopolymeric substrate, the essentially higher thickness h_s (about 100 nm) of the boundary layer was obtained [131] from experiments on the dynamic light scattering in a very thin wedge-like cell similar to the one shown in Figure 5.48. This value is essentially higher than analogous for boundary layers in the isotropic phase mentioned above (about some nanometers). It

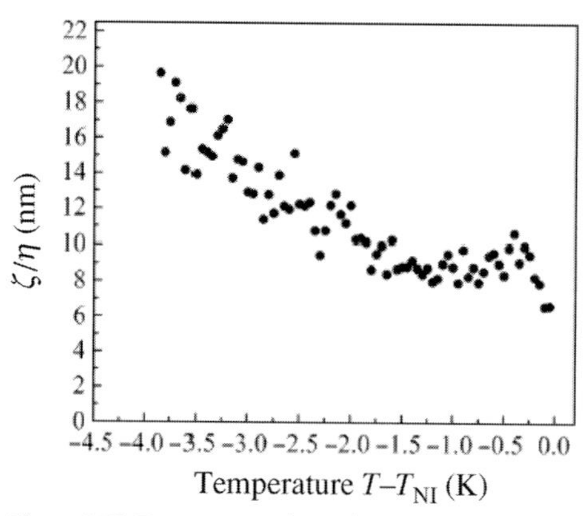

Figure 5.56 Temperature dependence of surface to bulk viscosity ratio η/ζ. (After Mertelj and Čopič [165].)

is reasonable to connect the existence of near surface layer in this case with the translational diffusion in the vicinity of the surface. It means that the boundary between this layer and the bulk nematic phase is not strictly localized and material parameters of this layer can continuously depend on the distance from the boundary. So, we have to consider the parameter h_s only as adjustable one for sublayer models [156, 162, 163]. The same is true for some other works, where extremely high values of h_s were reported [164, 166].

It is quite clear that fast surface dynamics has to be determined by physical properties of both liquid crystals and solid surface. This conclusion is even more important for slow surface dynamics induced by strong electric (or magnetic) fields and usually described in terms of "gliding" of an easy axis $\mathbf{n}_e$ referred to as the mean orientation of LC molecules absorbed by a surface.

First, this process was observed for the azimuthal reorientation of an easy axis for both lyotropic [153] and thermotropic [143] liquid crystals. In particular, it was found [141, 143, 154, 157, 159–161] that an extremely slow surface process resulting in overall change of preferred orientation $\mathbf{n}_b$ in the bulk of LC layer can be produced by strong electric (magnetic) field applied in the direction approximately orthogonal to the initial one imposed by preliminary surface treatment. Such phenomena are usually described in terms of slow "gliding" of an easy axis $\mathbf{n}_e$ referred to as the mean orientation of LC molecules absorbed by a surface. It was found that after turning the field off, the easy axis relaxes to the initial state for extremely long time (e.g., some hours).

A number of experimental studies of gliding imposed by both electric and magnetic fields have shown that this phenomenon is a general property of surfaces of different types including those with relatively strong anchoring (e.g., rubbed polyimide film). It takes place for the slow reorientation of absorbed molecules both in the plane of the LC layer (azimuthal gliding [143]) and out of the plane of the layer (zenithal gliding [160]). In all cases, slow surface dynamics shows some general features. For example, the increase in both the strength of applied field and its

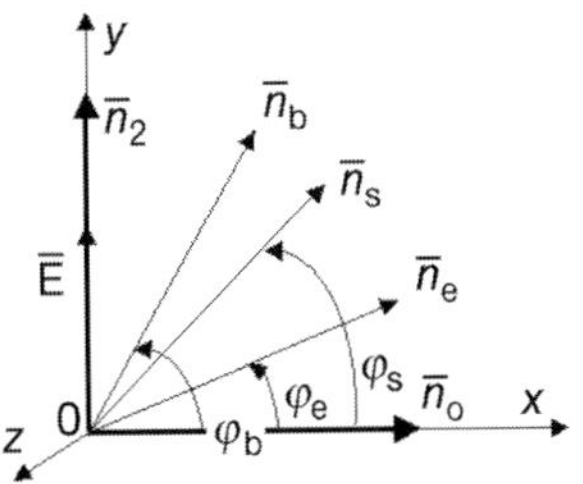

Figure 5.57 Modified phenomenological model of electrically induced gliding: $\mathbf{n}_b$ – bulk director, $\mathbf{n}_s$ – surface director, $\mathbf{n}_e$ – easy axis, $\mathbf{n}_0$ – unit vector of the initial surface direction produced by preliminary UV treatment, and n_2 – new preferable surface direction induced by both electric field and polarized light at combined effect under consideration. (After Pasechnik *et al.* [161].)

duration results in an increase in the characteristic relaxation times of the system to the initial state after turning the field off. So, using phenomenological models [143, 157–161] for the description of slow surface dynamics seems to be reasonable. At the same time, the physical background of such models has to be based on a clear understanding of molecular processes leading to the observed phenomena. In particular, the microscopic approach applicable to the description of dynamic process taking place both in surface solid layers [168] and in liquid crystal near-surface layers [116] seems to be very fruitful.

As an example of a phenomenological model, incorporating both fast and slow surface dynamics, we can consider a sublayer model proposed to explain experimental results of azimuthal gliding at a weak surface anchoring, induced by strong "in-plane" electric field [159] or by a combined action of electric field and light [161].

The model (see Figure 5.57) incorporates liquid crystal directors of the preferred surface orientation $\mathbf{n}_0$, provided by surface treatment, the preferred orientation of absorbed LC molecules (an easy axis $\mathbf{n}_e$), the mean orientation of molecules inside near-surface layer (a surface director $\mathbf{n}_s$), and the orientation of LC into the bulk of the layer (a bulk director $\mathbf{n}_b$). A strong electric field produces the twist-like deformation of the bulk director so the torque Γ_b transmitted via near-surface layer from bulk to the surface. It results in a slow azimuthal rotation (gliding) of an easy axis.

After turning the field off, the system relaxes to the initial state. These processes are described by two coupled torque balance equations (5.136) and (5.137) corresponding to torques acting on the surface director ($\mathbf{n}_s$) and on the easy axes ($\mathbf{n}_e$). In these equations, $\Gamma_{vs}(\Gamma_{ve})$ is viscous-like torque acting on the surface director and the easy axes, Γ_s is the elastic-like torque, transmitted from the surface to the surface director, K_{22} is the Frank modulus, $\xi \sim E^{-2}$ is the electric coherence length, which has to be replaced with the layer thickness d for backward relaxation ($E=0$), γ_s denotes the surface viscosity describing the phase shift between the surface director and the bulk one (such shift is essential only for fast processes), and γ_e is the analogous parameter referred to as viscous losses at rotation of an easy axis. The characteristic lengths L_s and L_e are connected (Equation 5.126) with anchoring strengths W_s and W_e, respectively, describing nonrigid anisotropic interaction of near-surface layer with the adsorbed layer of LC molecules and that of the LC-absorbed layer and the surface, h – an effective thickness of a near-surface layer characterized by the surface director $\mathbf{n}_S$.

$$\Gamma_b+\Gamma_s+\Gamma_{vs}=K_{22}[(\pi/2)-\varphi_s]/(\xi-h)-(K_{22}/2L_s)\sin 2(\varphi_s-\varphi_e)-\gamma_s(\partial\varphi_s/\partial t)=0 \quad (5.136)$$

$$\Gamma_b+\Gamma_{ve}=K_{22}[(\pi/2)-\varphi_s]/(\xi-h)-(K_{22}/2L_e)\sin 2\varphi_e-\gamma_e(\partial\varphi_e/\partial t)=0 \quad (5.137)$$

$$L_{e(s)}=K_{22}/W_{e(s)}. \quad (5.138)$$

It can be shown [159] that the simplest regime of a linear relaxation after turning the electric field off is described by two relaxation times τ_e^0 and τ_s^0 with considerably different values (Equation 5.139):

$$\tau_e^0=(\gamma_e L_e)/K_{22}\gg\tau_s^0=(\gamma_s L_s)/K_{22}. \quad (5.139)$$

In a stationary case, nonlinear effects become important at angles more than 10 degrees. For slow relaxation of an easy axis after turning off the field under study, one can obtain (by neglecting the torque transmitted from bulk LC) the torque balance equation for an easy axis:

$$(K_{22}/2L_e)\sin 2\varphi_e+\varphi_e(\partial\varphi_e/\partial t)=0. \quad (5.140)$$

So, the approximate nonlinear solution in this case can be written as

$$\text{tg}\varphi_e=\text{tg}\varphi_e^\infty\exp(-t/\tau_e^0), \quad (5.141)$$

where $tg\varphi_e^\infty$ is referred to as the stationary state of an easy axis before turning the electric field off. In a Mauguin regime, which takes place after turning the electric field off, motion of the easy axis can be easily detected by measuring the intensity of light I passing through the cell placed between crossed polarizers expressed as

$$I=I_0\sin^2\varphi_e=I_0\{[tg^2\varphi_e^\infty\exp(-2t/\tau_e^0)]/[1+tg^2\varphi_e^\infty\exp(-2t/\tau_e^0)]\}. \quad (5.142)$$

In the work [159], both the direct measurements of intensity of the laser light, passing through the narrow (about 100 μm) intraelectrode gap, and the digital snapshots processing were used to extract $I(t)$ dependences.

In the experiment, two types of a surface treatment that provided a weak planar anchoring were studied. In the first case of surface treatment (PS), a planar (or twist) orientation was achieved by rubbing a polystyrene film (the easy axis was oriented perpendicular to the rubbing direction). The UV photoalignment technique [117] was used in the second case. The main idea of the experiment was similar to that described earlier [143]. Namely, the liquid crystal cells with a strong planar surface anchoring and with a weak one on the opposite glass plates connected to a LC layer were used. The general scheme of LC cells and experimental geometry in many respects are similar to those shown in Figure 5.53.

Cells with a variable thickness of nematic layer (20–50 μm) and with a fixed one were studied. The first type simultaneously provides information on field-induced

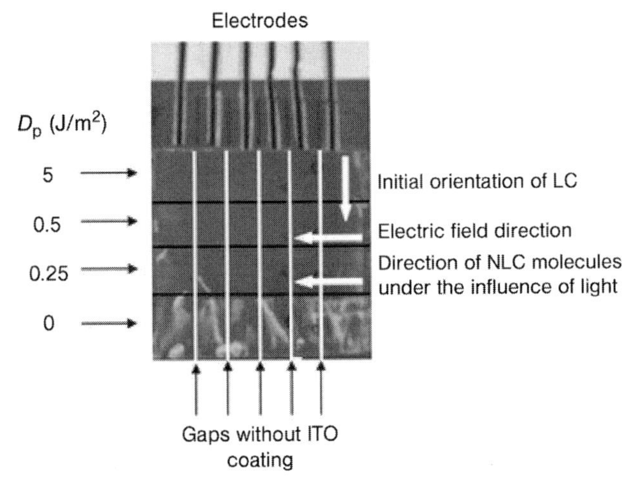

Figure 5.58 General scheme of the LC cell.

orientational changes at varying local thickness of the nematic layer. The cell with a fixed layer thickness ($d = 18\,\mu m$) was treated by UV light in a special way to obtain zones with different energies of illumination and so with different anchoring strengths (Figure 5.58).

A strong in-plane electric field ($f = 1$–$10\,kHz$) was formed in narrow gaps (of a width $g = 10$ and $100\,\mu m$) between transparent SnO_2 (ITO) electrodes placed on a glass plate with weak anchoring. It was also possible to apply polarized light (in combination with electric field) that stimulated the gliding process [161]. The gaps were visualized in a polarized light as bright stripes (for nematic mixture with a positive value of dielectric permittivity anisotropy) that showed memory effects after electric field (or electric field and light) were turned off. The brightness of the stripe (and the intensity of light) decreased very slowly when a proper choice of experimental conditions was made, showing some type of memory effects (Figure 5.59). It was established that slow relaxation of LCs after turning the field off and memory effects were suppressed with the increase in exposure energy that could be attributed to the increase in anchoring strength.

A strong combined effect was found due to the additional influence of polarized light on the electrically induced gliding in the case of a photoalignment surface treatment [161]. In particular, a pronounced rotation of an easy axis was registered upon the combined action of a relatively weak electric field (about 0.2 V/μm) and light. The strength of electric fields used in display technology and other LC devices is comparable to this value, so this fact has to be taken into account when using the PA technique. It was established that a number of control parameters (such as the dose D_p of a preliminary UV irradiation, the time of exposition t_{exp} of LCs under the combined action of light and field, the intensity of light I_2, and the field strength E) were responsible for slow surface dynamics.

By making a proper choice of control parameters, the total storage time of memorized images can extend up to some weeks (Figure 5.60). Nevertheless, in any

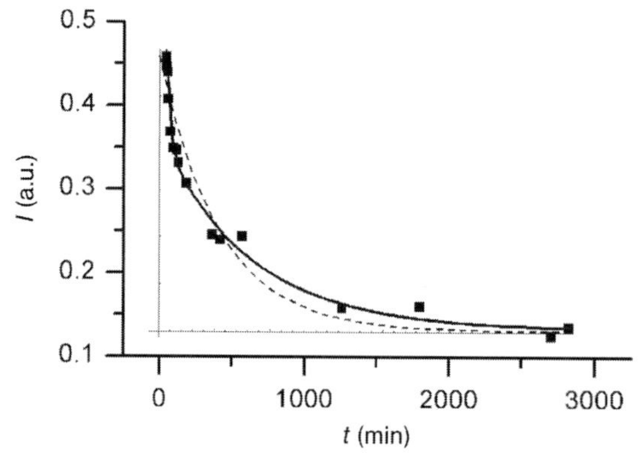

Figure 5.59 Dependence $I(t)$ after turning off the electric voltage U applied to the PA-treated surface for the exposition time, $t_{exp} = 17\,h$; $U/g = 300\,V/100\,\mu m/V$, $f = 5$ kHz; (- - -) theory (nonlinear simulation). (After Pasechnik *et al.* [159].)

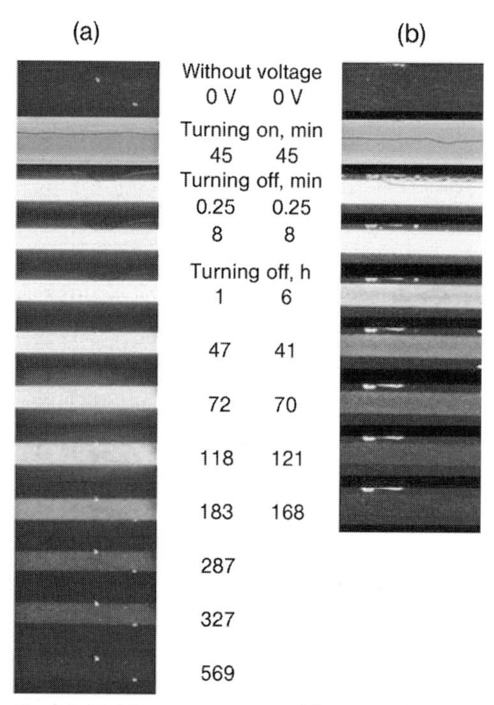

Figure 5.60 Microscopic images of the gap: before applying, after turning on strong the electric field ($U = 100$ V, $t_{exp} = 80$ min) and light ($I_2 = 74\,mW/m^2$) and after turning off the electric field and light at different doses (a) $D_p = 0.25\,J/cm^2$ and (b) $D_p = 5\,J/cm^2$ of the preliminary UV irradiation.

case, the optical images returned to the initial dark state without visible permanent memory. So, it means that this effect is very useful for temporary memory devices.

Only a slight modification of the original phenomenological model [159] was required to describe the main features of a very complicated surface dynamics under the combined action of field and light. Namely, the additional preferred surface direction $\mathbf{n}_2$ (Figure 5.57) with a corresponding anchoring strength W_2 has to be introduced to account for the dependence of the slow relaxation time τ_e on control parameters. In linear approximation, this dependence can be explicitly expressed as

$$\tau_e^0 = \gamma_e/(W_0 - \beta^{(E)} E^2 t \exp - 2\beta^{(L)} I_2 t_{\exp}), \tag{5.143}$$

where W_0 is the initial value of an anchoring strength W_e, provided by a preliminary UV treatment, $\beta^{(E)}$ and $\beta^{(L)}$ are coefficients, accounting for the changes in anchoring strengths W_0 and W_2 under the action of electric field and light. This expression was found to be in qualitative agreement with main experimental results. In particular, it explains the critical slowing down of gliding upon a proper choice of control parameters (Figure 5.60a) and suppression of gliding at the increase in the preliminary irradiation dose (Figure 5.60b). The estimated value $\gamma_e/W_0 = 880$ s can be considered as a timescale factor for slow surface dynamics.

It is worthwhile to note that the presented results are in qualitative agreement with those obtained at the study of magnetically induced azimuthal gliding at weak anchoring surfaces [153, 157].

In the case of relatively strong anchoring, the gliding phenomena become not so pronounced. Nevertheless, such effects were observed and studied by using a high-resolution technique [154, 160], even at a strongly anchoring surface (rubbed polyimide films, $W \sim 10^{-4}$ J/m^2). So, gliding can be considered as a general surface phenomenon resulting in long stability of LC devices.

Some conclusions concerning experimental methods applicable to the study of surface anchoring properties of liquid crystals and the results can be summarized as follows:

1. It is possible to describe a very complicated behavior of liquid crystals in near-surface layers in terms of restricted number of parameters such as an easy axis, a surface director, an anchoring strength, a surface viscosity, and so on. Contrary to the case of viscoelastic properties in bulk samples, these parameters reflect the interaction between liquid crystal media and solid one. So, they have to account for properties of both types of condensed matter.

2. A thorough preparation and control of properties of solid-like surfaces are of key importance for correct experimental determination of anchoring parameters. In particular, such new techniques as photoalignment and atomic force microscopy (AFM) are widely used for this purpose (see, for example, Refs [53, 120]). For example, photoalignment technique is very effective in surface pattering with well-defined anchoring properties (Table 5.2) that is of great importance [169, 170].

3. Usually, different surfaces connected to liquid crystals are referred to as strong anchoring or weak anchoring surfaces. Indeed, the overall behavior of liquid

Table 5.2 Effective azimuthal anchoring strength at different UV exposure conditions.

Period (μm)	First UV exposure (↕)		Second UV exposure (↔)		Azimuthal surface energy (10^{-5} J/m^2)	
	Average UV intensity (mW/cm^2)	τ_1 (s)	Average UV intensity (mW/cm^2)	τ_1 (s)	Inhomogeneous alignment ($W_{\varphi,\text{eff}}$)	Uniform alignment (W_φ)
2	1.94	24	2.95	6	0.25 ± 0.05	1.95 ± 0.04
1.67	2.02	16	2.99	5	0.15 ± 0.05	1.64 ± 0.04
0.78	0.45	110	0.46	43	0.08 ± 0.05	1.43 ± 0.04

Periodic micropattern structures with a given period were produced by intercepting two coherent UV beams. A two-stage procedure with different exposure times and UV intensities was used.

crystal layers confined by solid-like surfaces depends not only on anchoring strength but also on layer thickness. So, for the same surface, weak anchoring effects can be important (or not important) for relatively thin (thick) layers. The dimensionless anchoring strength w defined by expression (5.127) can be used to determine the type of anchoring.

4. Various experimental techniques used for measuring anchoring strength can be assigned to two main groups: field-off technique and field-on technique. In the first group, dynamic light scattering technique is most universal and applicable for determination of both azimuthal and polar anchoring strengths. Nevertheless, its application demands preparation of thin ($<2\,\mu$m) wedge-like cells of a well-controlled thickness and use of rather expensive experimental setups. A torque balance method is more suitable for routine laboratory measurements of azimuthal anchoring strength. The field-on technique can be realized by using both optical and electric capacity methods. The latter show some advantages at the determination of polar anchoring strength of strong anchoring surfaces ($W \sim 10^{-4}$–10^{-3} J/m^2) used in modern industry.

5. One can easily find essentially different values of the anchoring strength for the same boundary: liquid crystal–solid-like surface obtained by different authors (see, for example, Ref. [119]). So, a thorough account of possible errors and elaboration of new precise experimental technique for the study of surface anchoring is of great importance.

6. Near-surface layers of liquid crystals show a specific surface dynamics. In particular, it includes relatively fast (microseconds) processes responsible for the difference in dynamic behavior of bulk and surface layers and is described in terms of surface viscosity. The latter parameter can be hardly extracted from typical electrooptical experiments. Rare reliable data were obtained by dynamic correlation spectroscopy applied to liquid crystals under strong confinement.

7. The second type of specific surface dynamics is usually considered in terms of a slow rotation (gliding) of an easy axis under the action of strong magnetic (electric)

field. The characteristic times of such processes can widely vary (from minutes to weeks). The nature of such effects can be quite different depending on specific properties of surface-absorbed layers of liquid crystals and aligning solid-like layers. Moreover, recently the combined effects (electric field and polarized light) were established in the study of slow surface dynamics. Such effects can take place in LC devices and modify their long-term stability. Now, gliding phenomenon has been observed both for weak and for strong anchoring surfaces and is considered general a property of a boundary: liquid crystal–solid-like surface.

References

1 Stannarius, R. (1998) Elastic properties of nematic liquid crystals, in *Handbook of Liquid Crystals* (eds D. Demus, J. Goodby, G.W. Gray, H.-W. Spiess, and V. Vill), Wiley-VCH Verlag, Weinheim, pp. 60–84.

2 Dunmur, D.A. (2000) Measurements of bulk elastic constants of nematics, in *Physical Properties of Liquid Crystals: Nematics*, vol. 25 (eds D. Dunmur, A. Fukuda, and G. Luckhurst), EMIS Datareview Series, INSPEC, pp. 216–229.

3 Oswald, P. and Pieranski, P. (2005) *Nematic and Cholesteric Liquid Crystals*, Taylor and Francis, London.

4 Pieranski, P., Brochard, F., and Guyon, E. (1972) Static and dynamic behavior of a nematic liquid crystal in a magnetic field. Part I: Static results. *J. Phys. France*, **33**, 681.

5 Pieranski, P., Brochard, F., and Guyon, E. (1973) Static and dynamic behavior of a nematic liquid crystal in a magnetic field. Part II: Dynamics. *J. Phys. France*, **34**, 35.

6 Martins, A.F., Esnault, P., and Volino, F. (1986) Measurement of the viscoelastic coefficients of main-chain nematic polymers by an NMR technique. *Phys. Rev. Lett.*, **57**, 1745.

7 Martins, A.F. (2000) Measurement of viscoelastic coefficients for nematic mesophases using magnetic resonance, in *Physical Properties of Liquid Crystals: Nematics*, vol. 25 (eds D. Dunmur, A. Fukuda, and G. Luckhurst), EMIS Datareview Series, INSPEC, pp. 405–413.

8 Fréedericksz, V. and Zolina, V. (1933) Forces causing the orientation of an anisotropic liquid. *Trans. Faraday Soc.*, **29**, 919.

9 Fréedericksz, V. and Zwetkoff, V. (1934) Orientation of molecules in thin films of anisotropic liquids. *Phys. Z. Sowjet.*, **6**, 490.

10 Saupe, A. (1960) Die Biegungselastizität der nematischen Phase von Azoxianisol, *Z. Naturforsch.*, **15a**, 815.

11 Haller, I. (1972) Elastic constants of the neniatic liquid crystalline phase of *p*-methoybenzylidene-*p*-*n*-butylaniline (MBBA). *J. Chem. Phys.*, **57**, 1400.

12 Cladis, P.E. (1972) New method for measuring the twist elastic constant K_{22}/χ_a and the shear viscosity γ_1/χ_a. *Phys. Rev. Lett.*, **28**, 1629.

13 Leenhouts, F. and Dekker, A.J. (1981) Elastic constants of nematic liquid crystalline Schiff's bases. *J. Chem. Phys.*, **74**, 1956.

14 Kang, D., Mahajan, M.P., and Petschek, R.G.,*et al.* (1998) Orientational susceptibility and elastic constants near the nematic-isotropic phase transition for trimers with terminal–lateral–lateral–terminal connections. *Phys. Rev. E*, **58**, 2041.

15 Gruler, H. and Meier, G. (1972) Electric field-induced deformations in oriented liquid crystals of the nematic type. *Mol. Cryst. Liq. Cryst.*, **16**, 299.

16 Deuling, H.J. (1972) Deformation of nematic liquid crystals in an electric field. *Mol. Cryst. Liq. Cryst.*, **19**, 123.

17 Kresse, H. (1998) Dielectric properties of nematic liquid crystals, in *Handbook of Liquid Crystals* (eds D. Demus, J. Goodby, G.W. Gray, H.-W. Spiess, and V. Vill), Wiley-VCH Verlag, Weinheim, pp. 91–112.

18 Parry-Jones, L.A. and Geday, M.A. (2005) Measurement of twist elastic constant in nematic liquid crystals using conoscopici llumination. *Mol. Cryst. Liq. Cryst.*, **436**, 259.

19 Koyama, K., Kawaida, M., and Akahane, T. (1989) A method for determination of elastic constants K_1, K_2, K_3 of a nematic liquid crystal only using a homogeneously aligned cell. *Jpn. J. Appl. Phys.*, **28**, 1412.

20 Gray, G.W. (1999) Introduction and historical development, in *Physical Properties of Liquid Crystals* (eds D. Demus, J. Goodby, G.W. Gray, H.-W. Spiess, and V. Vill), Wiley-VCH Verlag, Weinheim, pp. 1–16.

21 de Gennes, P.G. (1968) Orientation of fluctuations and Rayleigh scattering in a nematic crystal. *C. R. Acad. Sci. Ser. B*, **266**, 15.

22 Orsay group (1969) Dynamics of fluctuations in nematic liquid crystals. *J. Chem. Phys.*, **51**, 816.

23 de Gennes, P. (1974) *The Physics of Liquid Crystals*, Oxford University Press, London.

24 Ho, J.T. (2001) Light scattering and quasielastic spectroscopy, in *Liquid Crystals: Experimental Study of Physical Properties and Phase Transitions* (ed. S. Kumar), Cambridge University Press, Cambridge, pp. 197–239.

25 van der Meulen, J.P. and Zijlstra, R.J.J. (1982) Optical determination of viscoelastic properties of nematic OHMBBA. *J. Phys. France*, **43**, 411.

26 Chen, G.-P., Takezoe, H., and Fukuda, A. (1989) Assessment of the method for determining the elastic constant ratios in nematics by angular dependence of scattered light intensity. *Jpn. J. Appl. Phys.*, **28**, 56.

27 Coles, H.J. and Sefton, M.S. (1986) Determination of splay elastic and viscotic constants using electric field dynamic light scattering. *Mol. Cryst. Liq. Cryst. Lett.*, **3**, 63.

28 Chigrinov, V.G. and Grebenkin, M.F. (1975) Determination of elastic constants K_{11} and K_{33} and viscosity coefficient γ_1 from orientational electrooptical effects. *Crystallogr. Rep.*, **20**, 1240.

29 Strömer, J.F., Brown, C.V., and Raynes, E.P. (2004) A novel method for the measurement of the nematic liquid crystal twist elastic constant. *Mol. Cryst. Liq. Cryst.*, **409**, 293.

30 Beresnev, G.A., Chigrinov, V.G., and Grebenkin, M.F. (1982) New method to determine K_{33}/K_{11} ratio in nematic liquid crystals. *Crystallogr. Rep.*, **27**, 1019.

31 Neubert, M.E. (2001) Chemical structure-property relationships, in *Liquid Crystals: Experimental Study of Physical Properties and Phase Transitions* (ed. S. Kumar), Cambridge University Press, Cambridge, p. 393.

32 Schadt, M., Petrzilka, M., Gerber, P.R., and Villiger, A. (1985) Polar alkenyls: physical properties and correlations with molecular structure of new nematic liquid crystals. *Mol. Cryst. Liq. Cryst.*, **122**, 241.

33 Saito, H. (2000) Elastic properties of nematics for applications, in *Physical Properties of Liquid Crystals: Nematics*, vol. **25** (eds D. Dunmur, A. Fukuda, and G. Luckhurst), EMIS Datareview Series, INSPEC, pp. 582–591.

34 de Jeu, W.H. (1980) *Physical Properties of Liquid Crystalline Materials*, Gordon and Breach, New York.

35 Grebenkin, M.F. and Ivashchenko, A.V. (1989) *Zhidkokristallicheskie Materiali (Liquid Crystal Materials)*, Chimiya, Moscow (in Russian).

36 Schadt, M., Buchecker, R., Leenhouts, F., Boller, A., Villiger, A., and Petrzilka, M. (1986) New nematic liquid crystals: influence of rigid cores, alkenyl side-chains and polarity on material and display properties. *Mol. Cryst. Liq. Cryst.*, **139**, 1.

37 Kneppe, H. and Schneider, F. (1998) Viscosity, in *Handbook of Liquid Crystals* (eds D. Demus, J. Goodby, G.W. Gray, H.-W. Spiess, and V. Vill), Wiley-VCH Verlag, Weinheim, pp. 142–169.

38 Belyaev, V.V. (2001) Physical methods for measuring the viscosity coefficients of nematic liquid crystals. *Phys. Usp.*, **44**, 255.

39 Gasparoux, H. and Prost, J. (1971) Détermination directe de l'anisotropie magnétique de cristaux liquides nématiques. *J. Phys. France*, **32**, 953.

40 Stewart, I.W. (2004) *The Static and Dynamic Continuum Theory of Liquid Crystals: A Mathematical Introduction*, Taylor and Francis, London.

41 Pasechnik, S.V., Larionov, A.N., Balandin, V.A., and Nozdrev, V.F. (1984) Etude acoustique de cristaux liquides nématiques sous champ magnétique pour différentes temperatures et pressions. *J. Phys.*, **45**, 441–449.

42 Bogdanov, D.L., Lagunov, A.S., and Pasechnik, S.V. (1980) Acoustical properties of liquid crystals in varying magnetic fields. *Primenenie Ultra-akoustiki k Issledovaniyou Vechtchestva*. **30**, 52 (in Russian).

43 Zwetkoff, V. (1939) Bewegung anisotroper Flüssigkeiten im rotierenden Magnetfeld. *Acta Physicochim. URSS*, **10**, 555.

44 Dorrer, H., Kneppe, H., Kuss, E., and Shneider, F. (1986) Measurement of the rotational viscosity γ_1 of nematic liquid crystals under high pressure. *Liq. Cryst.*, **1**, 573–582.

45 Kneppe, H. and Schneider, F. (1983) Determination of the rotational viscosity coefficient γ_1 of nematic liquid crystals. *J. Phys. E, Sci. Instrum. (UK)*, **16**, 512.

46 Leslie, F.-M., Luckhurst, G.R., and Smith, H.J. (1972) Magnetohydrodynamic effects in the nematic mesophase. *Chem. Phys. Lett.*, **13**, 368.

47 Wise, R.A., Olah, A., and Doane, J.W. (1975) Measurements of γ_1 in nematic CBOOA and (40-7) by NMR. *J. Phys.*, **36**, C1–C117.

48 van der Putten, D.D., Schwenk, N., and Spiess, H.W. (1989) Ultra-slow director rotation in nematic side-group polymers detected by N.M.R. *Liq. Cryst.*, **4**, 341.

49 Bock, F.-J., Kneppe, H., and Schneider, F. (1986) Rotational viscosity of nematic liquid crystals and their shear viscosity under flow alignment. *Liq. Cryst.*, **1**, 239.

50 Gerber, P.R. (1981) Measurement of the rotational viscosity of nematic liquid crystals. *Appl. Phys. A. Solids Surf.*, **26**, 139.

51 van Dijk, J.W., Beens, W.W., and de Jeu, W.H. (1983) Viscoelastic twist properties of some nematic liquid crystalline azoxybenzenes. *J. Chem. Phys.*, **79**, 3888.

52 Pasechnik, S.V., Shmeliova, D.V., Tsvetkov, V.A., Dubtsov, A.V., and Chigrinov, V.G. (2007) Liquid Crystal in rectangular channels: new possibilities for three-dimensional studies. *Mol. Cryst. Liq. Cryst.*, **479**, 59.

53 Dubtsov, A.V., Pasechnik, S.V., Shmeliova, D.V., Tsvetkov, V.A. and Chigrinov, V.G. (2009) Special optical geometry for measuring twist elastic module K_{22} and rotational viscosity γ_1 of nematic liquid crystals. *Appl. Phys. Lett.*, **94**, 181910.

54 Chigrinov, V.G., Kozenkov, V.M., and Kwok, H.S. (2008) *Photoaligning: Physics and Applications in Liquid Crystal Devices*, Wiley-VCH Verlag, Weinheim.

55 Zvenek, D. and Zumwer, S. (2001) Backflow-affected relaxation in nematic liquid crystals. *Liq. Cryst.*, **28**, 1389.

56 Gevorkian, E.V., Kashitsin, A.S., Pasechnik, S.V., and Balandin, V.A. (1990) The hysteresis of acoustic parameters and the bistability of smectic-C liquid crystal in a magnetic field. *Europhys. Lett.*, **12**, 353.

57 Beresnev, L.A., Blinov, L.M., Osipov, M.A., and Pikin, S.A. (1988) Ferroelectric liquid crystals. *Mol. Cryst. Liq. Cryst.*, **158**, 3150.

58 Woods, P.D., Stewart, I.W., and Mottram, N.J. (2004) Flow effect in director relaxation of bookshelf aligned smectic C liquid crystals. *Mol. Cryst. Liq. Cryst.*, **413**, 271.

59 Galerne, Y., Martinand, J.L., Durand, G., and Veyssie, M. (1972) Quasielectric

Rayleigh scattering in a smectic-C liquid crystal. *Phys. Rev. Lett.*, **29**, 562.

60 Blinov, L.M., Barnik, M.I., Baikalov, V.A., Beresnev, L.A., Pozhidaev, E.P., and Yablonsky, S.V. (1987) Experimental techniques for the investigation of ferroelectric liquid crystals. *Liq. Cryst.*, **2**, 1.

61 Pozhidaev, E.P., Osipov, M.A., Chigrinov, V.G., Baikalov, V.A., Blinov, L.M., and Beresnev, L.A. (1988) Rotational viscosity of smectic C* phase in ferroelectric liquid crystals. *Sov. Phys. JETP*, **94**, 125.

62 Pikin, S.A. (1991) *Structural Transformations in Liquid Crystals*, Gordon and Breach, New York.

63 Blinov, L.M. and Chigrinov, V.G. (1994) *Electrooptic Effects in Liquid Crystal Materials*, Springer, New York.

64 Clark, N.A. and Lagerwall, S.T. (1980) Submicrosecond bistable electrooptic switching in liquid crystals. *Appl. Phys. Lett.*, **36**, 899.

65 Barnik, M.I., Baikalov, V.A., Chigrinov, V.G., and Pozhidaev, E.P. (1987) Electrooptics of a thin ferroelectric smectic C* liquid crystalline layer. *Mol. Cryst. Liq. Cryst.*, **143**, 101–112.

66 Escher, C., Geelhaar, T., and Bohm, E. (1988) Measurement of the rotational viscosity of ferroelectric liquid crystals based on a simple dynamical model. *Liq. Cryst.*, **3**, 469.

67 Kimura, S., Nishiyama, S., Ouchi, Y., Takazoe, H., and Fukuda, A. (1987) Viscosity measurement in ferroelectric liquid crystals using a polarization switching current. *Jpn. J. Appl. Phys.*, **26**, L255.

68 Dahl, I., Lagerwall, S.T., and Skarp, K. (1987) Simple model for the poljarization reversal current in a ferroelectric liquid crystal. *Phys. Rev. A*, **36**, 4380.

69 Gouda, F., Skarp, K., Andersson, G., Kresse, H., and Lagerwall, S.T. (1989) Viscoelastic properties of the Smectic A* and Smectic C* phases studied by a new dielectric method. *Jpn. J. Appl. Phys.*, **28**, 1887.

70 Lagerwall, S.T. (1999) *Ferroelectric and Antiferroelectric Liquid Crystals*, Wiley-VCH Verlag, Weinheim.

71 Sako, T., Itoh, N., Sakaigawa, A., and Koden, M. (1997) Switching behavior of surface stabilized ferroelectric liquid crystals induced by pulse voltages. *Appl. Phys. Lett.*, **71**, 461.

72 Piecek, W., Rutkowska, J., Kedzierski, J., Perkowski, P., Raszewski, Z., and Dabrowski, R.S. (1998) Measurements of rotational viscosity of SmC* mixtures. *SPIE*, **3318**, 148.

73 Kiselev, A.D., Chigrinov, V.G., and Pozhidaev, E.P. (2007) Switching dynamics of surface stabilized ferroelectric liquid crystal cells: effects of anchoring energy asymmetry. *Phys. Rev. E*, **75**, 061706.

74 Hemine, J., Daoudi, A., and Legrand, C., *et al.* (2007) Structural and dynamical properties of the chiral smectic C phase of ferroelectric liquid crystals showing high. *Physica B*, **399**, 60.

75 Pozhidaev, E.P., Shevtchenko, S.A., Andreev, A.L., and Kompanets, I.N. (1999) Molecules rigid cores and aliphatic chains rotational viscosity in smectic C*-phase. Conference Summaries of 7th International Conference on Ferroelectric Liquid Crystals, Darmstadt, Germany, pp. 270–271.

76 Panov, V., Vij, J.K., and Shtykov, N.M. (2001) A field-reversal method for measuring the parameters of a ferroelectric liquid crystal. *Liq. Cryst.*, **28** (4), 615.

77 Rutth, J., Selinger, J.V., and Shashidhar, R. (1994) The electrostatic polarization of ferroelectric liquid crystals: a new interpretation of the triangle-wave technique. *Appl. Phys. Lett.*, **65**, 1590.

78 Garoff, S. and Meyer, B. (1979) Electroclinic effect at the A–C phase change in a chiral smectic liquid crystal. *Phys. Rev. A*, **19**, 338.

79 Belyaev, V.V. (2000) Relationship between nematic viscosities and molecular structure, in *Physical Properties of Liquid Crystals: Nematics* (eds D. Dunmur, A.

Fukuda, and G. Luckhurst), EMIS Datareview Series 25, INSPEC, pp. 414–428.

80 Kneppe, H. and Schneider, F. (1983) Rotational viscosity coefficients γ_1 for mixtures of nematic liquid crystals. *Mol. Cryst. Liq. Cryst.*, **97**, 219.

81 Haase, W., Ganzke, D., and Pozhidaev, E.P. (1999) Non-display applications of ferroelectric liquid crystals. *Mater. Res. Soc. Symp. Proc.*, **599**, 15.

82 Porter, R.S., Barrall, E.M. II, and Johnson, J.F. (1966) Some flow characteristics of mesophase types. *J. Chem. Phys.*, **45**, 1452.

83 Currie, P.K. (1979) Apparent viscosity during viscometric flow of nematic liquid crystals. *J. Phys. (France)*, **40**, 501.

84 Miesowicz, M. (1946) The three coefficients of viscosity of anisotropic liquids. *Nature*, **158**, 27.

85 Miesowicz, M. (1983) Liquid crystals in my memories and now – the role of anisotropic viscosity in liquid crystal research. *Mol. Cryst. Liq. Cryst.*, **97**, 1.

86 Chandrasekhar, S. (1992) *Liquid Crystals*, 2nd edn, Cambridge University Press, Cambridge.

87 Kneppe, H. and Shneider, F. (1981) Determination of the viscosity coefficients of the liquid crystal MBBA. *Mol. Cryst. Liq. Cryst.*, **65**, 3.

88 Gähwiller, C. (1973) Direct determination of the five independent viscosity coefficients of nematic liquid crystals. *Mol. Cryst. Liq. Cryst.*, **20**, 301.

89 Summerford, J.W., Boyd, J.R., and Lowry, B.A. (1975) Angular and temperature dependence of viscosity coefficients in a plane normal to the direction of flow in MBBA. *J. Appl. Phys.*, **46**, 970.

90 Beens, W.W. and de Jeu, W.H. (1983) Flow measurements of the viscosity coefficients of two nematic liquid-crystalline azoxybenzenes. *J. Phys. (France)*, **44**, 129.

91 Helfrich, W. (1969) Molecular theory of flow alignment of nematic liquid crystals. *J. Chem. Phys.*, **50**, 100.

92 Tsvetkov, V.A. (1980) in *Advances in Liquid Crystal Research and Applications* (ed. L. Bata), Pergamon Press, Oxford, pp. 567–572.

93 Pasechnik, S.V., Tsvetkov, V.A., Torchinskaya, A.V., and Karandashov, D.O. (2001) NLC under decay Poiseuille flow: new possibilities for measurements of shear viscosities. *Mol. Cryst. Liq. Cryst.*, **366**, 2017.

94 Narasimhan, M.N.L. and Eringen, A.C. (1974) Orientational effects in heat-conducting nematic liquid crystals. *Mol. Cryst. Liq. Cryst.*, **29**, 57.

95 Hennel, F., Janik, J., Moscicki, J.K., and Dãbrowski, R. (1990) Improved Miesowicz viscometer. *Mol. Cryst. Liq. Cryst.*, **191**, 401.

96 Janik, J., Moscicki, J.K., Czuprynski, K., and Dabrowski, R. (1998) Miesowicz viscosities study of a two-component thermotropic mixture. *Phys. Rev. E*, **58**, 3251.

97 Pasechnik, S.V., Chigrinov, V.G., Shmeliova, D.V., Tsvetkov, V.A., and Voronov, A.N. (2004) Anisotropic shear viscosity in nematic liquid crystals: new optical measurement method. *Liq. Cryst.*, **31**, 585.

98 Fischer, J. and Frederickson, A.G. (1969) Interfacial effects on the viscosity of a nematic mesophase. *Mol. Cryst. Liq. Cryst.*, **8**, 267.

99 Pasechnik, S.V., Krekhov, A.P., Shmeliova, D.V., Nasibullaev, I.Sh., and Tsvetkov, V.A. (2004) Orientational dynamics in nematic liquid crystal under decay Poiseuille flow. *Mol. Cryst. Liq. Cryst.*, **409**, 467.

100 Pasechnik, S.V. and Torchinskay, A.V. (1999) Behavior of nematic layer oriented by electric field and pressure gradient in the stripped liquid crystal cell. *Mol. Cryst. Liq. Cryst.*, **331**, 341.

101 Wahl, J. and Fisher, F. (1973) Elastic and viscosity constants of nematic liquid crystals from a new optical method. *Mol. Cryst. Liq. Cryst.*, **22**, 359.

102 Skarp, K., Carlsson, T., Lagerwall, S.T., and Stebler, B. (1981) Flow properties of nematic 8 CB. *Mol. Cryst. Liq. Cryst.*, **66**, 199.

103 Gähwiller, Ch. (1972) Temperature dependence of flow alignment in nematic liquid crystals. *Phys. Rev. Lett.*, **28**, 1554.

104 Pieranski, P. and Guyon, F. (1974) Two shear-flow regimes in nematic *p*-*n*-hexyloxybenzilidene-*p*′-aminobenzonitrile. *Phys. Rev. Lett.*, **32**, 924.

105 Cladis, P.E. and Torza, S. (1975) Stability of nematic liquid crystals in Couette flow. *Phys. Rev. Lett.*, **35**, 1283–1286.

106 Orsay Group (1969) Quasielastic Rayleigh scattering in nematic liquid crystals. *Phys. Rev. Lett.*, **22**, 1361.

107 van Eck, D.C. and Zijlstra, R.J. (1980) Spectral analysis of light intensity fluctuations caused by orientational fluctuations in nematics. *J. Phys. (France)*, **41**, 351.

108 Fellner, H., Franklin, W., and Christensen, S. (1975) Quasielastic light scattering from nematic *p*-methoxybenzylidene-*p*′-*n*-butylaniline (MBBA). *Phys. Rev. A*, **11**, 1440.

109 Hirakata, J.-I., Chen, G.-P., Toyooka, T., Kawamoto, S., Takezoe, H., and Fukuda, A. (1986) Accurate determination of K_1/η_{splay}, K_2/η_{twist} and K_3/η_{bend} in nematic liquid crystals by using photon correlation spectroscopy. *Jpn. J. Appl. Phys.*, **25**, L607.

110 Akiyama, R., Tomida, K., and Fukuda, A. (1986) Experimental confirmation of angular dependence of Rayleigh line spectral width in nematic liquid crystals. *Jpn. J. Appl. Phys.*, **25**, 769.

111 DiLisi, G.A. and Rosenblatt, C. (1992) Viscoelastic properties of a liquid-crystalline monomer and its dimmer. *Phys. Rev. A*, **45**, 5738.

112 Schmiedel, H., Stannarius, F.L., Grigutsch, M., Hirning, R., Stelzer, J., and Trebin, H.-R. (1993) Determination of viscoelastic coefficients from the optical transmission of a planar liquid crystal cell with low-frequency modulated voltage. *J. Appl. Phys.*, **74**, 6053.

113 Chigrinov, V.G. (1999) *Liquid Crystal Devices: Physics and Applications*, Artech House, Boston.

114 Takatoch, K., Hazegawa, M., Koden, M., Itoh, N., Hazekawa, R., and Sakamoto, M. (2005) *Alignment Technologies and Applications of Liquid Crystal Devices*, Taylor and Francis, London.

115 Sugimura, A. (2000) Anchoring energies for nematics in Physical properties of liquid crystals, in *Physical Properties of Liquid Crystals: Nematics* (eds D. Dunmur, A. Fukuda, and G. Luckhurst), EMIS Datareview Series 25, INSPEC, pp. 493–502.

116 Barbero, G. and Evangelista, R.E. (2006) *Absorption Phenomena and Anchoring Energy in Nematic Liquid Crystals*, Taylor and Francis, London.

117 Chigrinov, V.G., Kozenkov, V.M., and Kwok, H.S. (2003) *Optical Application of Liquid Crystals* (ed. L. Vicary), Top Publishing Ltd, pp. 201–240.

118 Faetti, S. (1991) Anchoring effects in nematic liquid crystals, in *Physics of Liquid Crystalline Materials* (eds I.C. Khoo and F. Simoni), Gordon and Breach Science Publishers, Philadelphia, PA, pp. 301–336.

119 Marusiy, T.Ya., Reznikov, Yu.A., Reshetnyak, V.Yu., and Chigrinov, V.G. (1990) Nematic liquid crystal anchoring energy with orienting substrates and methods of its measuring. *Poverhnost*, **7**, 5.

120 Rasing, Th. and Muševicš, I. (2004) *Surfaces and Interfaces of Liquid Crystals*, Springer, Berlin.

121 Deryagin, B.V., Altois, B.A., and Popovskiy, Yu.M. (1991) Orientational order of the saturated hydrocarbons layers on the quartz surface. *Doklady Akademii Nauk SSSR*, **137**, 130 (in Russian).

122 Sonin, A.A. (1995) *The Surface Physics of Liquid Crystals*, Gordon and Breach Publishers, Luxembourg.

123 Uchida, T. and Seki, H. (1991) Surface alignment of liquid crystals, in *Liquid*

Crystals Applications and Uses (ed. B. Bahadur), World Scientific, Singapore.

124 Barnik, M.I., Blinov, L.M., Korishko, T.V., Umansky, B.A., and Chigrinov, V.G. (1983) Investigation of NLC director orientational deformations in electric field for different boundary conditions. *Mol. Cryst. Liq. Cryst.*, **99**, 53.

125 Cognard, J. (1986) *Alignment of Nematic Liquid Crystals and Their Mixtures*, Gordon and Breach, London.

126 Myrvold, B. and Kondo, K. (1995) The relationship between chemical structure of nematic liquid crystals and their pretilt angles. *Liq. Cryst.*, **18**, 271.

127 Ryschenkow, G. and Kleman, M. (1976) Surface defects and structural transitions in very low anchoring energy nematic thin films. *J. Chem. Phys.*, **76**, 405.

128 Li, X.T., Pei, D.H., Kobayashi, S., and Iimura, Y. (1997) Measurements of azimuthal anchoring energy at liquid crystal/photopolymer interface. *Jpn. J. Appl. Phys*, **36**, L432.

129 Okubo, K., Kimura, M., and Akahane, T. (2003) Measurement of genuine azimuthal anchoring energy in consideration of liquid crystal molecular adsorption on alignment film. *Jpn. J. Appl. Phys.*, **42**, 6428.

130 Vilfan, M. and Cõpič, M. (2004) Surface anchoring coefficients measured by dynamic light scattering, in *Surfaces and Interfaces of Liquid Crystals* (eds Th. Rasing and I. Muševicš), Springer, Berlin, pp. 96–107.

131 Vilfan, M., Olenik, I.D., Mertelj, A., and Cõpič, M. (2001) Aging of surface anchoring and surface viscosity of a nematic liquid crystal on photoaligning poly-vinyl-cinnamate. *Phys. Rev. E*, **63**, 061709.

132 Vilfan, M., Mertelj, A., and Cõpič, M. (2002) Dynamic light scattering measurements of azimuthal and zenithal anchoring of nematic liquid crystals. *Phys. Rev. E*, **65**, 041712.

133 Vilfan, M. and Cõpič, M. (2003) Azimuthal and zenithal anchoring of nematic liquid crystals. *Phys. Rev. E*, **68**, 031704.

134 Škarabot, M., Osmanagic, E., and Muševic, I. (2006) Surface anchoring of nematic liquid crystal 8OCB on a DMOAPsilanated glass surface. *Liq. Cryst.*, **33**, 581.

135 Akiyama, R., Abe, S., Fukuda, A., and Kuze, E. (1982) Determination of tilt bias angles in nematic liquid crystal cells by observing angular dependence of Rayleigh line intensity. *Jpn. J. Appl. Phys.*, **21**, L266.

136 Akiyama, H. and Iimura, Y. (2002) Azimuthal anchoring properties of nematic liquid crystal on UV-exposed polyimide layers. *Jpn. J. Appl. Phys.*, **41**, L521.

137 Iimura, Y., Koboyashi, N., and Koboyashi, S. (1994) A new method for measuring the azimuthal anchoring energy of a nematic liquid crystal. *Jpn. J. Appl. Phys.*, **33**, L434.

138 Faetti, S., Gatti, M., Palleschi, V., and Sluckin, T.J. (1985) Almost critical behaviour of the anchoring energy at the interface between a nematic liquid crystal and a SiO substrate. *Phys. Rev. Lett.*, **55**, 1681.

139 Lee, S.H., Lee, G.-D., Yoon, T.-H., and Kim, J.C. (2004) Effect of azimuthal anchoring strength on stability in a bistable chiral splay nematic liquid crystal device. *Phys. Rev. E*, **70**, 041704.

140 Zhang, B., Sheng, P., and Kwok, H.S. (2003) Optical measurement of azimuthal anchoring strength in nematic liquid crystals. *Phys. Rev. E*, **67**, 041713.

141 Faetti, S. and Marianelli, P. (2005) Strong azimuthal anchoring energy at a nematic-polyimide interface. *Phys. Rev. E*, **72**, 051708.

142 Murauski, A., Chigrinov, V., Muravsky, A., Yeung, F.S.-Y., Ho, J., and Kwok, H.-S. (2005) Determination of liquid-crystal polar anchoring energy by electrical measurements. *Phys. Rev. E*, **71**, 061707.

143 Vorflusev, V.P., Kitzerow, H.S., and Chigrinov, V.G. (1995) Azimuthal anchoring energy in photoinduced anisotropic films. *Jpn. J. Appl. Phys.*, **34**, L1137.

144 Yokoyama, H., Kobayashi, S., and Kamei, H. (1987) Temperature dependence of the anchoring strength at a nematic liquid crystal-evaporated SiO interface. *J. Appl. Phys.*, **61**, 4501.

145 Pozhidaev, E., Chigrinov, V., and Li, X. (2006) Photoaligned ferroelectric liquid crystal passive matrix display with memorized gray scale. *Jpn. J. Appl. Phys.*, **45**, 875.

146 Chigrinov, V., Muravski, A., and Kwok, H.S. (2003) Anchoring properties of photoaligned azo-dye materials. *Phys. Rev. E*, **68**, 061702.

147 Yeung, F.S.Y. and Kwok, H.S. (2003) Truly bistable twisted nematic liquid crystal display using photoalignment technology. *Appl. Phys. Lett.*, **83**, 4291.

148 Crawford, G.P., Yang, D.K., Zumer, S., Finotello, D., and Doane, J.W. (1991) Ordering and self-diffusion in the first molecular layer at a liquid-crystal–polymer interface. *Phys. Rev. Lett.*, **66**, 723.

149 Korb, J.-P., Malier, L., Cros, F., Xu, S., and Jonas, D.J. (1996) Surface dynamics of liquids in nanopores. *Phys. Rev. Lett.*, **77**, 2312.

150 Derjaguin, B.Y., Popovskij, Yu.M., and Altoiz, B.A. (1983) Liquid-crystalline state of the wall-adjacent layers of some polar liquids. *J. Colloid Interf. Sci.*, **96**, 492.

151 Tsvetkov, V.A. (2005) Attempt of direct measuring of near surface shear viscosity. *Mol. Cryst. Liq. Cryst.*, **436**, 203.

152 Sebastião, P.J., Sousa, D., Ribeiro, A.C., Vilfan, M., Lahajnar, G., Seliger, J., and Žumer, S. (2005) Field-cycling NMR relaxometry of a liquid crystal above TNI in mesoscopic confinement. *Phys. Rev. E*, **72**, 061702.

153 Oliveira, E.A., Figueiredo, A.M., and Durand, G. (1991) Gliding anchoring of lyotropic nematic liquid crystals on amorphous glass surfaces. *Phys. Rev. A*, **44**, R825.

154 Faetti, S., Nobili, M., and Raggi, I. (1999) Surface reorientation dynamics of nematic liquid crystals. *Eur. Phys. J. B*, **11**, 445.

155 Rey, A.D. (2004) Line tension vector thermodynamics of anisotropic contact lines. *Phys. Rev. E*, **69**, 041707.

156 Tsoy, V.I. (1995) Freedericsz transition dynamics in a nematic layer with a surface viscosity. *Mol. Cryst. Liq. Cryst.*, **264**, 51.

157 Jánossy, I. and Kósa, T.I. (2004) Gliding of liquid crystals on soft polymer surfaces. *Phys. Rev. E*, **70**, 052701.

158 Alexe-Ionescu, A.L., Uncheselu, C., Lucchetti, L., and Barbero, G. (2007) Phenomenological model for the optically induced easy direction. *Phys. Rev. E.*, **75**, 021701.

159 Pasechnik, S.V., Chigrinov, V.G., Shmeliova, D.V., Tsvetkov, V.A., Kremenetsky, V.N., Zhijian, L., and Dubtsov, A.V. (2006) Slow relaxation processes in nematic liquid crystals at a weak surface anchoring. *Liq. Cryst.*, **33**, 175.

160 Joly, S., Antonova, K., Martinot-Lagarde, P., and Dozov, I. (2004) Zenithal gliding of the easy axis of a nematic liquid crystal. *Phys. Rev. E*, **70**, 050701.

161 Pasechnik, S.V., Dubtsov, A.V., Shmeliova, D.V., Tsvetkov, V.A., and Chigrinov, V.G. (2008) Effect of combined action of electric field and light on a gliding of easy axis in nematic liquid crystals. *Liq. Cryst.*, **35**, 569.

162 Kedney, P.J. and Leslie, F.M. (1998) Switching in a simple bistable nematic cell. *Liq. Cryst.*, **24**, 613.

163 Durand, G.E. and Virga, E.G. (1999) Hydrodynamic model for surface nematic viscosity. *Phys. Rev. E*, **59**, 4137.

164 Petrov, A.G., Ionescu, A.Th., Versache, C., and Scaramuzza, N. (1995) Investigation of flexoelectric properties of a palladium-containing nematic liquid crystal, Azpac, and its mixtures with MBBA. *Liq. Cryst.*, **19**, 169.

165 Mertelj, A. and Cǒpič, M. (1998) Surface-dominated orientational dynamics and surface viscosity in confined liquid crystals. *Phys. Rev. Lett.*, **81**, 28.

166 Khazimullin, M.V., Börzsönyi, T., Krekhova, A.P., and Lebedev, Yu.A. (1999) Orientational transition in nematic liquid crystal with hybrid alignment under oscillatory shear. *Mol. Cryst. Liq. Cryst.*, **329**, 247–254.

167 Pasechnik, S.V., Nasibullayev, I.Sh., Shmeliova, D.V., Tsvetkov, V.A., and Chigrinov, V.G. (2006) Oscillating Poiseuille flow in the hybrid cell with UV activated anchoring surface. *Liq. Cryst.*, **33**, 1153.

168 Chigrinov, V., Pikin, S., Verevochnikov, A., Kozenkov, V., Khazimullin, M., Ho, J., Huang, D.D., and Kwok, H.-S. (2004) Diffusion model of photoaligning in azo-dye layers. *Phys. Rev. E*, **69**, 061713.

169 Faetti, S., Mutinati, G.C., and Gerus, I. (2004) Measurements of the azimuthal anchoring energy at the interface between a nematic liquid crystal and photosensitive polymers. *Mol. Cryst. Liq. Cryst.*, **421**, 81.

170 Lu, X., Lee, F.K., Sheng, P., Kwok, H.S., Chigrinov, V., and Tsui, O.K.C. (2006) Substrate patterning for liquid crystal alignment by optical interference. *Appl. Phys. Lett.*, **88**, 243508.

6
Liquid Crystals for Display and Photonics Applications

This chapter is devoted to optimal rheological properties of liquid crystals (LCs) for applications in displays and photonics. We will provide a general insight into various electrooptical modes in LC with the purpose of (i) explaining the basic characteristics of the effects and their dependence on LC physical parameters; (ii) correlation of the LC rheological properties (elastic and viscosity constants, dielectric and optical anisotropy, type of LC alignment, and surface energy) with the application requirements. In this chapter, we will consider LC display (LCD) and photonics applications. There are few excellent books related to LC displays [1–4] and photonics [2], so we do not need to provide an extensive description of the quality criteria of the LC applications in these fields. An active matrix (AM) LCD based on thin film transistor (TFT) technology is the most important and up-to-date LCD development, which will soon dominate in the display market. However, passive matrix LCD (PM-LCD) still covers a substantial market share, and we will consider only this part of LCD trend. So, we are going to consider in this chapter both AM and PM-LCD applications. Low power consumption LCD with memory effects will be also highlighted. Finally, we will pay a special attention to a new trend in LC development in photonics [5]: passive optical elements for fiber optical communication systems (DWDM components).

6.1
Electrooptical Effects in Liquid Crystals

6.1.1
Electrically Controlled Birefringence

6.1.1.1 Static Director Distribution

The most important geometries of electrically controlled birefringence (ECB) are shown in Figure 6.1. A compromise between dielectric and elastic torques results in director reorientation from the initial alignment $\theta(z)$ with the maximum deviation θ_m at the center of the layer (the Fréedericksz transition). The effect occurs when the electric field exceeds a certain threshold value:

$$U_F = \pi(K_{ii}/\varepsilon_0\,\Delta\varepsilon)^{1/2} \quad \text{or} \quad B_F = (\pi/d)(\mu_0 K_{ii}/\Delta\chi)^{1/2}, \tag{6.1}$$

Liquid Crystals: Viscous and Elastic Properties
S. V. Pasechnik, V. G. Chigrinov, and D. V. Shmeliova

ISBN: 978-3-527-40720-0

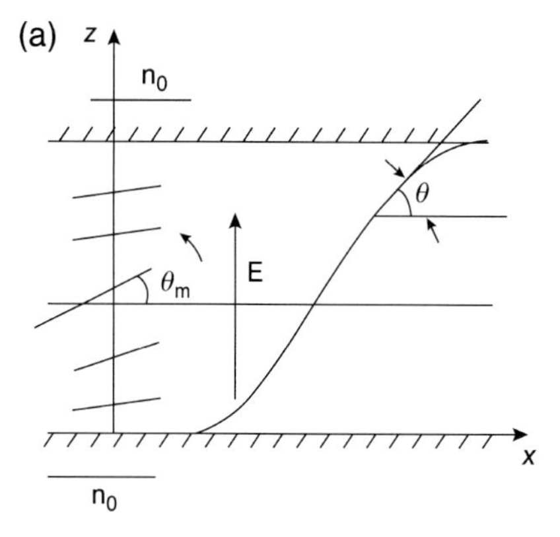

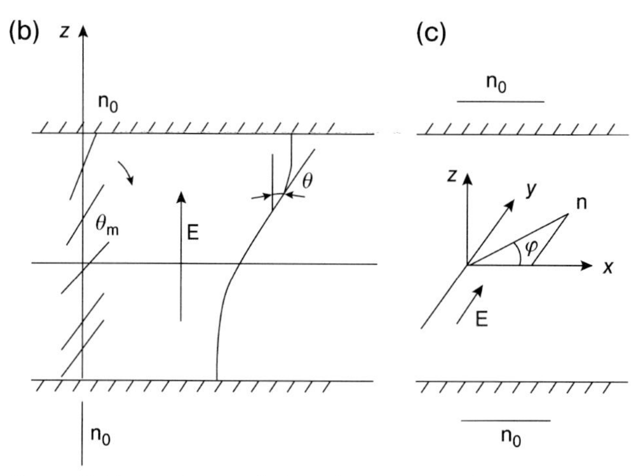

Figure 6.1 Electrically controlled birefringence mode for various LC configurations: (a) S-effect (homogeneous alignment, electric field is vertical or perpendicular to the substrates); (b) B-effect or VAN-effect (homeotropic alignment); (c) T-effect (homogeneous alignment, electrical field is horizontal or parallel to the substrates) [6].

where $K_{ii} = K_{11}$ or K_{33} for the splay (S) and bend (B) Fréedericksz transitions, respectively (Figure 6.1a and b). The initial director alignment is homogeneous ($\mathbf{E} \perp \mathbf{no}$) for the S Fréedericksz transition (S-effect) and homeotropic ($\mathbf{E} \| \mathbf{no}$) for the B-effect. As it follows from (2.40), the S-(B-)effect appears for the corresponding positive (negative) values of the dielectric anisotropy $\Delta\varepsilon$. The dielectrically isotropic point $\Delta\varepsilon = 0$ is stable for both types of the transitions.

Let us consider first the splay Fréedericksz transition or the S-effect. (All the expressions are also valid for the B-effect if the following exchange of parameters are made: $K_{11} \Leftrightarrow K_{33}$, $\varepsilon_{\|} \Leftrightarrow \varepsilon_{\perp}$, $n_{\|} \Leftrightarrow n_{\perp}$.) In this case, the director profile $\mathbf{n}$ takes the form (Figure 6.1a)

$$\mathbf{n}(z) = (\cos\theta(z), 0, \sin\theta(z)), \tag{6.2}$$

where the z-axis goes perpendicular to the substrates. Minimization of (2.42), together with the strong anchoring boundary conditions,

$$\theta(z=0)=\theta(z=d)=0,$$

allows us to obtain the following relationships for the director distribution $\theta(z)$ over the nematic layer:

$$2z/d=\int_0^{\psi(z)} G(x)\,\mathrm{d}x\Big/\int_0^{\pi/2} G(x)\,\mathrm{d}x,\quad G(x)=(G_k G_\gamma/G_{-1})^{1/2},$$
$$G_\alpha=1+\alpha\eta_\mathrm{m}\sin^2(x),\quad \alpha=k,\gamma,-1, \tag{6.3}$$
$$\psi(z)=\arcsin(\sin\theta(z)/\sin\theta_\mathrm{m}),\quad k=K_{33}/K_{11}-1,$$
$$\gamma=\varepsilon_\parallel/\varepsilon_\perp,\quad \eta_\mathrm{m}=\sin^2\theta_\mathrm{m}.$$

The director angle at the center of the layer θ_m is a function of the applied voltage U:

$$U/U_\mathrm{s}=2/\pi\,(1+\gamma\eta_\mathrm{m})^{1/2}\int_0^{\pi/2}(G_k/G_\gamma G_{-1})^{1/2}\,\mathrm{d}x, \tag{6.4}$$
$$U_\mathrm{s}=\pi\,(K_{11}/\varepsilon_0\,\Delta\varepsilon)^{1/2},\quad U\geq U_\mathrm{s}$$

As it follows from (6.4), for $\theta_\mathrm{m}=\eta_\mathrm{m}=0$ the voltage equals the threshold value $U=U_\mathrm{s}$.

In a real nematic cell, however, it is very difficult to satisfy the conditions (6.2). Moreover, to avoid degenerate solutions, when both clockwise and counterclockwise director rotations are possible (Figure 6.1), the boundaries are specially prepared with a tilted director orientation:

$$\theta(z=0)=\theta(z=d)=\theta_0\neq 0,\pi/2. \tag{6.5}$$

In this case, the threshold of the Fréedericksz transition disappears, that is, the deformation of the director alignment begins at infinitely small voltages [6, 7]. Figure 6.2 shows that for small values of θ_0 the corresponding electrooptic response of the nematic cell assumes a quasithreshold character [6].

The electric field E in a homogeneous liquid crystal layer satisfies Maxwell equation:

$$\mathrm{div}\,\mathbf{D}=Q, \tag{6.6}$$

where the space charge Q is equal to zero only if a nematic liquid crystal is considered as a pure dielectric. As the electric displacement vector $\mathbf{D}=[0,\ 0,\ D_z=\varepsilon_0\ (\varepsilon_\perp+\Delta\varepsilon\sin^2\Theta)]$ in case of a uniform director distortion, we have from (6.6) $D_z=\mathrm{const.}$ for $Q=0$. Consequently, large dielectric anisotropy leads to nonuniformity in the distribution of the field in a deformed liquid crystal layer.

6.1.1.2 **Effect of a Weak Anchoring at the Boundaries**

It should again be emphasized that in contrast to a tilted orientation where there is no threshold, the threshold for the Fréedericksz transitions shown in Figure 6.2

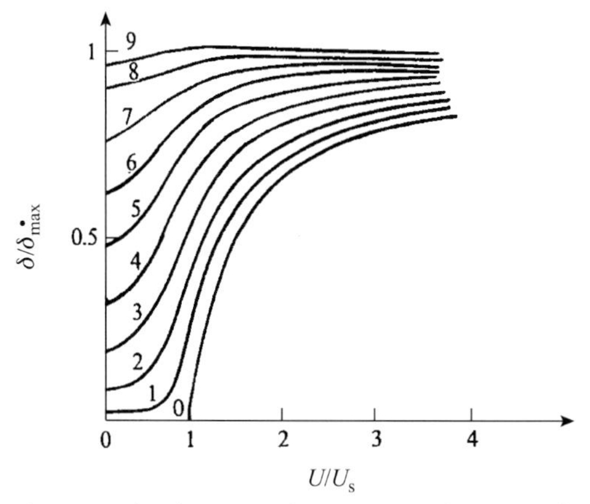

Figure 6.2 The electrooptical response (birefringence) of nematic cells with tilted director orientation at the boundaries (6.5) [6, 7]. δ/δ_{max} is the relative phase difference or birefringence of the cells. The director tilt angle is $\pi k/20$, where $k = 0, \ldots, 9$ is the number of the curve. The threshold is observed only for $\Theta_0 = 0$ ($k = 0$).

remains for a weak director boundary anchoring (2.37), but has a lower value. The corresponding equation is [6, 7]

$$\cot\text{an}\left[\pi U_F(W)/U_F(\infty)\right] = \pi K_{ii} U_F(W)/W d U_F(\infty), \tag{6.7}$$

where the effective elastic coefficient K_{ii} and the Fréedericksz transition threshold U_F are defined for finite (W) and infinite (∞) anchoring energies. For large anchoring energies, expression (6.7) reduces to

$$U_F(W) = U_F(\infty)(1-2K_{ii}/Wd), \quad Wd/K_{ii} \gg 1. \tag{6.8}$$

The dependence of the normalized Fréedericksz threshold on parameter $\lambda = \pi K_{ii}/Wd$ is shown in Figures 6.3 and 6.4 [8].

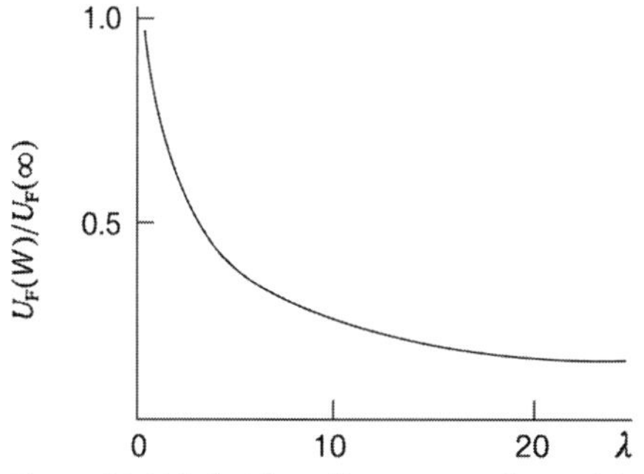

Figure 6.3 Fréedericksz effect in case of a weak boundary coupling of the LC director. Effective threshold as a function of the anchoring parameter $\lambda = \pi K_{ii}/Wd$ [8].

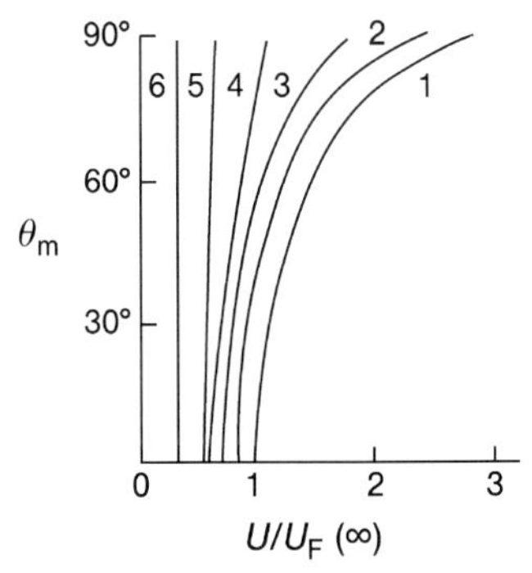

Figure 6.4 Fréedericksz effect in case of a weak boundary coupling of the LC director. Maximum deviation angle θ_m of the director at the center of the layer (Figure 6.1) for various anchoring parameters: (1) $\lambda = 0$; (2) $\lambda = 0.2$; (3) $\lambda = 0.5$; (4) $\lambda = 1$; (5) $\lambda = 2$; and (6) $\lambda = 10$ [8].

It is interesting to note that weak anchoring of the molecules to the walls results not only in a lower threshold but also in a stronger dependence of the angle of deviation of the director upon the external field. Curves of θ_m, the maximum angle of deviation at the center of the cell for different initial planar orientations, are given in Figure 6.4b. Physically, a steeper increase in the angle θ_m with the field is explained by a concerted rotation of the director throughout the depth of the layer, including the boundary zones and, hence, by a smaller reaction of elastic forces in these regions.

Methods of evaluating the anchoring energy and the corresponding experimental data are given in Chapter 5. We should note, however, that the value of the effective anchoring energy, defined from (5.124) and (5.125), depends on the form of the potential barrier $W(\theta)$. For sufficiently small values of anchoring energy, the electro-optical response of the nematic cell becomes infinitely steep, so that for a certain critical value of W a hysteresis and first-order Fréedericksz transition is possible [8].

In the case of finite anchoring, there also exists a saturation voltage for the total director reorientation, parallel (Figure 6.1a) or perpendicular (Figure 6.1b) to the field, when the boundary regions disappear [8]. Development of the nematic cells with a good and reliable control of anchoring energy is of great importance for applications, as very steep transmission–voltage curves (TVCs), memory states, and improved response times could be realized [6, 7].

6.1.1.3 **Dynamics of the Director Motion: Backflow Effect**

It is easiest to examine the kinetics of the director motion for a pure twist or T deformation (Figure 6.1c), which is not accompanied by a change in the position of centers of gravity of molecules. In this case, the electric or magnetic field is applied parallel to the substrates, exerting a torque, which causes a pure twist deformation of the initially homogeneous director alignment. In contrast, for S- and B-Fréedericksz transitions, the rotation of the director is accompanied by such a change in position, that is, by the movement of the liquid (backflow). In order to allow this backflow effect, the equation of the director motion is coupled to that of the fluid.

For a pure T deformation, the equation of the director motion expresses the balance between the torques due to the elastic and viscous forces and the external field (and does not contain the fluid velocity) [6, 7]

$$K_{22}\partial^2\varphi/\partial z^2 + \mu_0^{-1}\,\Delta\chi B^2 \sin\varphi\cos\varphi = \gamma_1\partial\varphi/\partial t. \tag{6.9}$$

This equation describes the director rotation in a magnetic field **B** with the inertia term $I\partial^2\varphi/\partial t^2$, which is omitted, $\gamma_1 = \alpha_3 - \alpha_2$ is a rotational viscosity, and α_i are Leslie viscosity coefficients. Equation 6.9 in the limit of small φ angles transforms to

$$K_{22}\partial^2\varphi/\partial z^2 + \mu_0^{-1}\,\Delta\chi B^2\varphi = \gamma_1\partial\varphi/\partial t \tag{6.10}$$

with the solution

$$\varphi = \varphi_m \exp(t/\tau_r)\sin(\pi z/d), \tag{6.11}$$

where $\tau_r = \gamma_1/(\mu_0^{-1}\,\Delta\chi B^2 - K_{22}\pi^2/d^2)$ is the reaction or switching on time. The solution (6.11) satisfies the strong anchoring boundary conditions

$$\varphi(z=0) = \varphi(z=d) = \varphi_m$$

and assumes a maximum value at the center of the layer $\varphi(z = d/2) = \varphi_m$.

Corresponding relaxation or decay times are found from (6.11) for $B=0$ in a similar way:

$$\tau_r = \gamma_1 d^2/K_{22}\pi^2. \tag{6.12}$$

Characteristic times remain the same when describing the dynamics of small T deformations in an electric field if the following substitution is made:

$$\mu_0^{-1}\,\Delta\chi B^2 \Rightarrow \varepsilon_0\,\Delta\varepsilon E^2. \tag{6.13}$$

Unlike T deformation, the director reorientation in S- and B-effects is always accompanied by a macroscopic flow of a nematic liquid crystal (backflow) with the velocity $\mathbf{V} = (V(z),\ 0,\ 0)$, where the z-axis goes perpendicular to the substrates and the deformations take place in the x, z-plane. The velocity V includes only the x-component because the z-component is zero according to the continuity equation (div $\mathbf{V} = 0$), and the y-component vanishes due to the symmetry of the problem (Figure 6.5).

For small variations of the angle θ, the characteristic times of the S- and B-effect (Figure 6.5) can be found from the solutions of the coupled dynamic equations of the nematic director $\theta(z)$ and velocity $V(z)$ in the following form:

$$\tau_r = \gamma_1^*/(\varepsilon_0\,\Delta\varepsilon E^2 - K_{ii}\pi^2/d^2), \qquad \tau_d = \gamma_1^*/(K_{ii}\pi^2/d^2), \tag{6.14}$$

where γ_1^* is the effective rotational viscosity. We have

$$\begin{aligned} \gamma_1^* &= \gamma_1 - 2\alpha_3^2/(\alpha_3 + \alpha_4 + \alpha_6), \qquad K_{ii} = K_{11} \quad \text{for S-effect},\\ \gamma_1^* &= \gamma_1 - 2\alpha_2^2/(\alpha_4 + \alpha_5 - \alpha_2), \qquad K_{ii} = K_{33} \quad \text{for B-effect}. \end{aligned} \tag{6.15}$$

The relationships (6.15) show that backflow considerably alters the response times and should be taken into account.

The relative difference $(\gamma_1 - \gamma_1^*)/\gamma_1$ becomes close to zero for voltages slightly above the threshold in the S-effect and very high voltages in the B-effect and can hardly

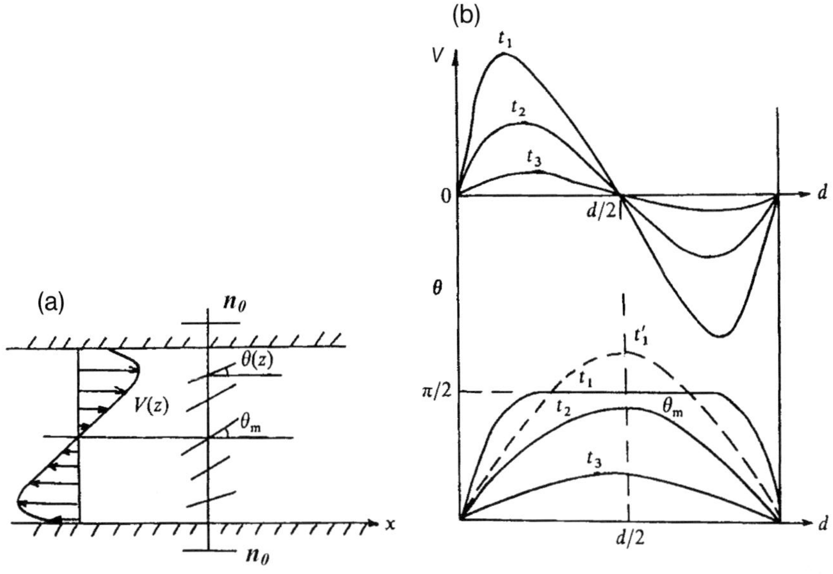

Figure 6.5 (a) Backflow effect in the process of the director reorientation. (b) The dependence of the velocity V and director angle θ on the coordinate z is given at different times: $t_1 < t_2 < t_3$. The dashed line shows possible nonmonotonous relaxation of θ at high voltages: $t_1 < t'_1 < t_2 < t_3$ [6].

exceed 50% within the whole voltage interval. The results of computer simulation were confirmed in experiment [6, 7].

For sufficiently high voltages $U/U_S > 4$ (U_S is the S-effect threshold), the director relaxation at the center of the layer goes in a nonmonotonous way (Figure 6.5). At the first stage of the relaxation, directions of θ rotations at the center of the cell and at the boundaries are opposite due to the large elastic energy accumulated in the boundary layers. This results in an increase in the maximum director angle θ_m up to $\theta_m \approx 3\pi/4$ at the first moment of relaxation, after which θ_m usually comes to zero (Figure 6.5).

Weak boundary anchoring decreases the rise time τ_r and increases the relaxation time τ_d. It can be easily understood if we make the substitution $K_{ii}\pi^2/d^2 \Rightarrow 2W/d$ in formulas (6.15), which proves to be correct for sufficiently low anchoring energies [6, 7].

6.1.1.4 **Optical Response**

To understand the optical characteristics of a liquid crystal layer in electrically controlled birefringence effect, let us consider the geometry of Figure 6.1a with the initial homogeneous director orientation along x-axis. If the applied voltage is below the threshold, the nematic liquid crystal layers manifest birefringence $\Delta n = n_e - n_o = n_{\|} - n_{\perp}$. When the field exceeds its threshold value, the director deviates from its orientation along the x-axis while remaining perpendicular to the y-axis. Therefore, the refractive index for the ordinary ray remains unchanged, $n_o = n_{\perp}$. At the same time, the refractive index for the extraordinary ray (n_e) decreases, tending toward n_o. It is simple to relate the magnitude of n_e to the orientation angle of the director $\theta(z)$ [6, 7]:

$$n_e(z) = n_e n_o \Big/ \sqrt{n_e^2 \sin^2\theta(z) + n_o^2 \cos^2\theta(z)}. \tag{6.16}$$

The phase difference between the extraordinary and the ordinary ray for a monochromatic light of wavelength λ is found by integrating over the layer depth (d):

$$\Delta\phi = 2\pi/\lambda \int_0^d \left(n_e(z) - n_o\right) dz = 2\pi d\langle\Delta n(z)\rangle/\lambda \tag{6.17}$$

The intensity of the light passing through the cell depends on the angle φ_0 between the polarization vector of the incident beam and the initial orientation of the director of the nematic liquid crystal

$$I = I_0 \sin^2 2\varphi_0 \sin^2(\Delta\Phi/2), \tag{6.18}$$

where I_0 is the intensity of the plane-polarized light incident on the cell. Hence, the external magnetic or electric field changes the orientation of the director, $\theta = \theta(E, z)$ and, consequently, the values of $\Delta n(E, z)$ and $\Delta\Phi(E)$. A change in the phase difference $\Delta\Phi$, in turn, results in an oscillatory dependence of the optical signal at the exit of the analyzer. The maximum amplitude of these oscillations corresponds to an angle $\varphi_0 = 45°$ and the maximum possible number of oscillations (e.g., the number of maxima during a complete reorientation of the director) is approximately $(n_{\parallel} - n_{\perp})d/\lambda$.

The characteristic curves of transmitted intensity in the ECB effect are shown in Figure 6.6. The contrast ratio I_{max}/I_{min} is not, in principle, limited (6.18), but is determined in experiment by the quality of the initial director orientation. The output light, in general, becomes elliptically polarized, so that its ellipticity e and the angle ψ, between the long ellipse axis and the polarizer, also depend on φ_0 and $\Delta\Phi$ [6, 7]:

$$\begin{aligned} e &= \tan[1/2 \arcsin(\sin 2\varphi_0 \sin \Delta\Phi)] \\ \tan 2\psi &= \tan 2\varphi_0 \cos \Delta\Phi. \end{aligned} \tag{6.19}$$

The experimental dependences of $e(U)$ and $\psi(U)$ on the applied voltage resemble those shown in Figure 6.6 for the intensity curve $I(U)$ [6, 7].

The effect of phase modulation for an initial planar orientation of the director (along x-axis) with positive dielectric or diamagnetic anisotropy ($\Delta\varepsilon, \Delta\chi > 0$) and with the field applied along the z-axis (Figure 6.1a) is called the S-effect [7], since the initial deformation is a splay deformation, even though a bend deformation is also induced above the threshold.

The applied field causes a phase difference to arise between the ordinary and extraordinary rays and the intensity of the light oscillates in accordance with (6.18). In case the bend deformation is now in the initial stages of its development (Figure 6.1b), the corresponding electrooptical effect is called B-effect, taking place for negative values of a dielectric anisotropy $\Delta\varepsilon < 0$. However, here the final orientation of the director is not defined (degenerate), and the sample does not remain monodomain and contains many specific defects (umbifics) [6, 7]. In

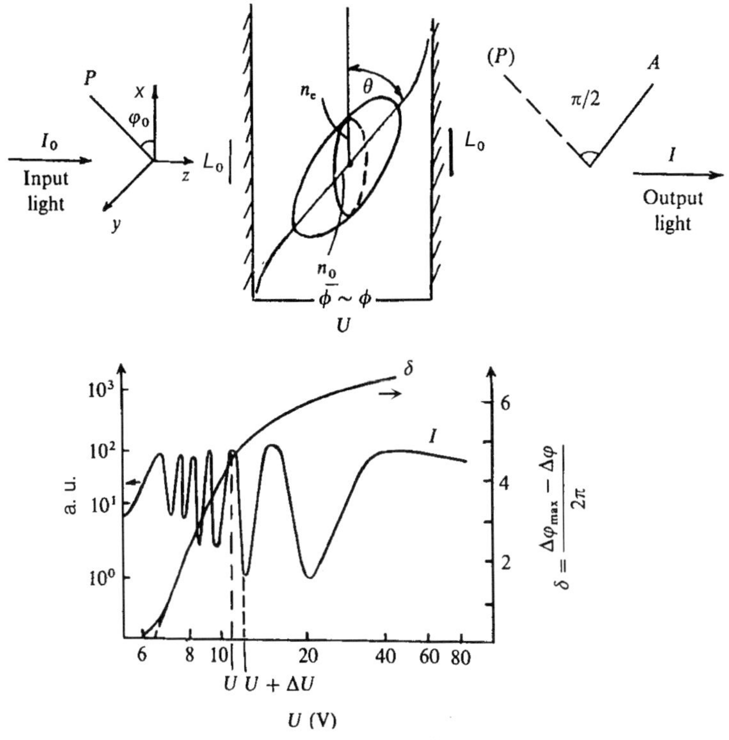

Figure 6.6 Electrically controlled birefringence in homogeneous liquid crystal cell (S-effect). Polarizer (P) is crossed with the analyzer (A) and set at the angle φ_0 with respect to the initial liquid crystal director. Below the intensity I of the transmitted light and phase difference δ are shown versus the applied voltage. The phase π-switching regime is indicated, which results in the variation of the intensity from I_{max} to I_{min} [6].

principle, the preferred direction of the final orientation of the director can be established, and these defects in the structure eliminated by special preparation of the surfaces with a slight pretilt.

The electrooptical (or magnetooptical) effect corresponding to the geometry in Figure 6.1c is called the T-effect. It is caused by a pure twist deformation and is amenable to theoretical analysis. Unfortunately, it is less suitable for experimental investigation since the field induces optical biaxiality in the nematic liquid crystal. The effect must not be confused with the twist effect, which includes combined T, S, and B deformations (to be discussed below).

An important problem of the ECB effect is to provide full switching from I_{max} to I_{min} with minimal response times. According to (6.18), such a switching is stipulated when the phase difference $\Delta\Phi$ is changed as much as by π. If a nematic cell is subjected to voltage U corresponding to a maximum intensity I_{max}, then to attain another state with I_{min} we have to supply an additional voltage $U_\pi \geq \Delta U$, where ΔU is the minimum possible value of U_π (Figure 6.6).

The main disadvantage of the S-effect for display applications is the strong dependence of the transmitted intensity on the light wavelength, and the nonuniform transmission–voltage characteristics at oblique light incidence. It is possible, however, to avoid them by placing a compensating birefringent plate between the liquid crystal cell and one of the polarizers or by using two nematic S-cells in series that have perpendicular initial directors [6, 7].

Special attention should be paid to the so-called π-cells, when the birefringent intensity is switched in the last fall of the oscillation curve (Figure 6.6) [6, 7]. In this case, switching is attained due to the very slight variation of the director distribution within the narrow regions near the boundaries, thus resulting in a very high response speed. The corresponding switching times can be estimated according to the formula [9]:

$$\tau \approx (\Delta/\pi/2\pi)^2 1/(1-\beta U_\pi/U_0)^2 \gamma_1 \lambda^2/K_{11}\,\Delta n^2, \tag{6.20}$$

where $\Delta/\pi \approx 1$ is a relative phase difference for the last intensity fall and $\beta \approx 1$ is the liquid crystal material constant. As we may see, the response time does not depend on the cell thickness (compare with (6.14)). Combining π-cells with phase retardation plates, both with a positive and with a negative phase shifts, it is possible to optimize a contrast and color uniformity of the liquid crystal cell [2]. B-effect in homeotropic or quasihomeotropic (slightly tilted) nematic samples remains attractive for applications, including displays with high-information content [7]. With good homeotropic orientation of a nematic liquid crystal, the B-effect is characterized by a steep growth in the optical transmission with voltage, that is, the threshold is very abrupt. This is due to the very weak light scattering of the homeotropically oriented layer, and the complete absence of birefringence in the initial state with crossed polarizers [6, 7]. Basically, the patterns observed experimentally are similar for the S- and B-effect, including dynamic behavior. Let us consider the transmission–voltage curves $I(U)$ for the B-effect at voltages slightly exceeding the threshold value, which are important for applications. As already mentioned, to avoid degeneracy in the director reorientation, a slight initial pretilt from the normal to the substrate is needed $\theta_0 \approx 0.5$–1° (Figure 6.1b). The electrooptical response is very sensitive to the θ_0 value, that is, small $\theta_0 < 0.5^\circ$ do not allow us to avoid defects, while larger angles, $\theta_0 > 1^\circ$, strongly reduce the contrast [6, 7]. According to (6.18), for the voltages slightly exceeding the threshold, the contrast is

$$C = I_{\text{on}}/I_{\text{off}} \approx \sin^2(A\langle\theta^2_{\text{on}}\rangle)/\sin^2(A\theta^2_{\text{off}}) \approx \theta^4_{\text{m}}/4\theta^2_0, \tag{6.21}$$

where $\langle\theta^2_{\text{on}}\rangle = d^{-1}\int_0^d \theta^2(z)\,dz \approx \theta^2_{\text{m}}/2$ for small θ and $\theta_{\text{off}} = \theta_0$ are the director angles in the switched-on and switched-off states and $A = \pi d n_\perp(1-n^2_\perp/n^2_\parallel)/2\lambda$ is a phase factor. As seen from (2.30), the contrast ratio C crucially depends on the pretilt angle θ_0. By using the expression similar to (2.14), we can obtain the following relationship for Θ^2_{m} in the B-effect:

$$\Theta_{\rm m}^2 = 4(U/U_{\rm B}-1)(K_{11}/K_{33}+\varepsilon_{||}/\varepsilon_{\perp})^{-1}, \tag{6.22}$$

where $U_{\rm B} = \pi\ [K_{33}/(\varepsilon_0|\Delta\varepsilon|)]^{1/2}$ is the B-effect threshold voltage. By using expression (6.18) for $\varphi_0 = \pi/4$, we derive the following relationship for the optical transmission $T = I/I_0$ in the ECB effect in a homeotropic nematic (B-effect):

$$\begin{aligned} T &= \sin^2 \pi d\,\Delta n \lambda^{-1} \langle \Theta_{\rm on}\rangle^2 \approx 4A^2(U/U_{\rm B}-1)^2(K_{11}/K_{33}+\varepsilon_{\perp}/\varepsilon_{||}-1)^{-2} \\ &\approx [2\pi d\lambda^{-1}(n_{||}-n_{\perp})/(K_{11}/K_{33}+\varepsilon_{\perp}/\varepsilon_{||}-1)]^2(U/U_{\rm B}-1)^2. \end{aligned} \tag{6.23}$$

As seen from (6.23), a steep electrooptical response of the cell is attained for sufficiently large values of the optical path difference $d(n_{||} - n_{\perp})$ and the elasticity anisotropy K_{33}/K_{11}, as well as for small dielectric anisotropy $|\Delta\varepsilon|/\varepsilon_{||}$. However, the values of d cannot be too large since the latter results in an increase in response times (6.14), while small values of the dielectric anisotropy lead to a growth in operating voltages. Thus, to develop a good ECB material, a compromise is needed. In order to obtain the steep electrooptical characteristics required for displays with high-information content, together with a fast response and uniformity of transmission for the oblique light incidence, the following parameters of nematic cells for the B-effect can be proposed [6, 7]: small thickness, $d < 5\ \mu$m; small pretilt angle $\theta_0 \approx 0.5$–$1°$; large optical retardation $d\Delta n \approx 1\ \mu$m; large K_{33}/K_{11} ratio; and small dielectric anisotropy $|\Delta\varepsilon|/\varepsilon_{||} < 0.5$. The regime of π-cell switching is shown in Figure 6.7.

Optically compensated π-cell is called optically compensated bend (OCB) cell and was first proposed by Uchida and Miyashita [10]. The mode comprises a bend-aligned liquid crystal layer and a biaxial phase retardation plate (Figure 6.8).

The special choice of a phase compensator, LC cell parameters, and surface pretilt angle made OCB mode very fast and attractive for applications in field sequential color (FSC) LCD, where R, G, and B images subsequently comes to the screen, thus (i) increasing the efficiency and resolution of the LCD due to the absence of color filters; (ii) improving the color coordinates of the image as flashing light emitting diodes (LEDs) are used for LCD backlighting system (Figure 6.9).

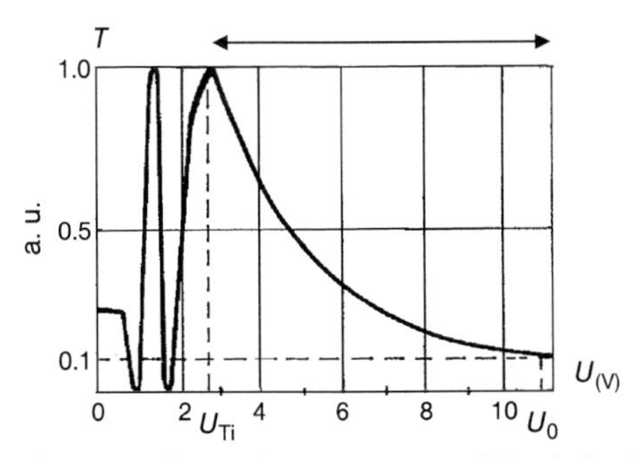

Figure 6.7 The switching regime in π-cell. For the last intensity fall, the controlling voltage has to change from U_π to U_0 [6, 7].

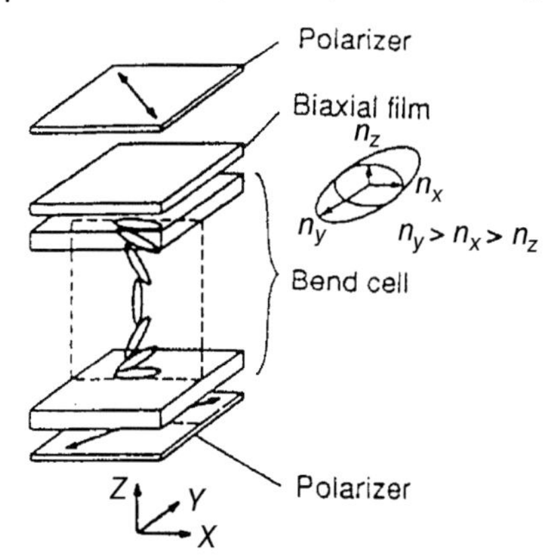

Figure 6.8 Optically compensated bend (OCB) mode [10].

High response speeds at lower switching voltages were achieved by adding diaryl-ether derivatives to negative- and positive-type nematic LC mixtures [11]. The nonplanar diaryl-ethers penetrated between the liquid crystal molecules, keeping them slightly apart, thus resulting in slightly reduced molecular interactions (Figure 6.10).

The effect of reduced molecular interactions was directly seen as a reduction in the rotational viscosities. Finally, the overall effect of reducing the rotational viscosity of liquid crystal mixtures with the use of anisotropic dopants results in improved switching speeds and decreased switching voltages. The described diaryl-ethers do not cause any change in the off state transmittance of the negative-type liquid crystals with initial homeotropic alignment, meaning there is no need to modify the alignment layers, and furthermore the contrast ratios remain unchanged. The voltage holding ratio (VHR), which characterized the ionic purity of LC mixture of the resulting mixtures, also remained almost unchanged.

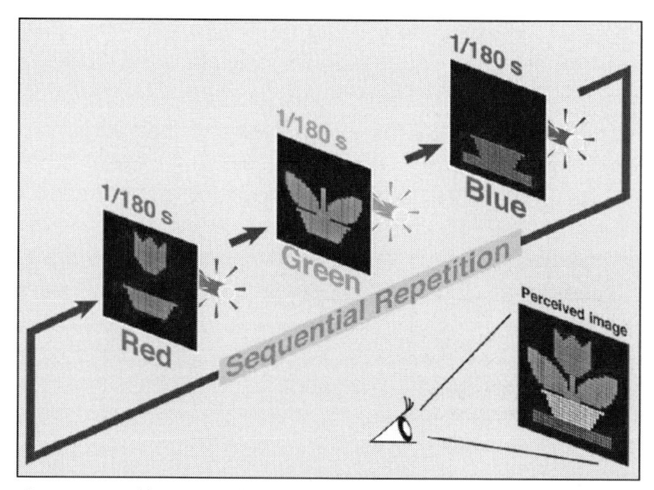

Figure 6.9 Field sequential color (FSC) LCD.

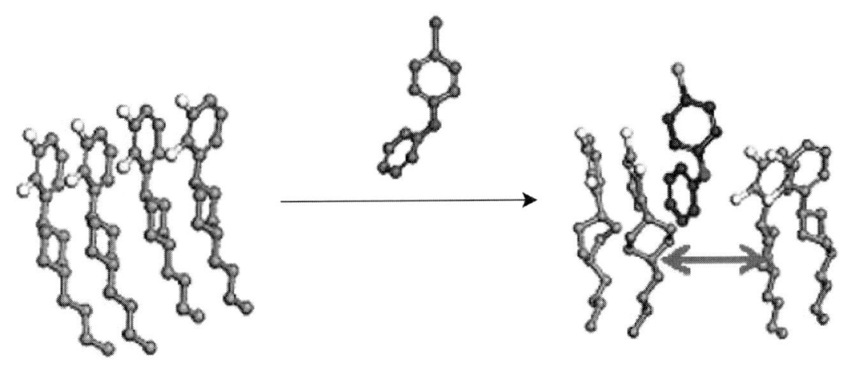

Figure 6.10 Decrease in the order parameter of LC mixture by adding diaryl-ether derivatives to negative- and positive-type nematic LC mixtures [11].

A dual fringe field switching (DFFS) mode was demonstrated to achieve a fast response time [12]. The LC employed has a positive dielectric anisotropy $\Delta\varepsilon > 0$ and the thickness of 10 μm to accumulate sufficient phase change (Figure 6.11). The electrodes on both substrates were in complementary positions to enhance the light efficiency and transmission uniformity.

DFFS mode was useful for wide view display and submillisecond response LC modulator applications.

Actually, there is no need for the application of the bias voltage to get OCB structure (no-bias bend LCD) if the high pretilt angles of the bend LC state can be initially generated (Figure 6.12) [13]. The properties and process conditions of the new nanostructured alignment layer were investigated for the purpose [13]. By forming nanodomains of two different kinds of polyimide (PI) with a "vertical" and

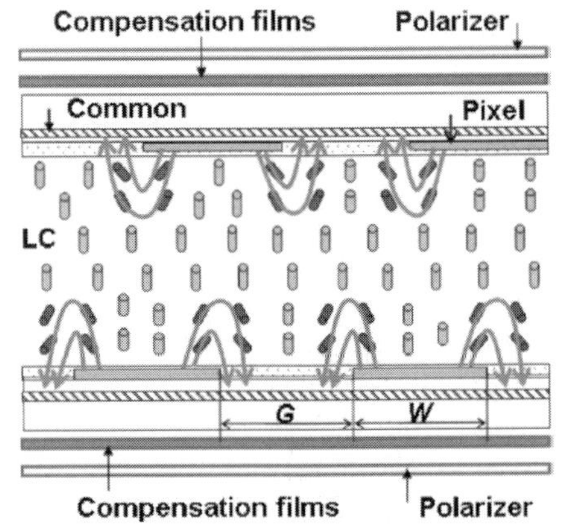

Figure 6.11 DFFS mode of homeotropically aligned LC with a positive dielectric anisotropy $\Delta\varepsilon > 0$ in a voltage on state [12].

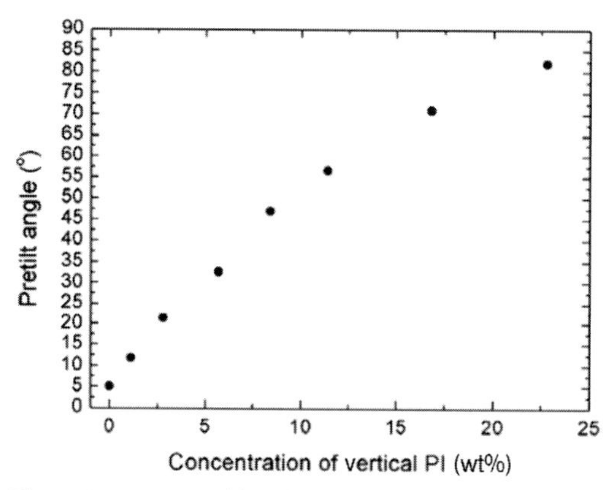

Figure 6.12 Measured liquid crystal pretilt angle as a function of concentration of the vertical polyimide [13].

"horizontal" LC alignment, it was possible to obtain any pretilt angle from 0° to 90°. These nanostructured alignment layers were robust, resilient, and reproducible. Even though the pretilt angle depends on many parameters, such as concentration of "vertical polyamide," baking temperature, anchoring energy, and so on, once these parameters are fixed, the pretilt angle produced is uniquely determined (Figure 6.12).

The combination of a nonaligning polymer and a conventional homogeneous polyimide can give homogeneous alignment with adjustable anchoring energies [13]. This is due to the averaging effect of strong and no anchoring of the two individual components. The concept of nanodomains of different materials can also be extended to include photoalignment [14].

6.1.2 Twist Effect

If x- and y-directions of the planar orientation of nematic liquid crystal molecules on opposite electrodes are perpendicular to each other and the material has a positive dielectric anisotropy $\Delta\varepsilon > 0$, then when an electric field is applied along the z-axis (Figure 6.13) a reorientation effect occurs that is a combination of the S, B, and T deformation [6, 7].

In the absence of the field, the light polarization vector follows the director and, consequently, the structure rotates the polarization plane up to an angle characterizing the structure $\varphi_m = \pi/2$ (Figure 6.13). This specific wave guide regime (the Mauguin regime) takes place when

$$\Delta nd/\lambda \gg 1. \tag{6.24}$$

When the applied voltage exceeds a certain threshold value

$$U_{tw} = \pi\{[K_{11} + (K_{33} - 2K_{22})/4]/\varepsilon_0\,\Delta\varepsilon\}^{1/2}, \tag{6.25}$$

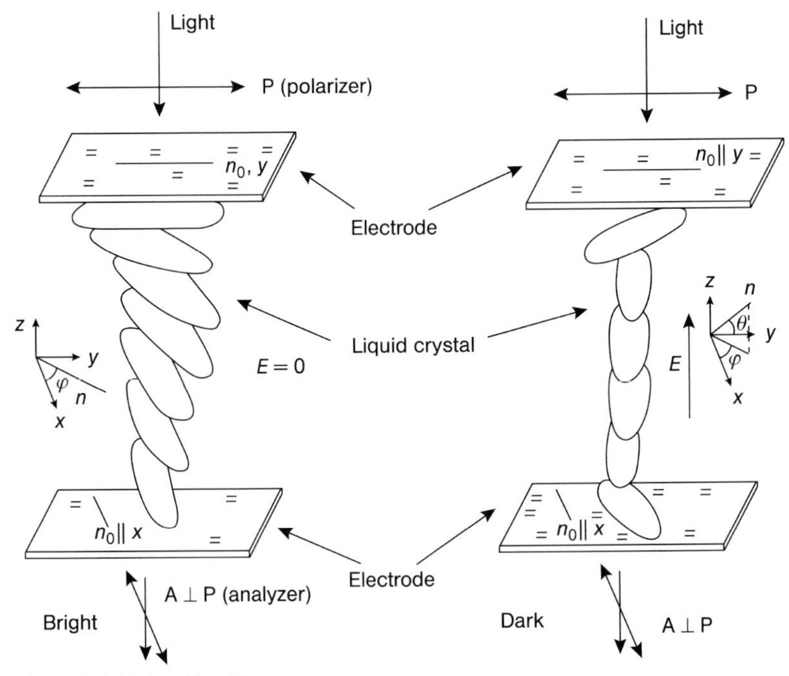

Figure 6.13 Twist effect [6].

the director **n** deviates from the initial orientation so that the linear dependence of the azimuthal angle $\varphi(z)$ disappears and the tilt angle $\theta(z)$ becomes nonzero (Figure 6.13). The qualitative character of the functions $\varphi(z)$ and $\theta(z)$ for different voltages is shown in Figure 6.14.

Since the director tends to orient perpendicular to the substrates, the effective values of Δn and d decrease and, for a certain voltage (optical threshold of the twist

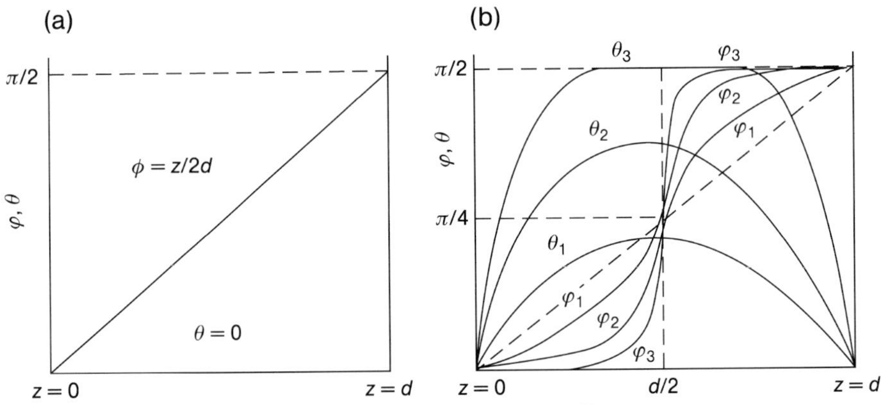

Figure 6.14 Distribution of the director angles $\varphi(z)$ and $\theta(z)$ in twist effect for different voltages: (a) $U \leq U_{tw}$ and (b) $U_{tw} < U_1 < U_2 < U_3$ [6].

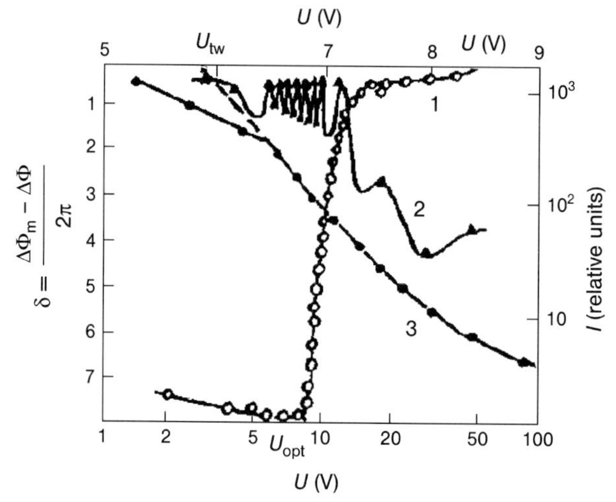

Figure 6.15 Optical response of the twist cell between parallel polarizers [15]: Curve 1: polarizers are parallel to the director on the input surface of the cell (conventional orientation); Curve 2: polarizers at the angle of 45° to the director on the input surface (maximum birefringence intensity); Curve 3: phase retardation (6.17) calculated from curve 2.

U_{opt}), the wave guide regime no longer remains. Let us note that despite the fact that the director starts to reorient at $U = U_{tw}$, a visible change in the twist-cell transmission is observed only for $U = U_{opt} > U_{tw}$ (Figure 6.15).

Figure 6.15 gives the dependence of the optical transmission of a twist cell for the conventional geometry $\mathbf{P}||\mathbf{n}(0)$ and when the polarizer $\mathbf{P}$ forms an angle of 45° with respect to the orientation of the director at $z = 0$ [15]. The deformation threshold, $U_{tw} = 6$ V, determined by extrapolating the linear section of the phase delay curve to $\delta(U) = 0$, coincides with that calculated from (6.25). The optical threshold for the twist effect increases with decreasing wavelength ($U_{opt} = 8.9$ and 10.2 V for $\lambda = 750$ and 450 nm, respectively) since the cutoff implied by the Mauguin condition occurs at higher voltages for shorter wavelengths (6.24).

6.1.2.1 Effect of the Cell Geometry and Liquid Crystal Parameters on the Steepness of the Transmission–Voltage Curve and its Correlation with the Information Content of Passively Addressed LCDs

6.1.2.1.1 Twist-Cell Geometry for Zero Voltage: Mauguin Conditions

A twist cell is usually formed by placing orienting glasses on top of each other. Then, twist directions at angles $\pi/2$ and $-\pi/2$ are equally probable. Regions with different signs of twist can be observed by interference methods [6, 7]. This degeneracy in the sign of the twist can be removed if small amounts of the optically active material are added to the nematic liquid crystal. In this case, the walls disappear and the twist cell has a uniform structure throughout its area. A nonzero tilt of the molecules to the cell

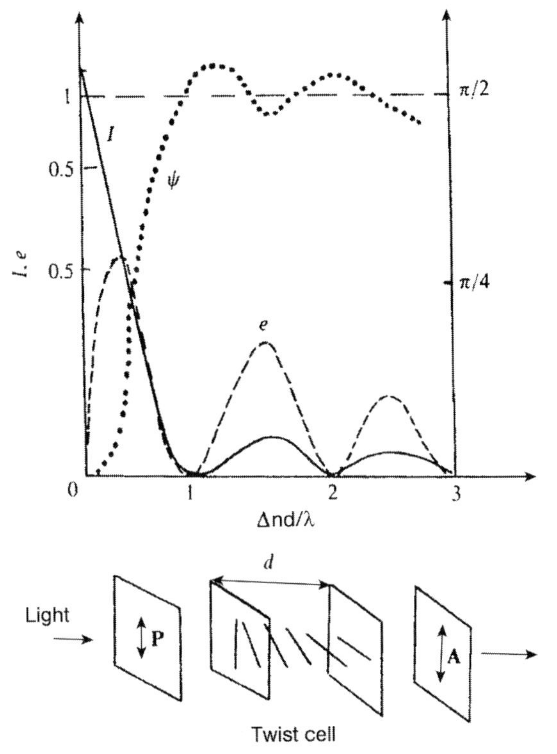

Figure 6.16 Optical characteristics of the twist cell in the absence of the field. *I*, intensity (solid line); *e*, ellipticity (dashed line); θ, rotation angle (dotted line) versus the parameter $\Delta nd/\lambda$. Without the twist cell ($d = 0$), the intensity between the parallel polarizer (P) and the analyzer (A) is taken equal to $I = 1$. The director at the first substrate of the twist cell is parallel to the polarizer (P) [6].

surface θ_0 also results in nonuniform twisting. In Ref. [16], it is suggested that the construction of twist cells with a certain pretilt angle will also eliminate the walls.

At zero field, the waveguide regime or Mauguin limit is violated for small values of $\Delta nd/\lambda$. Figure 6.16 shows the corresponding dependence of the transmitted light intensity (I), the ellipticity (e), and the rotation angle (ψ) of the long ellipse axis with respect to the polarizer P-axis on the parameter $\Delta nd/\lambda$ [16] (the director at the first substrate is parallel, or perpendicular, to the polarizer (P) and analyzer (A) (**P**||**A**). As seen from Figure 6.16, the exact Mauguin conditions

$$I = 0, \quad \psi = \pi/2, \quad e = 0 \tag{6.26}$$

take place for infinitely large $\Delta nd/\lambda$ values and also at some discrete points [6, 7]

$$\Delta nd/\lambda = (4m^2 - 1)^{1/2}/2, \quad m = 1, 2, 3, \tag{6.27}$$

usually called Mauguin minima.

We may consider the transmission of the twist cell for a white light, thus eliminating λ from the characteristic dependence of the twist-cell transmission. Writing $T(\Delta nd/\lambda)$ makes it possible to average $T(\lambda)$, together with the function of the sensitivity of the human eye $Y(\lambda)$, and the wavelength distribution of the illumination source $H(\lambda)$ [6, 7]. The corresponding optimal points, which provide the minimum transmission of the twist cell between parallel polaroids, are very close to those defined by (6.27) if we take the wavelength of the maximum sensitivity of the human eye in the range $\lambda \approx 550$–$580\,\text{nm}$.

6.1.2.1.2 Transmission–Voltage Curve for Normal Light Incidence

The typical transmission– voltage curve of the twist effect for normal light incidence is shown in Figure 6.17 for the twist cell placed between parallel polarizers. As already mentioned, the deformation of the director, measured by birefringence or capacity, begins for smaller voltages $U = U_{tw}$, then visible variations of the twist-cell transmission. In view of the Mauguin requirement (6.24), the optical threshold of the twist effect decreases for smaller values of the cell thickness, for the optical anisotropy of the liquid crystal, and for the director pretilt on the boundaries [6, 7]:

$$\theta = \theta_0 \quad \text{at}\, z = 0, d.$$

One of the most important parameters of a twist cell is the steepness of the TVC. Usually, the steepness parameter p is defined from the TVC of the twist cell, placed between crossed polarizers (Figure 6.17),

$$p_{50} = U_{50}/U_{90}-1, \quad p_{10} = U_{10}/U_{90}-1, \tag{6.28}$$

where U_{90}, U_{50}, and U_{10} correspond to 90, 50, and 10% levels of the optical transmission. As seen from (6.28), steeper TVCs correspond to smaller values of p_{50} and p_{10}.

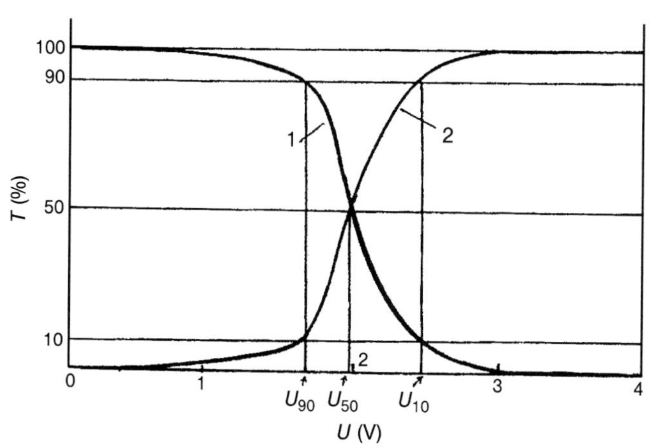

Figure 6.17 Optical transmission of the twist cell versus voltage for crossed (curve 1) and parallel (curve 2) polarizers [6]. The voltages U_{90}, U_{50}, and U_{10} correspond to the 90, 50, and 10% transmission level of curve 1, respectively.

The steepness of the TVC could be optimized for any specific case. There is no need to perform numerous experiments as the computer simulation could solve the problem with a high degree of accuracy. Many contemporary authors considered the problem of the calculation of electrooptical characteristics of twist cells, using both analytical estimations and computer programs [7, 17]. The procedure of the modeling of the twist-cell electrooptical behavior includes two steps. First, distributions of the director orientation are found for varying voltages (Figure 6.14). The second step consists of solving Maxwell equations for the light propagating in anisotropic liquid crystal media. To accomplish this, the liquid crystal is divided into a series of N equally thick uniaxial crystal sublayers, possessing a uniform optical axis direction within each sublayer. The direction of the optical axis changes by a small angle between adjacent sublayers, thus following the total director distribution in the cell (Figure 6.14).

Each sublayer is represented by a propagation matrix that alters the polarization state of the light passing through it, so that the total twist-cell transmission is found by multiplying all the matrices of sublayers. Computer methods for calculating the twist-cell properties were developed by Berreman, using the 4×4 propagation matrix [18]. Later on, computer procedures were improved, simplified by neglecting multiple reflections in sublayers, and made faster. The 2×2 matrix formalism, known as "Extended Jones Matrix Method," was developed [19]. Thus, the possibility of using the 2×2 Jones matrix that considerably accelerates the procedure without loss of accuracy was demonstrated. At present, several groups [20–23] use a set of commercially available computer programs for the simulation of electrooptical properties of twist and other nematic cells.

The attempts at optimization of the TVC steepness parameter show that the latter increases for the higher elastic ratios K_{33}/K_{11}, and increases for the lower values of the dielectric anisotropy $\Delta\varepsilon/\varepsilon_{\perp}$, while the effect of the optical path difference Δnd and the elastic ratio K_{33}/K_{22} is small [6, 7]. It was shown that the maximum steepness parameter is obtained when $\Delta nd \approx 2\lambda$, that is, near the second Mauguin minimum (6.27). If the director pretilt angle at the boundaries is not equal to zero, the TVC steepness parameter decreases. A phenomenological formula was proposed [24, 32] for the evaluation of the steepness parameter p_{50} based on experimental data for binary mixtures of compounds, belonging to 12 structurally and physically different liquid crystal classes,

$$p_{50} = 0.133 + 0.0266(K_{33}/K_{11} - 1) + 0.443\,(\ln \Delta nd/2\lambda)^2, \qquad (6.29)$$

which with a high accuracy correlates with the corresponding values measured in experiment.

A detailed analysis of TVC steepness p_{50} and p_{10}, based on computer simulation [17], was proposed in Ref. [7]. Let us note that the number of addressing lines N in the matrix liquid crystal displays, with a high-information content or multiplexing capability, sharply increases for steep TVCs, that is, low p-values. The precise dependence $N(p)$ is defined by the type of the driving scheme and will be discussed below.

The number of addressing lines N can be calculated from the relation [33]

$$N = [(1+p)^2+1]^2/[(1+p)^2-1]^2, \qquad N \approx 1/p^2 \quad \text{for } p \ll 1, \tag{6.30}$$

which is the result of the optimization of the driving conditions.

Our calculation show that decreasing K_{33}/K_{11} from 2 to 0.5 results in a considerable growth in the number of addressed lines of the passively addressed LCD. For low values of the ratio p ($p \approx 0.7$), the maximum of the multiplexing capability N is achieved for the layer thickness corresponding to the first Mauguin minimum; however, for $K_{33}/K_{11} \approx 1$, the second Mauguin minimum becomes optimal. The TVC steepness depends only on the product of Δnd, thus we can vary Δn and d independently keeping Δnd and the multiplexing capability constant.

The low-frequency dielectric anisotropy of nematic cells is also important. The twist-cell multiplexing capability depends differently on $\Delta\varepsilon/\varepsilon_\perp$, depending on the number of Mauguin minimum, m. For $m=1$, the number of addressed lines considerably decreases with $\Delta\varepsilon/\varepsilon_\perp$, while for $m=3$, on the contrary, the maximum values of N are obtained for largest $\Delta\varepsilon/\varepsilon_\perp$ ratios.

The geometry of the twist cell also affects the TVC steepness and, consequently, the multiplexing capability [7]. The TVC steepness increases for the lower values of the angle η_p between the polarizer and the analyzer. For $\eta_p = 70°$ (instead of the typical $\pi/2$ value, Figure 6.13), the number of addressed lines is, however, doubled with a small ($\approx$10%) decrease in the transmission in the off state. TVC steepness is also very sensitive to variations of the total twist angle φ_m and grows in more twisted cells. However, twist angles exceeding $\pi/2$ are unstable in pure nematic cells.

Larger values of the layer thickness lead to the higher operating voltages of the twist effect [7]. This is evident because the Mauguin parameter $\Delta nd/\lambda$ increases in the "off"state, and to break Mauguin's condition (6.24), smaller Δn values in the "on"state are needed, corresponding to the stronger director deformation in higher fields. As mentioned above, TVC in parallel and perpendicular polarizers are complementary (Figure 6.17). This is not true, however, for the contrast ratios as functions of applied voltages for a nonmonochromatic (white) light. The contrast ratio C is defined as the ratio of transmitted luminances in the on and off states [6, 7]:

$$C = B_{\text{on}}/B_{\text{off}}, \tag{6.31}$$

where

$$B = \int_\lambda H(\lambda)y(\lambda)I(\lambda)\,d\lambda \Big/ \int_\lambda H(\lambda)y(\lambda)\,d\lambda.$$

The intensities $I(\lambda)$ in the on and off states are averaged with the function of the sensitivity of the human eye $y(\lambda)$ and the energy distribution of imination source $H(\lambda)$ over the visible spectrum (380–780 nm). The electrooptic effect on the twist cell placed between parallel and crossed polarizers is called the "normally black" and "normally white" mode [6], in accordance with the appearance of the twist cell in the off state (dark or bright). Contrast ratios in the white mode are considerably higher than in the black mode as the transmission in the on state for a normally white mode

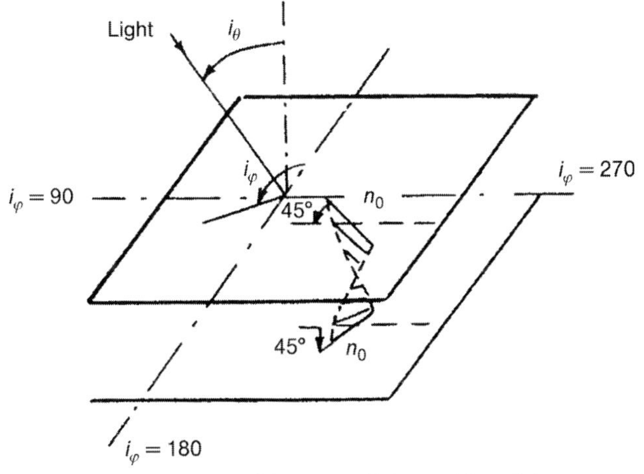

Figure 6.18 The geometry of the twist cell for an oblique light incidence. The angle i_θ is a polar angle of light incidence and the angle i_φ is the corresponding azimuthal angle measured from the symmetric plane of the twist structure [6].

could be very small, limited only by the quality of polarizers and liquid crystal orientation.

6.1.2.1.3 **Viewing Angle Dependences of Twist LCDs**

The twist-cell transmission at oblique incidence depends on the values of the polar i_θ and azimuthal i_φ angles of light incidence (Figure 6.18). This could be interpreted in terms of the corresponding Mauguin parameter $\Delta nd/\lambda$, which becomes a function of the light direction. The Mauguin parameter in the direction $\boldsymbol{e}$ is estimated by averaging the value of $p\Delta n/\lambda$ along $\boldsymbol{e}$, where $p = (\mathrm{d}\varphi/\mathrm{d}e)^{-1}$ is a local value of the pitch, that is, the distance of the total director azimuthal rotation by 2π (for e parallel to the twist axis $p = 4d$) and the effective optical anisotropy $\Delta n = (\sin^2\theta/n_{\|}^2 + \cos^2\theta/n_{\perp}^2)^{-1/2} - n_{\perp}$ with the polar θ and azimuthal φ angles of the director with respect to the e-axis.

The characteristics of transmission for oblique incidence $i_\theta \neq 0$ can be described in terms of the azimuthal dependence of the transmission $T(i_\varphi)$ for a given polar angle of incidence i_θ and the applied voltage U. For directors parallel to the boundaries the twist-cell transmission is symmetric with respect to the plane located at an angle of 45° to the director orientation on the boundaries [6]. However, for nonzero director pretilt angles, the symmetry is broken. In the T–V curves, obtained for oblique light incidence, there appears a minimum of transmission, which goes to the lower voltages for higher incidence angles (Figure 6.19). Indeed, for a certain voltage, the direction of light propagation may coincide with the director at the center of the layer, thus providing very low values of the Mauguin parameter, despite the fact that the director distribution, as a whole, is far from the homeotropic configuration. As seen from Figure 6.19, a twist cell for the oblique light incidence is the most sensitive to an external voltage for azimuthal angles $i_\varphi = 180°$, while $i_\varphi = 0°$ corresponds to quite an opposite case.

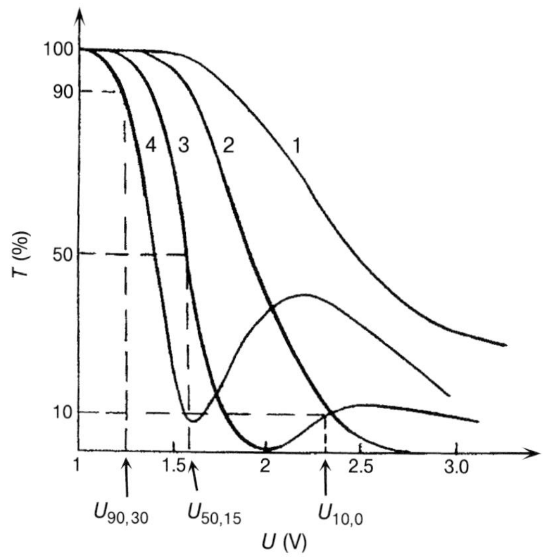

Figure 6.19 TVC of the twist cell for an oblique light incidence [6, 7]. Curve 1 corresponds to the light incidence angles $i_\theta = 15°$, $i_\varphi = 0°$; Curve 2: ($i_\theta = 0°$, $i_\varphi = 0°$); Curve 3: ($i_\theta = 15°$, $i_\varphi = 180°$); Curve 4: ($i_\theta = 30°$, $i_\varphi = 180°$).

The most crucial parameter that affects the uniformity of transmission is the Mauguin number m(6.27). For low $\Delta nd/\lambda$ values the anisotropy of transmission for an oblique incidence is weak [7]. According to this, liquid crystal mixtures with low Δn values and the first Mauguin minimum $\Delta nd/\lambda = \sqrt{3/2}$ as an operating point are preferable. We do not consider the influence of the parameters K_{33}/K_{11} and $\Delta\varepsilon/\varepsilon_\perp$ on the transmission characteristics of the twist cell, as it is very small compared to that of normal incidence mentioned above.

Let us note that coloration, that is, dependence of the light transmission on the wavelength, and the stronger temperature dependence of δn make the operation in the first Mauguin minimum less attractive. Sometimes, it seems more convenient to choose $\sqrt{3/2} < \Delta nd/\lambda < \sqrt{5/2}$ (between the first and second minima) even at the cost of the partial loss of the contrast.

The dependence of the T–V curves on the angles of incidence is taken into account as one of the characteristics of twist effect mixtures. Figure 6.19 shows that for a certain azimuth of an oblique incidence ($i_\varphi = 180°$), the optical threshold of the twist effect is lower than for the normal incidence. Thus, it is possible to consider new definitions of the switching-off voltage on a nonselected display element and switching-on voltage on the selected one. For instance, Figure 6.19 demonstrates that the optical threshold can be $U_{90,30}$ ($i_\theta = 30°$, $i_\varphi = 180°$, transmission 90%), while the selected voltage can be $U_{50,15}$ ($i_\theta = 15°$, $i_\varphi = 180°$, transmission 50%) or $U_{10,0}$ ($i_\theta = 0°$, transmission 10%). Similar limitations are imposed on the TVC steepness by the temperature dependence of the operating voltages. As a result, the multiplexing capability of the twist effect remains several times lower than that estimated from TVC at normal incidence and room temperature.

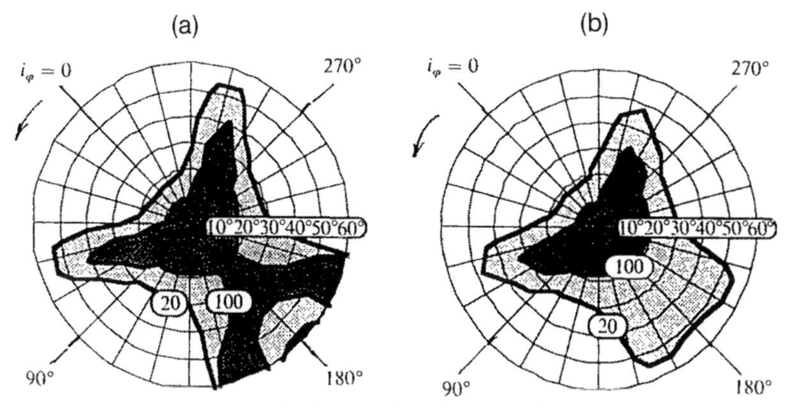

Figure 6.20 Isocontrast curves for the normally white mode. The first polarizer is (a) perpendicular and (b) parallel to the director on the input substrate [6, 7].

For applications, it is convenient to evaluate the angular dependence of the transmission by isocontrast curves, which show the levels of equal contrast ratio for different angles of incidence. An example of these curves for the normally white mode (twist cell between crossed polarizers in light) is given in Figure 6.20. (The contrast ratios are 20: 1 and 100: 1.) The radial coordinate in the isocontrast diagram defines the value of the incidence angle i_θ, while the azimuthal one defines the azimuthal incidence angle i_φ. It is seen from Figure 6.20 that the normally white mode provides wider viewing angles when the polarizer is perpendicular to the direct input substrate.

6.1.2.1.4 **Principles of Passive Matrix Addressing of Twist LCDs**

The principle of matrix addressing is shown in Figure 6.21 [6, 7]. Rows of the matrix are subsequently addressed in equal time subintervals by pulses of the amplitude U_S. In each time subinterval, all the columns of the matrix display are addressed simultaneously with pulses of the amplitude $\pm U_D$. The sign of the column pulse depends on whether an element of the matrix display (pixel) should be in the "off" or "on"state. Sometimes, "on" and "off"states of the pixel are called selected and nonselected states. Figure 6.21 demonstrates that the effective (root mean square voltage) on the selected $U_{sel}(A_{12})$ and nonselected $U_{nsel}(A_{11})$ elements are

$$U_{sel}^2 = (U_S + U_D)^2/N + U_D^2(1-1/N), \tag{6.32}$$

$$U_{nsel}^2 = (U_S - U_D)^2/N + U_D^2(1-1/N).$$

Excluding U_S from (6.32) and minimizing the obtained value of N with respect to U_D we have

$$N_{max} = [(U_{sel}^2 + U_{nsel}^2)/(U_{sel}^2 - U_{nsel}^2)]^2 = \{[(1+p)^2+1]/[(1+p^2)-1]\}^2 \tag{6.33}$$

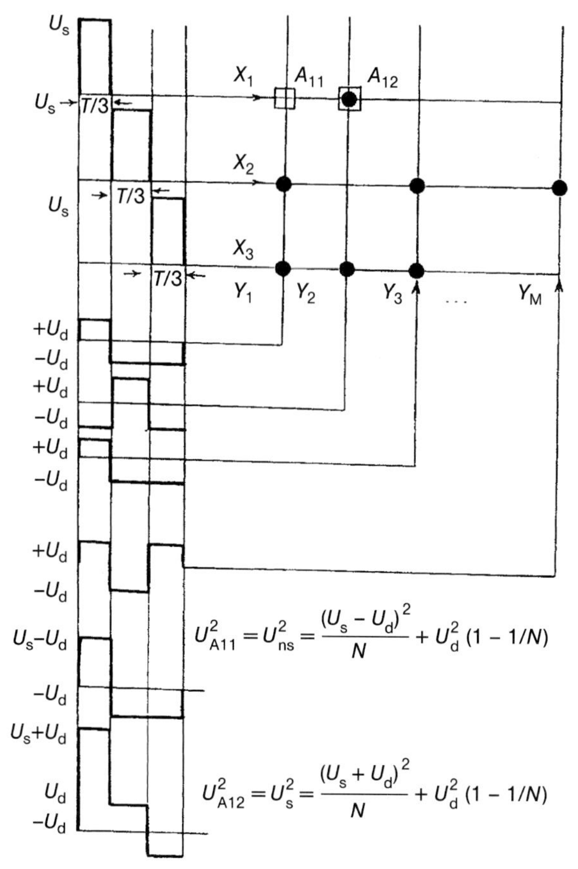

Figure 6.21 Principles of matrix addressing of LCD with N rows ($N = 3$) and M columns [6].

which coincides with (6.30) if we assume that

$$p = U_{sel}/U_{nsel} - 1.$$

Thus, the lower p-values enable us to address more rows of the matrix, that is, to increase their information content. Optimum amplitudes of voltage pulses U_S and U_D are written as

$$U_D = 1/2\sqrt{(U^2_{sel} + U^2_{nsel})},$$
$$U_S = 1/2(U^2_{sel} + U^2_{nsel})^{3/2}/(U^2_{sel} - U^2_{nsel}).$$

The advantages of multiplex driving is that $M \times N$ pixels can be accessed with just $M + N$ electrical contacts to the display, which considerably reduces a number of drivers. The disadvantage is the appearance of the so-called cross talk, as each pixel cannot be addressed independently, since an addressing pulse affects all the elements in the corresponding row.

The driving scheme, proposed in [25], allows us to address more rows than the classical 3 : 1 selection scheme [6, 7], where

$$U_S = 2U_{opt}, \qquad U_D = U_{opt}$$

with the optical threshold for the twist effect U_{opt}.

At present, there is no generally accepted system of parameters that characterizes the multiplexing capability of the twist cell. This is so because not only the quality of a mixture but also the number of technological conditions remains a very important characteristic of merit of the twist displays. As mentioned above, the multiplexing capability could be improved by a proper choice of the cell thickness, lower gap nonuniformity within the working area, better quality of nematic alignment, optimized angle between polarizers, driving scheme, and so on.

6.1.2.1.5 Dynamics of the Twist Effect

The definition of the rise t_r and decay t_{decay} times of the twist effect are clarified in Figure 6.22. Responding to the square wave voltage with the amplitude V liquid crystal twist cell between parallel polarizers increases its brightness from the initial value B_0 to the value of the optical threshold of the twist effect $B_0 + 0.1(B_m - B_0)$ during the so-called "delay" time t_d. The rise time t_r corresponds to the further increase in the brightness up to the value of 90% from the difference between the initial and the maximum B_m value. After switching off the voltage, the twist LCD brightness relaxes with a decay time t_{decay}. Qualitatively, the times for both the twist effect and the layer deformations are proportional to the viscosity and to the square of the thickness, and inversely proportional to an elastic constant of the liquid crystal material (6.14). Also, the rise times are inversely proportional to the difference $(V_2 - U_{tw}^2)$, where U_{tw} is a threshold voltage [6, 7].

For the higher applied voltages, the edge of the twist effect oscillograms develops a characteristic bounce (there is no such bounce on the front edge) (Figure 6.23). The

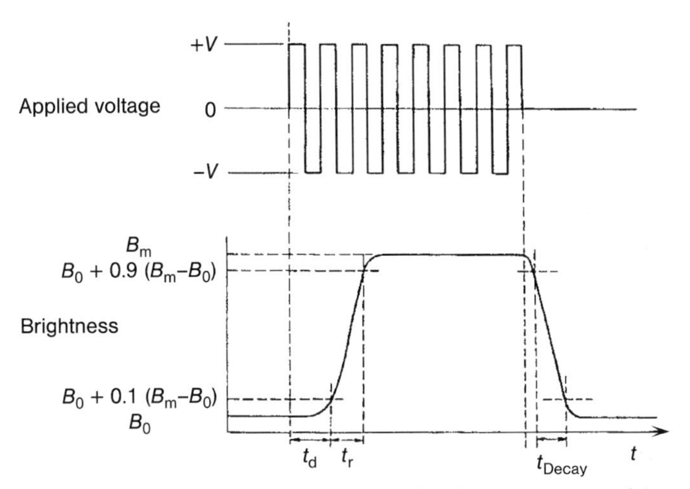

Figure 6.22 Typical oscillograms [6, 7] that define the response time of the twist effect.

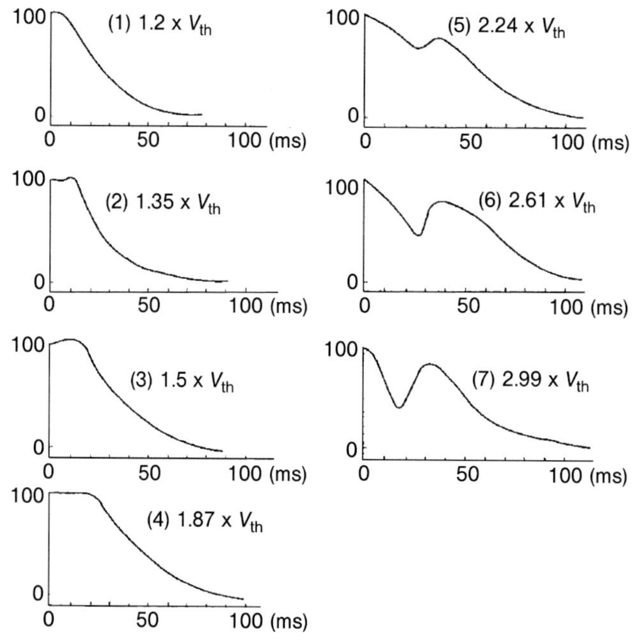

Figure 6.23 Occurrence of the bounce on the decay of the TVC curve [6, 7].

effect takes place if the driving voltage is sufficiently high, $V > 2U_{tw}$. This effect is accounted for theoretically by allowing the backflow [6, 7]. The point is that in the initial stages of the relaxation, the maximum elastic torque is found in the region of a rapid change of the director deformation θ with the z-coordinate in the vicinity of the walls as in the ECB effect (Figure 6.5). The highest rate of the director reorientation also occurs in this region, inducing the maximum associated flow of the liquid. This flow influences the director in the layer center in such a way that the director angle θ increases (to values above $\pi/2$). As relaxation proceeds, the backflow decreases, and the director can return to its initial state under the elastic forces. This corresponds to the maximum transmission of the cell (a bounce of the rear edge).

There exists a possibility of decreasing the twist effect relaxation times by applying a field with a frequency greater than that at which the dielectric anisotropy changes sign, as was discussed for ECB effect. However, higher operating voltages and a limited operating temperature range prevent the dual-frequency addressing scheme from finding commercial application.

6.1.2.1.6 New Developments

Let us analyze some new possible improvements of twist effect LCDs, which make it more attractive for electrooptical applications in LCDs.

Effect of Weak Boundary Anchoring: Reverse Twist Mode The effect of saturation, that is, total reorientation of the twist director alignment to the homeotropic one at finite voltage values, is possible. This effect could be used for the realization of memory states [6]. A considerable increase in the multiplexing capability of the twist cell at low anchoring energies is observed. The corresponding number of addressing lines is very sensitive to the form of the anchoring potential. For sufficiently small anchoring energies, the twist-cell response reveals bistability and hysteresis [8]. It should be pointed out that this is true only for cells with weak polar anchoring energies. Now the development of such systems is in progress. The azimuthal anchoring should be kept strong in all cases, otherwise the twist alignment can disappear at a certain critical thickness [26].

If the electrodes of a cell that have been pretreated by rubbing were further coated with a layer of surfactant (e.g., lecithin), the resulting orientation of the molecules of the nematic liquid crystal will be homeotropic and appears bright between parallel polarizers (Figure 6.13). In an electric field, a nematic liquid crystal with a negative dielectric anisotropy must reorient itself into a twist structure, and transmission through the cell decreases to zero (reverse twist effect) [7]. This effect has been observed experimentally in comparatively thick cells ($d \approx 50\,\mu m$) [7]. In cells with $d \approx 20\,\mu m$, the final twisted state proves to be unstable and the nematic liquid crystal layer gradually transformed into a planar structure. The addition of small quantities of cholesteric liquid crystals to the initial nematic mixture stabilizes the final twisted structure and improves the electrooptical characteristics of the device. The electrooptical response of electrically induced twist nematic cells includes intensity oscillations observed both in the switching-on and in the switching-off regimes [7]. These oscillations take place due to the variation of birefringence, which are not important in the usual twist effect.

Another reverse twist mode takes place when the boundary conditions favoring the twist configuration with a definite direction and the handedness of the chiral dopant in LC materials are opposite to each other [27]. Based on the LC molecular arrangement, low driving voltage [27] and fast response times [28] could be realized.

Improvement of Viewing Angles One possibility of improving the viewing angles of the twist effect is application of the so-called "plane switching" (IPS), when the electric field is applied parallel to the substrate's plane. In this case, it is possible to control the TVC of the twist cell mainly not by changing its phase retardation $\Delta\Phi$ but by affecting the angle φ_0 between the polarization plane of the incident light and the director axis (6.18). The dark state will correspond to the value of $\varphi_0 = 0$ and the angular dependence of the intensity will be absent as it is connected only with the phase retardation term $\Delta\Phi$ [29]. At the same time, the coloration, that is, the wavelength dependence of the transmission, is also absent due to the same reason [29]. However, certain problems of the IPS mode are yet to be solved, such as slow response (several times slower than in the usual twist mode) and the presence of defects.

Another possibility of getting the wider viewing angles in twist effect is the so-called multidomain alignment technique, that is, dividing each twist cell into two or four domains to observe the best side of the twist configuration, presented by each domain [30]. In some cases, the multidomain configuration can be obtained by the nonfixed (amorphous) liquid crystal orientation on one of the substrates of the cell [6, 7].

To improve the viewing angle dependence and contrast ratio of the twist cell, phase retardation plates are used [6, 7]. Sometimes, a second twist cell is placed after the first one with a 90° twist in the opposite sense [31]. The two cells optically compensate for each other when placed so that their directors are perpendicular on the facing surfaces. Thus, the double-layered twisted device appears black between crossed polarizers for all wavelengths and has better viewing characteristics at oblique incidence.

6.1.3
Supertwist Effects

6.1.3.1 Discovery of Supertwist Effect for LCDs: SBE Mode

In 1984, the display based on the "supertwist birefringent effect" (SBE) was proposed [32]. The geometry of a SBE display is typical for the supertwisted displays (Figure 6.24). A 270° supertwisted nematic layer is oriented with a 28° director tilt at the boundaries to prevent the appearance of light scattering structures. In an SBE cell, the thickness-to-pitch ratio $d/P = 0.75$, that is, three-quarters helix pitch, is fitted within the layer thickness d. It means that the second Grandjean zone for a 90° twisted director orientation has to be realized by a proper doping twist LCD cell with a chiral agent. When the field is switched on the director reorients to nearly homeotropic configuration (dielectric anisotropy $\Delta\varepsilon > 0$). Two polarizers used in the SBE display are located at angles β and γ with respect to the director projection on the input (n_1) and output (n_2) substrates.

The position of polarizers (β and γ) and the optical path difference Δnd define the color and contrast characteristics of the SBE device [33]. For a "yellow" mode with a positive contrast (purplish blue symbols on a bright slightly greenish yellow background), we have to choose $\beta = 32.5°$, $\gamma = 57.5°$, and $\Delta nd = 0.85\,\mu\text{m}$. For the "blue" mode, having a negative contrast with the colors, complementary to that of the "yellow" mode, the following optimum values were found: $\beta = 45°$, $\gamma = -45°$, and $\Delta nd = 0.79\,\mu\text{m}$. The transmission spectra for the blue (B) and yellow (Y) modes in "on" and "off"states are shown in Figure 6.25 [33]. The considerable wavelength dependence of the transmission of the SBE display imposes certain limitations to their applications.

The significant advantage of the supertwist mode is a steep transmission–voltage characteristic due to the sharp growth of the director deformation with the voltage (Figure 6.26). The first SBE display operated at $U_{\text{off}} = 1.6\,\text{V}$, $U_{\text{on}} = 1.75\,\text{V}$ that, according to (6.30), allows us to address more than 100 rows (or more than 200 rows for double addressing [32, 33]).

High contrast and uniformity of transmission characteristics at oblique incidence are also beneficial features of the SBE mode, which considerably improve the legibility of supertwist displays. Better viewing angles than that for the 90° twist structure seem to be a peculiar feature of highly twisted chiral nematics. Figure 6.27 demonstrates this for a 200° supertwist cell in comparison with the usual 90° twist cell [6, 7].

Thus, the supertwist birefringent effect was a step forward to the development of a new generation of matrix addressed displays with very high-information contents

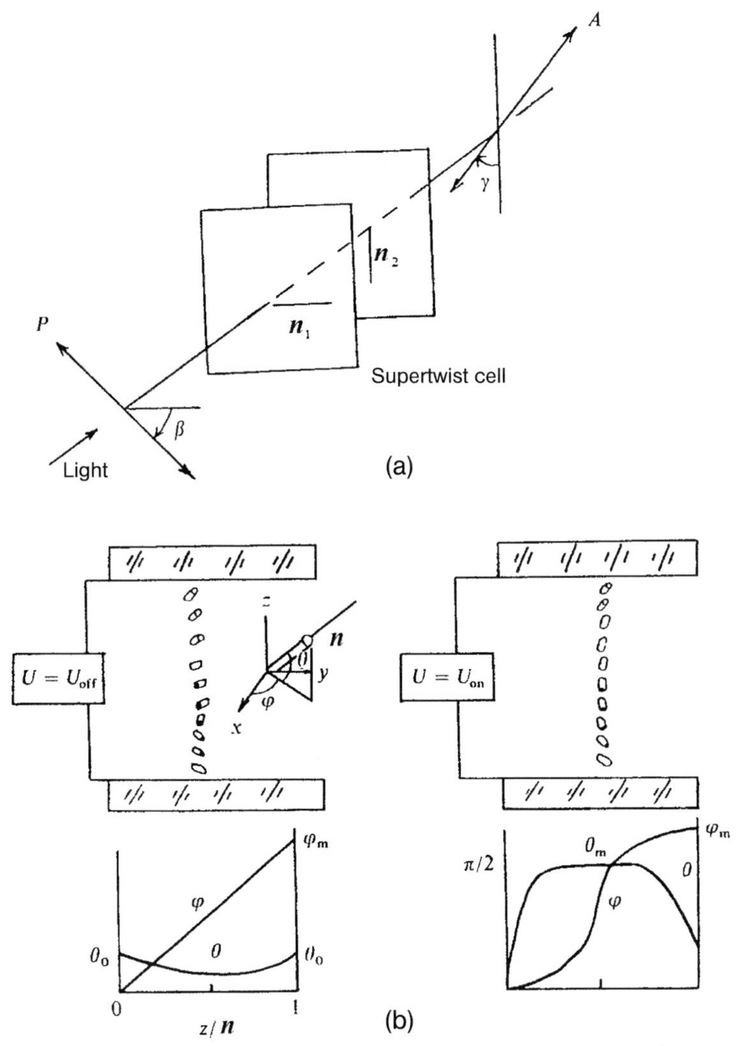

Figure 6.24 The geometry of the SBE display.
(a) Cell configuration, P: polarizer, A: analyzer, n_1 and n_2 the projections of the director onto the substrate plane. (b) Director distributions for the selected ("on") and nonselected ("off") voltages of the passively addressed LCD [6].

and excellent viewing characteristics. However, several characteristics of the SBE mode need further improvement. They are as follows [33]:

1. High sensitivity to the cell gap nonuniformity. More than 2% difference from the average thickness over the working area of the display results in the appearance of different colors.

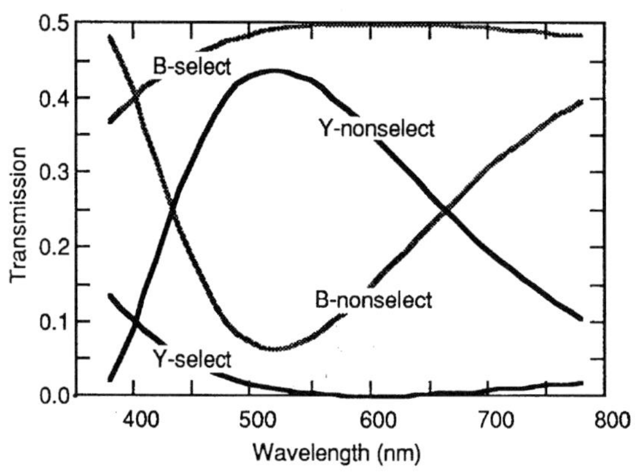

Figure 6.25 Transmission spectra of SBE display for yellow (Y_{on}, Y_{off}) and blue (B_{on}, B_{off}) modes in the on and off states, respectively [33].

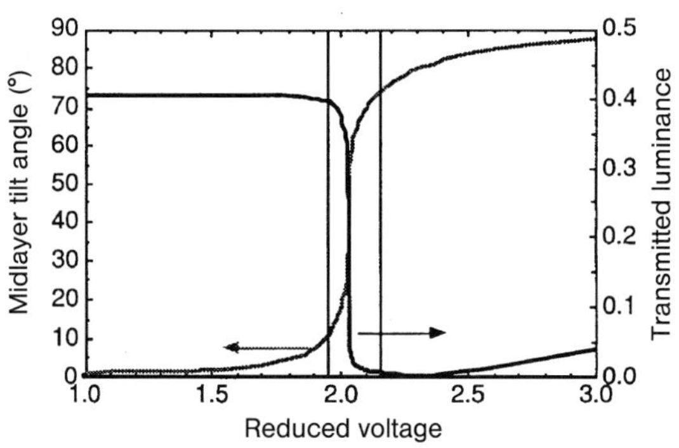

Figure 6.26 Electrooptical response (right) and director deformation (left) curves for SBE mode supertwist display [6].

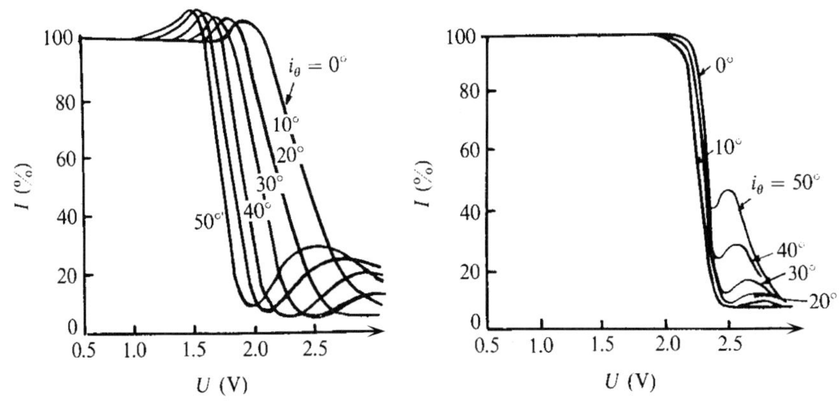

Figure 6.27 Transmission–voltage characteristics: (a) 90° twist structure and (b) a 200° supertwisted cell. The light incidence angles i_θ are shown on the curves, $i_\varphi = 180°$ [6].

2. Strict requirements to the quality of the tilted orientation at the substrates. The larger supertwist angles of the liquid crystal structure require the higher director tilt angles at the substrates to avoid parasitic domain structures.
3. Considerable wavelength dependence of the transmission spectra in the visible range (Figure 6.25).
4. Too large response times (300–400 ms).

New classes of supertwist displays appear to solve these problems.

6.1.3.2 **Various Supertwist Modes**

A general scheme of realization of highly informative supertwist displays is shown in Figure 6.28. Here, the scheme of supertwist display geometry shows the input and output orientations of the molecules or their projections (n_{in}, n_{out}) as well as the input and output orientation of the polarizers (P_{in}, P_{out}). The angle φ_m is the supertwist angle, β and η define the location of the polarizers, η is the angle between polarizers and β the angle of the first polarizer with respect to the director on the front substrate. As seen from Figure 6.28, various supertwist geometries could be obtained by altering the supertwist angle φ_m and polarizer angles β, η. Besides that, we can change such parameters of the supertwist mixtures as optical path difference Δnd and the director pretilt at the boundaries θ_0. Various realizations of supertwist structures for passively addressed high-information content LCDs are shown in Table 6.1 [6].

New methods for the realization of the electrooptical effects in supertwisted displays (supertwist nematic (STN) effect and optical mode interference (OMI) effect) avoid certain limitations and disadvantages that are observed in SBE displays. For example, the boundary tilt angles for preventing the appearance of domain structures are not as large as in SBE, and the requirement of thickness nonuniformity becomes softer. Consequently, manufacturing displays becomes easier. STN and OMI mixtures have the same or even better steepness as the SBE prototype, which in

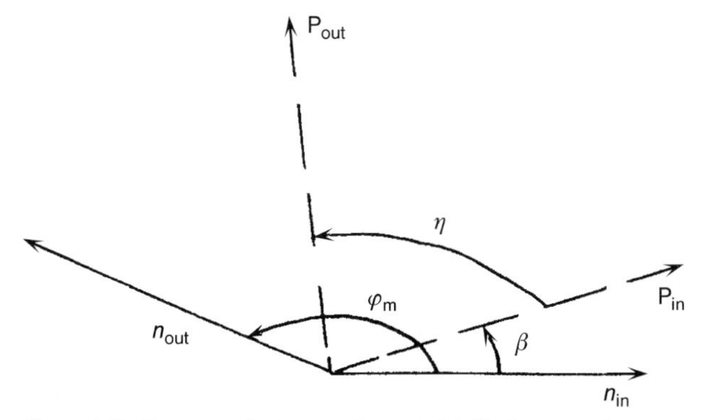

Figure 6.28 The general scheme of supertwist display geometry shows the input and output orientations of the molecules or their projections (n_{in}, n_{out}) as well as the input and output orientation of the polarizers (P_{in}, P_{out}). The angle φ_m is the supertwist angle, β and η define the location of the polarizers [6].

Table 6.1 Supertwist structures for passively addressed high-information content LCDs [6].

Supertwist mode	**Supertwist angle**	**Thickness-to-pitch ratio, d/P**	**Optical path length, Δnd (μm)**	**Polarizer angles β, η**	**Pretilt angles on the substrates, θ_0**
Supertwist birefringent effect [32, 33]	270°	0.75	0.85	−32.5°, 65°	28°
Supertwist nematic effect	240°	0.66	0.82	45°, 60°	4–6°
Optical mode interference effect [34]	180°	−0.3	0.46	0°, 90°	2–3°
Optical mode interference effect	240°	0.43	0.54		4–6°

accordance with (6.30) means addressing about 400 lines for XGA standard of passive LCD (1024 × 768 pixels for a dual scan drive) [6].

The optical mode inteference effect provides a weak wavelength dependence of the transmission in the visible region [34] (Table 6.1). The effect requires a low optical path difference, which leads to a strong interference of two polarization modes when propagating through an OMI cell (Mauguin's wave guide regime (6.24) is not valid). The transmission spectra of the OMI cell enables us to realize the black and white appearance of the two display states, which is impossible in STN or SBE cells (Figure 6.25). The electrooptical characteristics of the OMI cells are much more tolerant to the cell gap nonuniformity than that in STN and are less temperature dependent. However, one of the main disadvantages of the OMI displays is low brightness in the off state. For instance, if we take the brightness of two parallel polarizers equal to 100%, then the off states of the 90° twist cell and the 240° STN cell would correspond to 95 and 64%, respectively, while the brightness of the 180° OMI cell does not exceed 40% [33]. However, the brightness of OMI displays can be greatly improved up to 77% by increasing the twist angles up to 270° if we make an appropriate choice of Δnd and the angular position of polarizers [35]. Besides that, OMI displays possess the response times about 1.5 times lower than STN-LCDs for the same value of the contrast and multiplexing capability. Surprisingly, the viewing angles of OMI displays for high-information content screens are not much wider than that of STN, despite the lower values of Δnd. Thus, OMI displays are competitive with STN displays for high-information content passively addressed LCDs.

6.1.3.3 Dependence of TVC Steepness on the Material and Construction Parameters

The corresponding dependence of the steepness of the transmission–voltage curve on physical parameters of the chiral nematic mixture and geometry of the supertwist display is shown in Table 6.2. Correlations mentioned in Table 6.2 are used in the development of supertwist mixtures for highly informative displays.

Table 6.2 Effect of material parameters and cell geometry on the electrooptic response of supertwist LCD [6].

Parameter	Effect
Thickness-to-pitch ratio, d/P_0, with equilibrium pitch, P_0	Lower d/P_0 values result in a decrease in driving voltages and make them steeper. Response times decrease. The supertwist structure exists for $\varphi_m/2\pi - 0.25 < d/P_0 < \varphi_m/2\pi + 0.25$
Pretilt boundary angle, θ_0	Higher pretilt angles shift the TVCs to the lower voltages and slightly increase its steepness at the cost of the lower contrast ratios. Parasitic domain structures do not appear for the sufficiently large θ_0
Dielectric ratio, $\Delta\varepsilon/\varepsilon_\perp$	Lower $\Delta\varepsilon/\varepsilon_\perp$ ratios increase the steepness on account of the higher driving voltages, but domains may appear. Response times improve
Supertwist angle, φ_m	Larger angles φ_m increase TVC steepness up to the appearance of a hysteresis on the TVC curve. Driving voltages and response times increase

According to Table 6.2, the problem of developing supertwist materials with a steep transmission–voltage curve is always a matter of compromise. Steeper TVCs are either accompanied by higher driving voltages or the appearance of domains and hysteresis [6]. Besides that, the parameters of newly developed supertwist mixtures should have a weak temperature dependence, as the necessary driving voltages are very sensitive to temperature. Another problem is to provide a qualitative oblique alignment over the large surface area of the display. This problem can be solved by using rubbed polyimide layers, which provide the pretilt angles between 1° and 10°, and new methods, such as the photoaligning procedure [13, 14].

One of the most crucial problems in the development of supertwist materials is decreasing response times. A fast response, which enables to operate in the TV standard mode, should not be worse than 50 ms. To improve response times of supertwist displays, it is preferable to use low viscosities, large Δn values, larger ratios of K_{33}/K_{11}, low d/P ratios, smaller relative dielectric anisotropy $\Delta\varepsilon/\varepsilon_\perp$, and so on [6]. The TV standard has not yet been achieved, but response times are small enough to use supertwisted LCDs in personal computers both in portable and in desktop variants. We should note that the large size of STN-LCD screen with a high-information content (e.g., XGA standard) cannot be fast in principle as the steeper TVCs inevitably lead to the slower response times.

6.1.3.4 Supertwisted LCDs with Improved Characteristics: STN-LCDs with Phase Retardation Plates

One of the main goals is to attain a black–white switching of a supertwist cell with a high contrast ratio and an acceptable brightness. The possible solution of the problem is to use phase retardation plates in combination with an STN display. The idea is to compensate phase difference induced by a supertwisted structure. Supertwisted displays with one and two phase retardation plates were reported [36]. The orientation of the retardation plates and their optical path differences are optimized to provide both black and white switching and wide viewing angles [17]. The double plates provide better achromatic appearance and contrast ratio than a single plate, especially when they are placed on both sides of the STN cell. Thin polymer films (polycarbonate, polyvinyl alcohol, etc.) are used now, as phase retardation plates. Biaxial compensator films and optically negative polymeric films composed of discotic molecules were also developed [37]. The contrast ratios of the STN-LCD with negative birefringence films achieve 100 : 1 for the normally incident light. However, it is very difficult to realize such a film with a uniform phase retardation over a large surface area.

6.1.3.5 Double-STN-Cell (DSTN) Configuration: Triple STN Subtractive Color System

The main problem with the STN-LCD using the phase retardation plate is a limited temperature range over which a good compensation is possible due to the different temperature dependence of Δn in liquid crystal and polymer layers [33]. This disadvantage is overcome when, instead of the phase compensator, another supertwist cell of the same thickness is used that has no electrode and is twisted in the opposite direction. Thus, the second passive layer optically compensates the active STN layer in the off state and the light passed through two cells becomes linearly polarized perpendicular to the input polarizer direction. As a result, this light is absorbed by the analyzer, which is crossed with the polarizer (Figure 6.29) [6]. The device consisting of two supertwist cells is called double-layer STN-LCD or DSTN-LCD and in the off state looks dark for all wavelengths in the visible region. DSTN-LCDs demonstrate both higher contrast ratios and wider viewing angles than STN-LCD with phase retarders. Due to the double-cell construction, the requirement of the gap nonuniformity in DSTN-LCD is more stringent than in STN-LCD. Other

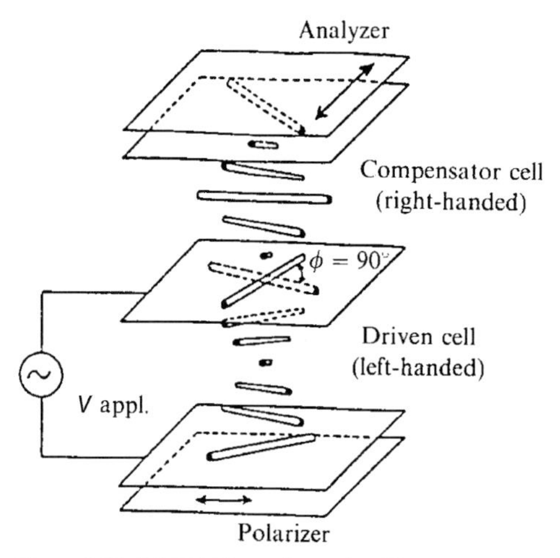

Figure 6.29 DSTN-LCD configuration [6].

drawbacks include increased display thickness and weight, a problem especially for applications in portable computers. The TVC of DSTN cell possesses the same steepness as TVC of STN-LCD, but the viewing angles are considerably wider [33]. Of course, supertwist LCDs in general have much higher contrast ratios and better viewing angles than TN-LCDs.

6.1.3.6 Multiline Addressing: Shadowing

In case of sufficiently fast response of the liquid crystal material, the time distribution of the controlling voltage over the pixel of the STN-LCD screen begins to play an important role. In particular, the usual Alt-Pleshko scheme (Figure 6.21) does not provide an appropriate behavior of the LCD response due to the fast decay of the transmission within one addressing period of the screen (conventional frame rate is about 60 Hz). The phenomenon is known as frame response (Figure 6.30) and can be avoided only if the controlling voltage is distributed more uniformly over the frame period [33]. Several methods are known to overcome this drawback: active addressing [33] and multiline selection (MLS) [38]. Both methods need additional circuitry for the generation of the orthogonal row addressing function and calculation of column addressing functions on their basis. However, the additional cost due to the more complex addressing scheme is not a big problem [38]. Multiline addressing technique has been realized for STN-LCDs with a high information capacity and low power consumption [39].

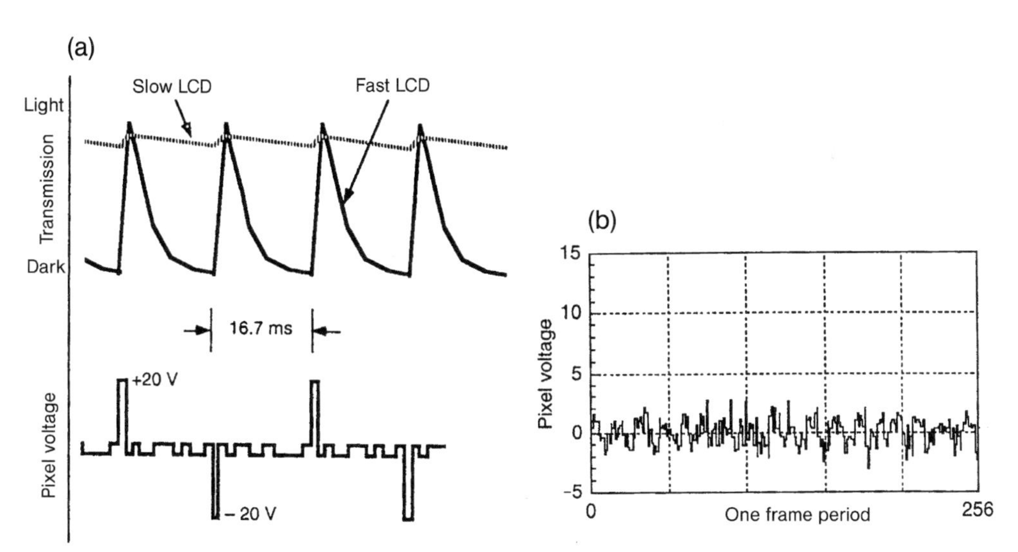

Figure 6.30 The phenomenon of frame response, resulting in fast decrease in the transmission of the STN-LCD pixel in case of strongly nonuniform distribution of the controlling voltage over the pixel (a) frame response in case of conventional addressing [33] and (b) voltage distribution over the pixel in case of active addressing [33, 38, 39].

Another problem in driving STN-LCD is the so-called shadowing, that is, distortion of the switching of the driving waveform. The shadowing occurs more frequently when the number of scanning lines increases. The distortion can be omitted by using low-on-resistance drivers, low-resistant terminal contacts, and ITO electrodes, and improved liquid crystal materials with reduced capacity (low dielectric constant).

6.1.4 Electrooptical Modes in Cholesterics

6.1.4.1 Selective Reflection Band

Various cholesteric textures (planar, focal conic, and homeotropic) are shown in Figure 2.23. Furthermore, we will consider the electrooptic switching between these textures more carefully. Now, let us summarize the basic optical properties of a cholesteric planar texture for the light normally incident on this texture (Figure 6.31) [6].

1. There exists one resonance band (selective reflection band), that is, a full reflection of light with the circular polarization having the same handedness as the pitch of the helix and a corresponding total transmission of the light with an opposite circular polarization (Figure 2.34a). The band has rather flat top with a center at

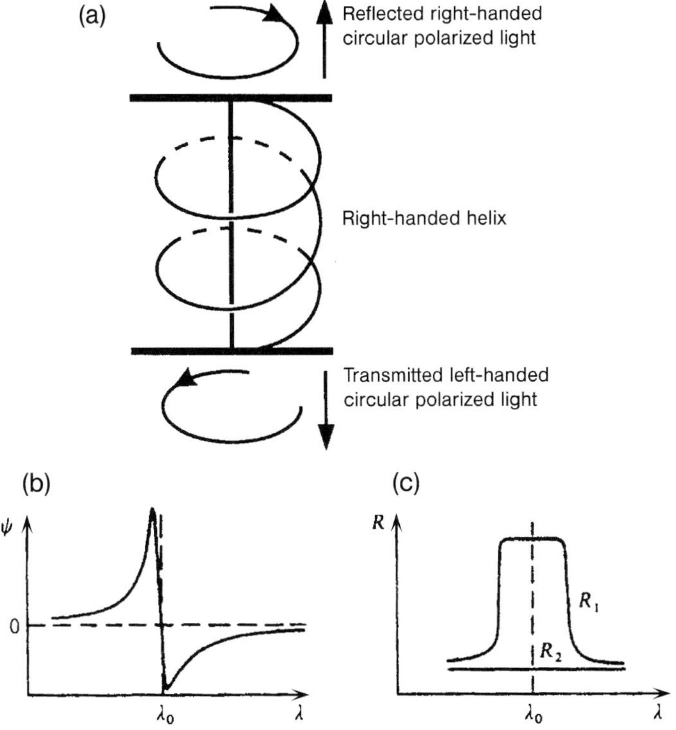

Figure 6.31 Optical properties of a cholesteric planar texture. (a) selective reflection of a circularly polarized light; (b) optical rotatory power; (c) circular dichroism [6].

$$\lambda_0 = P_0\langle n\rangle, \tag{6.34}$$

where $\langle n\rangle = (n_{||} + n_{\perp})/2$. The higher orders of reflection are forbidden.

2. The spectral width of this reflection $\Delta\lambda$ is proportional to the optical anisotropy $\Delta n = n_{||} - n_{\perp}$ of a cholesteric liquid crystal

$$\Delta\lambda = P_0\,\Delta n. \tag{6.35}$$

3. On each side of the selective reflection band, there are regions of strong rotation of a plane polarized light.

A schematic representation of wavelength dependence of optical rotatory power and circular dichroism in cholesteric systems is provided in Figure 6.31b and c, respectively.

6.1.4.2 Unwinding of a Cholesteric Helix

Due to the nonzero dielectric anisotropy $\Delta\varepsilon \neq 0$ (2.9) and flexoelectric coupling, the cholesteric LC interacts with the external electric field resulting in the two main processes: unwinding of the cholesteric helix and linear flexoelectric switching.

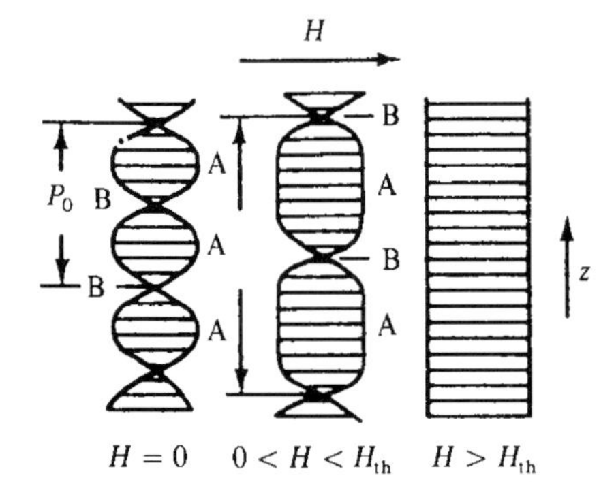

Figure 6.32 Unwinding of the cholesteric helix in a magnetic field H perpendicular to the helix axis [6].

The transformation (unwinding) of the cholesteric helix in a magnetic field is shown in Figure 6.32. The process consists of (i) increasing helix pitch in intermediate field values $0 < H < H_{th}$ and (ii) divergence of helix pitch at the threshold value $H = H_{th}$. The corresponding magnitude of the electric field [6]

$$E_{th} = \pi^2/P_0(K_{22}/\varepsilon_0\,\Delta\varepsilon)^{1/2}. \tag{6.36}$$

The dynamics of helix unwinding in cholesterics is described by the characteristic rise time τ_{rise} and decay or relaxation time τ_{rel} according to the following relations:

$$\tau_{rise} = \gamma_1/(\varepsilon_0\,\Delta\varepsilon E^2/2 - K_{22}q^2), \quad \tau_{rel} = \gamma_1/K_{22}q^2. \tag{6.37}$$

As follows from (6.37), the relaxation times of the helix diverge near the threshold field:

$$q = 2\pi/P \Rightarrow 0 \quad \text{for } P \Rightarrow \infty \text{ at } E = E_{th}.$$

6.1.4.3 Linear Flexoelectric Effect

We should also allow flexoelectric polarization (Figure 6.33)

$$\mathbf{P}_f = e_{11}\mathbf{n}\,\mathrm{div}\,\mathbf{n} - e_{33}(n \times \mathrm{curl}\mathbf{n}), \tag{6.38}$$

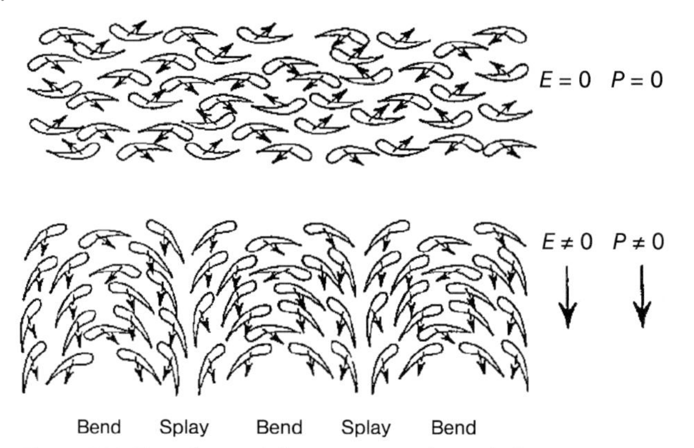

Figure 6.33 Flexoelectric deformation in a planar cholesteric structure.

where e_{11} and e_{33} are flexoelectric moduli, which gives rise to the following term g_f in the density of the nematic free energy:

$$g_f = -(\mathbf{P}_f\mathbf{E}). \tag{6.39}$$

The flexoelectric term is connected with the interaction of the LC director deformation and is complementary to the usual elastic term, which does not depend on the electric field (2.15). If a cholesteric axis is parallel to the plane of the cell, the flexoelectric coupling leads to the reversible distortion of the structure with the characteristic time on the order of hundreds of microseconds [40]. For values of the electric field sufficiently smaller than the threshold $E \ll E_{th}$, the deformation amplitude of the helix axis is

$$\varphi(E) = \arctan(e_f E P_0/2\pi K), \tag{6.40}$$

that is, changes its sign simultaneously with the field E. (Here e_f and K are average flexoelectric and elastic coefficients, respectively, and P_0 is the helix pitch.) It results in corresponding modulations of the intensity of the light on the output of LC cell placed between crossed polarizers. The problem is not to disturb the helix pitch too much as the electrooptic response can deviate from the linear character due to the helix unwinding effect [41]. We will not consider here the appearance of dielectric, flexoelectric, and other modulated scattering structures in cholesterics, which are clearly described elsewhere [7]. Such structures are parasitic and must be avoided by a special choice of LC parameters, cell construction (e.g., increased tilt angle), and driving regime.

6.1.4.4 **Reflective Cholesteric Structures**

The reflective cholesteric textures were intensively investigated in Kent State University [42, 43]. The basic structures were the so-called polymer-stabilized cholesteric textures (PSCTs) with stabilization by dispersing a low concentration of polymer inside the cholesteric liquid crystal. The electric field can switch between the three main PSCT configurations [6] (Figure 6.34).

The planar cholesteric structure is prepared by UV-curing of the mixture of a cholesteric liquid crystal, a monomer, and a photoinitiator with a simultaneous

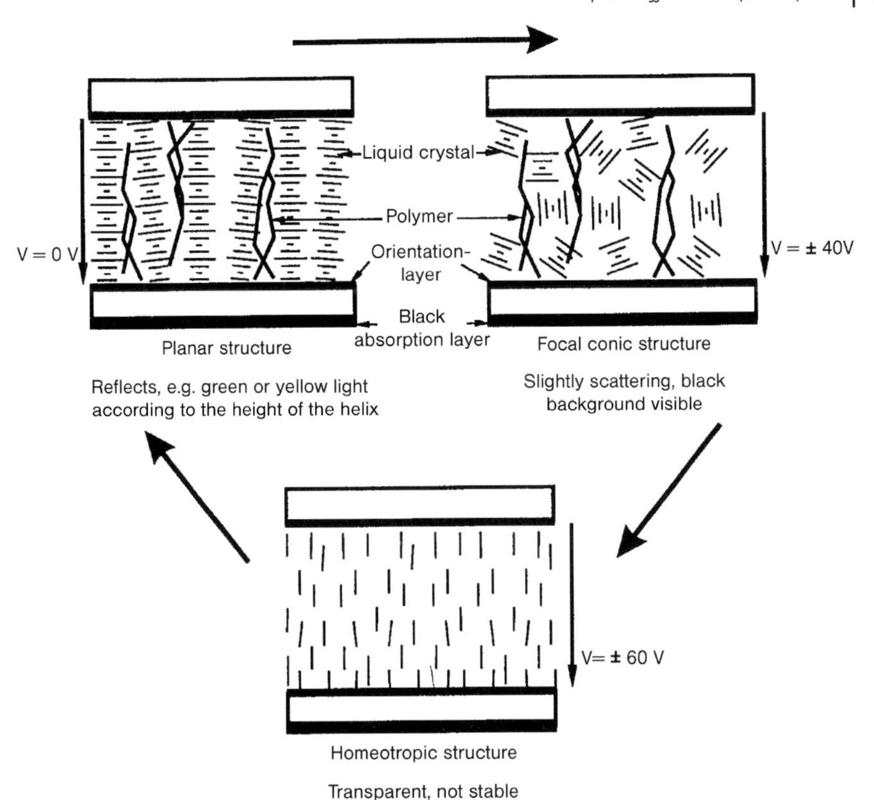

Figure 6.34 Switching between three cholesteric textures in PSCT configuration [6].

application of electric field to achieve the homeotropic orientation of LC molecules together with a monomer for forming a proper polymer network [42]. The competition between the polymer network, which favors the homeotropic LC alignment, and the surfaces treated for a homogeneous LC orientation results in a multidomain PSCT structure (Figure 6.34). The cell is then coated with a black absorbing layer to enhance the contrast between the two PSCT states.

The planar PSCT structure exhibits Bragg reflection of green or yellow light in accordance with the formula (6.34). The texture is stable in the absence of the applied field, so no power is needed to keep the image when it is written (intrinsic memory effect). The image is readable in both indoor room lighting and outdoor sunlight conditions. The contrast is very good for wide viewing angles and no polarizers are needed. The wavelength of the reflected light is easily controlled by developing chiral nematic mixtures with the temperature-stabilized pitch P, polymer network, and/or surface alignment layers.

The PSCT planar structure with a positive dielectric anisotropy $\Delta\varepsilon > 0$ can be electrically driven to a focal conic one by applying a short pulse of a sufficiently high voltage (about $\pm 40\,\text{V}$ for a 7-μm thick sample). In the focal conic structure, cholesterics become slightly scattering making visible black background, especially

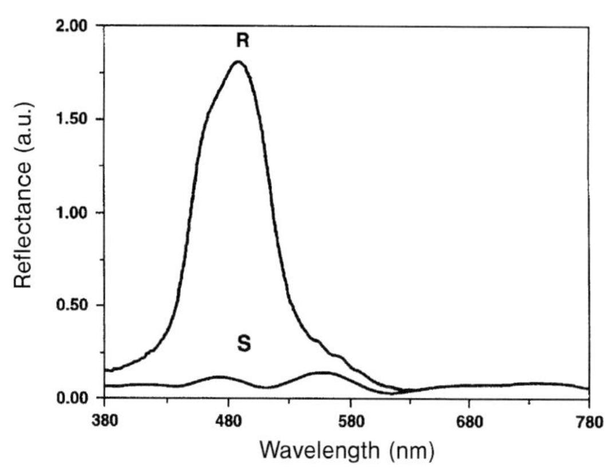

Figure 6.35 Spectral and viewing angle characteristics of PSCT-LCDs. Spectral characteristics in the two different states [6].

prepared for the purpose (Figure 6.34). The contrast and viewing angles of the PSCT cells are defined by a Bragg color reflection in the planar state and slightly scattered state near the dark background (Figure 6.35). The viewing angle in the focal conic state is large partially due to the random orientation of the helixes and different size of domains (Figure 6.34). The superiority of the PSCT-LCD viewing angle characteristics over reflective STN-LCDs was clearly shown [42, 43].

In order to switch from the focal conic state to the planar state an even larger voltage of ± 60 V is shortly applied for about 20 ms. This induces the transparent unstable homeotropic alignment in Figure 6.34 transforming into the planar helical texture if the voltage is quickly turned off. The scattering state is reached if the voltage is slowly switched off.

The addressing voltage V across the LC cell and the corresponding reflectance in different textures is shown in Figure 6.36. It is important to note the scenario of a cell switching between the states [42]. In case the initial state is planar (circles), we have

$V < V1$ no change;
$V1 < V < V2$
grey scales of reflected light (change in domain sizes and helix orientation);
$V2 < V < V3$ cell switched to focal conic state (black);
$V > V4$ cell switches from the focal conic to planar state.

For a cell in the focal conic state, the voltage $V > V4$ switches to the planar state (diamonds in Figure 6.36). The matrix addressing scheme of the PSCT-LCDs was developed with a special form of a column and row pulses, taking into account the peculiarities of the PSCT switching [42]. Shortcomings of the PSCT-LCD include slow response (hundreds of milliseconds), high addressing pulses, and sensitivity to pressure. The high resolution (100 dpi) and a large size (14″ in diagonal) are already realized.

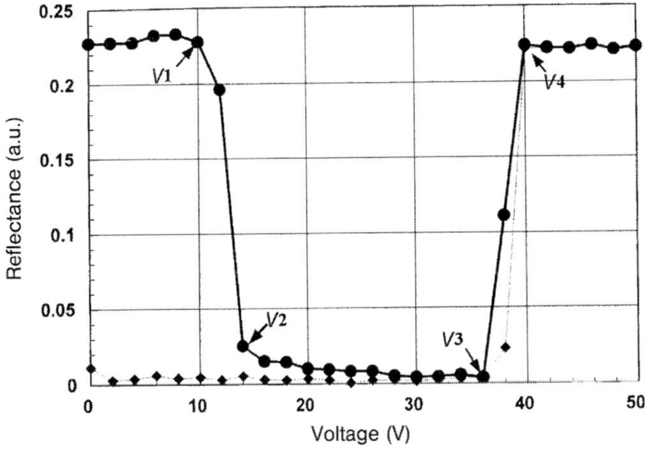

Figure 6.36 Electrooptical response of PSCT-LCD. The circles represent the reflectance evolution from the quasiplanar state, while the diamonds represent the reflectance evolution from the focal conic state [6].

The new developments in reflective cholesteric displays include photosensitive cholesteric displays [44]: when irradiated, the resulting conformation variation of the dye in the LC bulk affects the helical twisting power and changes the pitch of the cholesteric, thereby changing the peak reflected wavelength of the display.

6.1.5 Electrooptic Effects in Ferroelectric LC

6.1.5.1 Basic Physical Properties

6.1.5.1.1 Structure and Symmetry

The structure of ferroelectric liquid crystals (FLCs), methods of preparation of FLC mixtures, and the most important problems in their alignment were considered in Chapter 2. Here, we will consider the optical and electrooptical properties of FLC cells, taking into account their main physical parameters (such as polarization, viscosity, tilt angle, etc.) and experimental conditions (initial orientation, helical structure, anchoring energy, polarizers location, driving voltage, etc.). Special attention will be paid to the most important one for applications of electrooptic modes in FLCs.

As already mentioned, the symmetry of the ferroelectric smectic C^* phase corresponds to the polar symmetry group C_2, Figure 2.6, so that when going along the z-coordinate parallel to a helix axis and perpendicular to the smectic layers, the director $\mathbf{n}$ and the polarization vector $\mathbf{P}$, directed along the C_2 axis, rotate such as

$$\mathbf{n}(z+R) = \mathbf{n}(z), \quad \mathbf{P}(z+R) = \mathbf{P}(z), \tag{6.41}$$

that is, the helix pitch R is equal to a spatial period of the FLC structure. In the absence of external fields, the FLC equilibrium helix pitch is R_0 and the average polarization of the FLC volume is equal to zero (Figure 2.6).

6.1.5.1.2 Main Physical Parameters

The main physical parameters that define FLC electrooptical behavior are:

(i) tilt angle, θ;
(ii) spontaneous polarization, P_s;
(iii) helix pitch, R;
(iv) rotational viscosity, γ_φ;
(v) dielectric anisotropy, $\Delta\varepsilon$;
(vi) optical anisotropy, Δn;
(vii) elastic moduli; and
(viii) anchoring energy of the director with a solid substrate.

Let us briefly characterize each of these parameters.

6.1.5.1.3 Tilt Angle

The value of the tilt angle could vary from several degrees to $\theta \approx 45°$ in some FLCs. Usually, in electrooptic FLC materials, $\theta \approx 22.5°$ in an operating temperature range. However, for some electrooptical applications it is desirable to have the value of θ as high as possible. The temperature Tdependence near the phase transition point T_c to the more symmetric LC phase (smectic A, nematic, or chiral nematic)

$$\theta \approx (T_c - T)^{1/2} \tag{6.42}$$

is typical of second-order phase transitions.

6.1.5.1.4 Spontaneous Polarization

The value of the spontaneous polarization depends on the molecular characteristics of an FLC substance itself and the achiral dopant introduced into the matrix, and can vary from 1 to more than 200 nC/cm^2. The value of the spontaneous polarization is one of the main FLC characteristics, which define FLC electrooptic response.

6.1.5.1.5 Rotational Viscosities

We have already partially discussed the rotational viscosities of FLC in Chapter 5. The switching time in electrooptical effects in FLCs is defined by the rotational viscosities γ_φ and γ_θ that characterize the energy dissipation in the director reorientation process. According to the FLC symmetry, two viscosity coefficients should be taken into account, γ_θ and γ_φ, which determine the corresponding response rates with respect to the director angles θ and φ (Figure 2.6). In some FLC materials characterization, this viscosity is called the **n** and **c** director viscosity, where **n** is a real FLC director and **c** director is its projection onto the plane perpendicular to the FLC layers normal (Figure 6.37). The relevant dynamic equations take the form [6, 7]

$$\gamma_\theta d\theta/dt + A\theta = 0, \qquad \tau_\theta = \gamma_\theta / A, \tag{6.43a}$$

$$\gamma_\varphi d\varphi/dt - P_S E \sin\varphi = 0, \qquad \tau_\varphi = \gamma_\varphi / P_S E, \tag{6.43b}$$

where $A = 2a(T_c - T)$ and $a > 0$ is an effective elastic coefficient for the tilt, τ_θ and τ_φ are the characteristic response times for the θ and φ angles of the FLC director. The

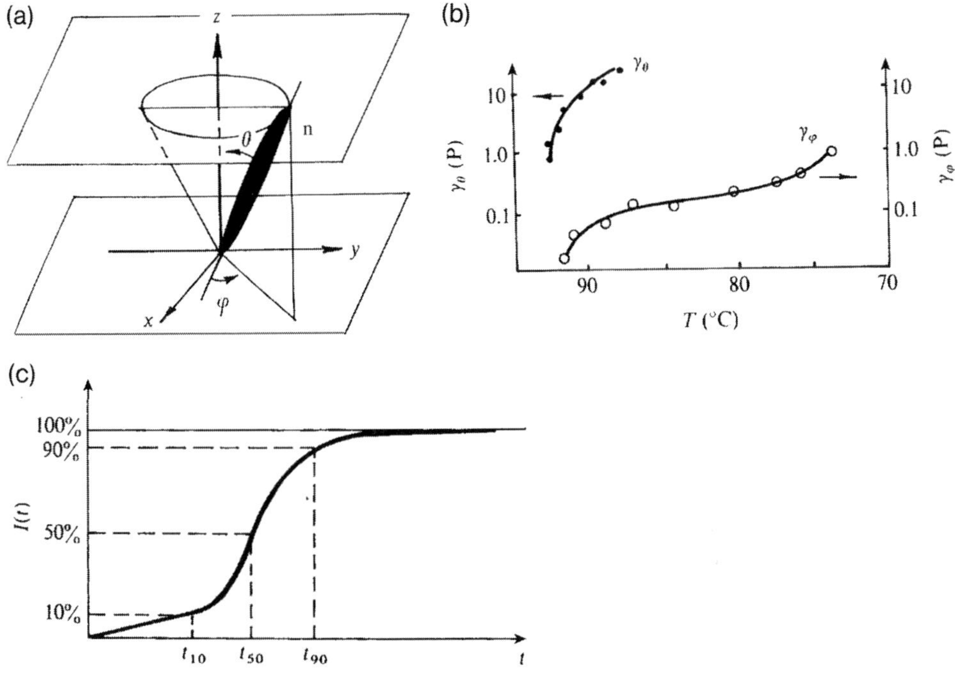

Figure 6.37 Characteristic viscosities γ_θ and γ_φ, which define the characteristic times of electrooptic response in FLC cell. (a) FLC director **n**; (b) temperature dependences of the viscosities γ_θ and γ_φ in DOBAMBC (Figure 2.6) [45]; (c) dynamics of FLC cell transmission.

viscosity coefficient γ_φ can be rewritten as

$$\gamma_\varphi = \gamma'_\varphi \sin^2\theta, \tag{6.44}$$

where γ'_φ is independent of the angle θ. According to (6.44), $\gamma_\varphi \Rightarrow 0$ for $\theta \Rightarrow 0$, that is, γ_φ is very low for small tilt angles θ.

Far from the phase transition point T_c, it is reasonable to consider only the azimuthal director angle φ because the angle θ is frozen. (Variations of the angle θ result in a density change, which is energetically unfavorable.) However, near T_c we can change the θ angle, for example, by applying electric field E; this effect is known as electroclinic effect [6, 7].

The rotational viscosity γ_φ could be estimated from the experimental dependence of the electrooptical response as follows [45, 46]:

$$\gamma_\varphi = P_S E \tau_\varphi$$

$$\tau_\varphi = (t_{90} - t_{50})/\ln\sqrt{5}, \tag{6.45}$$

where t_{90} and t_{50} are the corresponding times for 90 and 50% transmission from the maximum level (Figure 6.37c).

One of the possible ways to increase the switching rate in FLCs includes minimizing the viscosity with a simultaneous rise in polarization. However, this method is not very promising because (i) the increase in polarization is hardly compatible with the lower values of rotational viscosity [45] and (ii) the repolarization component of the AC current grows rapidly with the value of polarization, that is,

$$i_P \approx P_S/\tau_\varphi \approx P_S^2/\gamma_\varphi, \tag{6.46}$$

which is always undesirable for applications. The temperature dependence of the viscosity γ is shown in Figure 6.37b for DOBAMBC (Figure 2.6). In accordance with (6.42) and (6.44) at the phase transition point $T = T_c$, we have $\gamma_\varphi = 0$.

6.1.5.1.6 **Helix Pitch**

The helix pitch value R_0 is easily controlled by varying the concentration of a chiral dopant in a smectic C matrix. To provide variation of the helix pitch from 0.1 to 100 μm and more, we should have a chiral dopant with a high twisting power and good solubility. At the same time, the chiral dopant should not depress the smectic C temperature range [7]. Moreover, the growth of polarization should be more pronounced than that of the rotational viscosity γ_φ with an increasing concentration of the chiral dopant. The inverse pitch of the FLC helix, for reasonably large concentrations of the helix of the chiral dopant C_D, is proportional to the concentration

$$R_0^{-1} \propto C_D. \tag{6.47}$$

6.1.5.1.7 **Dielectric and Optical Properties**

Dielectric properties of FLCs strongly depend on the temperature (especially near phase transition points) and frequency of the field. The dielectric susceptibility of an FLC could be defined as

$$\chi = (\varepsilon - 1)/4\pi = \lim \langle P \rangle / E \quad \text{for } E \Rightarrow 0. \tag{6.48}$$

As mentioned in Chapter 5, *two modes* contribute to the value of the averaged value of polarization $\langle P \rangle$. The first of them (*soft mode*) is induced as a result of the amplitude change in the polarization, that is, variation of the tilt angle θ. This mode is the most important near the phase transition point of the FLC phase and results in the electroclinic effect [47]. The second mode, called the *Goldstone mode,* is responsible for the variation of the polarization phase, that is, the azimuthal director angle φ. The contributions of both modes to FLC dielectric properties were considered in Chapter 5 ((5.63) and (5.64)).

It is important for applications to have negative dielectric anisotropy $\Delta\varepsilon = \varepsilon_{||}$ $\varepsilon_\perp < 0$, which stabilizes the relevant director structures [6, 7]. Usual values of the negative dielectric anisotropy $\Delta\varepsilon$ in FLC mixtures are between -0.5 and -2 in kilohertz region decreasing at the higher frequencies due to the relaxation of the Goldstone and soft modes [6, 48].

Let us point out a specific property of FLCs that makes its dielectric behavior very different from nematic liquid crystals. Dielectric susceptibility, defined as $\chi = \partial P/\partial E$, strongly depends on the field amplitude and for $E < E_u$ (E_u is the field of FLC helix unwinding) can reach values of 10^2–10^3 and even higher. This should be taken into account in applications, for instance, in photosensitive FLC light valves cells, which require impedance matching between photoconductor and liquid crystal layers [7].

The effect of nonzero dielectric anisotropy can be taken into account, inserting the additional term into Equation (6.43b) of the azimuthal director motion in the electric field E [6, 7]:

$$\gamma_\varphi \, d\varphi/dt = P_S E \sin\varphi + (\varepsilon_0 \, \Delta\varepsilon) E^2 \sin^2\theta \sin\varphi \cos\varphi. \tag{6.49}$$

A real dielectric FLC ellipsoid has three components $\varepsilon_{||}$ (along the director **n**), ε_s (along the C_2 axis), and ε_t (in the direction perpendicular to both C_2 and **n**), this should be taken into account when considering the FLC interaction with external field (Figure 6.38) [49].

For instance, let the field be parallel to the y-axis, where θ and φ are the director angles in the FLC layer. Then, the contribution to FLC free energy in case of *dielectric biaxiality* gives [49]

$$\begin{aligned} \mathbf{ED}/2 &= -\mathbf{ED}_y/2 = -\varepsilon_0(\varepsilon_{||} \sin^2\theta \sin^2\varphi + \varepsilon_s \cos^2\varphi + \varepsilon_t \sin^2\varphi \cos^2\theta) E^2/2 \\ &\approx -\varepsilon_0[(\varepsilon_{||} - \varepsilon_\perp)\sin^2\theta \sin^2\varphi + (\varepsilon_t - \varepsilon_s)\sin^2\varphi] E^2/2, \end{aligned} \tag{6.50}$$

where we have omitted the constant term $\varepsilon_0 \varepsilon_s E^2/2$. The relationship (6.50) shows that the dielectric contribution exists in FLC free energy even in the case of very small tilt angles θ, if the biaxiality is essential, $\varepsilon_s \neq \varepsilon_t = \varepsilon_\perp$. As shown in [49], the dielectric biaxiality $|\varepsilon_s - \varepsilon_t|$ is comparable in some cases with the dielectric anisotropy $|\varepsilon_{||} - \varepsilon_t|$ of an FLC, and should be taken into account as a factor of the dielectric stabilization of a director configuration.

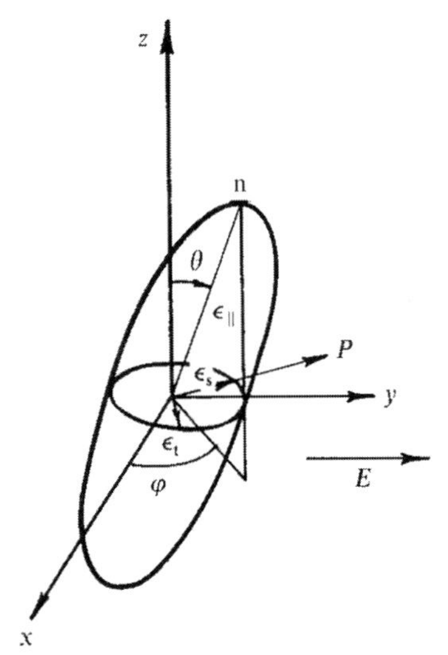

Figure 6.38 Dielectric biaxiality in FLC. P: polarization, **n**: director, E: electric field, z-FLC layers normal. The biaxiality is equal to $\varepsilon_t - \varepsilon_s$, where ε_s is measured along the direction of P [6].

All three principal values of the dielectric tensor could be determined, measuring the FLC tilt angle θ and the dielectric constants in different geometry, with the FLC layers parallel and perpendicular to the substrates, in the helical and unwound state (the FLC layer pretilt must also be taken into account) [49, 50]. The value and sign of the dielectric biaxiality $\varepsilon_s - \varepsilon_t$ was found to be frequency dependent with the inversion frequency of about 1 kHz.

The dielectric tensor of an FLC at *optical frequencies* could be regarded as uniaxial [6, 7]. Then, the FLC possesses only two refractive indices, $n_{||}$ along the director and $n_{\perp}$ perpendicular to it. This approximation is confirmed by direct measurements of the FLC optical properties. As a rule, *the biaxiality* does not exceed 10^{-1} and tends to zero at the phase transition point T_c [6, 7].

The electrooptical behavior of an FLC is mainly defined by the optical anisotropy $\Delta n = n_{||} - n_{\perp}$. The birefringence value Δn can be obtained from electrooptical measurements, where the dependence of the transmitted intensity on the phase factor $\Delta nd/\lambda$ (d is the cell thickness and λ is the light wavelength) is used [6, 7]. The dispersion law for the optical birefringence follows the well-known Cauchy formula [6, 7]:

$$n(\lambda) = \Delta n(\infty) + C/\lambda^2, \quad C = \text{const.} \tag{6.51}$$

The temperature dependence of $\Delta n(\lambda)$ in FLCs is less pronounced than in nematics. Within the whole temperature range of the smectic C* phase, its variation does not exceed 0.01. The spatial period of the dielectric tensor in FLCs is equal to the helix pitch R_0 (not to $R_0/2$, as in cholesterics). This results in additional diffraction orders in reflection when light is propagating obliquely with respect to the helical axis. Electrooptical effects in FLCs are usually observed for $R_0 \gg \lambda$ (surface-stabilized FLC structure (SSFLC) [48]) or for $R_0 \ll \lambda$ (deformed helix ferroelectric (DHF) effect [51]), that is, over the limits of the diffraction region.

6.1.5.1.8 Elastic Properties and Anchoring Energy

The elastic properties of FLCs are usually discussed using the density of the elastic energy g_{el} as follows [52]:

$$g_{el} = \frac{1}{2}\left[K_{11}(\text{div}\,\mathbf{n})^2 + K_{22}(\mathbf{n}\,\text{curl}\,\mathbf{n} - t)^2 + K_{33}(\mathbf{n} \times \text{curl}\,\mathbf{n} - \mathbf{b})^2\right], \tag{6.52}$$

where t and $\mathbf{b}$ denote spontaneous twist and bend of the FLC director $\mathbf{n}$, respectively, and K_{ii} are FLC elastic moduli. The value of $t > 0$ and $t < 0$ for the right- and left-handed FLCs, respectively. The general continuum theory of FLC deformations must take into account not only deformations of the FLC director $\mathbf{n}$ but also possible distortions of the smectic layers, which can be described in terms of the variation of the layer normal $\mathbf{v}$.

The FLC free energy should also include the surface terms, which are polar $w_p = -W_p(\mathbf{P}\mathbf{v})$ and dispersion $w_d = -W_d(\mathbf{P}\mathbf{v})^2$ contributions, where W_p and W_d are the corresponding anchoring strength coefficients, $\mathbf{v}$ is the layer normal, and $\mathbf{P}$ is the FLC polarization [53]. Thus, the total free energy F_d of the FLC director deformations

$$F_d = \int_V g_{el}\,d\tau + \int_S (w_p + w_d)\,d\sigma, \tag{6.53}$$

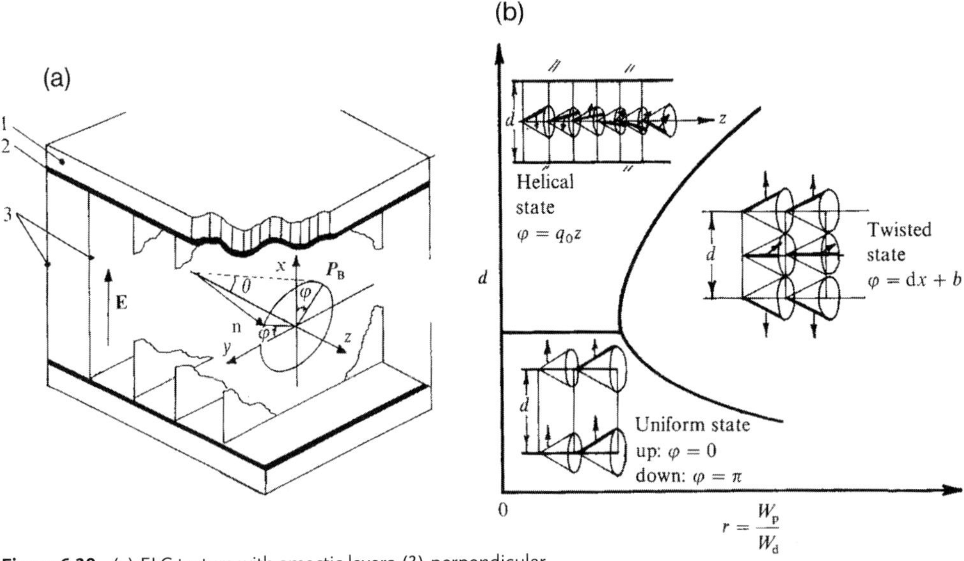

Figure 6.39 (a) FLC texture with smectic layers (3) perpendicular to the substrates (1) with current-conducting ITO layers (2) in "ideal" defectless configuration. (b) Different FLC states in zero field for a various cell thickness d and anchoring ratio $r = W_p/W_d$ [6].

that is, consists of a sum of the volume (v) and surface (s) terms. Various textures of FLC layers with a normal parallel to the substrates are shown in Figure 6.39 as functions of the layer thickness d and the ratio of the linear-to-quadratic anchoring strength coefficients $r = W_p/W_d$. The following characteristic states are observed [6, 7]: helical state in sufficiently large thickness $d \gg R_0$, where R_0 is a helix pitch; uniform states with $R_0 > d$ due to the parallel orientation order effect on the substrates ("up" and "down" states differ by the direction of the FLC polarization **P**) and twisted state for large values of $r \gg 1$, which favors antiparallel surface aligning of the FLC director.

In the presence of the external field, the total free energy includes the energy of the director deformations F_d and the energy F_E of the interaction of the ferroelectric phase with the field **E**:

$$F = F_d + F_E = F_d + \int_v [-(\mathbf{PE}) - \mathbf{DE}/2] d\tau. \tag{6.54}$$

The last term in (6.54) is already written (6.50) taking into account the FLC biaxiality. Anchoring energy of FLC can be determined either (i) by measuring the width of the coercetivity loop ΔV in the $P(E)$ dependence in a static field according to the relation $\Delta V = 8W_d/P_s$ or (ii) by measuring the free relaxation times τ_r of the FLC director to bistable states $\tau_r = \gamma_\varphi d/4W_d$, where γ_φ is FLC rotational viscosity for the **c** director (Figure 6.37) [54]. Results of these two methods coincide with each other with an accuracy of 30%. The dispersion anchoring energy of an FLC for different conducting surfaces was found to be on the order of 10^{-4} J/m^2 [7]. Near the transition point T_c to the smectic A phase, the anchoring energy goes to zero according to the law: $W_d \propto (T_c - T)$ [54]. FLC polar and dispersion energy strongly depends on the molecular content of FLC mixture and boundary conditions [55, 56].

6.1.5.1.9 Aligning and Textures

Contrary to nematic liquid crystals, there is no standard method for preparing uniform defect-free FLC orientation. In the smectic C* phase, both the director and the smectic layers should be oriented in a proper way. We are going to discuss the various techniques of FLC alignment, which provide defect-free samples with high optical quality.

One of the standard techniques often used for this purpose is the preparation of a perfectly aligned smectic A sample. A liquid crystal often undergoes the following phase sequence: isotropic phase ⇒ (nematic) ⇒ smectic A ⇒ smectic C. The nematic phase (if it exists) could be oriented by traditional methods, such as the rubbing of polymer films or the evaporation of inorganic films at an oblique angle. The quality of orientation improves for lower values of the specific heat at the nematic ⇒ smectic A transition and the minimum amount of crystallization centers.

If an FLC material does not possess the nematic phase, the smectic A phase could be oriented directly from the isotropic state. In thick (>100 μm) cells, and even in thin ones with weak anchoring of the director to boundaries, it is possible to grow well-oriented smectic A layers using either a magnetic field or a temperature gradient. In both cases, the preferred alignment direction is predetermined by a rubbed polymer layer, or by an edge of the spacer fixing the cell thickness and playing the role of a crystallization center [7].

FLCs can be aligned using a parallel shift of one cell substrate with respect to the other. The smectic layers tend to orient perpendicular to the substrates ("bookshelf geometry", Figure 6.39a), when mechanical shear is applied perpendicular to the rubbing direction [57]. This is one of the most reliable techniques and allows us to orient the smectic A phase in a broad range of cell thickness over an area of several square centimeters [48]. FLC mixtures with no smectic A phase can be oriented by cooling down the nematic phase in an AC electric field that defines the direction of a smectic layer, while the director is oriented by rubbing the polymer films on the substrates [7].

One of the important aligning techniques, which allows us to avoid zig-zag defects between two adjacent smectic layers bent in opposite directions (chevrons), remains the oblique evaporation of silicon monoxide [48] (Figure 6.40). Zig-zag defects are avoided by the promotion of only one possible bend or tilt of the smectic layers due to a specific oblique director orientation at the boundaries (Figure 6.40). Samples with the antiparallel evaporation direction contain uniformly tilted layers. Those with parallel directions exhibit the chevron structure of tilted layers free of zig-zag defects.

The chevron structure and zig-zag defects were shown to disappear if a low-frequency electric field of sufficient amplitude is applied to an FLC cell [58]. The voltage required for the reorientation decreases with increasing spontaneous polarization. The texture appearing resembles the "bookshelf" geometry of smectic layers perpendicular to the substrates. It has minimum response time, maximum

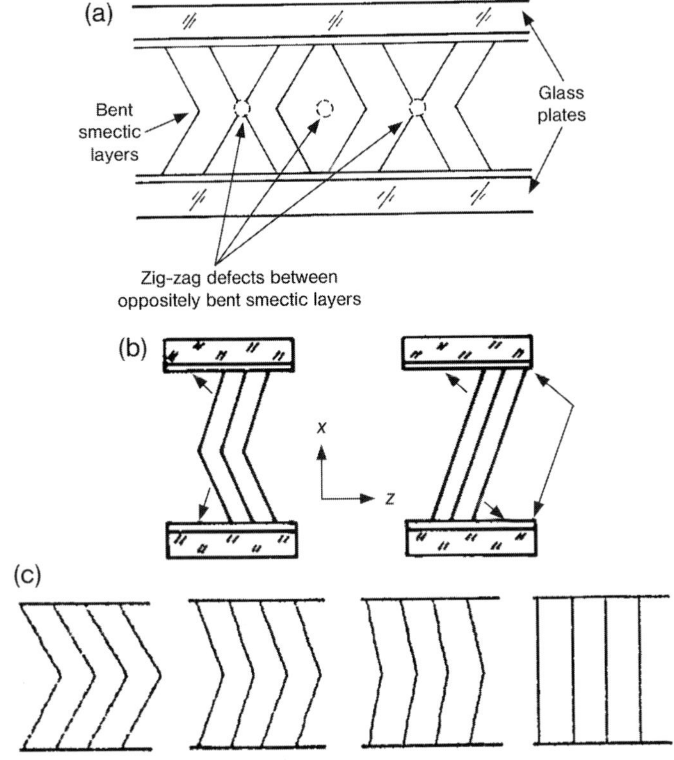

Figure 6.40 Zig-zag defects in FLC cells between two oppositely bent layers (a); avoiding zig-zag defects by preparing uniform layer tilt (b); electric field-induced transition to "quasibookshelf" structure (c) [6].

transmission difference between both stable states, and a relatively uniform viewing area. Therefore, this texture, referred to as "quasibookshelf geometry," seems to be very promising for display applications [6, 7].

The typical texture for FLCs with a short pitch R_0 compared to the layer thickness d ($d \gg R_0$) is shown in Figure 6.41. The structure occurs when an AC electric field (15 V/μm, 10 Hz) is applied during cooling from the isotropic phase down to the smectic A phase and further to the ferroelectric phase [59]. This stripe texture means a periodic modulation of the refractive index perpendicular to the lines, while the lines themselves are parallel to the rubbing direction. The spacing between lines increases with increasing cell gap. The stripes seem to suppress the formation of the FLC helix.

The optical quality of short-pitch FLC (SPFLC) cell could be improved if only one substrate is rubbed, and FLC layers are allowed to slide along the other untreated substrate with a low anchoring energy. The final structure will be a tilted-bookshelf smectic layer structure (Figure 6.40b) with a high optical quality.

Photo-induced alignment for FLC cells based on the azo dye films was investigated [60, 61]. The alignment quality of FLC cells depended on the asymmetric

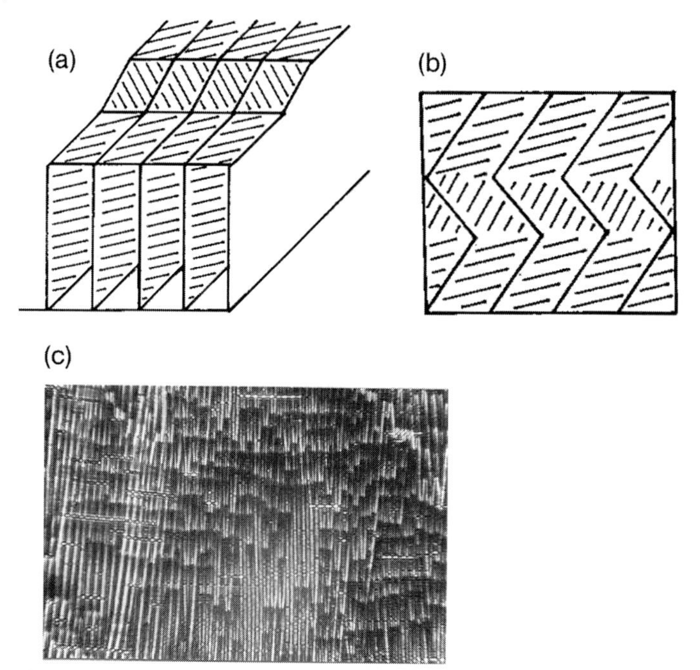

Figure 6.41 Typical "horizontal-chevron" texture for short pitch FLCs. (a) SPFLC cell; (b) top view; (c) photograph of the texture in crossed polarizers [6]. (Courtesy of M. Schadt.)

boundary structure and treatment of ITO layer, including photoaligning substrate and planar substrate. Asymmetric boundary conditions may be useful to avoid a competition in aligning action of solid surfaces of FLC cells. High exposure energy resulted in high quality of photoaligned azo dye layer and hence a perfect electro-optical response of FLC display. An optimal (about 3–5 nm) azo dye layer thickness that provided both the highest multiplex operation steadiness and the best contrast ratio of FLC display cells was found [60] (Figure 6.42). A prototype of passively addressed FLC display based on the photoalignment technique has been developed with perfect bistable switching in the multiplex driving regime [60].

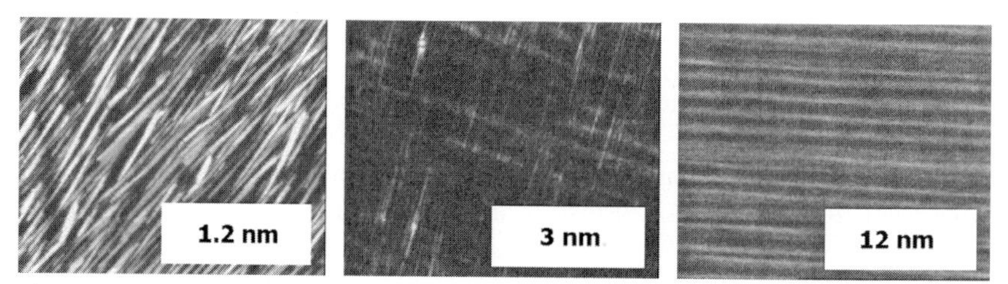

Figure 6.42 Improving FLC quality by a photosensitive azo dye layer with an optimal thickness. The optical quality of FLC cell is the best for the azo dye layer thickness equal to 3 nm [60].

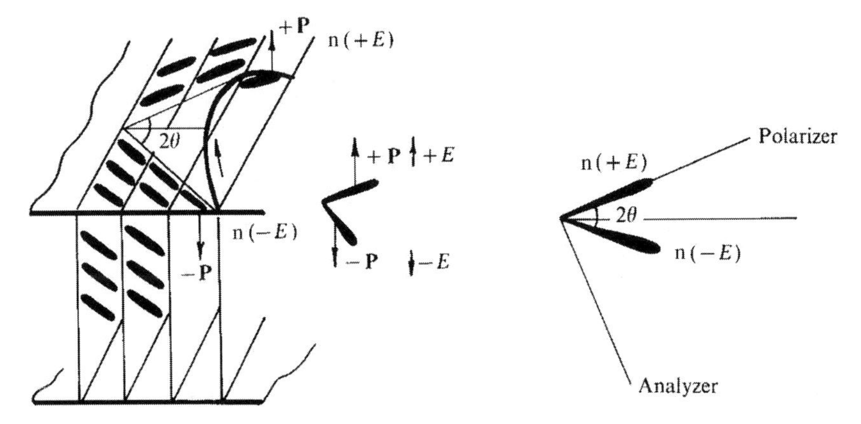

Figure 6.43 Clark–Lagerwall effect in FLC cell [6].

6.1.5.1.10 Electrooptic Effects in FLC Cells

Clark–Lagerwall Effect Let us consider the main electrooptical phenomena in FLCS. The best known is the Clark–Lagerwall effect [57], which results in director reorientation from one bistable state to the other when an external electric field changes its sign (Figure 6.43). In this case, the FLC layers are perpendicular to the substrates and the director moves along the surface of a cone whose axis is normal to the layers and parallel to the cell substrates. In each final position of its deviation, the director remains parallel to the substrates, thus transforming the FLC cell into a uniaxial phase plate. The origin of electrooptical switching in the FLC cell is the interaction of the polarization **P** perpendicular to the director with the electric field **E**. The maximum variation of the transmitted intensity is achieved when the FLC cell is placed between crossed polarizers, so that an axis of the input polarizer coincides with one of the final director positions. The total angle of switching equals the double tilt angle θ (Figure 6.43). The Clark–Lagerwall effect is observed in the so-called surface-stabilized FLC structure. In SSFLC cells, $d \ll R_0$ and we come to the situation where the existence of the helix is unfavorable, that is, the helix is unwound by the walls (Figure 6.39b).

The variation of the azimuthal director angle φ in the Clark–Lagerwall effect (Figures 6.39 and 6.43) is described by the equation of the torque equilibrium (6.49), which comes from the condition of the minimum of the FLC free energy (6.53). The total model of FLC reorientation in a "bookshelf" geometry (Figure 6.39) with appropriate boundary conditions is

$$\begin{aligned} &\gamma_{\varphi} d_{\varphi}/\mathrm{d}t = P_s E \sin\varphi + (\varepsilon_0\,\Delta\varepsilon) E^2 \sin^2\theta \sin\varphi \cos\varphi, \\ &K \partial\varphi/\partial x + W_p \sin\varphi \pm W_d \sin 2\varphi|_{z=0,d} = 0, \end{aligned} \tag{6.55}$$

where allowing the biaxiality (6.50) we have $\Delta\varepsilon = (\varepsilon_{\parallel} - \varepsilon_{\perp})\sin^2\theta + \varepsilon_t - \varepsilon_s$.

For typical values of polarizations $P_S \approx 20\,\mathrm{nC/cm^2}$, driving fields $E \approx 10\,\mathrm{V/\mu m}$, and the dielectric anisotropy $\Delta\varepsilon \approx 1$, we have

$$|\varepsilon_0\,\Delta\varepsilon E| < P_s, \tag{6.56}$$

and, consequently, the dielectric term in (6.55) could be omitted. If the inequality (6.56) is invalid, what occurs for sufficiently high fields, then for $|\varepsilon_0 \Delta\varepsilon E| \propto P_s$ the response times of the Clark–Lagerwall effect sharply increase for positive $\Delta\varepsilon$ values [62]. Experiments show that for negative values of $\Delta\varepsilon$ the slope of the FLC dynamic response increases, that is, the corresponding switching times become shorter. This is especially important for practical applications because it promotes an increase in the information capacity of FLC displays. If the driving field increases, FLC response time passes through a minimum, then grows, passes through a maximum at $|\varepsilon_0 \Delta\varepsilon E| \propto P_s$, and finally decreases again.

For $|\varepsilon_0 \Delta\varepsilon E| \gg P_s$, the FLC switching times τ are approximately governed by the field squared, $\tau \propto \gamma_\varphi/(\varepsilon_0 \Delta\varepsilon E^2)$, as in the ECB effect in nematics (6.14). FLC mixtures with negative dielectric anisotropy $\Delta\varepsilon < 0$ are also used for the dielectric stabilization of the initial orientation in FLC displays [48].

As shown above (6.43b), response times in the Clark–Lagerwall effect are determined by $\tau_\varphi \propto \gamma_\varphi/P_s E$. Values of $t_{90} - t_{10}$ and $t_{90} - t_{50}$, measured in experiment (Figure 6.37c), are fairly close to τ_φ. Comparing the response times in nematics (N)

$$\tau_r^{(N)} \propto \gamma_1/(\varepsilon_0 \Delta\varepsilon E^2)\tau_d^{(N)} \propto d^2\gamma_1/K\pi^2 \tag{6.57}$$

and ferroelectric liquid crystals

$$\tau_r^{(FLC)} \propto \tau_d^{(FLC)} \propto \tau_\varphi \propto \gamma_\varphi/P_s E, \tag{6.58}$$

we come to the conclusion that the electrooptical switching in the Clark–Lagerwall effect in FLCs is much faster than in nematics. Slower response of nematic LCs is mainly due to relatively large decay times $\tau_d^{(N)}$, which in FLC case can be very short in sufficiently high electric fields E.

The optical transmittance I in the Clark–Lagerwall effect is calculated as follows [57, 62]:

$$I = \sin^2 4\theta \sin^2 \Delta\Phi/2, \tag{6.59}$$

where $\Delta\Phi = \Delta n d/\lambda$ the phase difference, $\Delta n = n_{||} - n_\perp$. As it follows from (6.59), the maximum contrast is obtained for $\theta = \pi/8$ (22.5°), $\Delta n d/\lambda = 1/2$, which, for $\Delta n = 0.125$, $\lambda = 0.5\ \mu m$, gives $d = 2\ \mu m$. We should note that the variation of the cell thickness $\Delta d = \lambda/8\ \Delta n$ from the optimum value $d = \lambda/2\ \Delta n$ results in considerable difference in the FLC electrooptical response (Figure 6.44) [6, 7]. The practical criteria of an FLC display quality, however, require more precise limitations $d = 2 \pm 0.2\ \mu m$ [6, 7].

The existence of two or more thermodynamically stable states with different optical transmission is a very important feature of FLC cells in Clark–Lagerwall effect. Figure 6.45 shows various cases of experimentally observed electrooptic switching of FLC cell under the action of sign-alternating periodic voltage pulses. The perfect bistable switching is given by the solid curve in Figure 6.45b, while the cases (c) and (d) present imperfect and quasibistable switching, respectively. The monostable switching without optical memory is given in Figure 6.45e. Reliable reproduction of bistability conditions is very crucial for technological applications.

The practical realization of a perfect bistability in Clark–Lagerwall FLC cell is a complicated problem. First of all, we have to point out that chevron geometry, even in

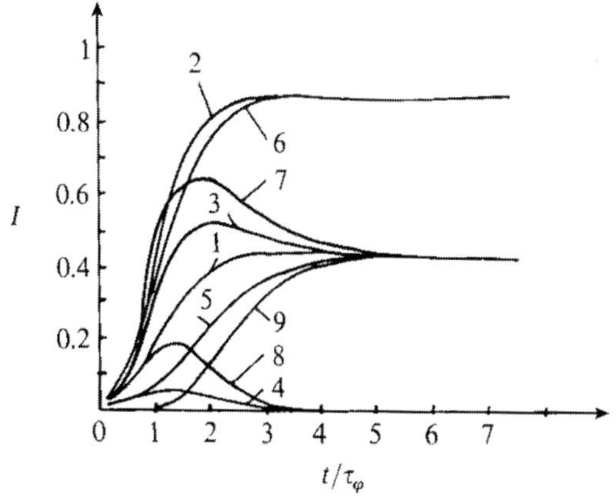

Figure 6.44 Electrooptical response of the FLC cell in Clark–Lagerwall effect for different phase factors $d\Delta n/\lambda = k/4$, $k = 1, \ldots, 9$ [6].

zig-zag free samples, hinders the occurrence of bistability. Chevrons and, more generally, the tilt of smectic layers (Figure 6.40) should be eliminated from the cell by any means, such as the use of a low-frequency AC field to arrange a "quasibookshelf" structure or developing FLCs with a very short helix pitch preventing chevron appearance [48]. But even in the favorable case of smectic layers perpendicular to the substrates, it is important to control different FLC physical parameters, such as

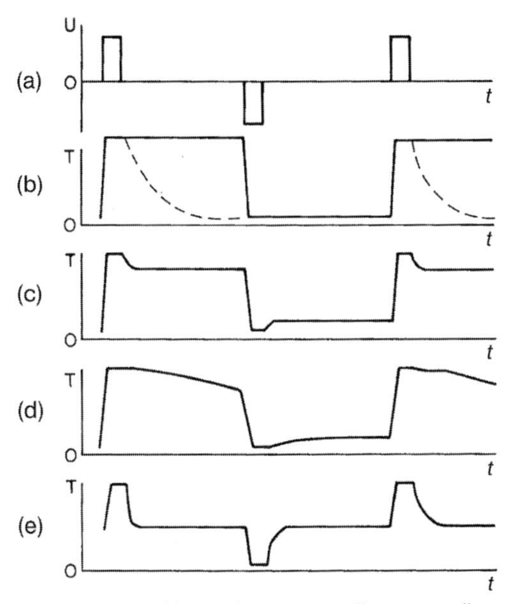

Figure 6.45 Bistable switching in FLC cell. (a) Controlling voltage pulses; (b–d) various realizations of bistable switching with two optically different FLC states; (e) monostable FLC switching without optical memory [6].

the value of polarization, the helix pitch, the elastic moduli, and the anchoring energies [7]. For instance, sufficiently large cell thickness d and linear (polar) anchoring energies W_p do not promote bistability because the helical and twisted states become more favorable (Figure 6.39b). The bistable switching takes place above a certain threshold field $E_{th} \propto W_d/K^{1/2}$, where K is an average elastic constant and W_d is a dispersion anchoring energy and polar anchoring energy is taken equal to zero $W_p = 0$ [7]. Thus with increasing anchoring we have to increase the switching amplitude of the electric field. As the energy of switching electric torque is proportional to the product of $P_s E$, the bistability threshold is inversely proportional to the value of the FLC spontaneous polarization P_s.

The problems we face when using the Clark–Lagerwall effect include not only severe restrictions to the optimum layer thickness and requirements of defectless samples but also difficulties in realizing a perfect bistability or optical memory switched by the electric field and providing the gray scale. The last is one of the most crucial problems because it is very inconvenient to provide the gray scale using either complicated driving circuits or increasing the number of working elements (pixels) in an FLC display [48]. This problem arises in the Clark–Lagerwall effect because the level of transmission is not defined by the amplitude of the driving voltage pulse U but by the product $U\tau$, where τ is the electric pulse duration.

The space charge accumulated due to the spontaneous polarization also influences the switching properties of the Clark–Lagerwall effect [63]. If we take into account the charge density $\sigma(t)$, which is accumulated at the interface between the orienting layer and the FLC medium,

$$\sigma(t) = P_s(\cos\varphi(t) - \cos\varphi(0)), \tag{6.60}$$

then the electric field strength $E(t)$ within the FLC cell is given by [63]

$$E(t) = V_0\varepsilon'/(2d'\varepsilon + d\varepsilon') - 2\sigma(t)d'/(2d'\varepsilon + d\varepsilon'), \tag{6.61}$$

where (d, ε) and (d', ε') are the thickness and dielectric constant for the FLC and the orienting layer, respectively (Figure 6.46). The formula (6.61) is easily obtained if we

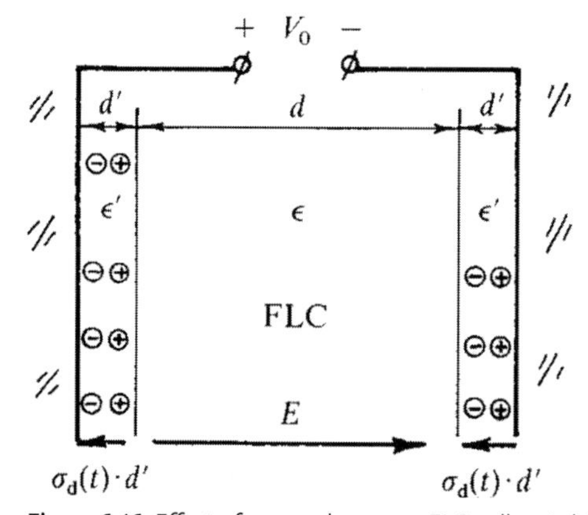

Figure 6.46 Effect of space charge on FLC cell switching in electric field [6].

take into account that the effective voltage value V'_0 across the FLC medium is less than the applied voltage V_0 as part of the voltage drops across the orienting layer. In view of (6.61), we come to the conclusion that, in order to avoid screening the switching field by charges, we have to use thin orienting layers with high dielectric permittivity and sufficiently thick FLC cells with low spontaneous polarization.

Charge-controlled switching was suggested as one of the possible methods to obtain a gray scale in SSFLC devices [64]. In this case, in order to switch totally a FLC cell with the surface area A we need to apply a charge equal to $Q = 2P_sA$ and then electrically isolate the cell. In case the charge is equal to $q < Q$, the FLC cell is only partially switched and thus a gray scale is obtained. Thus, controlling the amount of charge by addressing each pixel of FLC screen individually via a special active element, for example, thin film transistor, we have an analogue gray scale in FLC display, which cannot be obtained in usual passively addressed ferroelectric LCD.

A surface-stabilized FLC structure with bistable switching cannot provide an intrinsic continuous gray scale, unless a time- or space-averaging process is applied [65]. The inherent physical gray scale of passively addressed FLC cells can be obtained if the FLC possesses multistable electrooptical switching with a sequence of ferroelectric domains, which appear if the spontaneous polarization P_s is high enough [61]. Ferroelectric domains in a helix-free FLC form a quasiperiodic structure with a variable optical density as it appears between crossed polarizers [61] (Figure 6.47). The bookshelf configuration (Figure 6.40) of smectic layers is preferable for the observation of these domains. If the duration of the electric pulse applied to a helix-free SSFLC layer containing ferroelectric domains is shorter than the total FLC switching time, the textures shown in Figure 6.47 are memorized after switching this pulse off and short-circuiting the FLC cell electrodes. The domains appear as a quasiregular structure of bright and dark stripes parallel to the smectic layer planes. The bright stripes indicate the spatial regions with a complete switching of the FLC director,

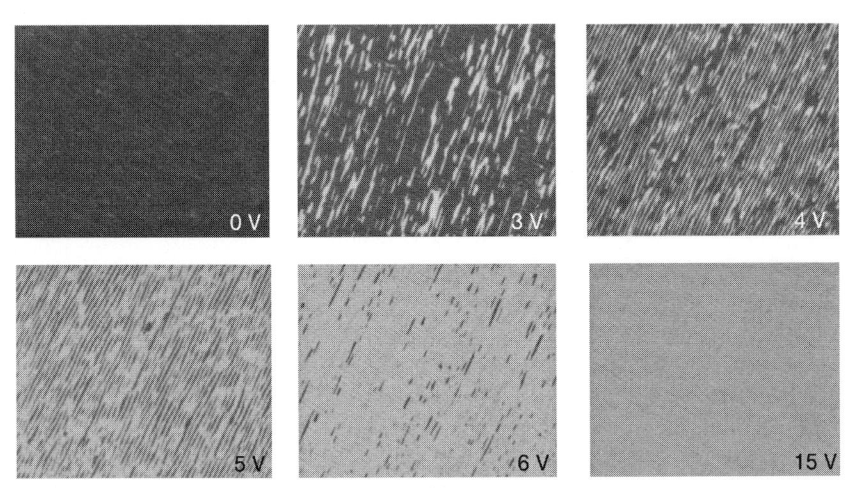

Figure 6.47 Continuous variation of the width of ferroelectric domains with a change in the applied voltage of the FLC layer between crossed polarizers [65].

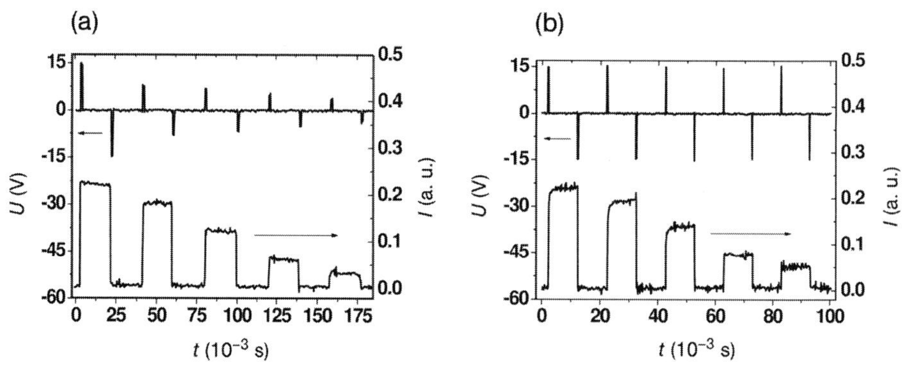

Figure 6.48 Light transmission (bottom curves) memorized by the multistable FLC cell on (a) the amplitude of 1-ms alternating driving pulses and (b) the duration of alternating driving pulses ranging from 250 to 50 μs [65].

while the dark stripes indicate the regions that remain in the initial state. The sharp boundaries between the black and white domain stripes seem to illustrate the fact that only two stable director orientations exist. The variation of the occupied area between bright and dark stripes depends on the energy of the applied driving pulses. The total light transmission of the structure is the result of a spatial averaging over the aperture of the light passed through the FLC cell and is always much larger than the period of the ferroelectric domains. Both the amplitude and the duration of the driving pulses can be varied to change the switching energy, which defines the memorized level of FLC-cell transmission in a multistable electrooptical response (Figure 6.48). Therefore, any level of the FLC-cell transmission, intermediate between the maximum and the minimum transmissions, can be memorized after switching the voltage pulses off and short-circuiting the cell electrodes (Figure 6.48).

Conditions necessary for multistable switching modes are (i) sufficiently high FLC spontaneous polarization $P_s > 50$ nC/cm^2 and (ii) a relatively low energy of the boundaries between the two FLC states existing in FLC domains (Figure 6.47), which is usually typical for the antiferroelectric phase [6, 7]. The multistability is responsible for three new electrooptical modes with different shapes of the gray scale curve that can be either S-shaped (double or single depending on the applied voltage pulse sequence and boundary conditions) or V-shaped depending on boundary conditions and FLC cell parameters (Figure 6.49).

Deformed Helix Ferroelectric Effect The geometry of the FLC cell with a deformed helix ferroelectric effect is presented in Figure 6.50 [51, 59]. The polarizer (P) on the first substrate makes an angle with the helix axis and the analyzer (A) is crossed with the polarizer. The FLC layers are perpendicular to the substrates and the layer thickness d is much higher than the value of the helix pitch R_0 (see also Figure 6.39b):

$$d \gg R_0. \tag{6.62}$$

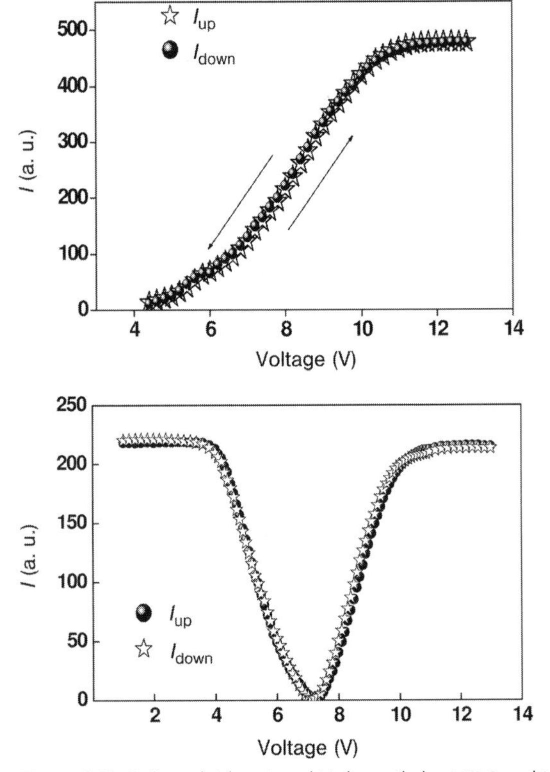

Figure 6.49 S-shaped (above) and V-shape (below) FLC multistable switching [65].

The light beam with the aperture $a \gg R_0$ passes parallel to the FLC layers through an FLC cell placed between the polarizer and analyzer. In an electrical field, the FLC helical structure becomes deformed so that the corresponding dependence of the director distribution cos φ, as a function of coordinate $2\pi z/R_0$, oscillates symmetrically in $\pm E$ electric fields (Figure 6.50). These oscillations result in a variation of the effective refractive index, that is, electrically controlled birefringence appears. The effect takes place till the fields of FLC helix unwinding

$$E_U = \pi^2/16K_{22}q_0^2/P_s, \tag{6.63}$$

where K_{22} is FLC twist elastic constant and $q_0 = 2\pi/R_0$ is the helix wave vector.

The characteristic response times τ_c of the effect in small fields $E/E_u \ll 1$ are independent of the FLC polarization P_s and the field E, and are defined only by the rotational viscosity γ_φ, and the helix pitch R_0:

$$\tau_c = \gamma_\varphi/K_{22}q_0^2. \tag{6.64}$$

The dependence (6.64) is valid, however, only for very small fields E. If $E \leq E_u$, the FLC helix becomes strongly deformed and $\tau_c \propto E^{-\delta}$, where $0 < \delta < 1$ [66]. If E is close to the unwinding field E_u the helical pitch R sharply increases $R \gg R_0$. Consequently,

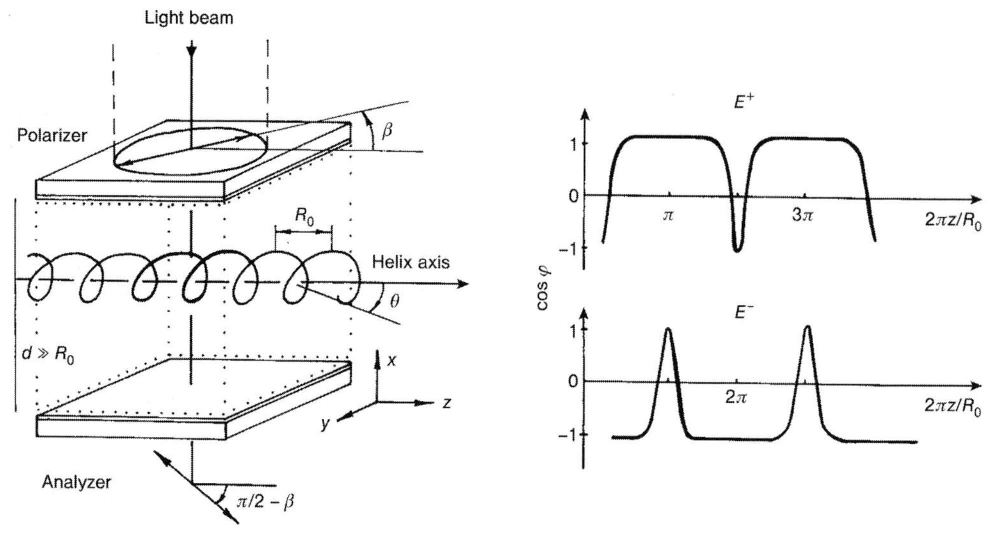

Figure 6.50 Deformed helix ferroelectric (DHF) effect [6].

the times of the helix relaxation τ_d to the initial state also rise

$$\tau_d/\tau_c \propto R^2/R_0^2, \tag{6.65}$$

that is, for $E \approx E_u$, it is possible to observe the memory state of the FLC structure [51]. In this regime, the electrooptical switching in the DHF effect reveals a pronounced hysteresis especially for $E \Rightarrow E_u$ (Figure 6.50).

However, if the FLC helix is not deformed too much, fast and reversible switching in the DHF mode could be obtained [59]. Figure 6.51 shows one of the possible realizations of the driving scheme with the remarkable decrease of switching time down to the 10 μs range [59]. The diodes were used to provide the necessary amplitude of the blocking voltage to keep the FLC cell in the off state in the absence of an additional "information" field, which is necessary in matrix-addressed displays. Figure 6.51 also shows that addressing pulses are followed by the stabilizing voltage of a small amplitude $E < E_u$, which provides a fast return of the helix to the initial state and compensates charge-screening fields (Figure 6.46).

The optical transmission of the DHF cell could be calculated as follows:

$$I = \sin^2(\pi\,\Delta n(z)d/\lambda)\sin^2(2(\beta - \alpha(z)), \tag{6.66}$$

where β is the angle between the z-axis and the polarizer (Figure 6.50),

$$\alpha(z) = \arctan(\tan\theta\cos\varphi(z)) \tag{6.67}$$

is the angle between the projection of the optical axis on the y, z-polarizer plane and the z-axis, and $\Delta n(z) = n_{\text{eff}}(z) - n_\perp$ is the effective birefringence:

$$n_{\text{eff}} = n_{||}n_\perp/[n_\perp^2 + (n_{||}^2 - n_\perp^2)\sin^2\theta\,\sin^2\varphi]^{1/2}. \tag{6.68}$$

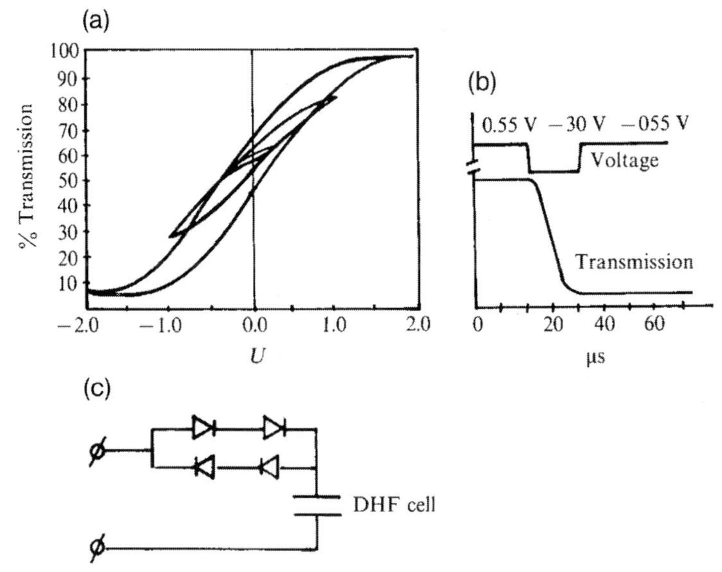

Figure 6.51 Electrooptical switching in the DHF effect. (a) Typical hysteresis in TVC; (b and c) addressing circuit [6].

Diffraction of light on the helical structure is avoided because the helical pitch of the FLC mixture $R_0 \propto 0.1–0.3\,\mu m < \lambda = 0.5\,\mu m$ is in the visible range.

In the case of small angles $|\theta| \ll 1$, the transmission in (6.66) can be expanded in θ series

$$I \propto (\sin^2 2\beta - 2\theta \sin 4\beta \cos\varphi + 4\theta^2 \cos 4\beta \cos^2\varphi)\sin^2 \pi\, \Delta n d/\lambda. \quad (6.69)$$

As shown in Ref. [6, 7] for the small values of the applied field $\cos\varphi \propto E/E_u$ and changes its sign for the field reversal $E \Rightarrow -E$ (Figure 6.50). Thus, according to (6.69) for $\sin 4\beta = 0$, we have a quadratic gray scale, that is,

$$\Delta I \propto \theta^2 \cos^2\varphi \propto \theta^2 E^2/E_u^2 \quad (6.70)$$

and for other values of θ, the gray scale is linear. For $\cos 4\beta = 0$, the quadratic component in the modulated intensity I is absent, that is,

$$\Delta I \propto \theta \cos\varphi \propto \theta E/E_u. \quad (6.71)$$

If $E(t) = E_0 \cos wt$, then in the case of (6.70) we come to the modulation regime, which doubles the frequency of the applied field. The relationships (6.70) and (6.71) were confirmed in experiment [51]. Using a "natural" gray scale of the DHF mode, many gray levels have been obtained with fast switching between them [59]. New ferroelectric mixtures with the helix pitch $R_0 < 0.3\,\mu m$ and tilt angle $0 > 30°$ have recently been developed for the DHF effect [67]. The helix unwinding voltage was about 2–3 V. Short-pitch FLC mixtures could be also used to obtain pseudobistable switching in FLC samples.

Let us point to certain advantages of the DHF electrooptical effect for applications compared to the Clark–Lagerwall mode.

1. High operation speed is achieved for low driving voltages. This takes place because a slight distortion of the helix near the equilibrium state results in a considerable change in the transmission. Consequently, the instantaneous response of the FLC cell is provided without the so-called delay time inherent to the Clark–Lagerwall effect.
2. The DHF effect is also less sensitive to the surface treatment and more tolerant to the cell gap inhomogeneity. As it follows from experiment and qualitative estimations [51, 59, 67], the effective birefringence value Δn_{eff} is approximately twice as low as $\Delta n = n_{||} - n_{\perp}$ in the Clark–Lagerwall effect. The insensitivity of DHF cells to the surface treatment enables us to successfully use the same aligning technique that has been developed for nematic liquid crystals.
3. The DHF effect allows the implementation of a "natural," that is, dependent on voltage amplitude, gray scale both linear and quadratic in voltage. Moreover, at $E \approx E_u$ long-term optical memory states are possible.

6.1.5.1.11 **Addressing Principles of Passive Ferroelectric LCDs**

The passive addressing of FLC displays is made similar to TN or STN-LCDs (Figure 6.21), that is, select pulses address sequentially all the rows of the display matrix, while the data pulses address all the columns simultaneously in each select row time. As previously, the pixel voltage $V_{ij} = V_i - V_j$, that is, the difference between a row V_i and column V_j voltage pulses [6, 7].

A basic difference between passive addressing of TN (STN) and FLC displays is the possibility of using an optical memory in the latter case. This means that the sharp transmission–voltage curve is no longer needed, as we do not have to refresh the voltage on the pixels. Ones switched, the pixel of FLC screen will remain in this state. The problem is to make gray scale because usually only two transmission levels are memorized. The gray levels can be achieved by rather artificial means either by time-control technique, that is, making several sequential frame pictures of one image or space-control, that is, dividing one pixel of the image to several subpixels, so that the transmission of the pixel can vary due to the number of switched subpixels [48]. Let us consider for simplicity the FLC passive addressing scheme without gray scale [6]. One of the schemes, called Seiko scheme [6] is shown in Figure 2.66. Four voltage signals, the so-called *driving waveforms*, are needed to address FLC matrix: two row signals, selecting V_s or nonselecting V_{ns}, and two column signals controlled by the picture data (column white V_{on} and column black V_{off}). On the selected row, the effective pixel voltages $V_s - V_{on}$ and $V_s - V_{off}$ must produce switching of the pixel to one or another state within the selection time, which is called the line addressing time. At the same time, the voltages $V_{ns} - V_{on}$ and $V_{ns} - V_{off}$ must not affect the state of the pixel. In Seiko scheme, the requirements of zero average voltage over pixel is taken into account to avoid FLC degradation and the height of the writing pulses $U_R + U_D$ is three times as much as the height of AC stabilization pulses U_R. Here, two scans are used to write one picture with a small interrupting time between them – one subframe to write

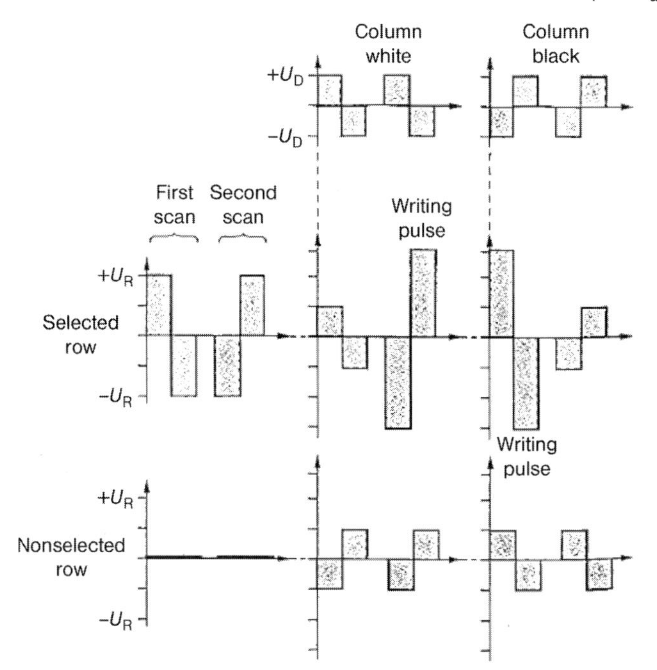

Figure 6.52 Seiko multiplexing scheme for passive addressing of FLC displays [6].

white pixels and another to write black ones. Total time to write one row in Seiko scheme is $N\tau$, where τ is the time required to achieve one or another bistable FLC state using the pulses with the amplitude of $U_R + U_D$ and $N = 2 + 2 = 4$ – a number of intervals. The scheme is not optimized as the value of N is too large. Minimization of the value of N is the purpose of the newly developed FLC addressing schemes [6, 7].

The multiplexing ratio M of passively addressed FLC display is defined by the frame time τ_{fr} and the row writing time (or line address time) $N\tau$ according to the formula (Figure 6.52)

$$M = \tau_{fr}/N\tau. \tag{6.72}$$

Intrinsic gray scale generation and stabilization in ferroelectric liquid crystal display (FLCD) have been proposed and investigated [68]. A study of the FLC samples shows that the switching process takes place through the formation and evolution of domain formation (Figure 6.47). Dynamic current and electrooptical response in FLC testing cell are discussed as a criterion of memorized gray scale generation for FLC display. Gray scale stability for different cross-talk effects under a passive multiplex driving scheme was demonstrated.

Based on the driving scheme, 160×160 passive matrix addressing FLC display was developed. For the FLC display, the dark state is fixed here as a basic one, while the gray levels depend on the writing voltage with a fixed duration time [68]. The bright state can also be selected as a reference state of the FLCD. Figure 6.53 shows the original image and the same image on our bistable reflective FLCD with four memorized gray scale levels. Owing to a birefringence for a 5 μm cell gap, the

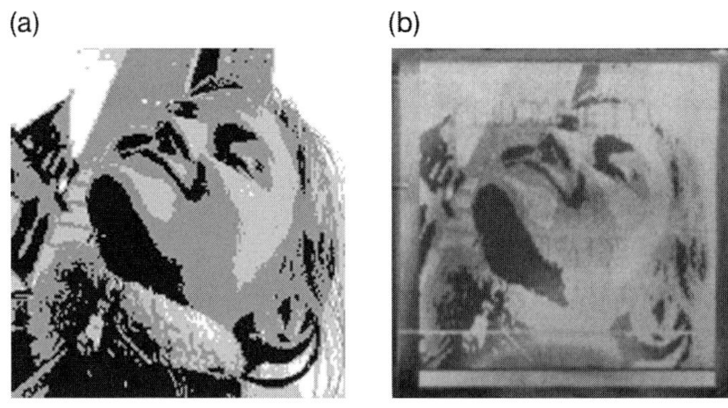

Figure 6.53 (a) Original image and (b) the same image on our bistable reflective FLCD with four memorized gray scale levels [68].

nonselected and selected states are, respectively, green and dark for this FLC display (Figure 6.53) [68]. Images can be saved for infinite time without any power supply. The contrast ratio could even be increased if better FLC mixtures, smaller cell gap, and compensation films are used.

A prototype of matrix type panel for electric label system was developed using FLC on SiO oblique evaporation as the alignment method to achieve best memory effect [69]. The low power consumption, the stable memory states, and the dot matrix enable not only complicated information display but also a readable complicated two-dimensional barcode pattern (Figure 6.54).

Polymer stabilization of FLCs exhibiting a continuous gray scale V-shaped switching, free from zig-zag defects producing a high contrast ratio, and a high-speed response below 400 μs, called polymer-stabilized FLC displays (PSV-FLCDs) was developed by Kobayashi and coworkers [70]. The advantages for PSV-FLCDs are (i) a continuous gray scale capability for full color display by dissolving the bistability; (ii) uniaxal alignments causing a pure dark state; and (iii) capability of unwinding

Figure 6.54 (a) Citizen watch (U-100) with FLC display and (b) prototype panel for electrical label system [69].

helical structure in smectic C* phases by polymer stabilization. Materials used in PSV-FLCDs include mixtures composed of conventional and newly synthesized FLC mixtures and several newly synthesized photocurable monomers. PSV-FLCDs are fabricated with a unique process combined with a UV-light exposure for the formation of polymeric nanostructures stabilizing FLCs alignment and applying AC voltage for creating uniaxial orientation exhibiting a dark state free from zigzag defects. By this process, a biaxial orientation based on bistable switching in SSFLC displays changes into a uniaxial orientation exhibiting V-shaped switching to be possible to produce a continuous gray scale [70]. To accommodate the polymer-stabilized FLCDs to active matrix (thin film transistor) driving, the reduction of operating voltage within 10 V was provided, as well as the suppression of temperature variation in operating voltage, within 3 mV/°C in the range from −5 to 50 °C, resulting in successful demonstration of a TFT driving field sequential full color type featured by a high-resolution of SVGA (800 × 600 pixels) in a small size of 4 in. diagonal display without color filter (Figure 6.55). The novel PSV-FLCDs show the fast response times below 400 μs at 25 °C even in a gray scale. In the lower temperature of operation, PSV-FLCs exhibits a fast response time less than 1000 μs even at −5 °C. New FLC materials with a lower birefringence should be developed to

Figure 6.55 Top: V-shape response of polymer-stabilized FLC display (PSV-FLCD). Bottom: A photograph of displayed images on the screen of the prototype field sequential full color LCDs using PSV-FLCD [70].

increase the FLC cell gap for achromatic black/white switching required for color sequential FLC displays (see, e.g., (6.59)).

6.2
Liquid Crystal Display Optimization

6.2.1
Various LCD Addressing Schemes

Display addressing is the means by which an image, represented by data in a video stream or data stored in memory, is processed into electrical impulses and applied to the display, where the image is made visible. In liquid crystal displays, an image is made visible by altering the liquid crystal orientation in certain regions, creating an optical contrast through the anisotropy in the liquid crystal's refractive index and/or optical absorption. LCD modifies the light passing through them, unlike emissive displays such as light emitting diode and field emission display (FED), which generate light of their own. The electrical addressing of LCD can be subdivided into active matrix LCD and passive matrix LCD. In most active matrix LCDs, a discrete nonlinear device, such as a diode or a transistor, is associated with each pixel of the display, which is regularly arranged in a rectangular array or matrix. Figure 6.56 shows a diagram of the most common active matrix LCD, the thin film transistor display.

There are three types of electrodes present in this display: control electrodes, signal electrodes, and pixel electrodes. For TFTs, control electrodes operate the gates of transistors to momentarily switch the transistors into their conducting states so that the voltage present in the signal electrodes can be channeled to appropriate pixel electrodes at proper times. The pixel is defined by the region of overlap between the pixel electrode and a single, common counter electrode positioned on the opposite side of the liquid crystal layer. In a passive matrix display, there are no control electrodes since there are no nonlinear devices to control; and the signal electrodes

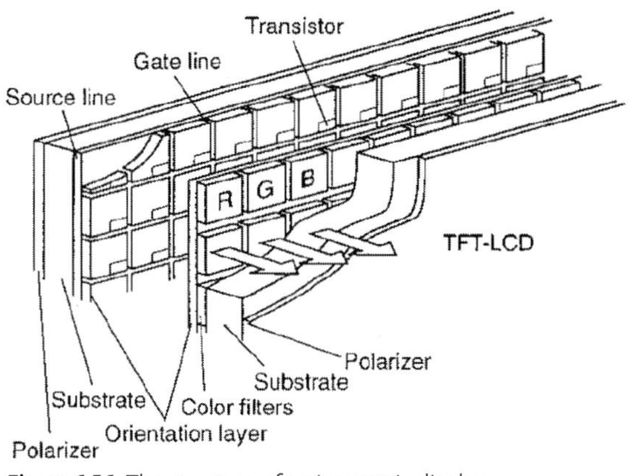

Figure 6.56 The structure of active matrix display.

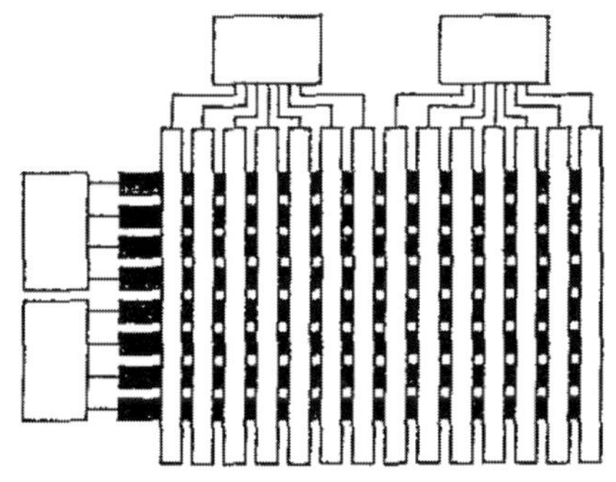

Figure 6.57 Common passively addressed matrix LCD. The row drivers (left) and column drivers (above) are used to address every pixel of LCD matrix.

are the same as the pixel electrodes. Every pixel is uniquely determined by the region of overlap of a row and column electrode, making it possible to access $N \times M$ pixels with only $N + M$ electrodes. For a passive matrix display, it follows that the voltage across the liquid crystal, which determines the optical state of the pixel, is the potential difference between the voltages applied to the corresponding row and column electrodes (Figure 6.57). We have provided Figure 6.21 below, which indicates the principles of passive matrix LCD.

It is important to realize that a drive waveform applied to a column electrode influences the pixel waveforms appearing across every pixel in that column. In order to display a desired pattern of pixels, it is therefore necessary to ascertain the characteristics of the pixel waveform the electrooptical effect responds to so that this can be exploited by the addressing technique. In the surface-stabilized ferroelectric liquid crystal display (SSFLCD), for example, the polarity of the last applied pulse whose amplitude–time product lies above a certain threshold value determines the optical state of the pixel (Figure 6.52). For the reflective cholesteric display, the magnitude of a bipolar selection pulse is used to switch between the highly contrasting planar and focal conic textures that then remain stored at a lower, nonselecting voltage or even at zero volts (Figure 6.34). Other electrooptic effects such as the TN, STN, and LCD effects respond, over a wide range of frequencies, only to the root mean square (rms) value of the pixel waveform (Figure 6.21).

One of the new prospective methods of LCD addressing is color sequential LCD. The principle of color sequential LCD is shown in Figure 6.58. The driving of such color sequential LCD includes three subframes. The response time of the LCD needs to be very fast in order to fit the field sequential driving method. Suppose there are 60 frames/s, then there is 16.67 ms/frame. Each subframe will have 5.5 ms. Again, each subframe has to further divide into three parts. The first one is the data loading time that is about 0.5 ms. Second, it comes to the LC response time that is about 2.1 ms. Finally, the total LED illumination time is around 2.9 ms (Figure 6.58).

Let us consider the LCD quality factors as a function of LC mixture parameters and its anchoring properties for various types of LCD described above with different addressing.

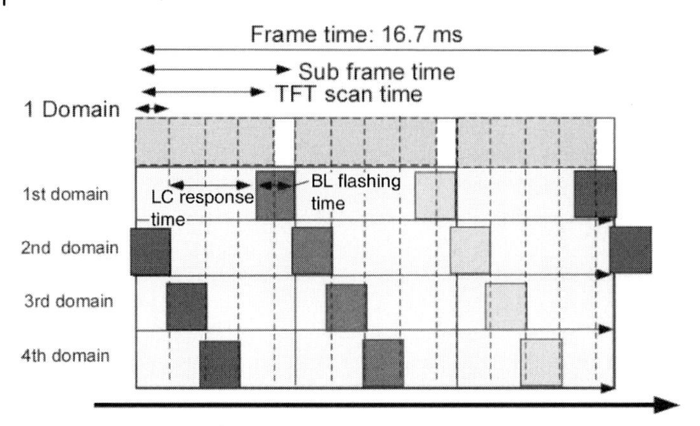

Figure 6.58 Principle of LCD color sequential switching.

6.2.2
Passive Matrix Displays

One of the basic limitations of passive matrix LCD (Figure 6.57) is the so-called "iron law," which states that the resolution of the LCD strongly correlates with the steepness of the transmission–voltage/reflection–voltage (TVC/RVC) curves (6.30). As mentioned above, the steepness of the iron law considerably restricts the quality of PM-LCD, especially for a sufficiently high resolution (Figure 6.59). The selection ratio, shown in Figure 6.59

$$V_{on}/V_{off} = 1 + p, \tag{6.73}$$

where p is a steepness parameter (6.28)–(6.30), V_{on} and V_{off} correspond to the switched-on and switched-off state of PM-LCD, respectively.

The dependence of the steepness parameters and electrooptical characteristics of TN and STN-LCD on the properties of LC mixture and cell configurations are listed Tables 6.3 and 6.2, respectively. All elastic, dielectric, and optical parameters of LC mixtures, as well as LC cell configuration (pretilt angle, twist angle, and anchoring energy), are very important [6, 7]. Results useful for applications are published elsewhere [1–7]. All the properties of highly multiplexed PM-TN and STN-LCD such as contrast ratio, brightness, response time, temperature range, and so on are, however, getting worse with increasing resolution (number of addressing lines); so, to keep a high steepness of TVC curves is a challenging task.

The LCD modeling software, such as MOUSE-LCD [17] shown above [17–23], provides an efficient tool for modeling and optimization of PM-LCD electrooptical performance based on TN, STN, and other electrooptical modes in LC mixtures depending on the LC mixture parameters and cell configuration. Some results of PM-LCD optimization based both on calculations and on experiment for TN and STN effects are summed up as follows:

1. A proper choice of "backflow" viscosities considerably shortens PM STN-LCD response time (Figure 6.60).

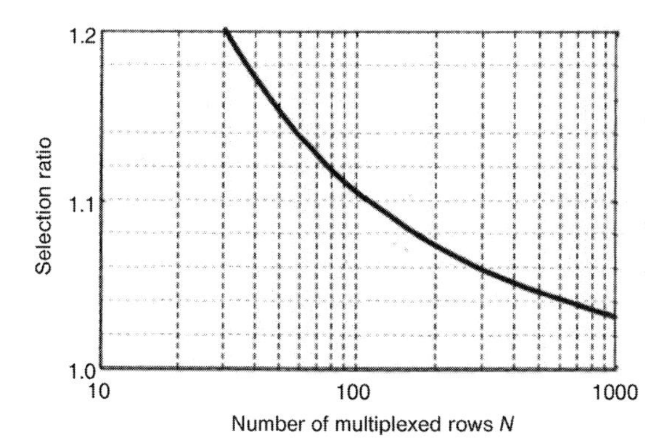

Display resolution	Selection ratio, single	Selection ratio, dual scan
160x160	1.0824	1.1188
VGA (640x480)	1.0467	1.0667
SVGA (800x600)	1.0417	1.0595
XGA (1024x768)	1.0368	1.0524
SXGA (1280x1024)	1.0318	1.0452
UXGA (1600x1200)	1.0293	1.0417

Figure 6.59 Iron law of passive matrix addressing LCD: the number of the addressing lines strongly correlates with the selection ratio, V_{on}/V_{off}, where V_{on} and V_{off} corresponds to the switched-on (switched-off) state of PM-LCD, respectively.

Table 6.3 The steepness of the transmission–voltage curve as a function of LCD parameters [6].[a]

The LCD parameter	Effect on the steepness and information capacity
Elastic ratio, K_{33}/K_{11}	Low values (<0.7) improve the steepness. The effect is considerably pronounced for the small optical anisotropy Δn (first Mauguin minimum)
Dielectric ratio, $\Delta\varepsilon/\varepsilon_{\perp}$	Low values (<1) improve the steepness for the small optical anisotropy Δn. The situation is quite reverse for the larger values Δn
Cell thickness, d	The effect is the same as for the optical anisotropy Δn
The angle between the polarizer and analyzer	The angles smaller than 90° improves the steepness on account of the lower transmission in the off state
Twist angle in the LC cell	The steepness deteriorates for the smaller twist angles. The angles more than 90° are not stable in nematic LC cells without chiral dopants

[a]According to (6.30), the information content of the passively addressed LCD increases with the steepness of the TVC.

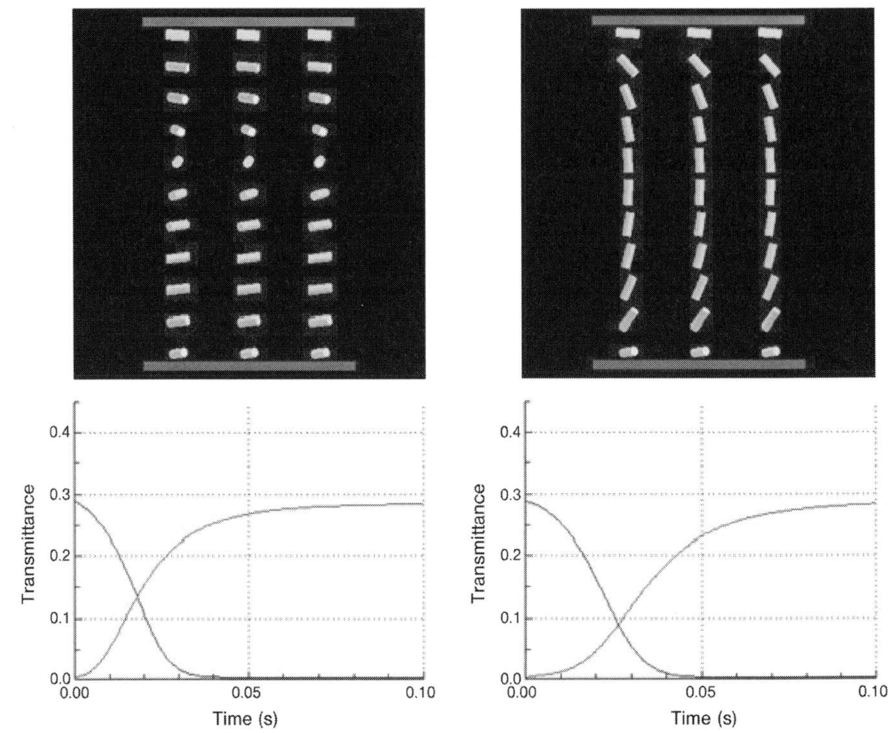

Figure 6.60 Simulation of PM STN-LCD. 20% decrease in response time for a proper choice of a shear (backflow) LC viscosities (see Equation 2.28). Top: off (left) and on states (right) of STN-LC cell. Bottom: STN-LCD response: optimized (left) and common (right).

2. Viewing angles of PM TN-LCD are better in a normally black mode (Figure 6.61).
3. Special "stress" configuration of TN-LCD can both decrease the driving voltage (power consumption) and improve the response time (Figure 6.62) [27, 28]. We have mentioned the "stress" TN configuration above. The twist sense of TN-LCD is determined by the direction of pretilt angles on both alignment layers. In commercially available TN-LCDs, the chiral material is used to stabilize the LC twist sense. The LC materials for TN-LCDs possess the twisting property of the same direction as the one determined by the combination of pretilt angle directions. On the contrary, by adding the chiral reagent in which twist direction is opposite to the one determined by the combination of pretilt angle directions, the splayed twist state is formed (Figure 6.62). The stability of "stress" TN state structure depends largely on the pitch length and pretilt angle. The larger pitch length and pretilt angle can stabilize the new mode.

The interest in PM-LCD has decreased recently as all high-resolution displays for most applications, such as TV, desktop and PC monitors, mobile phones, and PDAs are active matrix LCD (AM-LCD).

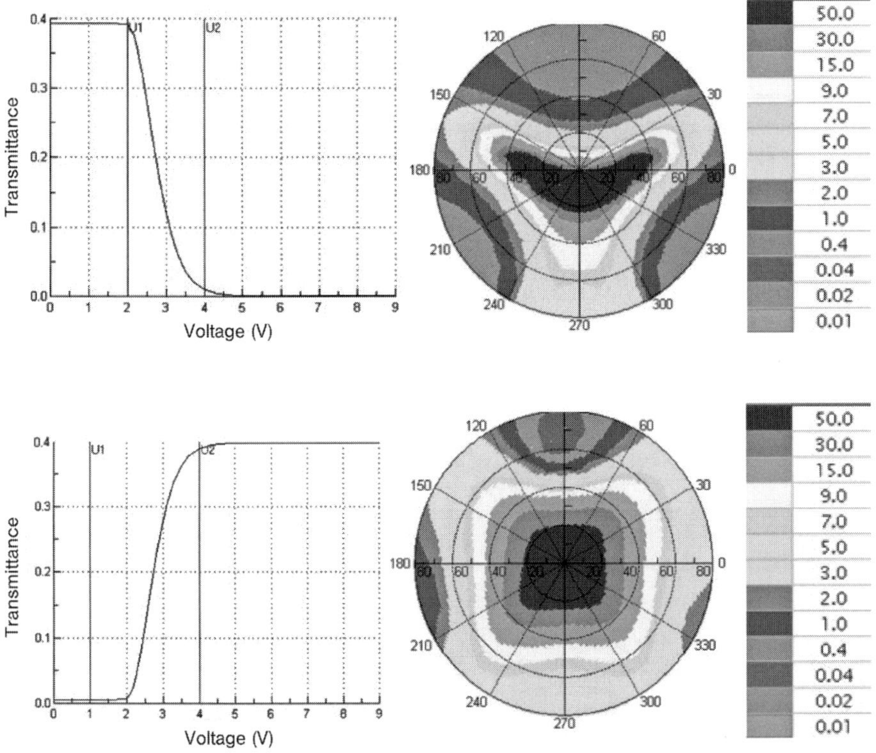

Figure 6.61 Viewing angles of normally white (above) and normally black (below) TN-LCD. The TVC curves of the modes are shown in the left. The viewing angles are defined by contrast ratios at different viewing angles (see also Figure 6.20).

6.2.3 Active Matrix Displays

Thin film transistors (TFT-LCDs) were considered to be the basic structure for active matrix addressed LCDs (Figure 6.56). TFT operates as a very good conductor when the gate (row) voltage is large and a very poor conductor in case of small gate voltages [6] (Figure 6.63). When the row voltage is zero, no matter how high the column voltage is, the current will not flow through the LC pixel element. Opening the gate voltage by a selected row pulse makes it possible to deliver to the LC pixel element a definite space charge, proportional to the required contrast.

The LC pixel is connected to the ground and, as a result, the space charge is allowed to relax during the frame interval, with the characteristic time defined by the capacity C_{LC} and resistance R_{LC} of the LC layer (Figure 6.63). A zero DC voltage in LC element is provided by a corresponding sign alternation of the column voltages during the subsequent frame times. The operation is allowed due to the symmetry of the TFT conductance in both directions. There is no need to change polarity of the row-addressing voltages, as their only function is to open the gate (Figure 6.63). A small

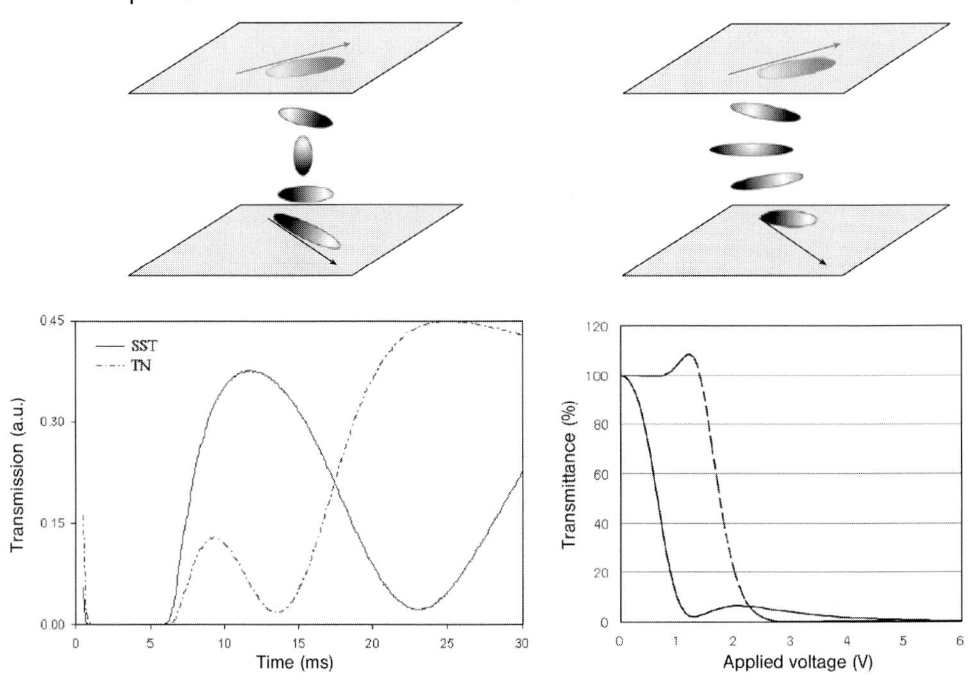

Figure 6.62 Stress splay twist (SST) [28] or reverse TN (RTN) [27] mode. Top: common TN (left), SST/RTN mode (right). Bottom: comparison of response time (left) [28] and applied voltage (right) [27] between common TN and SST/RTN mode.

capacitor C_s is added to the LC pixel capacity C_{LC} to retain the contrast during the frame time.

TFTs are certainly advantageous since (i) the cross talk is practically absent, because the column voltage hardly affects LC pixels during the nonaddressed period, and (ii) grayscale generation is much easier and resembles that of directly addressed LCD elements. However, the aperture ratio of TFT AM-LCDs (i.e., the ratio of the dead area on pixel), where TFT is located, to the total active area is lower than the total area of the pixel. Requirements for a TFT are derived from the number N of rows and the frame time T in which a picture is written in. The row is addressing during the interval

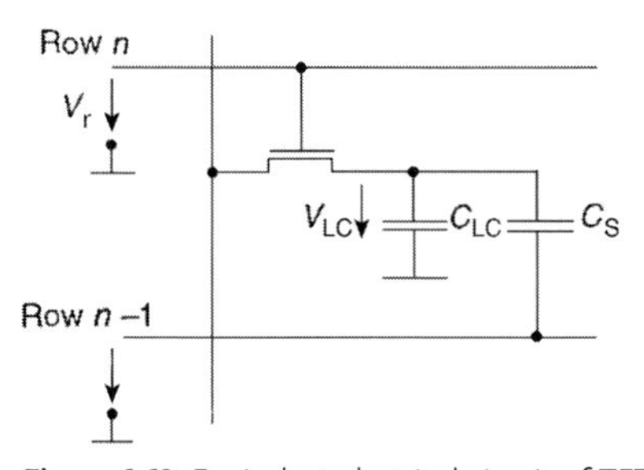

Figure 6.63 Equivalent electrical circuit of TFT AM-LCD pixel [6].

$\tau = T/N$. In order to guarantee a proper charging of capacitors $C_{LC} + C_S$ during the frame time T, the following condition for time intervals must be met [1, 6]:

$$R_{on}(C_{LC} + C_S) < 0.2\tau = 0.2T/N, \qquad (6.74)$$

where the factor 0.2 means 1% error in charging voltage [1].

The stored charge has to be held during the flame time T to avoid flickering of the image. This enables us to write (1% error in holding voltage):

$$R_{off}(C_{LC} + C_S) \geq 100T. \qquad (6.75)$$

Taking into account (6.74) and (6.75), we have for the TFT resistance ratio:

$$R_{off}/R_{on} \geq 500N. \qquad (6.76)$$

If $I_{off} \sim 10^{-12}$ A and $N = 1000$ rows, we must have $I_{on} = I_{off}R_{off}/R_{on} \geq 0.5\,\mu A$.

One of the important characteristics of the TFT-LCD is the voltage holding ratio. The VHR is defined as the voltage at TFT-LCD pixel at the end of the frame period divided by the voltage initially set to it during the row addressing time (Figure 6.64).

$$\mathrm{VHR}\,(\%) = \frac{\int_0^{16.67\ \mathrm{ms}} V(t)\mathrm{d}t}{5\ \mathrm{V} \cdot 16.67\ \mathrm{ms}} \times 100\%. \qquad (6.77)$$

The VHR strongly depends on the decay time of the voltage on the pixel element to the frame time T. If the VHR is too small, not only will the pixel voltage become smaller it will also cause flickering. Thus, the brightness of the pixel changes periodically. This is highly undesirable.

The voltage holding ratio is a critical electrical parameter for LCDs. It is a measure of the amount of ions in the bulk of LCs. In TFT-LCD configuration, the VHR shows how fast the charge will relax on LC pixel, while the other LCD rows are being addressed. Too fast relaxation results in the flicker effect. The VHR value can be measured as follows (Figure 6.64): A voltage pulse of 5 V was applied to the LC cell for 64 μs (260 lines in 16.67 ms frame time, Figure 6.64) and then the voltage $V(t)$

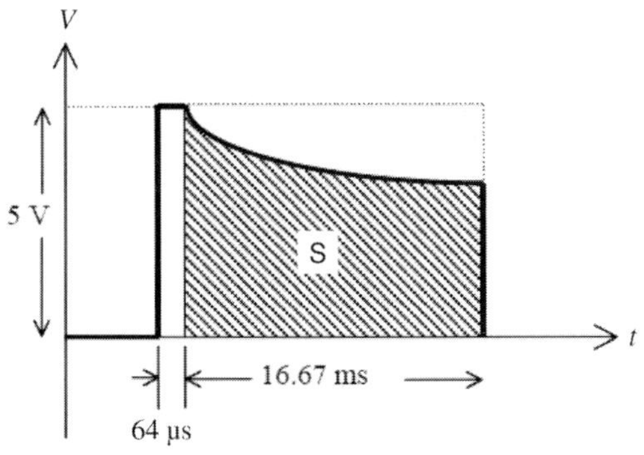

Figure 6.64 Measurements of VHR: voltage pulse of 5 V is applied during 64 μs to LC cell. Using this addressing time 260 lines of AM-LCD can be addressed during the frame time of 16.67 ms, that is, about 60 Hz or 60 frames/s.

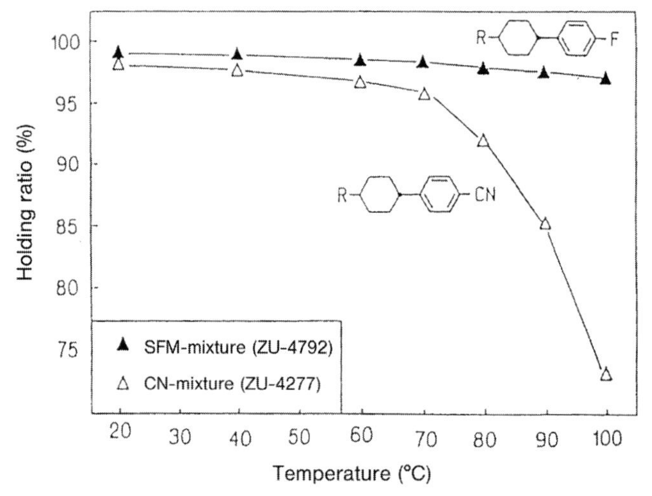

Figure 6.65 Holding ratio of TFT AM-LCD and its correlation with the LC molecular structure [6]. Temperature dependence of VHR for two different LC structures.

dropped down during the period of $T = 16.67$ ms (frame frequency about 60 Hz) after the pulse was switched off at $t = 0$ (Figure 6.64). The acceptable value of the VHR for TFT-LD applications is more than 99% for both room temperature and 80 °C.

The residual DC (RDC) property for photoaligned LCD cells can also be measured. RDC values can be measured using two methods: flicker-minimizing method and DC voltage method [71]. In the flicker-minimizing method, the test cell is driven by a square wave 30 Hz voltage with a varied DC offset, and the output signal with flicker could be observed [71]. The RDC is defined by the offset value, where minimum flicker is achieved.

DC voltage method can also be applied for the measurement. A voltage of 5 V is applied to a cell for 10 min, and then the cell is shorted for 1 s. The RDC value is defined as the voltage measured after 10 min. This method is simpler but has less accuracy than the flicker-minimizing method. The acceptable RDC value for TFT-LCD applications is less than 50 mV [71].

Proper values of VHR and RDC require a high purity of LC mixtures for TFT-LCD operations. The holding ratio must be high (>99%) and temperature-stable, which depends on LC compounds used in AM-LCDs. This is possible only for well-purified LC mixtures containing fluorinated compounds (Figure 6.65) [6]. LCs with a large dielectric anisotropy are also difficult to apply as the LC capacitance significantly diminishes during the reorientation of the LC director in electrooptical effect, thus deteriorating the holding ratio.

By all means, the number of ions both in the bulk (evaluated by VHR) and on the surface of LC cell (evaluated by RDC) should be minimized to provide an effective AM-LCD switching without flicker effect. The LC for AM-LCD should be of very high resistivity to hold the charges between frames. Actually, LC switching off time

$$\tau_{LC,off} \sim \varepsilon\rho, \tag{6.78}$$

where ε is the LC average dielectric constant (see Section 2.1.4) and ρ is specific resistivity that should be at least $10^{11}\,\Omega\,\text{cm}$. For AM-LCD, LC with ρ of $10^{12}\,\Omega\,\text{cm}$ is not uncommon.

VHR and RDC values also correlate with the so-called image sticking effect, which takes place when LCD displays the same image for some time after the image was switched off [72]. Investigations in this field have shown that high VHR and low RDC prevent the image sticking effect in LCD [72]. The other factor that correlates with an image sticking effect is LC anchoring energy, which must be sufficiently high to avoid the sticking phenomenon [73].

AM-LCDs are mostly based on the two electrooptical modes in LC: VAN mode, that is, ECB effect in homeotropic LC cells (Figure 6.1b), and IPS (in-plane switching mode, Figure 6.66). The advantage of VAN mode is a very high contrast

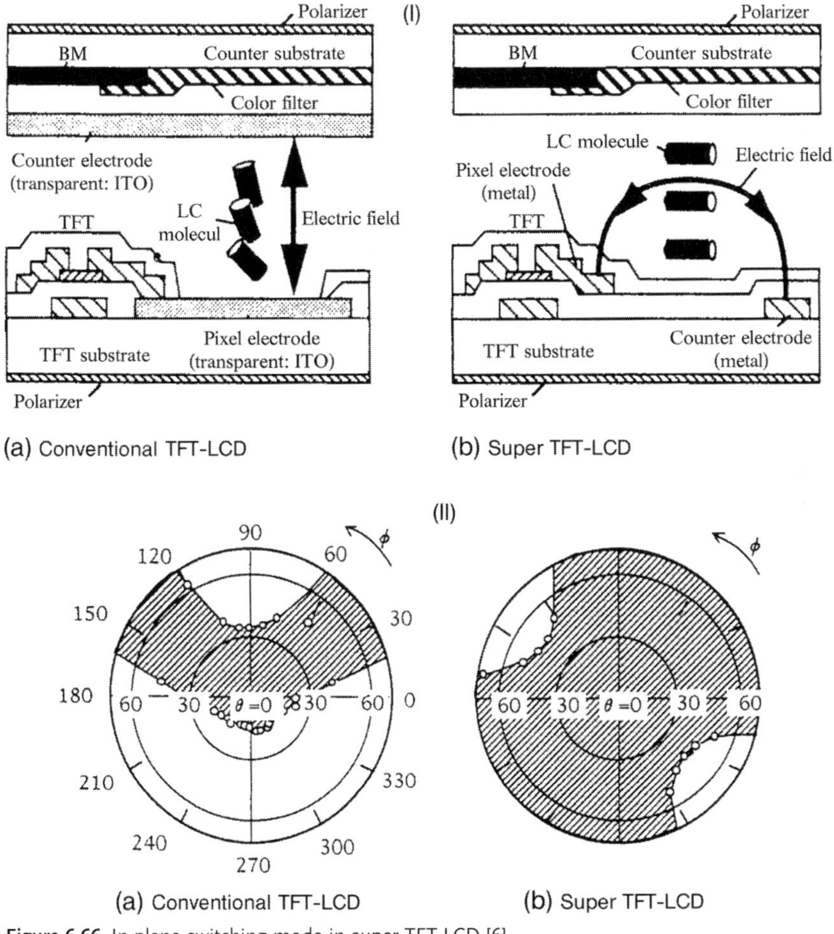

Figure 6.66 In-plane switching mode in super TFT-LCD [6]. (I) Geometry of realization of super TFT-LCD and (II) comparison of viewing angle dependences with conventional TFT-LCD.

ratio, which is mostly defined by the quality of the polarizers and pretilt angle θ_0 (1–3° from the initial homeotropic alignment, see e.g., (6.21)). IPS mode is characterized by a special geometry of electric field application parallel to the cell substrates (Figure 6.66). In this case, the variation of the LC director polar angle θ (6.2) and consequently the change of the LC cell phase retardation $\Delta\Phi$ (6.17) is minimal and all the intensity changes are due to the LC twist angle φ variations (Figure 6.1c). As it follows, for example, from (6.18) the intensity variations with the angle of light incidence are small in IPS mode, so IPS TFT-LCD viewing angles are very broad (Figure 6.66) [1, 2].

6.2.4
Low Power Consumption LCD with Memory Effects

We have already described electrooptical effects in cholesteric and ferroelectric LC as well as bistable and multistable cholesteric and ferroelectric LCD with a low power consumption and memory. We are going to provide some more examples of these "Electronic paper" LCD below.

6.2.4.1 Surface Bistability

6.2.4.1.1 BTN with 0 ⇔ π Twist Angle Switching

A bistable nematic twisted (BTN) with $0 \Leftrightarrow \pi$ twist angle switching bistable device using monostable surface anchoring switching was demonstrated [74]. The two bistable textures were planar and π twisted alignments (Figure 6.67). The switching to one or another bistable state from the initially homeotropic configuration, obtained in a strong

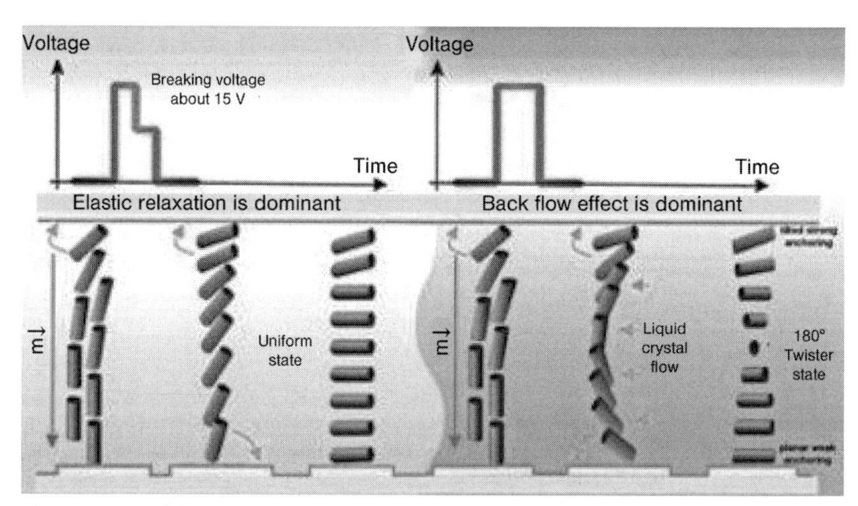

Figure 6.67 Bistability in nematic LC [6, 74]. Slow decreasing field down to zero results in a planar state, while instantaneously turning the field off forces the creation of a bend state, which spontaneously transforms to π twisted state.

electric field, depends on the rate of the field decrease and is associated with an anchoring breaking on the "slave" substrate (Figure 6.67). Slow decreasing field down to zero results in a planar state, while turning the field off instantaneously initiates a strong backflow effect (Figure 6.5), thus forcing the transformation to π twisted state.

The cell gap of BTN LCD ($d \sim 1.5\,\mu\text{m}$) is thinner than that of conventional LCDs. The top and bottom substrates are rubbed in the same direction (Figure 6.67). One substrate (slave substrate) is coated with a special alignment layer, which gives a nearly planar anchoring with moderately strong polar anchoring energy ($W_\theta \approx 3.10^{-4}\,\text{J/m}^2$). The other substrate ("naster" substrate) is coated with a conventional polyimide, which gives tilted anchoring with a strong polar anchoring energy ($W_\theta \approx 10^{-3}\,\text{J/m}^2$). Both azimuthal anchoring energies are kept strong enough ($W_\varphi \approx 10^{-4}\,\text{J/m}^2$) to maintain the azimuthal orientation of the liquid crystal on the alignment layer. Two stable textures U and T exist. The U texture is 0° twisted while the T texture is 180° (π) twisted. A chiral dopant is added to the liquid crystal mixture to adjust the free pitch about four times higher than the LC cell thickness to equalize the energies of U and T. U and T appear, respectively, black and white in the standard reflective configuration [74]. The big problem of BTN-LCD is to find a proper surface with a moderate polar, but still large azimuthal anchoring energy, which is not common. Typical BTN-LCDs are shown in Figure 6.68.

Recently, a new switching mechanism for bistable nematic twisted displays (BTN) was proposed [75]. By using the bidirectional alignment surface, which is fabricated by a photopolymerizable polymer, $0 \Leftrightarrow \pi$ twist angle switching can be controlled by changing the voltage levels. Such switching avoids any anchoring breaking, hence large cell gap can be possible. Furthermore, since the switching depends on the voltage level, it is completely compatible with ordinary low cost STN drivers [75].

Figure 6.68 Typical BTN-LCD. (Courtesy of Nemoptic.)

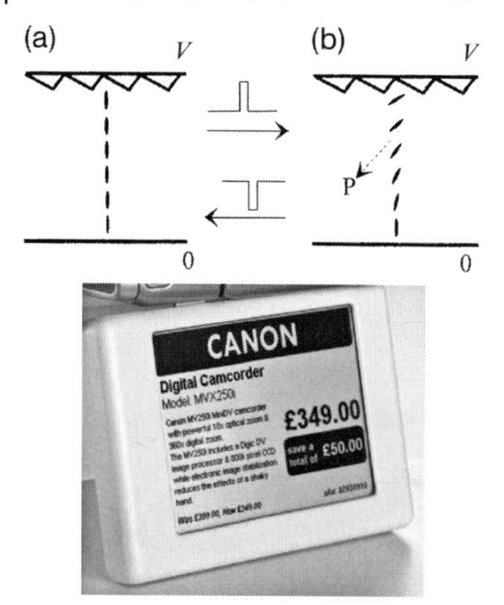

Figure 6.69 Bistable device made by special grating surface. (a) Homeotropic configuration and (b) hybrid homeotropic to tilted alignment.

6.2.4.1.2 **Zenithal Bistability (ZBD)**

A grating surface treated with a homeotropic surfactant was found to produce two bistable pretilt LC configurations when the groove's depth-to-pitch ratio is optimum [76]. Bistability results for a range of grating shapes between stable shallow grating and stable deep grating [76]. Current production devices use a grating with 0.8 μm pitch and 0.9 μm amplitude.

Switching is done by favoring one of the states having a nonzero flexoelectric polarization with a voltage pulse of the same polarity. (The opposite polarity pulse supports the homeotropic configuration (Figure 6.69).) The grating surface was made by a photolithography technique and coated by a surfactant to induce a homeotropic LC orientation. There exists a possibility to optimize flexoelectric coefficients, anchoring energy, and the grating profile in the bistable geometry, as well as the optical characteristics of the device. Typical response times of the matrix-addressed LCDs were found to be in a submillisecond range at $U = 10\,\text{V}$ [76].

The potential of the ZBD device has begun to be realized with its first commercial application. New technological advances are necessary to understand the relationship between grating shape and performance, the reproducible mass production of deep, submicron features on the inner surfaces of LCDs, and new liquid crystal material and addressing schemes that use the flexoelectric effect in nematic liquid crystals. The disadvantages of ZBD device are rather high switching voltage and the problem of memorizing the intermediate levels, so only black/white bistable optical states become possible (Figure 6.69).

6.2.4.1.3 **Optically Rewritable LCDs**

Optical rewritable technology (ORW) [77, 78] is a modified method of azo dye photoalignment [60, 78] that possesses traditional high azimuthal anchoring energy,

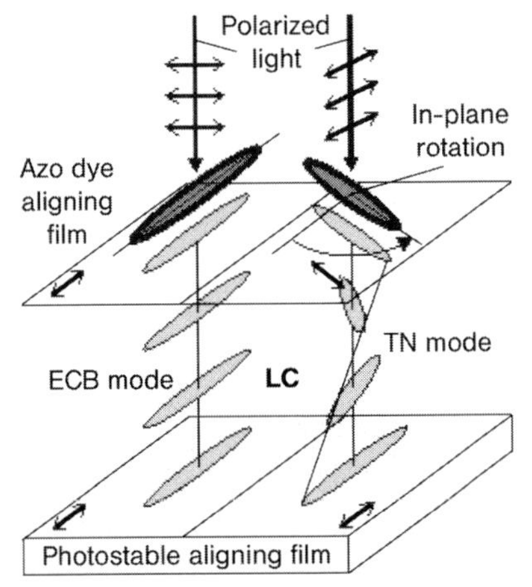

Figure 6.70 Operation principle of LC ORW cell.

up to 2×10^{-4} J/m^2, and has a unique feature of reversible in-plane aligning direction reorientation, that is, rotation perpendicular to the polarization of an incident light. An ORW-LC cell consists of two substrates with different aligning materials (Figure 6.70).

One aligning material is optically passive and keeps aligning direction on one substrate. The other aligning material is optically active and can change its alignment direction being exposed with polarized light through the substrate. In comparison with electrically controlled plastic display, ORW can be significantly thinner and requires no ITO photolithography and etching on plastic substrate because no electrodes are needed. Switching and continuous gray scale are achieved by control of aligning direction of photoaligning azo dye layer, which is insoluble in liquid crystal. By this means, one can obtain a specified twist angle in the ORW-LC cell that

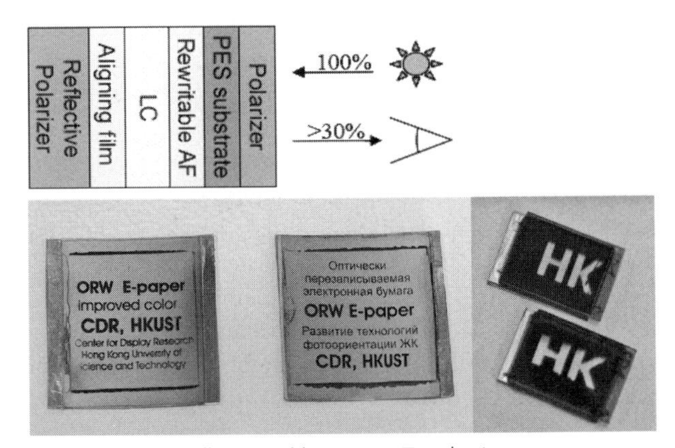

Figure 6.71 Optically rewritable E-paper. Top: basic structure; bottom: typical ORW image on flexible substrate [77, 78]. ORW image based on LC mixtures with fluorescent dyes is also shown (bottom right corner).

corresponds to the transmission level defined by the initial polarizer configuration (Figure 6.71). ORW is very tolerable to the cell gap variation as even 50% change in the cell gap will not cause any noticeable change in LC transmission value, while achromatic switching of all ORW gray levels can be obtained [77, 78]. Every transmission level is stable and visualizes information with zero power consumption for a long time. Optical rewritable LC alignment is a good base for new types of LC displays, especially on plastic substrates, where no ITO films and electrodes are needed (Figure 6.71) [77, 78].

No backlight is required since a reflective type polarizer is used as the bottom substrate. The image is truly stable, can be written to gray level with saturation, and rewritten a number of times with high reproducibility of properties. The overall reflection is about 30% (Figure 6.71), while the aspect ratio is 1. The unique property of being cell gap nonsensitive allows to maintain proper performance even when the device is bent. Fluorescent dye dopant of liquid crystal partly absorbs blue light and re-emits green light improving photopic reflection and enhancing color of the ORW E-paper (Figure 6.71). The ORW E-paper's potential applications are light printable rewritable paper, labels, and plastic card displays. Other ORW papers have also been recently developed, including (i) doping of photoaligning azo dye to the bulk of cholesteric LC cell [79]; (ii) photosensitive chiral dopants in cholesteric LC capable of changing the twisting power under UV exposure [44]; (iii) cholesteric LC layer with an organic photoconductor layer [80]. However, these versions of ORW devices require sufficiently high electric field in combination with a writing light, which is not convenient.

6.3
LC Applications in Photonics: Passive Optical Elements for Fiber Optical Communication Systems

Photonics LC applications include passive LC elements for fiber optical communication systems (DWDM components) based on liquid crystal cells, which can successfully compete with other elements used for the purpose, such as microelectromechanical (MEM), thermooptical, optomechanical, or acoustooptical devices [5]. The known LC applications in fiber optics enable one to produce switches, filters, attenuators, equalizers, polarization controllers, phase emulators, and other fiber optical components. Application of nematic and ferroelectric LC for high-speed communication systems, producing elements that are extremely fast, stable, durable, of low loss, operable over a wide temperature range, and requiring low operating voltages and extremely low power consumption. Good robustness due to the absence of moving parts and compatibility with VLSI technology, excellent parameters in a large photonic wavelength range, whereas the complexity of the design and the cost of the device are equivalent to regular passive matrix LC displays, makes LC fiber optical devices very attractive for mass production. The quality criteria for photonics fiber optical LC applications may also be very different for various applications. For instance, the LC variable optical attenuators (VOA) are characterized by (i) cross talk attenuation;

(ii) insertion loss; (iii) operation wavelength; (iv) working temperature range; (v) switching power; and (vi) response time. The photonics LC devices should be standardized for the same criteria as usual fiber optical components for photonics applications. We will consider various rheological factors, which can affect photonics applications of LC devices.

6.3.1
LC Switches

Switches for optical fiber networks are increasingly important. LC switches show certain advantages in comparison with MEM switches, commonly used for the same purpose, such as (i) fast switching time; (ii) low controlling voltages and power consumption; and (iii) higher reliability and working time [5]. However, wavelength dependence of the response times and thermal drift of the characteristics of LC switches should be avoided [5]. There are two main techniques, which can be used in LC optical switches, working in nonpolarized light. In the first, the nonpolarized light from the input fiber is decomposed by polarizing beam splitters (PBSs) on the two orthogonal polarizations, which are then independently rotated by the SLM and collected on the output fiber, thus transferring the input nonpolarized light to the proper output channel. The fast switching of optical channels in nonpolarized light was proposed in Ref. [81] (Figure 6.72).

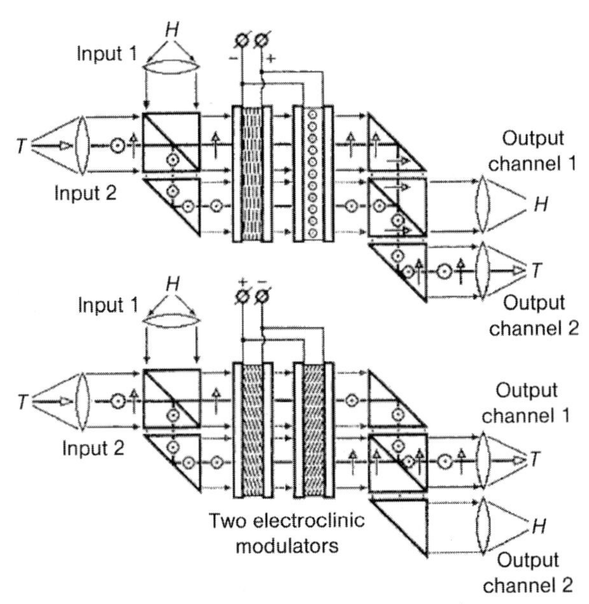

Figure 6.72 FLC switch of nonpolarized light images between two input and two output channels based on two polarization beam splitters and two FLC electroclinic modulators, mounted in series. The transfer of the images (letters "H" and "T") is shown in two different cases [81].

The switching device includes ferroelectric LC modulators and polarization beam splitters, which may control the polarization of four polarized input beams, transferring the input images (letters "H" and "T") from the input 1 or 2 to the corresponding outputs (channels 1 and 2). The transfer of the images $1 \Rightarrow 1$ and $2 \Rightarrow 2$ or $1 \Rightarrow 2$ and $2 \Rightarrow 1$ depends on the amplitude and sign of the voltage applied to the modulators (Figure 6.72). The advantage of the switch is very fast response time (5 μs at the voltage ±30 V). The device has an intrinsic gray scale and very small wavelength dispersion, making it possible to redistribute the input light intensities between the outputs within the whole visible range. An integrated LC switch has also been proposed and designed that works independent of the polarization state of the input light [82]. The proposed structure consists of a polarization splitter, two polarization converters, and a polarization combiner. The light propagation patterns within the whole device in the off and on states for both TE and TM modes have been demonstrated.

In the second, the voltage controllable diffraction can be used for the optical switch [83, 84]. The LC optical switching can also be arranged by changing the boundary refractive index for the fiber using LC cell [85, 86]. Another LC optical switches can be used using the principle of Mach Zehnder interferometer with a switchable ferroelectric LC cladding [87] or Fabry–Perot LC infiltration cell integrated in high index contrast silicon-on-insulator waveguide [88].

The following electrooptical effects in liquid crystals can be used in optical switching:

1. Electroclinic effect in smectic LCs [6] (Figure 6.72). Electroclinic effect is a phase transition between smectic A (Figure 2.4b) and smectic C phase (Figure 2.4c) arranged by electric field [6, 7]. The switching time of 5 μs was reported for the controlling voltages of ±30 V (Figure 6.72) [81]. The effect has a weak dependence on the light wavelength. However, the switching must be provided in the double-cell configuration (one cell does not allow 90° rotation of the light polarization).

2. Deformed helix ferroelectric effect in ferroelectric LCs (Figure 6.50). The switching time, less than 10 μs at the controlling voltage of ±20 V, can be provided, which is temperature independent over the broad temperature range [89]. DHF-LC cells are the basis of fast responding modulators and optical filters, which are applicable in fiber optical communication systems. The fast ferroelectric LC shutters (deformed helix ferroelectric effect) with the response time less than 1 μs in a broad temperature range from 20° to 80 °C were also developed [90] (Figure 6.73). We believe, that DHF-FLCs are the fastest electrooptical mode in LC cells for photonics applications.

3. Nematic LC cells with 270° twist can provide switching times of less than 5 ms [6]. The prototypes of LC modulators are already known. The modulators can be used as LC fiber optical switch. The LC switches can use the effect of total internal reflection (TIR) in nematic LC [86, 91] or selective reflection effect in cholesteric LC [92]. The total internal reflection switch operates only for one light polarization (TE mode) and the most promising is VAN configuration (Figure 6.1b) [91]. The

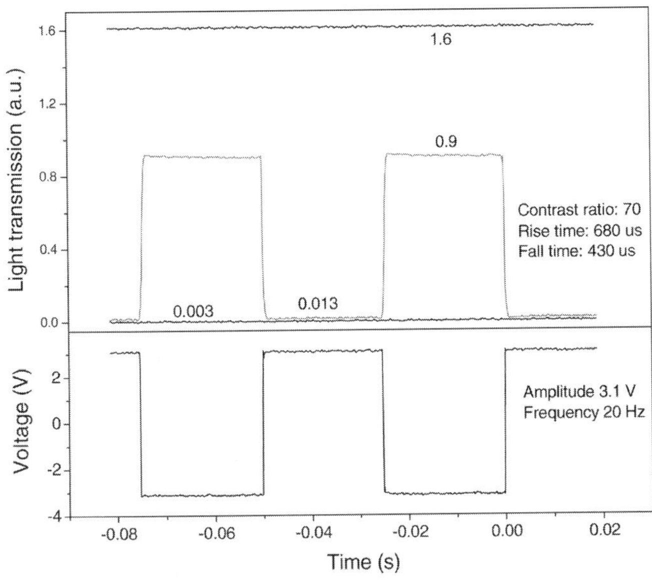

Figure 6.73 Switching time of deformed helix ferroelectric (DHF) LC cells [90].

switching time of 1 ms can be easily obtained in this case for the switching pulse amplitude of 5 V [91] (Figure 6.74).

4. Bistable nematic switches are also possible using BTN cells [93]. In this case, the switching power is reduced as the LC cell keeps the switched state without voltage.

5. The bypass optical switch based on two nematic liquid crystal cells with a switching time less than 200 μs was demonstrated using two temperature-stabilized nematic LC birefringent cells [94] (Figure 6.75). Two subsequent NLC cells with crossed optical axes compensate for the relaxation of NLC birefringence if turned off simultaneously. Thus, the switching speed of two-cell switch can be as high as NLC cell turn-on times. The cells can be specified to a certain fiber wavelength by adjusting the cell gap thickness.

6. LC switch can be made to control light beams in a plane of LC layers (Figure 6.76) [95]. It was shown experimentally that the direction of the light beam can be considerably changed by refraction and reflection of light at the sharp boundaries between the regions with different orientations [95]. LC switching can be controlled by electric field. Certain ways were proposed for optimization insertion loss and cross talk of the 1×2 switcher for real photonic applications (Figure 6.76).

Using different ITO templates, it is possible to create $N \times M$ switch and other different optical processing data elements, for example, attenuators. There are a number of ways to optimize such types of LC devices including application of fast operating ferroelectric liquid crystal layers, which can provide operation times in a microsecond range.

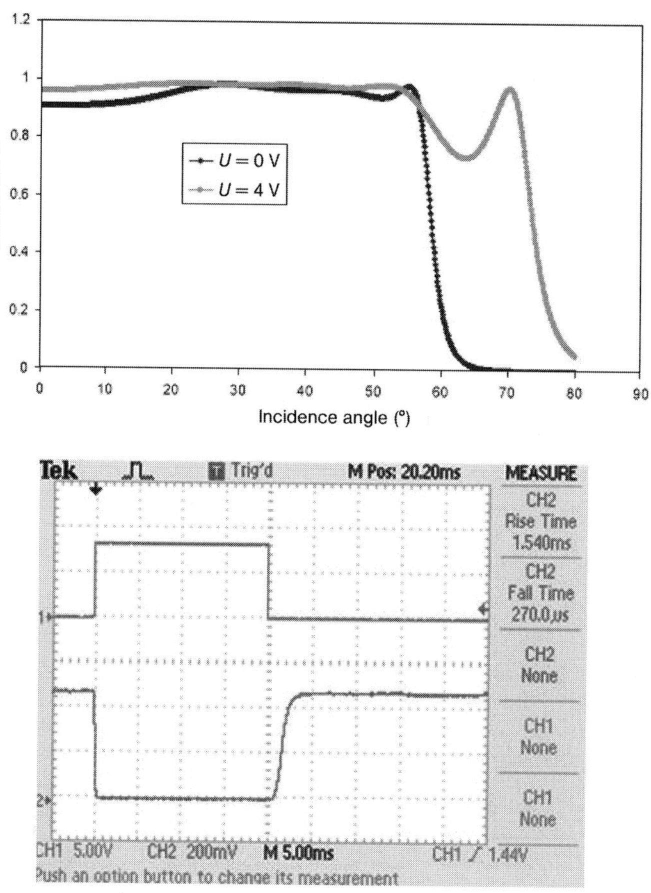

Figure 6.74 Total internal reflection in nematic LC [91]. Left: calculations of TIR transmission versus light incidence angle for two voltages $U = 0\,V$ and $U = 4\,V$. Right: experimental data of switching dynamics of TE mode in VAN configuration (Figure 6.1b).

6.3.2 Other LC Passive Elements for Photonics Applications

Polarization controllers are the elements that can transfer any input state of polarization (SOP) to the desired one, thus controlling the unpredictable polarization change or drift, which comes from the polarization-dependent components of the fiber optical system. These elements can be made on the basis of the three subsequently placed LC cells, which exhibit the effect of electrically controlled birefringence with a homogeneous initial orientation [96]. Typical switching times are dependent on LC material and are on the order of 10 ms for the wavelength $\lambda = 1.3\,\mu m$ (Figure 6.77) [96].

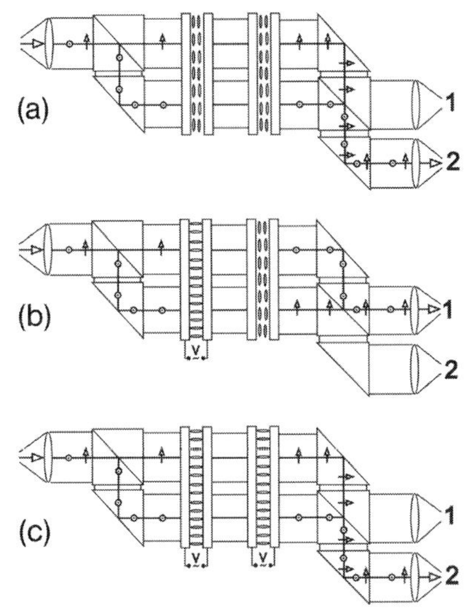

Figure 6.75 A bypass two-cell optical switch based on nematic liquid crystals [94].

Polarization rotator is an LC element that can rotate the linear polarized light in the output of the LC system to any desirable angle still keeping the linear polarization at the output [97, 98]. A novel configuration for a TN-LCD between two-quarter wave plates can operate as a controllable polarization rotator [97]. The plane of polarization of the transmitted light can be rotated as a function of the phase introduced by TN-LCD for different voltages [97]. The other LC

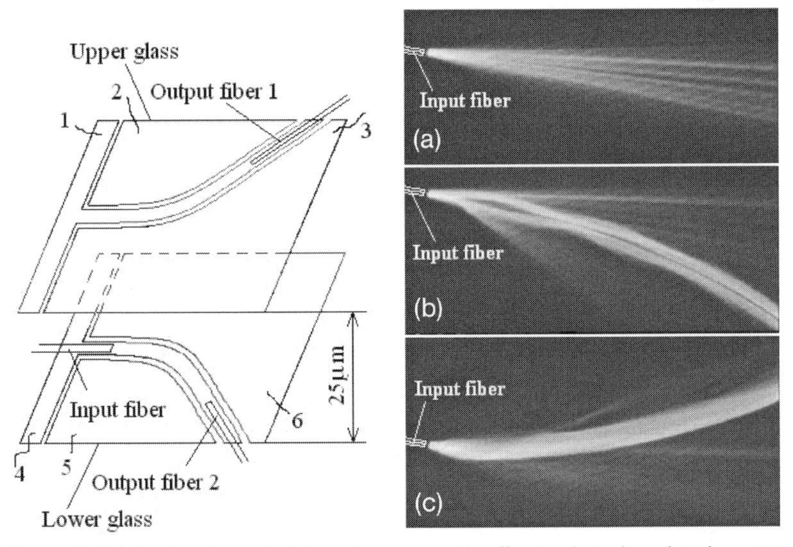

Figure 6.76 A 1 × 2 optical switch based on waveguide effect in photoaligned LC layer [95].

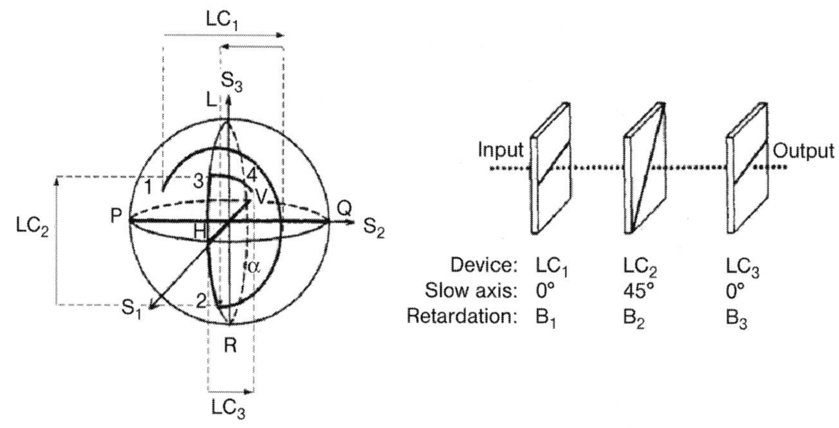

Figure 6.77 A polarization controller constructed with a cascade of three LC cells [96]. This configuration allows a fast arbitrary-to-arbitrary transformation of state of polarization (SOP) on the Poincare sphere (left). The diagram on the left shows a transformation of SOP at point 1 to a final SOP at point 4 following two intermediate transformation at point 2 and point 3.

configuration consists of the polarizer and two LC homogeneous cells placed at 45° with respect to each other (one with the a voltage controllable phase change and the other fixed as a quarter wave plate, Figure 6.78) [98]. The configuration can rotate the light polarization state at any angle between 0° and 90° depending on voltage.

Variable optical LC attenuators (*VOA*) have the typical attenuation range of 30 dB for the driving voltage of 12 V at 1525–1575 wavelength range [99–102] with the response time of about 10–30 ms. Some of them are based on the light scattering of LC cell

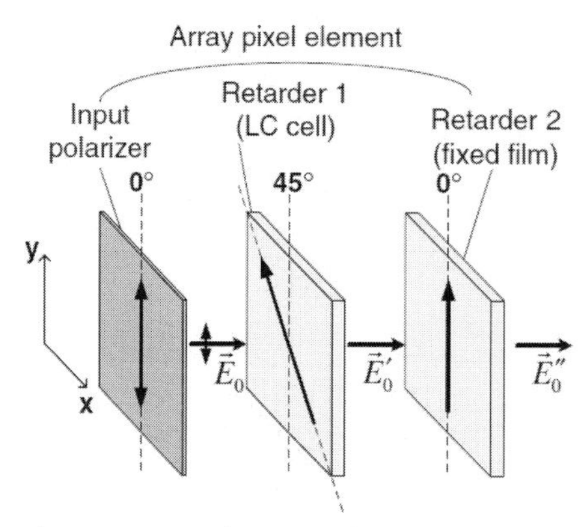

Figure 6.78 LC polarization rotator, consisted of a polarizer and two-phase retarders, one of which is controlled by voltage. The rotation of the polarization plane is between 0° and 90° [98].

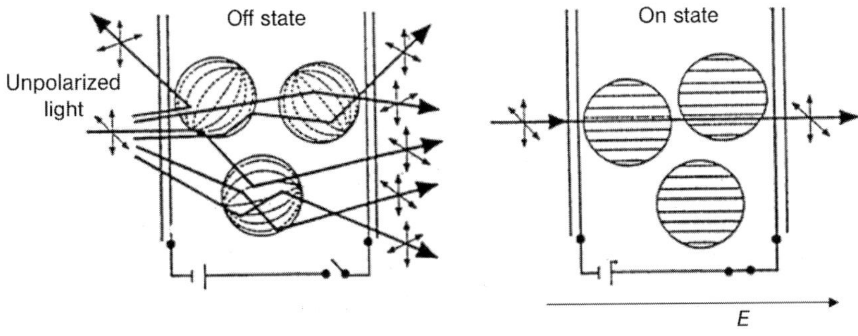

Figure 6.79 The structure and electrooptic properties of PN-LC films, attenuated nonpolarized light. Left: field off state, scattering unpolarized light; right: transparent for the normally incident light field on state [6].

filled with a polymer network (PN-LCD) [6] (Figure 6.79). The light from the input fiber was scattered as a result of refractive index mismatching of the LC and the polymer when no voltage was applied, and it was passed through the PN-LC layer as a result of the refractive index matching of the LC and the polymer, when a voltage was applied (Figure 6.79).

6.3.3 Photonic Crystal/Liquid Crystal Structures

Another promising technique for making passive fiber optics components is the use of photonic crystal (PC) materials [5]. Photonic crystals are artificial media with 1D, 2D, or 3D periodic structures that exhibit both unidirectional and omnidirectional photonic band gaps. The band gap defines a range of frequencies at which electromagnetic radiation striking the crystal is reflected by the crystal rather than being permitted to propagate through it. In these crystals, the dimensions of the lattice structures and the dielectric elements are selected to produce band gaps having desired center frequencies and bandwidths. Electromagnetic radiation at a frequency within the band gap is reflected from the structure via the well-known Bragg reflection phenomenon.

Electrically tunable photonic band gap (PBG) structures hold the potential to become a versatile and compact backbone for optical signal processing. Electrical tuning of two-dimensional PBG structures infiltrated with liquid crystals was reported [103–107]. The structures can be operated with any electrooptic materials and lead to fast and efficient modulators, routers, and tunable filters [103]. The effect of LC alignment is very important, in particular, a high-quality LC alignment parallel to the photonic hole surface is highly desirable [5]. The example of voltage controllable photonic crystal/liquid crystal (PC/LC) structure is given in Figure 6.80 [106]. Liquid crystals are materials that present a variation in their effective refractive index if an electric field is applied (6.16). In addition to this, LCs are cost-effective and easy to manipulate [6, 7]; that is the reason for selecting this material instead of others

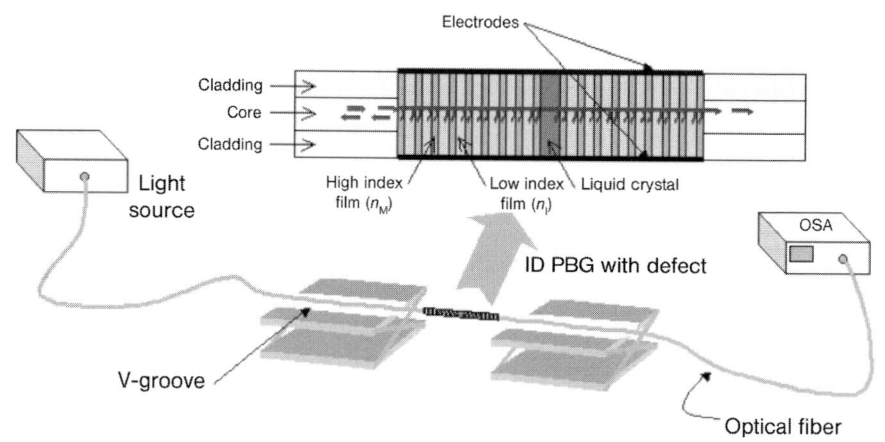

Figure 6.80 Structure of the 1D PBG wavelength filter consisting of 2 Bragg mirrors of 30 layers and a cavity (defect) [106].

dependent on the temperature or a mechanical force, for instance. By changing the voltage applied to the structure, the defect state that constitutes the filter will sweep a range of wavelengths [106]. Other important applications of liquid crystals in the middle of one-dimensional PC/LC structures are electrooptical switches and modulators [107].

Controllable PC/LC devices became a "hot" topic of research. Photonic crystals are known to provide wavelength-dependent filters, beam splitters, and mirror components. As photonic crystals are on the order of several wavelengths, in each of two major dimensions, and are made with microprocessing techniques, optical processing systems employing photonic crystals can be very small, thereby permitting extremely high bit density and high rates of data processing. Filling the interstices of the photonic crystal with the properly aligned LC material and subjecting the LC to a varying electric field can produce a tunable photonic crystal element. We will consider in the following section photoaligning materials to align LC mixtures in small cavities, such as holes and tubes of photonic crystals, with a size of 1 μm and less, and obtained excellent LC orientation inside the tubes by photoalignment. New PC/LC passive elements of fiber optical systems include optical switches, fiber optical isolators, optical filters, wavelength selective mirrors for fiber lasers, polarization controllers, attenuators, and so on [5].

6.3.4 Photoalignment Technology for LC Photonics Devices

Photoalignment possesses obvious advantages in comparison with the usually "rubbing" treatment of the substrates of LCD cells. Possible benefits for using this technique in photonics LC devices include [60, 78]

(i) new advanced applications of LC in fiber communications, optical data processing, holography, and other fields, where the traditional rubbing LC alignment is not possible due to the sophisticated geometry of LC cell and/or high spatial resolution of the processing system;

(ii) ability for efficient LC alignment on curved and flexible substrates; and

(iii) manufacturing of new optical elements for LC technology, such as patterned polarizers and phase retarders, tunable optical filters, variable optical attenuators, and so on.

The photoaligning materials developed by us are based on photopolymerized and cross-linked dye photosensitive layers that enable [60, 78] (i) high-order parameter more than 0.8; (ii) excellent alignment quality of nematic and ferroelectric LC materials in various modes; (iii) temperature and UV stability due to the polymerization and cross-linking effect in dye layers; (iv) perfect adhesion and anchoring energy comparable to rubbed polyimide layers; (v) excellent sensitivity with a minimum exposure energy; and (vi) have the ability to align LC materials in curved surfaces and photonic holes.

Photonic crystal fiber is a glass or polymer fiber with an array of microscopic air holes running along the length of the fiber. The waveguide properties of such a fiber can be controlled by introducing an additional material into the air holes [103–105]. LC is suitable for this purpose because its refractive index can be easily tuned by electric field or temperature. The technique of photoconfigurable alignment of LC in glass microtubes and in photonic crystal fiber was developed (Figure 6.81) [108]. Good homogeneous alignment was detected with polarized microscopy and FTIR spectroscopy methods. The presented technique of alignment is based on properly developed photoaligning azo dye material and is promising as a noncontact method for LC orientation in complex photonic crystal structures [108]. Figure 6.81 presents the glass tube of inner diameter 4 μm treated with the photoaligning sulfonic azo layer (SD1) and filled with uniform nematic LC orientation without point defects or linear disclinations. The order parameter S of LC has been obtained from FTIR spectroscopy data and has demonstrated good alignment quality ($S = 0.63$ [108]). The presented technique can be used as noncontact method of LC alignment in complex photonic crystal structure.

The uniform azo dye photoalignment on the profiled 3D surface (substrate with bulk relief) was demonstrated [109]. The three-step exposure process for uniform surface

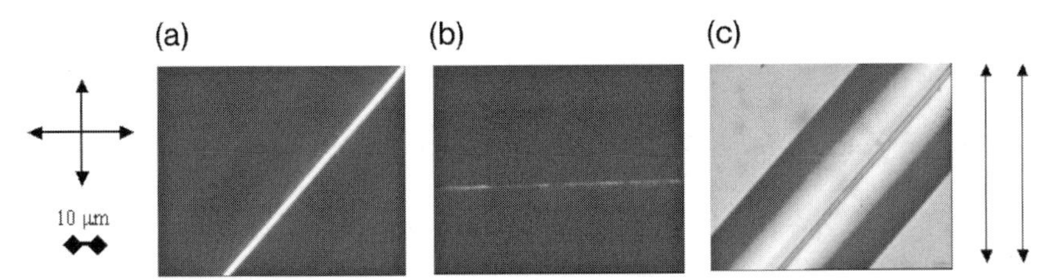

Figure 6.81 Photoalignment in microtube [108], crossed polarizers: (a) angle between polarizer and tubes axis is 45°, (b) angle between polarizer and tubes axis is 0°; parallel polarizers: (c) angle between polarizer and tube's axis is 45°.

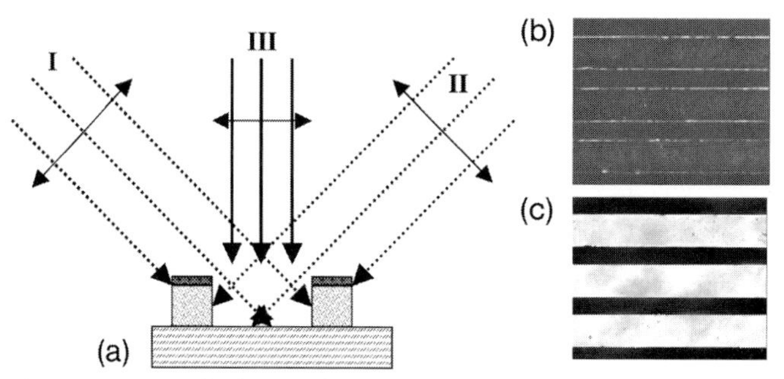

Figure 6.82 Optical scheme of three-step 3D substrate illumination (a); profiled substrate LC cell between crossed (b) and parallel (c) polarizers [5].

alignment was developed (Figure 6.82). Patterned exposure of azo materials is a very useful procedure for LC photonic devices as a high-quality LC alignment on profiled surfaces can be obtained.

LC alignment on submicrometer-sized rib waveguides on silicon chips was studied [5, 78, 109]. Experiments using nematic liquid crystal cladding on silicon waveguides and microrings coated with photoalignment layer, and covered with a vertically aligned polyimide rubbed glass, revealed a defect-free hybrid aligned liquid crystal cell on silicon substrate (Figure 6.83). An electrically tunable microresonator using photoaligned liquid crystal as cladding layers, where a photoalignment layer on the device surface defined the orientation of the liquid crystal molecules and the transmission property of the waveguide-coupled microresonator was electrically tuned by varying the cladding refractive index under an applied electric field in the vertical direction, was demonstrated [5, 78].

The liquid crystal cladding refractive index was then varied according to the applied voltage and subsequently the microresonator resonance wavelengths were tuned. Based on our initial measurements, the FSR (free spectral range) wavelength shift within the range of 20 nm was obtained, which is comparable with a thermooptic

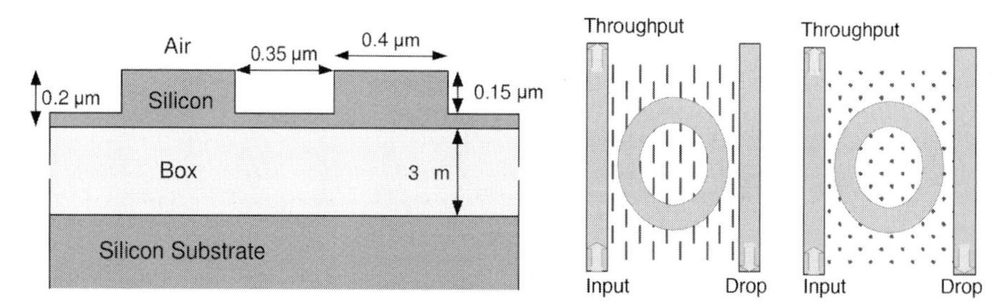

Figure 6.83 From left to right: schematic profile of submicrometer-sized rib Si microring system; planar and homeotropic LC configurations on photoaligned cladding azo dye layer [5, 78].

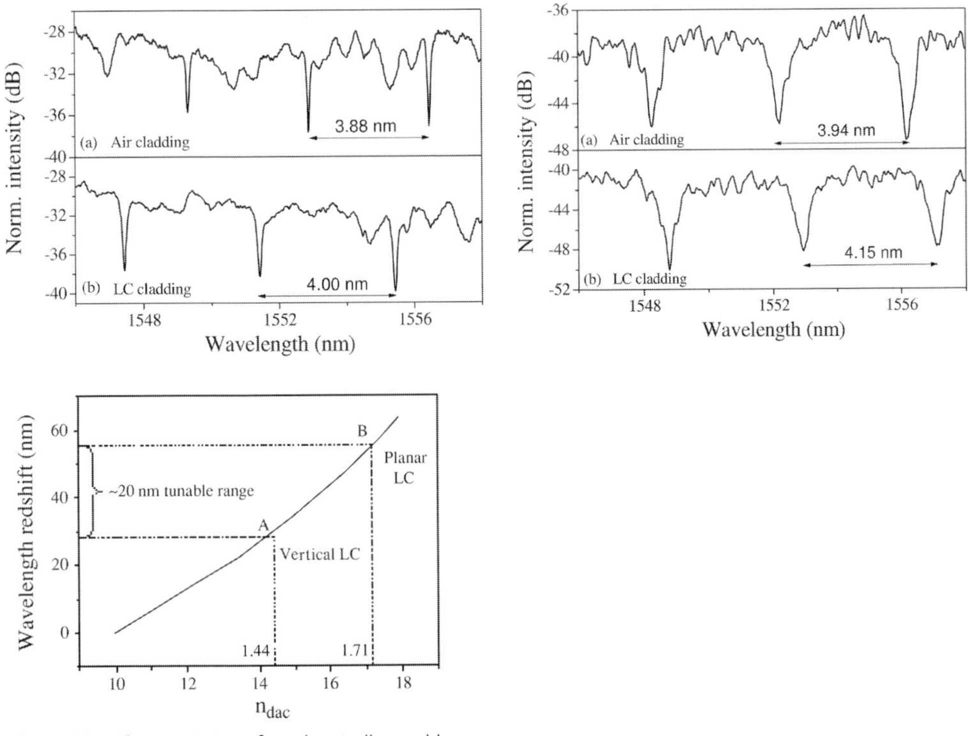

Figure 6.84 Characteristics of an electrically tunable microresonator using photoaligned liquid crystal as cladding layers. From left to right: FSR shift, when air cladding is changed to planar-oriented LC; FSR shift, when air cladding is changed to homeotropic oriented LC; wavelength redshift as a function of the effective refractive index of the LC layer [5].

effect (Figure 6.84). New voltage controllable Si-based add drop filters are envisaged based on this principle.

Remnant high-efficiency polarization gratings were created in nematic liquid crystal cells by layers of azo dye molecules deposited on the cell substrates and exposed to "interfering" beams with opposite circular polarizations [110]. The diffraction picture was controlled by an electric field applied across the LC cell (Figure 6.85). Obtained polarization gratings (PGs) can be used for electrically controlled discrimination and detection of polarized components of light. All molecules of LC are reoriented to a uniform homeotropic state at high voltage, and there is no more modulation of LC alignment in the cell. Applications in LC optical switches are envisaged.

The achromatic polarization grating exhibiting only three diffracted orders (0, ± 1) can also be produced using photoalignment [111]. The achromatic PG offered high efficiencies of thick (Bragg) gratings over nearly the entire range of visible light. When used as an optical element of displays, the achromatic PG can be implemented as a switchable LC grating for modulator applications by allowing one of the grating

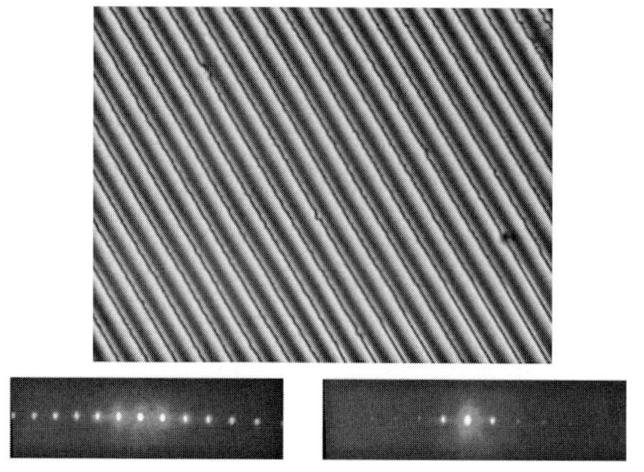

Figure 6.85 Tunable polarization gratings using liquid crystals [110].

twists to be implemented as a switchable nematic LC and by placing the entire structure between the substrates with electrodes [111].

More work in the field of passive liquid crystal elements for fiber optical communication systems is needed. High-quality LC switches, variable optical attenuators, voltage controllable filters, based on Si photonics devices, polarization controllers, and rotators have to be developed. New working prototypes ready for packaging are highly desirable. Using photonic crystals and photoalignment technique makes it possible to develop new LC fiber components. There exist certain problems in wavelength dependence, too large losses, thermal drift, and high cross talk. New LC materials for the IR region used for fiber optical communications have to be tested. Specific LC materials with a high birefringence and low rotational viscosity have to be developed for IR region. The combination of photonic crystals and liquid crystals can provide effective switching of the refractive index of diffraction grating in the microsecond range with low controlling voltages, as the thickness of the LC layer in the hole can be less than one micron. LC aligning in the microtubes of photonic crystals can be produced by the photoaligning technique. Fast PC/LC efficient switches and tunable filters are envisaged.

References

1 Lueder, E. (2001) *Liquid Crystal Displays: Addressing Schemes and Electrooptical Effects*, John Wiley & Sons, Ltd, Chichester.

2 Yang, D.-K. and Wu, S.-T. (2006) *Fundamentals of Liquid Crystal Devices*, John Wiley & Sons, Ltd, Chichester.

3 Armitage, D., Underwood, I., and Wu, S.T. (2006) *Introduction to Microdisplays*, John Wiley & Sons, Ltd, Chichester.

4 den Boer, W. (2005) *Active Matrix Liquid Crystal Displays: Fundamentals and Applications*, Elsevier.

5 Chigrinov, V.G. (2007) Liquid crystal devices for photonics applications. *Proc. SPIE*, **6781**, 67811.

6 Chigrinov, V.G. (1999) *Liquid Crystal Devices: Physics and Applications*, Artech House.

7 Blinov, L.M. and Chigrinov, V.G. (1994) *Electrooptical effects in liquid crystalline materials*, Springer, Berlin.

8 Nehring, J., Kmetz, A.R., and Scheffer, T.J. (1976) Analysis of weak boundary coupling effects in liquid-crystal displays. *J. Appl. Phys.*, **47**, 850.

9 Wu, S.T. and Wu, C.S. (1991) High speed nematic liquid crystal modulators. *Mol. Cryst. Liq. Cryst.*, **207**, 1.

10 Uchida, T. and Miyashita, T. (1995) Optically compensated bend mode with wide viewing angle and fast response time for AM-LCD. IDW'95 Digest, p. 39.

11 Kilickiran, P., Masutani, A., Roberts, T., Hollfelder, N., Schüller, B., Nelles, G., and Yasuda, A. (2008) Towards faster liquid crystals at lower driving voltages. *J. SID*, **16/1**, 63.

12 Jiao, M., Ge, Z., Wu, S.-T., and Choi, W.-K. (2008) Submillisecond response nematic liquid crystal modulators using dual fringe field switching in a vertically aligned cell. *Appl. Phys. Lett.*, **92**, 111101.

13 Kwok, H.-S. and Yeung, F.S.Y. (2008) Nano-structured liquid-crystal alignment layers. *J. SID*, **16/9**, 911.

14 Ho, J., Chigrinov, V., and Kwok, H.S. (2007) Variable liquid crystal pretilt angles generated by photoalignment of a mixed polyimide alignment layer. *Appl. Phys. Lett.*, **90**, 2435061.

15 Grebenkin, M.F., Seliverstov, V.A., Blinov, L.M., and Chigrinov, V.G. (1975) Properties of nematic liquid crystals with positive dielectric anisotropy. *Kristallografiya*, **20**, 984.

16 Gooch, C.H. and Tarry, H.A. (1975) The optical properties of twisted nematic liquid crystal structures with twist angles $\leq 90°$. *J. Phys. D: Appl. Phys.*, **8**, 1575.

17 Chigrinov, V., Yakovlev, D., and Kwok, H.S. (2004) Optimization and modeling of liquid crystal displays. *Inform. Display*, **20**, 26.

18 Berreman, D.W. (1983) Numerical modeling of twisted nematic devices. *Phil. Trans. Roy. Soc. London*, **309**, 203.

19 Yeh, P. and Gu, C. (1999) *Optics of Liquid Crystal Displays*, John Wiley & Sons, Inc., New York.

20 Wohler, H. and Becker, M.E. (2002) Numerical modelling of LCD electro-optical performance. *Opto-Electron. Rev.*, **10**, 23.

21 http://www.autronic-melchers.com/.

22 http://www.shintech.jp/.

23 http://www.sanayisystem.com.

24 Schadt, M. and Gerber, P.R. (1982) Class specific physical properties of liquid crystals and correlations with molecular structure and static electrooptical performance in twist cells. *Z. Naturforsch.*, **37a**, 165.

25 Alt, P.M. and Pleshko, P. (1974) Scanning limitations of liquid crystal displays. *IEEE Trans. Electron. Dev.*, **ED-21**, 146.

26 Vorflusev, V.P., Kitzerow, H.S., and Chigrinov, V.G. (1995) Azimuthal anchoring energy in photoinduced anisotropic films. *Jpn. J. Appl. Phys, Part 2*, **34**, L1137.

27 Takatoh, K., Akimoto, M., Kaneko, H., Kawashima, K., and Kobayashi, S. (2008) The low driving voltage, the molecular arrangement and the stability of reverse TN mode. IDW'08 Digest, p. 1567.

28 Li, Y.W., Tan, L., and Kwok, H.S. (2008) Field-sequential-color LCDs based on transient modes. SID'08 Digest, p. 32.

29 Kim, K.J., Park, S.J., Choi, S.H., Kim, J.H., and Shin, H.H. (2008) A novel LCD using S-IPS technology for the application of public display. IDMC'08 Digest, p. 1649.

30 Seiberle, H. and Schadt, M. (2000) Photoalignment and photo-patterning of planar and homeotropic liquid-crystal-display configurations. *J. SID*, **8/1**, 67.

31 Mak, H.Y., Chigrinov, V.G., and Kwok, H.S. (2007) Transflective twisted-nematic liquid-crystal display with double twisted-nematic cell and twisted liquid-crystal retarder. *J. SID*, **15/7**, 527.

32 Scheffer, T.J. and Nehring, J. (1021) A new, highly multiplexable liquid crystal display. *Appl. Phys. Lett.*, **45**, 1984.

33 Scheffer, T.J. and Nehring, J. (1991) Twisted nematic and supertwisted nematic mode LCDs, in *Liquid Crystals: Applications and Uses* (ed. B. Bahadur), World Scientific.

34 Schadt, M. and Leenhouts, F. (1987) Electro-optical performance of a new black-white and highly multiplexable liquid crystal display. *Appl. Phys. Lett.*, **50**, 236.

35 Li, J., Kelly, J., Hoke, Ch., and Bos, P.J. (1996) Optimization of luminance for optical-mode-interference (OMI) displays. *J. SID*, **4/2**, 95.

36 Morrissy, J.H. (2000) Will traditional TN/FSTN LCDs dominate in low power, reflective display applications. SID'2000 Digest, p. 80.

37 Mori, H., Nagai, M., Nakayama, H., Itoh, Y., Kamada, K., Arakawa, K., and Kawata, K. (2003) Novel optical compensation method based upon a discotic optical compensation film for wide-viewing-angle LCDs. SID'2003 Digest, p. 1058.

38 Hirai, Y. and Kitamura, M. (1997) STN-LCDs enhanced by multiple line addressing (MLA): present and future applications. SID'97 Digest, p. 401.

39 Ruckmongathan, T.N. (2008) Low power by energy multiplexing in liquid crystal displays. IDRC'08 Digest, p. 63.

40 Broughton, B.J., Clarke, M.J., Blatch, A.E., and Coles, H.J. (2005) Optimized flexoelectric response in a chiral liquid crystal phase device. *J. Appl. Phys.*, **98**, 034109.

41 Rudquist, P., Komitov, L., and Lagerwall, S.T. (1994) Linear electro-optic effect in a cholesteric liquid crystals. *Phys. Rev. E*, **50**, 4735.

42 Yuan, H. (1996) Bistable reflective cholesteric displays, in *Liquid Crystals in Complex Geometries Formed by Polymer and Porous Networks* (eds G. Crawford and S. Zumer), Taylor and Francis, London, pp. 265–281.

43 Doane, J.W., Davis, D., Khan, A., Montbach, E., Schneider, T., and Shiyanovskaya, I. (2006) Cholesteric reflective displays: thin and flexible. IDRC'06 Digest, p. 9.

44 Venkataraman, N., Magyar, G., Montbach, E., Khan, A., Schneider, T., Doane, J.W., Green, L., and Li, Q. (2008) Thin flexible photosensitive cholesteric displays. IDRC'08 Digest, p. 101.

45 Pozhidaev, E.P., Osipov, M.A., Chigrinov, V.G., Baikalov, V.A., Blinov, L.M., and Beresnev, L.A. (1988) Rotational viscosity of smectic C* phase of ferroelectric liquid crystals. *Sov. Phys. JETP*, **67**, 283.

46 Barnik, M.I., Baikalov, V.A., Chigrinov, V.G., and Pozhidaev, E.P. (1987) Electrooptics of a thin ferroelectric smectic C* liquid crystal layer. *Mol. Cryst. Liq. Cryst.*, **143**, 101.

47 Garoff, S. and Meyer, B. (1979) Electroclinic effect at the A-C phase change in a chiral smectic liquid crystal. *Phys. Rev. A*, **19**, 338.

48 Lagerwall, S.T. (2004) Ferroelectric and antiferroelectric liquid crystals. *Ferroelectrics*, **301**, 15.

49 Jones, J.C. and Raynes, E.P. (1992) Measurement of biaxial permittivities for several smectic C host materials used in ferroelectric liquid crystal devices. *Liq. Cryst.*, **11**, 199.

50 Itoh, N., Koden, M., Miyoshi, Sh., and Akahane, T. (1995) Novel frequency dependence of dielectric biaxiality of surface stabilized ferroelectric liquid crystals. *Liq. Cryst.*, **18**, 109.

51 Beresnev, L.A., Chigrinov, V.G., Dergachev, D.I., Pozhidaev, E.P., Funfschilling, J., and Schadt, M. (1989) Deformed helix ferroelectric liquid crystal display: a new electrooptic mode in ferroelectric chiral smectic C liquid crystal. *Liq. Cryst.*, **5**, 1171.

52 Tsuchiya, T., Takezoe, H., and Fukuda, A. (1986) Importance of controlling material constants to realize bistable uniform states in surface stabilized ferroelectric liquid crystals. *Jpn. J. Appl. Phys.*, **25**, L27.

53 Chilaya, G.S. and Chigrinov, V.G. (1993) Optics and electrooptics of chiral smectic C liquid crystals. *Phys.-Usp.*, **36**, 909.

54 Chigrinov, V., Panarin, Yu., Vorflusev, V., and Pozhidaev, E. (1996) Aligning properties and anchoring strength of ferroelectric liquid crystals. *Ferroelectrics*, **178**, 145.

55 Huang, D.D., Pozhidaev, E.P., Chigrinov, V.G., Cheung, H.L., Ho, Y.L., and Kwok, H.S. (2004) Photo-aligned ferroelectric liquid crystal displays based on azo-dye layers. *Displays*, **25**, 21.

56 Kiselev, A.D., Chigrinov, V.G., and Pozhidaev, E.P. (2007) Switching dynamics of surface stabilized ferroelectric liquid crystal cells: effects of anchoring energy asymmetry. *Phys. Rev. E*, **75**, 061706.

57 Clark, N.A. and Lagerwall, S.T. (1980) Submicrosecond bistable electrooptic switching in liquid crystals. *Appl. Phys. Lett.*, **36**, 899.

58 Hikmet, R.A.M. (1995) *In situ* observation of smectic layer reorientation during the switching of a ferroelectric liquid crystal. *Liq. Cryst.*, **18**, 927.

59 Funfschilling, J. and Schadt, M. (1994) Physics and electronic model of deformed helix ferroelectric liquid crystal displays. *Jpn. J. Appl. Phys.*, **33**, 4950.

60 Chigrinov, V.G., Kwok, H.-S., Hasebe, H., Takatsu, H., and Takada, H. (2008) Liquid-crystal photoaligning by azo dyes. *J. SID*, **16/9**, 897.

61 Pozhidaev, E.P. and Chigrinov, V.G. (2006) Bistable and Multistable States in Ferroelectric Liquid Crystals. *Cryst. Rep.*, **51**, 1030.

62 Xue, J.-Z., Handschy, M.A., and Clark, N.A. (1987) Electrooptical switching properties of uniform layer tilted surface stabilized ferroelectric liquid crystal devices. *Liq. Cryst.*, **2**, 707.

63 Yang, K. and Chieu, T. (1989) Dominant factors influence the bistability of SSFLC devices. *Jpn. J. Appl. Phys.*, **28**, L1599.

64 Yeoh, C.T.H., Mosley, A., and Nicholas, B.M. (1993) Effect of alignment structure of charge controlled switching on ferrolectric liquid crystals. *Ferroelectrics*, **149**, 333.

65 Pozhidaev, E., Chigrinov, V., Hegde, G., and Xu, P. (2009) Multistable electro-optical modes in ferroelectric liquid crystals. *J. SID*, **17/1**, 53.

66 Panarin, Yu., Pozhidaev, E., and Chigrinov, V. (1991) Dynamics of controlled birefringence in an electric field deformed helical structure of a ferroelectric liquid crystal. *Ferroelectrics*, **114**, 181.

67 Hegde, G., Xu, P., Pozhidaev, E., Chigrinov, V., and Kwok, H.S. (2008) Electrically controlled birefringence colours in deformed helix ferroelectric liquid crystals. *Liq. Cryst.*, **35**, 1137.

68 Li, X.H., Murauski, A., Muravsky, A., Xu, P.Z., Cheung, H.L., and Chigrinov, V. (2007) Gray scale generation and stabilization in ferroelectric liquid crystal display. *J. Display Technol.*, **3**, 273.

69 Amakawa, H. and Kondoh, S. (2008) Developments of ferroelectric liquid crystal devices and applications utilizing memory effect. IDW'08 Digest, p. 1559.

70 Fujisawa, T., Hatsusaka, K., Maruyama, K., Nishiyama, I., Takeuchi, K., Takatsu, H., and Kobayashi, S. (2008) V-shaped E-O properties of polymer stabilized (PSV-) FLCD free from conventional surface stabilization: advanced color sequential LCDs. IDW'08 Digest, p. 1563.

71 Huang, H.C., Cheng, P.W., and Kwok, H.S. (2000) On the minimization of flicker and image retention in silicon light valves. SID'00 Digest, p. 248.

72 Choi, N., Ahn, H., and Shin, S. (2008) Developing the new evaluation method of the image sticking. IDW'08 Digest, p. 43.

73 Woo, J.W., Shin, D.C., Lim, E.J., Park, H.J., and Shin, H.H. (2008) UV photoalignment technology comparable to the rubbing alignment method in terms of the azimuthal anchoring energy. IDW'08 Digest, p. 37.

74 Angelé, J., Elyaakoubi, M., Guignard, F., Jacquier, S., Joly, S., Martinot-Lagarde, P., Osterman, J., Laffitte, J.D., and Leblanc, F. (2007) An A4 BiNem displays with 3.8 mega pixels. IDW'07 Digest, p. 265.

75 Li, Y.-W., Lee, C.Y., and Kwok, H.S. (2008) Permanent bistable twisted nematic displays using bi-directional alignment surface. SID'08 Digest, p. 1026.

76 Cliff Jones, J. (2007) The zenithal bistable device: from concept to consumer. SID'07 Digest, p. 1347.

77 Muravsky, A., Murauski, A., Chigrinov, V., and Kwok, H.-S. (2008) Optical rewritable electronic paper. *IEICE Trans. Electron.*, **E91-C**, 1576.

78 Chigrinov, V.G., Kozenkov, V.M., and Kwok, H.S. (2008) *Photoalignment of Liquid Crystalline Materials: Physics and Applications*, John Wiley & Sons, Inc., New York.

79 Cheng, K.-T., Liu, C.-K., Ting, C.-L., and Fuh, Y.-G. (2008) Optical addressing in dye-doped cholesteric liquid crystals. *Opt. Commun.*, **281**, 5133.

80 Sato, M., Ishii, T., Hiji, N., Tomoda, K., Yamamoto, S., and Baba, K. (2008) High resolution electronic paper based on LED print head scanning exposure. SID'08 Digest, p. 923.

81 Beresnev, L. and Haase, W. (1998) Ferroelectric liquid crystals: development of materials and fast electrooptical elements for non-display applications. *Opt. Mater.*, **9**, 201.

82 Wang, Q. and Farrell, G. (2007) Integrated liquid-crystal switch for both TE and TM modes: proposal and design. *J. Opt. Soc. Am. A*, **24**, 3303.

83 Crossland, W., Wilkinson, T., Manolis, I., *et al.* (2002) Telecommunications applications of LCOS devices. *Mol. Cryst. Liq. Cryst.*, **375**, 1.

84 Crossland, W.A., Clapp, T.V., Wilkinson, T.D., Manolis, I.G., Georgiou, A.G., and Robertson, B. (2004) Liquid crystals in telecommunications systems. *Mol. Cryst. Liq. Cryst.*, **413**, 363.

85 Beccherelli, R., Bellini, B., Donisi, D., and d'Alessandro, A. (2007) Integrated optics devices based on liquid crystals. *Mol. Cryst. Liq. Cryst.*, **465**, 249.

86 Zhang, A., Chan, K.T., Demokan, M.S., Chan, V., Chan, P., Kwok, H.S., and Chan, A. (2005) Integrated liquid crystal optical switch based on total internal reflection. *Appl. Phys. Lett.*, **86**, 211108.

87 Hoshi, R., Nakatsuhara, K., and Nakagami, T. (2006) Optical switching characteristics in Si-waveguide asymmetric Mach-Zehnder interferometer having ferroelectric liquid crystal cladding. *Electron. Lett.*, **42**, 635.

88 Riboli, F., Daldosso, N., Pucker, G., Lui, A., and Pavesi, L. (2005) Design of integrated optical switch based on liquid crystal infiltration. *IEEE J. Quant. Electron.*, **41**, 1197.

89 Pozhidaev, E.P. (2001) Electrooptical properties of deformed-helix ferroelectric liquid crystal display cells. *SPIE Digest*, **4511**, 92.

90 Presnyakov, V., Liu, Z., and Chigrinov, V.G. (2005) Fast optical retarder using deformed-helical ferroelectric liquid crystals. *Proc. SPIE*, **5970**, 426.

91 Xu, P., Chigrinov, V., and Kwok, H.S. (2008) Optical analysis of a liquid-crystal switch system based on total internal reflection. *J. Opt. Soc. Am.*, **25**, 866.

92 Semenova, Yu., Panarin, Yu., Farrell, G., and Dovgalets, S. (2004) Liquid crystal based optical switches. *Mol. Cryst. Liq. Cryst.*, **413**, 385.

93 Zhuang, Z., Kim, Y., and Patel, J. (1999) Bistable twisted nematic liquid crystal optical switch. *Appl. Phys. Lett.*, **75**, 3008.

94 Muravsky, Al. and Chigrinov, V. (2005) Optical switch based on nematic liquid crystals. IDW'05 Digest, p. 223.

95 Maksimochkin, A.G., Pasechnik, S.V., Tsvetkov, V.A., Yakovlev, D.A., Maksimochkin, G., and Chigrinov, V.G. (2007) Electrically controlled switching of light beams in the plane of liquid crystal layer. *Opt. Commun.*, **270**, 273.

96 Zhuang, Z., Suh, S.-W., and Patel, J.S. (1999) Polarization controller using nematic liquid crystals. *Opt. Lett.*, **24**, 694.

97 Moreno, I., Martínez, J.L., and Davis, J.A. (2007) Two-dimensional polarization rotator using a twisted-nematic liquid-crystal display. *Appl. Opt.*, **46**, 881.

98 Muravsky, A., Murauski, A., Chigrinov, V., and Kwok, H.-S. (2008) Light printing of grayscale pixel images on optical rewritable electronic paper. *Jpn. J. Appl. Phys.*, **47**, 6347.

99 http://www.lightwaves2020.com.

100 Du, F., Lu, Y., Ren, H.,*et al.* (2004) Polymer-stabilized cholesteric liquid crystal for polarization-independent variable optical attenuator. *Jpn. J. Appl. Phys.*, **43**, 7083.

101 Wu, Y.-H., Liang, X., Lu, Y.-Q., Du, F., Lin, Y.-H., and Wu, S.-T. (2005) Variable optical attenuator with a polymer-stabilized dual-frequency liquid crystal. *Appl. Opt.*, **44**, 4394.

102 Hirabayashi, K. and Amano, C. (2004) Liquid-crystal level equalizer arrays on fiber arrays. *IEEE Photon. Technol. Lett.*, **16**, 527.

103 Haurylau, M., Anderson, S., Marshall, K., and Fauchet, P. (2006) Electrical modulation of silicon-based two-dimensional photonic bandgap structures. *Appl. Phys. Lett.*, **88**, 061103.

104 Alagappan, G., Sun, X., Yu, M., Shum, P., and Engelsen, D. (2006) Tunable dispersion properties of liquid crystal infiltrated into a two-dimensional photonic crystal. *IEEE J. Quant. Electron.*, **42**, 404.

105 Weiss, S., Ouyang, H., Zhang, J., and Fauchet, P. (2005) Electrical and thermal modulation of silicon photonic bandgap microcavities containing liquid crystals. *Opt. Express*, **13**, 1090.

106 Del Villar, I., Matías, I.R., and Arregui, F.J. (2003) Analysis of one-dimensional photonic band gap structures with a liquid crystal defect towards development of fiber-optic tunable wavelength filters. *Opt. Express*, **11**, 430.

107 Sireto, L., Coppola, G., Abatte, G., Righini, G.C., and Otón, J.M. (2002) Electro-optical switch and continuously tunable filter based on a Bragg grating in a planar waveguide with liquid crystal overlayer. *Opt. Eng.*, **41**, 2890.

108 Presnyakov, V., Liu, Z., and Chigrinov, V. (2005) Infiltration of photonic crystal fiber with liquid crystals. *Proc. SPIE*, **6017**, 102.

109 Poon, A., Li, C., Ma, N., Lau, S.L., Tong, D., and Chigrinov, V. (2004) Photonics filters, switches and subsystems for next-generation optical networks. *HKIE Trans.*, **11**, 60.

110 Presnyakov, V., Asatryan, K., Galstian, T., and Chigrinov, V. (2006) Optical polarization grating induced liquid crystal micro-structure using azo-dye command layer. *Opt. Express*, **14**, 10558.

111 Oh, C. and Escuti, M.J. (2007) Achromatic diffraction using reactive mesogen polarization gratings. SID'07 Digest, p. 1401.

7
Liquid Crystal Sensors

The main aim of this chapter is to demonstrate a principal possibility of elaboration of liquid crystal sensors of mechanical perturbations based on the results of basic investigations of linear and nonlinear phenomena in shear flows of liquid crystals (Chapter 3).

This chapter is organized as follows. First, we will define the physical background of sensor applications of liquid crystals and discuss the main parameters of mechanooptical effects in Poiseuille flows of liquid crystals, important for practice and optimization of these parameters. Then, we will concentrate on some types of liquid crystal sensors proposed for sensing of steady and low-frequency mechanical forces and motion. In particular, we will consider the sensors of the pressure difference, acceleration, vibration, and inclination. We will consider the possible construction of such devices and estimate their technical parameters. The use of liquid crystal sensors for the control of liquid and gas flows will be discussed too. Finally, we will describe some experimental results of acoustooptical effects on liquid crystals at additional action of electric and magnetic fields. The latter can be important for the elaboration of electrically controlled liquid crystal sensors, which can be used in detection and visualization of high-frequency (ultrasonic) vibrations.

7.1
Liquid Crystals as Sensors of Mechanical Perturbations: Physical Background and Main Characteristics

Liquid crystals can be considered a very promising material for sensor applications. It is determined by high sensitivity of liquid crystal media to the action of different factors of physical (or chemical) nature. In general, a sensor is often defined as a device that receives and responds to a signal or stimulus [1]. Stimulus is the quantity, property, or condition that is sensed and converted into definite (mostly electrical) signal [1–3]. Sensing and visualization of temperature fields via cholesteric liquid crystals can be considered as the most traditional sensor application of liquid crystals. It is based on a strong sensitivity of the selective reflection spectra to temperature variations that makes it possible to visualize temperature fields [4]. Such possibility is traditionally used in biomedical applications of liquid crystals [5, 6] for the diagnostics

Liquid Crystals: Viscous and Elastic Properties
S. V. Pasechnik, V. G. Chigrinov, and D. V. Shmeliova

ISBN: 978-3-527-40720-0

of different kinds of pathologies. It was also applied for registration, visualization, and measurement of air flows that modify the temperature fields [7–9]. Cholesterics and twisted nematics were also proposed for the fiber optic pressure sensors [10] operating at high pressures (up to 100 MPa). In addition, more exotic LC pressure sensors such as those based on freely suspended ferroelectric smectic films [11] or lyotropic liquid crystals [12] can be found in literature. We also can mention the use of LC as chemical and biological sensors that are attracting a growing interest [13–16].

In this chapter, we will describe the physical background and particular technical decisions that make it possible to propose liquid crystal sensors of mechanical forces, position, and motion based on the physical phenomena in flows of liquid crystals (see Chapter 3). Such devices have to be referred to as optical sensors as they initially convert the mechanical stimulus into an optical response. A lot of optical sensors of pressure, strain, flow, vibration, and so on based on different physical phenomena are used in modern industry [1, 2]. At present, interest in the elaboration of new types of optical sensors (e.g., fiber optic sensors [3]) is growing.

In liquid crystal sensors under consideration, the flow-induced optical effects play a key role. It is worthwhile to note that flow effects such as backflow phenomena are also important for display applications (Chapter 6).

In accordance with the global aim of this chapter, we will avoid to describe in detail other types of above-mentioned liquid crystal sensors based on different physical effects.

Results of investigations of flow phenomena in nematic liquid crystals (Chapter 3) show that such materials can be considered very interesting from the point of view of sensors applications. Indeed, there are a number of phenomena resulting in linear and nonlinear deformations of orientational structure that can be well registered by optical methods. It opens new ways for practical applications of liquid crystals as optical sensors of different types of mechanical perturbations.

For practical realization of such devices, it is important to find answers to some main questions.

- What particular mechanooptical effect provides potential advantages of the sensors?
- What are the there ways for optimization of technical parameters of the sensors?
- What is the best technical decision for a particular application of the sensors?

We will try to answer these questions for liquid crystal sensors based on mechanooptical effects in Poiseuille flows of nematics. In such type of sensors, the flow of an anisotropic liquid inside a channel induces changes in optical characteristics. It is more useful for practical applications compared to the relative motion of substrates producing Couette flow.

As it was mentioned above, the main peculiarity of nematic liquid crystal interesting for sensor applications is the ability to change its orientation (and so optical properties) under the action of extremely weak flows induced by a pressure gradient. So, the main potential advantage of LC sensors is connected with high sensitivity. Results presented in Chapter 3 have shown that flow-induced changes in birefringence at a homeotropic boundary orientation demonstrate highest sensitivity to the pressure gradient in comparison to other mechanooptical effects (e.g., flow-induced light scattering). So, the changes in optical parameters such as the optical phase delay and the polarized light intensity in the linear regime of a director motion is of prime interest for the elaboration of highly sensitive liquid crystal sensors. It is important to note that such changes can be quantitatively described in the framework of the simple linear model considered in Chapter 3 at least for a steady and low-frequency Poiseuille flow. So, construction of this type of LC sensor has to include LC cell with the initially homeotropic layer of a nematic liquid crystal placed between two polarizers. The main specific feature of the transfer function describing transformation of the input pressure gradient G into output electric signal U from opto-electric transducer is determined by a nonmonotonous dependence of a light intensity I on the flow-induced phase δ delay (as it was shown in Chapter 3, $I = I_0 \sin^2(\delta/2)$). The latter parameter is proportional to the squared pressure gradient at the linear regime of a director motion in the flow plane (Chapter 3). So, the pressure gradient can be easily determined by measuring the light intensity I at low values of the phase delay ($\delta < 0$). Contrary to it, such determination becomes more complicated in the case of relatively high value of the phase delay ($\delta > \pi$) as different values of a pressure gradient can correspond to the same value of the light intensity (Figure 7.1).

So, different technical decisions and algorithms of optical signal processing are needed for the two cases mentioned above.

(a) $\delta < \pi$

In the simplest case, it is enough to place LC cell of a constant gap with the initial homeotropic orientation between two polarizers, oriented at 45° relative to the flow direction to achieve the maximal sensitivity to the pressure gradient acting on a liquid crystal. The measurement of flow-induced changes in monochromatic light intensity

$$\Delta I = I - I_d$$

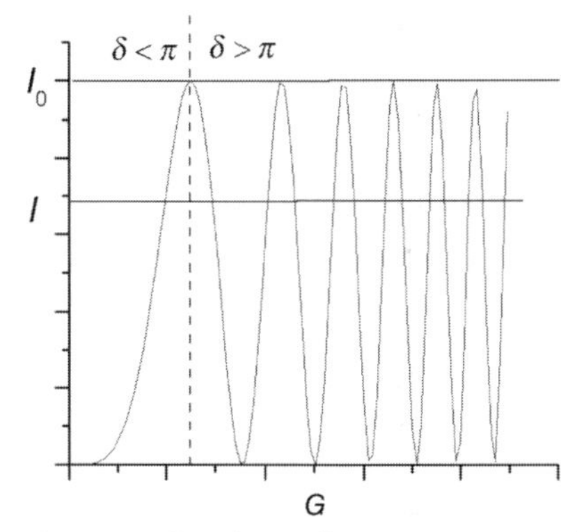

Figure 7.1 The schema illustrating the principle of the inverse transformation of the polarized light intensity I into the pressure gradient G for low ($<\pi$) and high ($>\pi$) values of the optical phase delay δ.

is needed to extract the information about a pressure gradient. The parameter I_d in expression (7.1), corresponding to the dark state of the system, arises due to the declination of real optical characteristics of a light beam, polarizers, and a layer orientation from the ideal ones ($I_d = 0$). For example, thermal fluctuations of a director in thick layers can be responsible for light scattering and average declination of a director from the uniaxial crystalline structure.

As it was shown in Chapters 3 and 5 for an ideal case ($I_d = 0$), the dependence of the optical phase delay on the pressure gradient can be expressed in the same manner for both steady and oscillating flows:

$$\delta = AG^2. \tag{7.1}$$

In the case of a steady or a quasistationary (decay) flow in the absence of fields, the parameter A is easily expressed from Equation 5.96:

$$A = \frac{1}{15120} \frac{\pi d}{\lambda} \frac{n_o (n_e^2 - n_o^2)}{2n_e^2} \left(\frac{\alpha_2 d^3}{K_{33} \eta_1} \right)^2. \tag{7.2}$$

In the general case of oscillatory flow of frequency ω in the presence of electric field, the parameter $A = A(\omega, \omega_0, \omega_E)$ depends on three characteristic frequencies (ω, ω_0, ω_E) and can also be calculated by using Equations 3.34–3.41. Taking into account the above-mentioned connection between light intensity I and phase delay δ, it is simple to derive the next general expression for reverse transformation $G = G(I)$:

$$G = \sqrt{\frac{2}{A} \arcsin \sqrt{\frac{I}{I_0}}}, \tag{7.3}$$

where I_0 is input light intensity. In practice, the intensity of light registered by a photoreceiver (e.g., photodiode) is decreased by absorption of light into polarizers and reflection of light from a number of boundaries with different refraction indexes. To exclude these factors, the ratio I/I_0 has to be determined by measuring the output light intensities in crossed (I) and parallel (I_0) polarizers as it was done in experiments described in Chapter 3.

It was shown in Chapter 3 that hydrodynamic theory correctly describes the optical response of a homeotropic layer of LC on the applied pressure gradient. It opens a good possibility to predict the technical parameters of LC sensors. In particular, it helps calculate the lower (G_{min}) and the upper (G_{max}) limits of the dynamic diapason of LC sensor like functions of a number of parameters (the thickness of LC layer, the electric field intensity, the frequency of oscillations, and so on). Contrary to the case of destabilizing action of electric field (Fréedericksz transition in homeotropic layer of LC with a negative sign of $\Delta\varepsilon$), there is no physical threshold for a pressure gradient that induces the distortion of a homeotropic layer. So, the minimal value of a pressure gradient G_{min} is defined mostly by technical parameters of light sources and receivers. The value of I/I_0 has to be small but reasonable to estimate G_{min}. In the calculations presented below, we will use the value of this ratio equal to 10^{-2}.

The upper limit G_{max} corresponding to the first maximum on the oscillating curve (Figure 7.1) is defined by the condition $I = I_0$. So, the expressions for parameters G_{min} and G_{max} can be written as

$$G_{min} = \sqrt{\frac{0.2}{A}} \tag{7.4a}$$

$$G_{max} = \sqrt{\frac{\pi}{A}}. \tag{7.4b}$$

It is worthwhile to note that the ratio G_{max}/G_{min} of two parameters defined in such a way does not depend on the value of the parameter A and is equal to about 4. This value corresponds to the rather narrow dynamic range of the LC-sensor realizing the simplest technical decision and algorithm of a signal processing. It restricts the use of such type of sensors by cases where the high threshold sensitivity is more important than a dynamic diapason. Now, we will consider the ways of optimization of this important parameter.

(b) $\delta > \pi$

One of the possible ways to increase the dynamic diapason of LC-sensors is to use measurements in the region of a nonmonotonous dependence of the light intensity on the phase delay and therefore on the applied pressure gradient, as shown in Figure 7.1. Indeed, the total number N_t of the interference maxima arising at the complete reorientation of the layer from the homeotropic state to the planar one is described by expression

$$N_t = (d\,\Delta n)/\lambda. \tag{7.5}$$

For relatively thick LC layers ($d = 100$–$200\,\mu m$) and a typical value of the refractive index anisotropy $\Delta n \sim 0.2$, the parameter N_t will be varied in the range 30–60 at $\lambda = 0.63\,\mu m$ (red light). A major part of this number will be realized at relatively small values of the angles θ of a director rotation from the initial homeotropic orientation when the linear approximation is valid. It means, in accordance with Equation 7.2, that the upper limit of the dynamic range G_{max} of the sensor is increased approximately $N_t^{1/2}$ times compared to that defined by Equation 7.4 corresponding to the first maximum ($N = 1$). So, the dynamic range in the case (b) can be extended to approximately 5–7 times compared to a rather narrow range realized in the case (a). Though such possibility seems to be attractive, its practical realization is not so simple as in the case (a).

First, one has to find ways to identify the number of the maximum, corresponding to the current state of the layer. The use of the cell of a variable gap (like the one described in Chapter 3) together with some optical channels can be considered as a possible decision for such a problem. The more complicated algorithm of the data processing will also be needed in this case.

Second, there are a number of hydrodynamic instabilities arising in a homeotropic layer at corresponding threshold gradients (see Chapter 3). In particular, the first homogeneous instability is connected with an escape of a director from the flow plane. The dynamics of such instability is more complicated and slower compared to the simple linear dynamics observed for a director motion in the flow plane. Upon increasing further the pressure gradients, extremely slow quasimemory effects take place (Chapter 3). Such phenomena are quite undesirable from the point of view of sensor application. For steady and infrasound frequency flows, the situation can be partly improved by the application of stabilizing electric field. In conclusion, we can say that in practice, case (b) can be realized first for the pressure gradient varying with a well-defined low frequency. Some of such examples will be considered below.

7.2 Technical Parameters of Liquid Crystal Sensors of the Pressure Gradient: Ways of Optimization

The problem of optimization of LC sensors can be solved in different ways. In particular, it can be done by using the cells of a variable gap, as low-frequency optical response of a homeotropic layer critically depends on this parameter (see Chapter 3). Then, the use of stabilizing electric fields makes possible an effective control of modes and technical characteristics of sensors. As we will show below the combination of these two ways is very promising for sensor application. At last, the realization of a particular optical geometry can be effective too. We will begin a description of the optimization problem with this topic.

7.2.1 Decreasing Threshold Pressure Gradient via Choice of Optimal Geometry

One of the simplest ways to increase the threshold sensitivity of LC cell to the action of a pressure gradient is related to the use of the tilted optical geometry instead of the ordinary orthogonal geometry. In principle, it can be done by using initial tilted orientation of a director. The latter is realized by oblique evaporation of SiO on a glass surface or by using hybrid, homeotropic–planar boundary conditions (see Chapter 3). An alternative way to choose the optimal optical geometry for a homeotropic layer that was proposed and experimentally confirmed [17–19] for a case of low-frequency pressure gradient acted on the homeotropic layer of a liquid crystal.

The geometry of the experiment is shown in Figure 7.2.

The homeotropic layer of the nematic mixture (ZhK654, NIOPiK production) with a positive value of $\Delta\varepsilon$ was exited by a low-frequency ($f = 0.3$ Hz) pressure difference applied along a 4 cm-long channel. The cell was rotated on the angle Y relative to the initial orthogonal state as shown in Figure 7.2.

Strong variations in the dynamic optical response were registered at relatively small changes in angle Y (Figure 7.3). The dependence of an amplitude value J_a of the optical response on angle Y is shown in Figure 7.4.

It is seen that the cell demonstrates the maximal sensitivity at $Y = \pm 17°$. In this case, the amplitude J_a is approximately proportional to the pressure difference

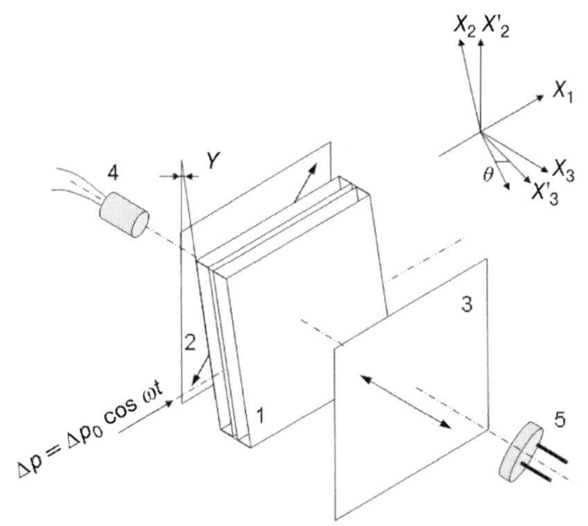

Figure 7.2 Geometry of experiment: (1) LC cell; (2) polarizer; (3) analyzer; (4) light emitting diode; (5) photodiode.

amplitude, and very small values of a pressure gradient (about 2.5 Pa/m) can be detected (Figure 7.5).

The observed effect also remains when an electric voltage is applied to the layer. It provides good opportunity to control the sensitivity of LC cell (Figure 7.6). This way of optimization will be discussed in more detail below.

Qualitatively, the same result was obtained [20] for another nonorthogonal geometry (the cell was rotated on angle θ around x_2-axis). However, in the latter case, increase in sensitivity was not such pronounced.

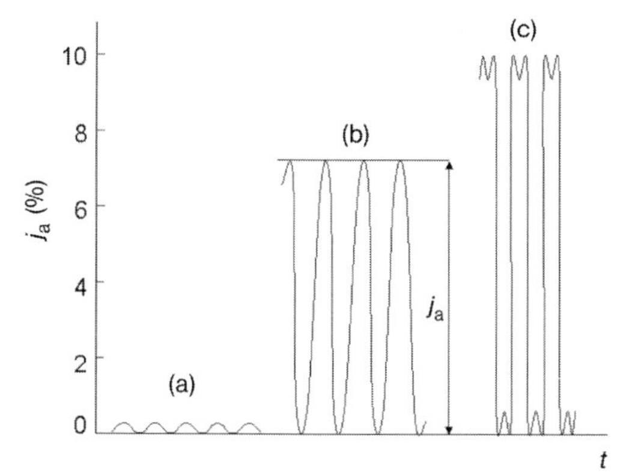

Figure 7.3 The pressure-induced variations of transmitted light intensity $J = \Delta I/I_0$ as a function of time. I_0 is the transmitted intensity in the absence of a pressure gradient ($\Delta P_0 = 0$). (a) $\Delta P_0 = 1.0$ Pa, $Y = 0$; (b) $\Delta P_0 = 1.0$ Pa, $Y = 17°$; (c) $\Delta P_0 = 10$ Pa, $Y = 17°$.

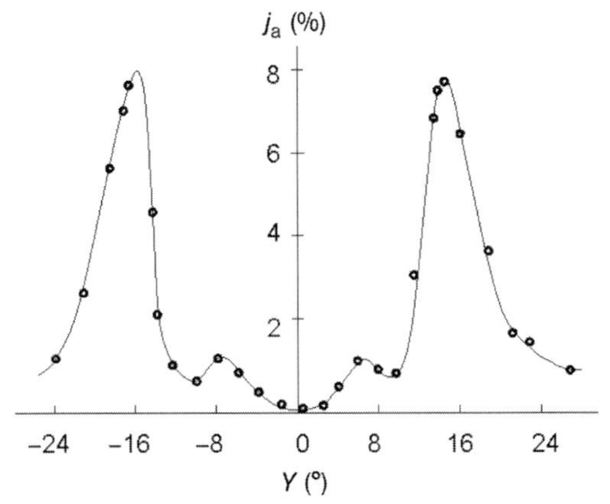

Figure 7.4 Variation of the amplitude J_a of optical response with the angle Y; $\Delta P_0 = 1.0$ Pa.

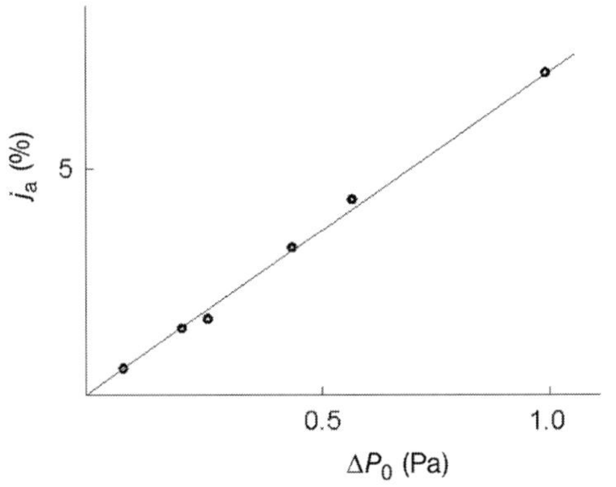

Figure 7.5 Dependence of the amplitude J_a of the optical response on the pressure difference amplitude ΔP_0; $Y = 17°$.

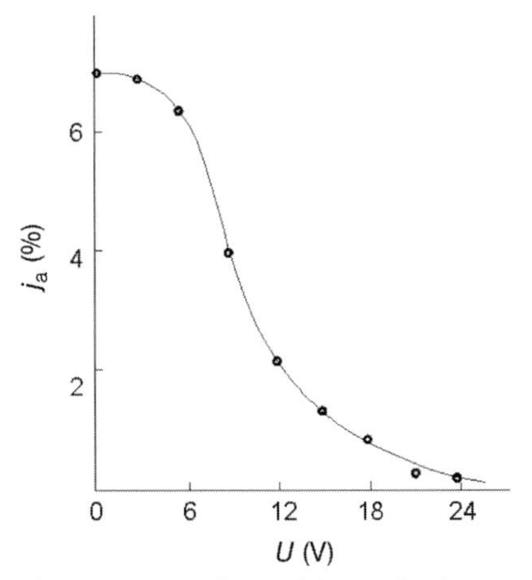

Figure 7.6 Dependence of the amplitude J_a on the applied voltage; $Y = 17°$, $\Delta P_0 = 1.0$ Pa.

7.2.2
Use of Electric Fields

Electric fields can be considered an optimal tool for the control of technical parameters of liquid crystal sensors. In Chapter 3, it was shown that a linear response of a homeotropic layer of MBBA on the applied pressure gradient depends on the applied voltage. Nevertheless, the possibility of an electric control of this response in the case of LC with a negative sign of a dielectric permittivity anisotropy $\Delta\varepsilon$ (such as MBBA) is restricted by the emergence of electrically induced instability (Fréedericksz transition) at relatively low voltages (about 4 V for MBBA). Of course, there are some nonlinear effects connected with the combined action of destabilizing electric fields and shear flows considered in Chapter 7 that can be of interest for sensor application. In particular, the decreasing (down to zero) of the threshold pressure gradient corresponding to the homogeneous instability (see Chapter 3) makes it possible to propose high-sensitive and electrically controlled LC sensors. The same is also true for the effect of the ordering of electrically induced convective rolls by weak Poiseuille flows, also described in Chapter 3. The optical phenomena resulted via such effects are very pronounced, which makes them attractive for practical use. Nevertheless, it is difficult to receive fast operating times in these cases due to slowing down of an orientational motion at the structural transformations mentioned above. So, the probable use of such devices is possible only for detecting slow varying pressure gradients.

The quite different situation is realized for liquid crystals with a positive sign of $\Delta\varepsilon$. In this case, it appears possible both to realize a linear mode of a director motion for a wide-enough dynamic range of registered pressure gradient and to improve the inertial properties of LC sensors.

The detailed experimental research on the optical response of LC layer to the pressure gradient described in Chapter 3 was conducted for a standard nematic – MBBA. It provided to draw the key conclusion about correspondence of the linear theory and experimental results. Nevertheless, for practical realization of liquid crystal sensors, stable nematic mixtures with a wide temperature range are desirable. Though there are a lot of liquid crystal materials of such type, only some of them are studied in detail to provide reasonable theoretical estimations. In the experiments described below [21–26], two liquid crystal mixtures ZhK616 and ZhK654 were under investigation. They were produced by NIOPiK on the basis of well-studied binary nematic mixture ZhK440 [27] doped by a polar substance (cyanophenyl ether of heptylbenzoic acid [28]) of a different concentration (12 and 40% for ZhK616 and ZhK654, respectively). Thus, the two mixtures had different positive values of $\Delta\varepsilon$ (3.4 for ZhK616 and 10.5 for ZhK654) that provided the stabilizing action of an electric field on the initial homeotropic orientation used in these experiments. All measurements were carried out in a dielectric regime of electric field influence to avoid possible EHD instabilities.

Two types of sandwich-like liquid crystal cells with open edges were used in the described experiments. In the cell of the first type [21, 25, 26] shown in Figure 7.7, the rectangular channel of a constant thickness ($h = 105\ \mu m$) was formed and one of

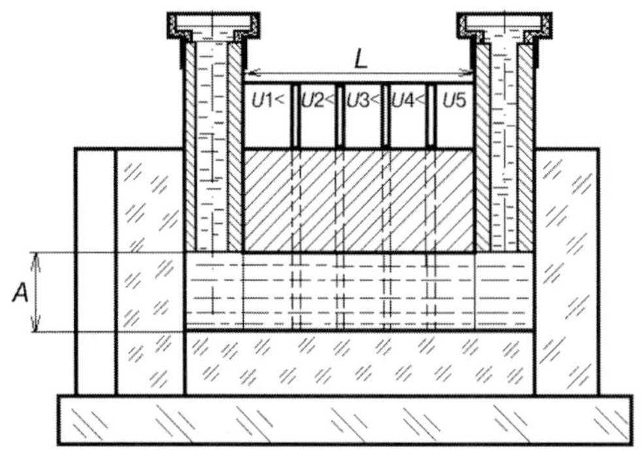

Figure 7.7 Liquid crystal cell with stripped electrodes; different voltages U_i applied to stripped electrodes provide variation of the cell sensitivity. (After Pasechnik and Torchinskaya [21].)

the inner electrodes was subdivided into different electrically isolated stripped electrodes. This made it possible to apply different electric voltages to different parts of the cell to induce a step-like nonhomogeneity in the direction of the flow. The cell of a variable gap (30–210 μm) described in Chapter 3 was also used as the second type of cells to study the optical response at different voltages and variable layer thicknesses. It was characterized by nonhomogeneity in the direction normal to the flow. This cell can be approximated by a number of channels of a different thickness acted upon by the same pressure gradient $G(t) = \Delta P(t)/L$, where L is the length of the channel (this parameter was equal to 2.5 and 1.0 cm for the cells of the first and of the second type, respectively). Both cells can be considered real prototypes of sensitive elements of liquid crystal sensors with a number of optical channels of different sensitivity controlled via electric field (the first type) or via thickness in combination with electric field (the second type).

The setups used in experiments were essentially the same as that described in Chapter 3. They provided registration of changes of intensity of polarized light induced by a low-frequency pressure gradient $G = G_0 \sin(2\pi f t)$. In some experiments with the cell of the first type, the optical response was simultaneously recorded from different optical channels, corresponding to the different stripes. So, it was possible to compare the phase shift between optical signals obtained at different voltages.

As far as the experiments described in Chapter 3 are concerned, the applied pressure gradient G induces the oscillating Poiseuille flow with a quasiparabolic velocity profile and distortions of an initial homeotropic structure described by the angle $\theta(z, t)$. Electric voltage applied to the cell acts as a factor stabilizing the initial orientation.

The typical time dependence of the intensity of a polarized light passing through the LC cell of the second type under the action of low-frequency pressure variations $\Delta P(t)$ in the presence of electric field is shown in Figure 7.8.

The nonlinear character of $I(t)$ dependencies is connected with an interference between an extraordinary ray and ordinary one and can be excluded [25, 26] by transforming $I(t)$ into $\delta(t)$ dependencies as it was described in Chapter 3. So, it is

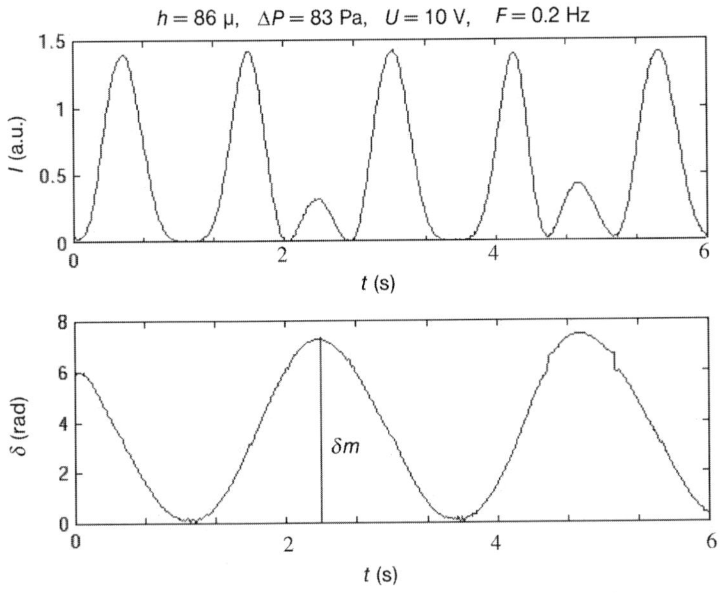

Figure 7.8 Time dependencies of light intensity $I(t)$ and phase difference $\delta(t)$; ZhK616, LC cell of the second type. (After Pasechnik *et al.* [25].)

reasonable to use $\delta(t)$ dependencies to compare experimental results with linear theory predictions. An example of such transformation is presented in Figure 7.8. Contrary to the $I(t)$ dependence, the time dependence of the phase difference $\delta(t)$ can be described by a simple harmonic like a law, which reflects the linear regime of motion of a director. Indeed, our experiments have shown that the general result of a linear theory, namely, $\delta_m \sim (\Delta P)^2$ (see Equations 7.1 and 7.2), is in an accordance with experimental dependencies obtained at variations of a pressure amplitude, frequency, and layer thickness both in the absence and in the presence of electric field.

Theoretical analysis performed in the framework of a linear hydrodynamic model (Chapter 3) has shown that the electric field application to the LC layer with $\Delta\varepsilon > 0$ leads to two main consequences. First, it decreases the amplitude of the director distortion, induced by a pressure gradient. Second, it effectively suppresses the phase shift between the orientation and pressure oscillations, so a quasistationary regime of motion can be realized [23, 26]. One can see it in Figure 7.9 by comparing the calculated profiles of "in-phase" (θ_r) and "antiphase" (θ_i) parts of function θ (z, t) describing a dynamic linear response of the LC layer (see Equations 3.35 to 3.37). The similar effects take place under local thickness variations [23].

The case shown in Figure 7.9b ($\theta_i \ll \theta_r$) corresponds to the quasistationary regime of the director motion. One can use the inequality $m\omega \ll \omega_E$ (the parameters m and ω_E are defined in Section 3.4.3.2) to estimate the electric voltage needed to realize the quasistationary regime at the given frequency of oscillation ω. An example of such an estimate is given in Figure 7.10, where the frequency ω_E of a director relaxation in a strong electric field is shown as a function of electric voltage U. The quasistationary

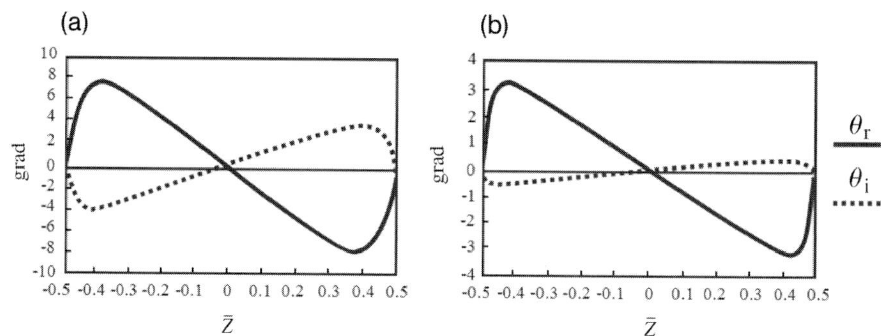

Figure 7.9 Profiles of "in-phase" (θ_r – solid lines) and "antiphase" (θ_i – dotted lines) parts of θ (z, t) function. ZhK654, $\Delta P_0 = 33$ Pa, $f = 0.286$ Hz; (a) $U = 6.3$ V, (b) $U = 12.6$ V; $\bar{\mathbf{z}} = \mathbf{z}/\mathbf{h}$.

regime will be achieved at $U \gg U'$. The boundary voltage U' increases with the frequency of oscillations as shown in Figure 7.10.

As it was shown in [21, 25] in the case of a quasistationary linear oscillating flow, the dependence of the phase difference δ on electric voltage can be described by the universal function $F(\mathrm{pl})$:

$$\delta \sim \langle \theta^2 \rangle \sim F(\mathrm{pl}), \tag{7.6}$$

where $\langle \theta^2 \rangle$ the average value of $\theta^2(z, t)$ and the universal function $F(\mathrm{pl})$ for the case $\Delta\varepsilon > 0$ can be written as

$$F(\mathrm{pl}) = \frac{1}{(\mathrm{pl}^6)} \left\{ \frac{\mathrm{pl}}{4\mathrm{sh}^2(\mathrm{pl})} [\mathrm{sh}2\mathrm{pl} - 2\mathrm{pl}] + 2\left[1 - \mathrm{pl}\frac{\mathrm{ch}(\mathrm{pl})}{\mathrm{sh}(\mathrm{pl})}\right] + \frac{(\mathrm{pl})^2}{2} \right\}, \tag{7.7}$$

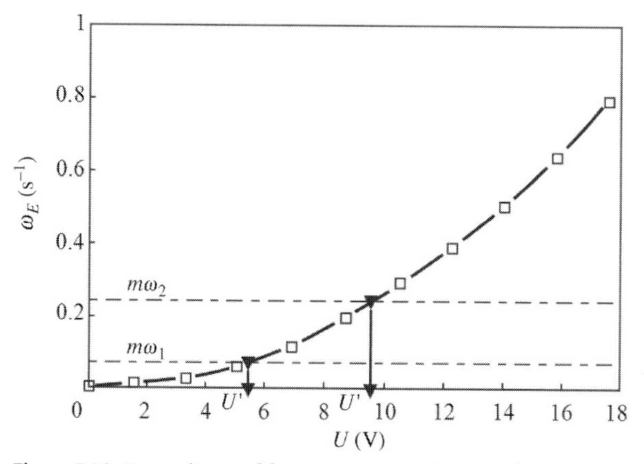

Figure 7.10 Dependence of frequency ω_E on electric voltage U. The boundary voltage U' of the quasistationary regime is determined by intersection of this dependence and the straight line corresponding to the given value $m\omega$. ZhK654 ($f_1 = 0.092$ Hz s, $f_2 = 0.286$ Hz).

where

$$\text{pl} = \frac{U}{2}\sqrt{\frac{\varepsilon_0\,\Delta\varepsilon}{k_{33}}} = \frac{\pi}{2}\left(\frac{U}{U_F}\right), l = h/2.$$

From the view point of possible practical applications of LC cells as sensors of mechanical vibrations controlled by electric field, it is important to note that the function $F(\text{pl})$ does not depend on viscosity coefficients. It gives the opportunity to calculate the degree of field influence on the mechanooptic response of LC layer in a very simple way (one has to know only the $k_{33}/\Delta\varepsilon$ value for such calculation). An analysis of experimental results obtained has shown that this universal behavior really takes place for low-frequency oscillations. The comparison between the theory predictions and the experimental results, obtained for ZhK616 (LC cell of the second type) and ZhK654 (LC cell of the first type) at varying experimental parameters is shown in Figures 7.11 and 7.12. One can see that $\delta_m(U)$ dependencies are satisfactorily described by the universal function mentioned above.

The existence of a quasistationary regime of a director motion plays an essential role in the application of LC cells for determining instant values of time-dependent pressure gradients $G(t)$ by processing the dynamic optical response $I(t)$ in the framework of simple equations such as (7.3).

Experimentally, such possibility was confirmed by the results of investigations of mechanooptical effects in the layer of ZhK654 stabilized by electric field [25, 26]. An example of the dynamic optical phase delay δ extracted from $I(t)$ dependence as a function of the current pressure difference $\Delta P(t)$ applied to the stripped cell and simultaneously registered in experiments by the independent pressure sensor is shown in Figure 7.13. In these experiments, the optical response $I(t)$ was obtained from the third segment of the cell stabilized by voltage U_3. The existence of simple dependence $\delta(t) \sim \Delta P^2(t)$ makes it possible to fulfill the inverse transformation

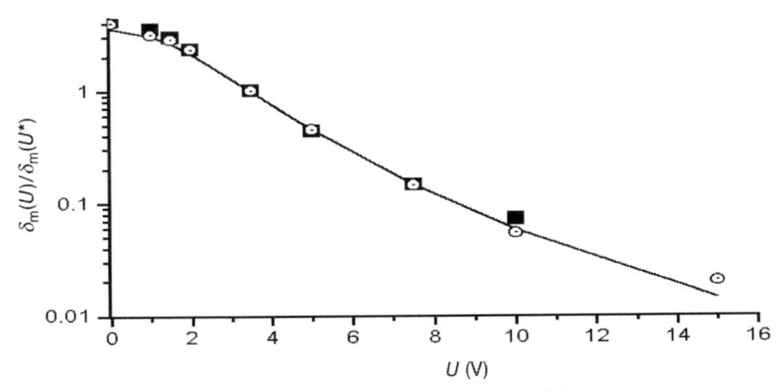

Figure 7.11 Universal dependence of the normalized phase delay $\delta_m(U)/\delta_m$ ($U^* = 3.5$ V) on voltage for ZhK616: (⊙) $G = 2230$ Pa/m; (■) $G = 1170$ Pa/m; $f = 0.04$ Hz.; (solid line) theoretical dependence. (After Pasechnik *et al.* [25].)

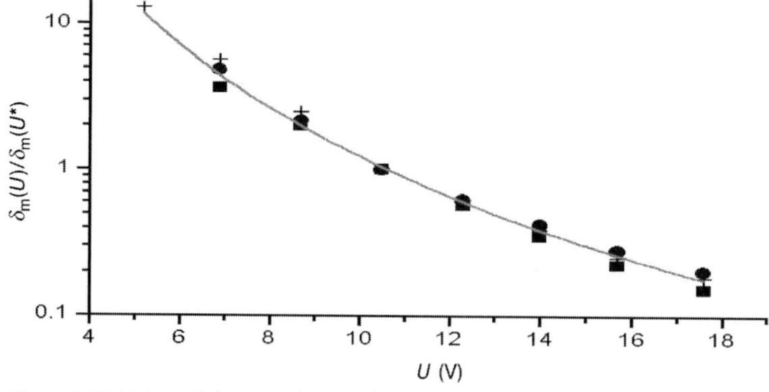

Figure 7.12 Universal theoretical dependence and experimental values of $\delta_m(U)/\delta_m(U^* = 10.5\,\text{V})$ (for ZhK654: (●) $G = 2000\,\text{Pa/m}$, $f = 0.29\,\text{Hz}$; (■) $G = 3200\,\text{Pa/m}$, $f = 0.29\,\text{Hz}$; (+) $G = 3200\,\text{Pa/m}$, $f = 0.092\,\text{Hz}$. (After Pasechnik *et al.* [25].)

$\Delta G(t) = \Delta P(t)/L = F[\delta(t)]$. Such transformation does not depend on frequency in the range corresponding to the quasistationary regime.

Breaking off a quasistationary regime results in a hysteresis loop, shown in Figure 7.14a, which arises due to a phase shift between the optical response and the pressure oscillations. This shift can be excluded from the account as shown in

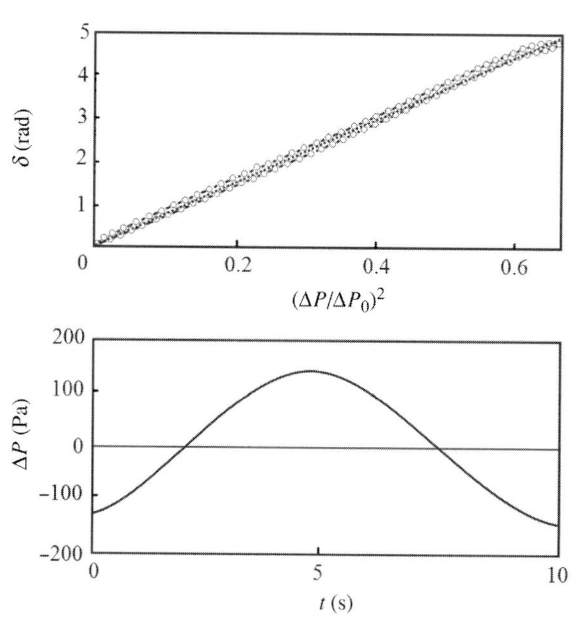

Figure 7.13 Dependence of optical phase delay $\delta(t)$ on the instant pressure difference $\Delta P(t)$ (ΔP_0 is an amplitude value of $\Delta P(t)$) (a) and time dependence of the pressure difference (b) registered by the independent pressure sensor; ZhK654, $U_3 = 15.7\,\text{V}$. (After Torchinskaya [26].)

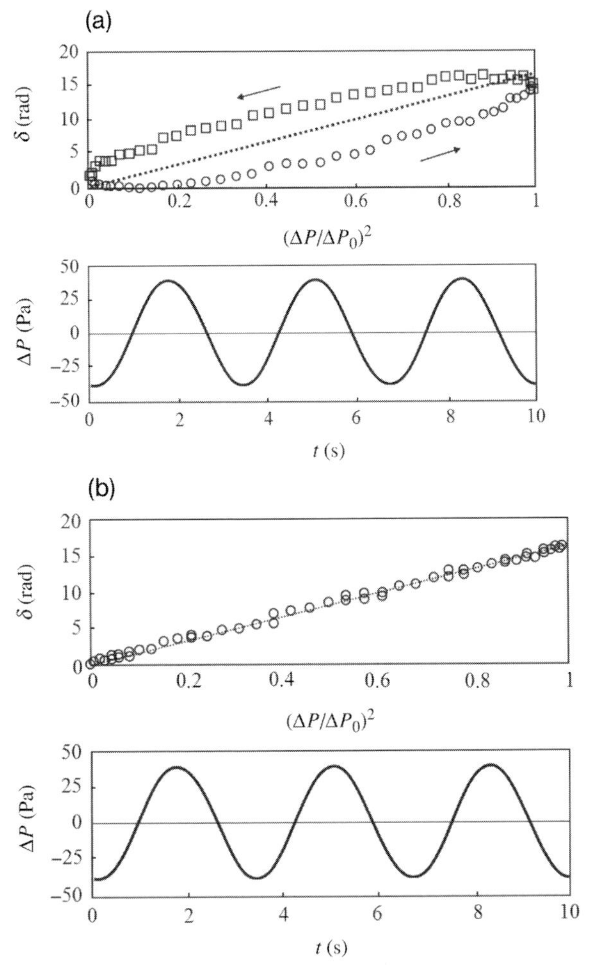

Figure 7.14 Optical phase delay $\delta(t)$ as a function of the instant pressure difference $\Delta P(t)$; ZhK654, $U_3 = 5.2$ V, $f = 0.286$ Hz; (a) results obtained via direct measurements demonstrate hysteresis at increasing and decreasing pressure difference shown by arrows; (b) the same results after phase shifting of $\delta(t)$ dependence. (After Torchinskaya [26].)

Figure 7.14b. In practice, such procedure is available for periodic processes with a well-defined frequency. On the contrary, the restoration of the instant pressure gradient is possible via experimental calibration of the cell at different frequencies and fields. Results of such calibration providing determination of the dynamic range of the pressure gradient registered via the stripped cell mentioned above are presented in Table 7.1.

One can see that using stabilizing electric field makes it possible to increase the ratio G_{max}/G_{min} to more than 50 corresponding to a rather wide dynamic range. An additional use of stabilizing electric field can also essentially improve the inertial

Table 7.1 Limits G_{min} and G_{max} of the dynamic range of pressure gradients at different frequencies and voltages calculated in accordance with Equations 7.3 and 7.4.

	$f = 0.286$ Hz		$f = 0.092$ Hz	
U (V)	G_{min} (Pa/m)	G_{max} (Pa/m)	G_{min} (Pa/m)	G_{max} (Pa/m)
1.6	101.9	404.0	52.52	208.2
3.4	130.5	517.1	93.13	369.1
5.2	173.8	688.6	144.6	573.2
6.9	234.9	930.9	218.5	866.1
8.7	321.3	1273	326.6	1294
10.5	478.1	1895	496.0	1966
12.3	586.7	2325	603.0	2390
14	816.5	3236	675.7	2678
15.7	969.0	3841	894.4	3545
17.6	1236	4897	1155	4577

Data are obtained via experimental calibration of the stripped cell filled with ZhK654.

properties of LC cell (even in the region of relatively high values of local thickness). Moreover, it is useful for the suppression of hydrodynamic instabilities that can take place in the studied geometry (see Chapter 3).The additional variation of the layer thickness seems to be the best variant for the elaboration of LC sensors of an extremely wide dynamic range (G_{max}/G_{min} is about 1000).

7.3
Liquid Crystal Sensors of Pressure, Acceleration, Vibrations, and Inclination

The optical response of liquid crystals to the applied pressure gradient and its dependence on electric field described above can be used in electrically controlled LC sensors of different types. Below we will consider basic principles of operation and possible constructions of such devices. Estimates of some important characteristics of the sensors will also be presented.

7.3.1
Liquid Crystal Sensors of Differential Pressure

Sensors of differential pressure based on different physical phenomena are extremely useful in various branches of modern industry [1–3]. The main peculiarities of LC sensors [18, 29, 30], which make them differ from analogous devices, are a high threshold sensitivity and electrically controlled characteristics. It is obvious that liquid crystal cells with open edges described above can be considered the simplest example of differential pressure sensors. In particular, they are proposed to register the air pressure difference applied directly to free surfaces of liquid crystals in open tubes connected to the channel. The optical response of such sensors is determined by a number of geometrical parameters (such as the size of channel and open tubes) and the applied voltage U. These parameters can be varied to optimize the technical characteristics of the sensor depending on a particular application. For example, in accordance with results presented in (7.2), one way to achieve the maximal threshold sensitivity of the sensor is to increase the gap (h) of the channel and (or) decrease its length (L). The increase in h is restricted by breaking the initial single domain homeotropic structure of LC layer. In practice, a perfect homeotropic orientation is realized in layers with a thickness smaller than 200–300 μm.

The decrease in the channel length is restricted by two factors. First, there are some transitional regions of the length ζ_t with a nonstabilized velocity profile. For a channel of a rectangular cross section, this length is expressed as [31]

$$\zeta_t \approx 4h' \cdot Re, \tag{7.8}$$

where h' – hydrodynamic radius-ratio of flow section square to perimeter, for rectangular channel = h/2; Re – the Reynolds number is expressed as

$$Re = \frac{\langle v \rangle 4h'}{\eta_k}. \tag{7.9}$$

For the case under consideration, the average velocity $\langle v \rangle$ is proportional to the pressure gradient:

$$\langle v(t) \rangle = \frac{h'^2}{12 \cdot \eta_3} \frac{\Delta P(t)}{L}, \tag{7.10}$$

where kinematic (η_k) and dynamic (η_s) viscosities are connected in the usual manner ($\eta_s = \rho\eta_k$, ρ is a density). The estimates made in Ref. [31] for a 200 μm thick MBBA layer under the action of pressure gradient 10^4 Pa/m have shown that the transitional length ζ_t is about 0.13 mm. The channel length has to essentially exceed this value to be sure that simple expressions presented above can be applied to the differential pressure sensor.

The second reason that restricts the length of the channel is related to inertial properties of liquid crystals. Indeed, a director in each element of flowing LC has to get enough time to change it s orientation. It results in the next inequality

$$\tau_n = \omega_n^{-1} \ll \tau_v = (L/\langle v \rangle), \tag{7.11}$$

where the characteristic frequency of a director motion ω_n is defined in Section 3.4.3.2 and the time of flow through the channel τ_v can be estimated using Equations 5.78, 5.79 and 7.10. In the absence of the field and typical thickness $h = 100$ μm, the inequality (7.11) will be valid for channels longer than 1 cm at a pressure difference equal to 1 Pa. The electric field application results in decrease in characteristic time and possibly the length of the channel. It is also accompanied by an increase in the threshold sensitivity and extension of the dynamic range of registered pressure. At the same time, the increase in the pressure difference and the layer thickness results in a corresponding increase in L. Estimates have shown that the channel length of about 1 mm can be considered as the minimal reasonable value at registration of the pressure difference in the range 0.1–10 Pa even when using stabilizing electric field.

The correct estimate of the frequency range of the registered pressure difference is also of great importance from the point of view of practical applications. Such estimate can be made on the basis of simple models and calculations presented in Chapter 3. In particular, a low-frequency limit (f_{min}) of the range is defined by inequality (3.30) obtained for a wedge-like cell. It can be rewritten as

$$T \ll \frac{48\pi^2 D^2 L}{A(h_{max}+h_0)(h_{max}^2+h_0^2)}\left(\frac{\eta_s}{\rho g}\right), \tag{7.12}$$

where $T = 1/f$ is the period of pressure oscillations, h_{max} (h_0) is the maximal (minimal) layer thickness, A is the width of the channel, and D is the diameter of open tubes with the free surface of LCs.

It is obvious from (7.12) that a low-frequency limit can be easily regulated with a proper choice of the geometrical size of the channel and tubes. For example, for the size $A = 3$ cm, $L = 3$ mm, $h_{max} = 200\,\mu$m, $h_0 = 50\,\mu$m and typical values of the shear viscosity ($\eta_s = 0.1$ Pa s) and the density ($\rho = 10^3$ kg/m^3), one can get $T \ll 4500$ s and $T \ll 180$ s for $D = 1$ cm and $D = 2$ mm, respectively. So, it is possible to apply LC sensors for the registration of weak low-frequency pressure oscillations as the sensitivity of LC layer is maximal in a quasistationary regime simply realized in this case. Further decrease in the low-frequency limit can be achieved by using the relatively narrow channel of a constant gap. For example, at $A = L = 3$ mm, $D = 1$ cm, $h_{max} = h_0 = 100\,\mu$m, the maximal period of oscillation has to be lower than 10^5 s. In the last case, a sensor can operate both as the sensor of extremely low-frequency pressure difference and as the sensor of a steady pressure that maintains its characteristics for a long time. It can be of great importance for some particular applications of LC sensors as will be shown below.

The upper limit of the frequency range of LC sensors is mostly determined by the threshold sensitivity needed for a particular application. A general feature of such devices is the decrease in the latter parameter with frequency (see, for example, Table 7.1). The phase shift between the optical response and the applied pressure difference arising at high frequencies due to a break in the quasistationary regime of a director motion also prevents the use of LC sensors. A decrease in the layer thickness or application of strong electric fields needed to restore a quasistationary regime results in corresponding increase in the low limit of the registered pressure difference. So, the optimization of the sensor characteristics is possible in the given field of their application.

In many cases, a separation of a liquid crystal from the surrounding environment is needed. It can be done with the help of both limp diaphragms and elastic ones. A possible approximate construction of the electrically controlled differential pressure sensor of the latter type is shown in Figure 7.15.

This technical decision leads the above-mentioned possibilities of controlling the technical parameters of LC sensors. Namely, the layer of a nematic liquid crystal of a variable thickness is used to get a broad dynamic range of the sensor. It also makes possible the visual registration of the applied pressure difference due to the sharp

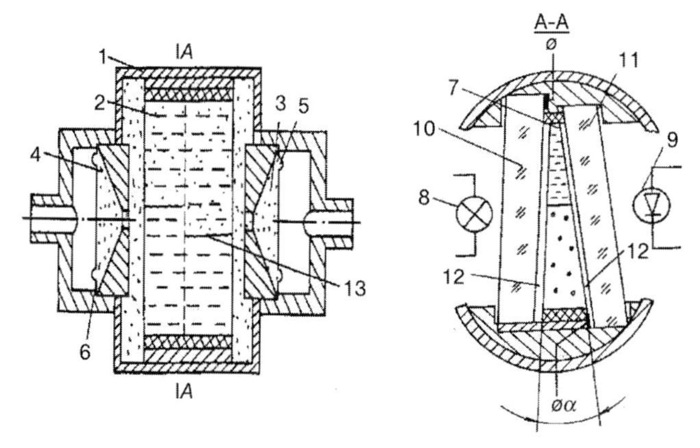

Figure 7.15 The possible construction of the liquid crystal sensor of pressure: (1) a body; (2) nematic liquid crystal; (3, 4) chambers; (5, 6) membranes; (7) a capillary; (8) a polarized light source; (9) a photodetector; (10, 11) glass plates; (12) current-conducting coating; (13) a boundary between distorted and undistorted regions of a liquid crystal.

boundary separating the undistorted region of LC layer and the region where the orientational changes are strong enough to induce variation of local optical properties (e.g., birefringenece). The additional control of the boundary position (or the local optical response) is possible via application of electric voltage. In the construction shown in Figure 7.15, different voltages can be applied to the separate electrodes into two parts of the cell. It provides a fine regulation of the optical response to the applied pressure difference.

Contrary to the case of simple liquid crystal sensors described above, technical parameters of differential pressure sensors are determined by a number of additional parameters. In particular, elasticity of the membranes separating a liquid crystal and surrounding environment has to be small enough to transmit the pressure difference to the channel filled with LCs. In experiments, the best results were obtained specially for membranes made of a chemically stable rubber. They provided transmission of a pressure difference as small as 0.1 Pa at the membrane diameter of about 1 cm. The sensitivity of the sensor can be essentially increased by using membranes with larger diameters. Nevertheless, it also leads to an increase in the sensor sizes and LC amount. Other ways of optimization are essentially the same as those described above for the simpler LC sensors.

Taking into account the estimates, we can conclude that using liquid crystals helps elaborate on extremely sensitive and electrically controlled differential pressure sensors. It is possible to achieve the minimal value of the registered pressure difference of about 0.1 Pa, which is two orders of magnitude lower than for the most industrial sensors of comparable sizes used for low-pressure measurements (see, for example, the sensor MPXV 5004, Motorolla, which provides measurements in the range 0–4 kPa with 1.5% accuracy).

7.3.2
Liquid Crystal Sensors of Acceleration, Vibrations, and Inclination

The differential pressure sensors described above are based on the optical response of LC layer to a pressure gradient arising due to the applied pressure difference. Indeed, such gradient can be produced by accelerated mechanical motion of LC cell due to inertia forces. The principle of operation and the possible construction of the linear acceleration sensor are shown in Figure 7.16.

The main peculiarity of the sensor construction in comparison with the simplest pressure gradient sensors considered above is the existence of two channels (1 and 2) of different length (L_1 and L_2) and gap (h_1 and h_2). It is done to increase the sensitivity of the sensor to the linear acceleration a that induces the flow of LC in the channels due to inertia forces. As shown in Chapter 5, the hydrodynamic resistance R of the channel is proportional to h^3L. So, with a proper choice of h_1 and h_2, the resistance of the first channel R_1 can be made essentially smaller than that of the second channel R_2 even in the case $L_2 = L_1$ shown in Figure 7.16. In this case, a pressure gradient G arising in channel 2 due to inertia forces is expressed as

$$G = a\rho\frac{L_1}{L_2}. \tag{7.13}$$

The minimal value of the acceleration a_{min} registered by the sensor is determined by a low-limit G_{min} of the pressure gradient range considered above. The estimates made in accordance with (7.13) give for $G_{min} = 1$ Pa/m and $L_1/L_2 = 10$ acceleration $a_{min} \approx 10^{-4}$ m s$^{-2} \approx 10^{-5}g$, where g is acceleration of gravity. It is much better than for the most accelerometers operating in an infrasound frequency range. For example, the piezo sensor 626 AX*4 (PSB, USA) with a mass of 0.635 kg used for monitoring of buildings provides a range of acceleration $\pm 0.5g$ at frequencies 0.03–500 Hz. It corresponds to a low limit of the range about $5 \times 10^{-3}g$ at a typical accuracy of 1%. So, a highly sensitive linear acceleration sensor can be realized on the basis of liquid crystals. Maximal linear sizes of such sensor are determined by the length L_1 equal to 3 cm at $L_2 = 3$ mm in the considered example. Of course, both electric field and channel 2 of a variable gap can be used to optimize the technical characteristics of the sensor.

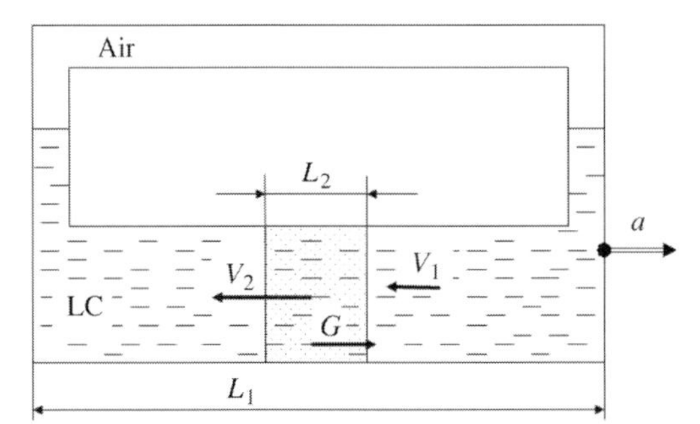

Figure 7.16 Linear acceleration sensor: principle of operation and construction.

The described sensor is also applicable for the registration of low-frequency vibration. The amplitude a_0 of the acceleration produced by periodic vibrations is connected with the amplitude of displacement X_0 ($a_0 = X_0\omega^2$). The minimal value of the registered amplitude of vibrations can be estimated as 100 μm at frequency $f = 0.15$ Hz ($\omega \approx 1$) and the parameters of the channel mentioned above. This value essentially depends on frequency and can be decreased by increasing the ratio L_1/L_2.

The technical decision illustrated in Figure 7.16 also elaborates on highly sensitive electrically controlled sensors of relatively small inclinations. In the case of small angles of inclination α, the pressure gradient G is expressed as

$$G = g\rho \frac{L_1}{L_2} \alpha. \tag{7.14}$$

Estimates made in accordance with (7.14) show extremely high sensitivity of such sensors to inclination (the minimal registered angle is of order 2″ at $G_{min} = 1$ Pa/m, and $L_1/L_2 = 10$). It is essentially better than the typical value of industrially produced inclonometers of comparable sizes (see, for example, ZEROTRONIC inclination sensors). The typical optical response of LC inclinometers and the dependence of the inclination angle on the optical phase delay are presented in Figures 7.17 to 7.19.

This dependence confirms rather high sensitivity of the LC inclinometer. At the same time, the dynamic range of registered inclination angles is rather narrow, and a number of interference maxima arise with an increase in α. So, using stabilizing electric field seems to be very useful. In particular, this makes it possible to provide a simple shape of $I(t)$ dependence as shown in Figure 7.18.

Calibration curves for the inverse dependences $\alpha(\delta)$ obtained by processing the optical response in the absence and presence of electric field are shown in Figures 7.19 and 7.20.

It is obvious that electric control of the sensor calibration curve becomes effective even when a relatively low electric voltage is applied. It affords good prospects for the elaboration of wide range LC inclination sensors.

It is worth noting that the operating time of the sensors of inclination is restricted by the characteristic time T_{fl}, which can be estimated in the same manner as a low-frequency limit of the pressure sensor expressed by inequality (7.12). Indeed, the initial pressure difference,

$$\Delta P_0 = g\rho L_1 \alpha, \tag{7.15}$$

arising at the initial inclination of the sensor will be slowly decreased due to the flow of LCs in the channel. The value of this decreasing ΔP_{fl} can be expressed as

$$\Delta P_{fl} = 2g\rho \frac{Q}{S} T_{fl}, \tag{7.16}$$

where Q is the instantaneous rate of flow, proportional to the pressure gradient (7.14). The operating time of the sensor is defined by inequality

$$\Delta P_{fl} \ll \Delta P_0. \tag{7.17}$$

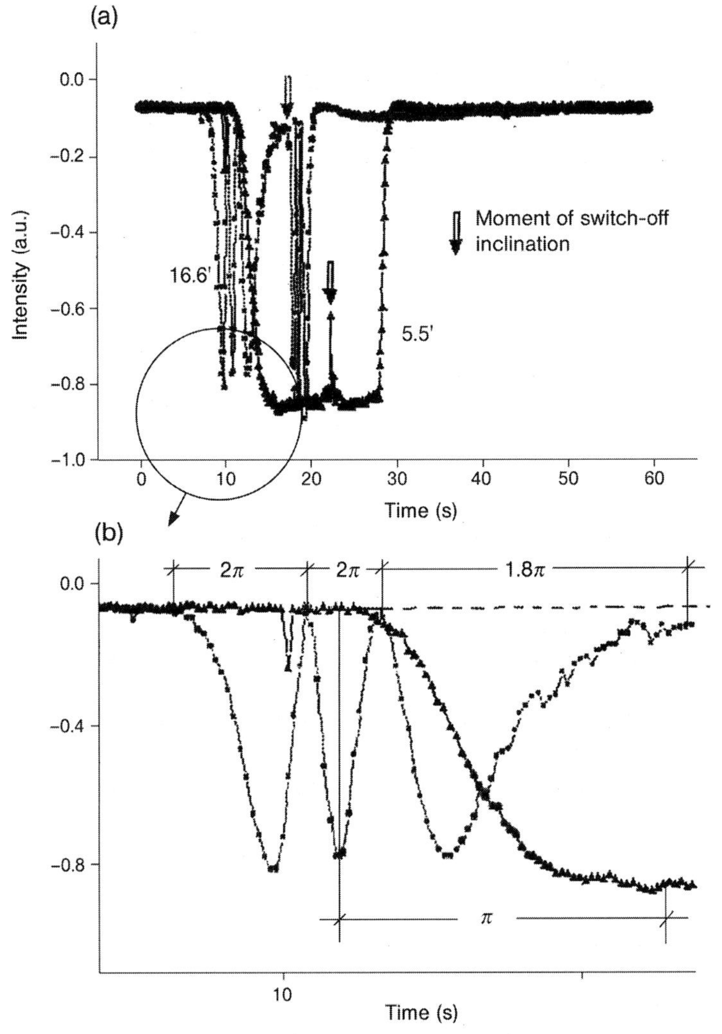

Figure 7.17 Dynamic response $I(t)$ of the inclination sensor on an abrupt increase in the inclination angle α (up to values 5.5′ and 16.6′) and the forthcoming abrupt decreasing of α to zero; the data were obtained in the absence of electric field.

By using (7.14) to (7.17) and the corresponding expression for Q presented in Chapter 3, it is simple to estimate the operating time of the sensor. In particular, for channel 2 of a constant gap h_2, we will obtain

$$T_{\mathrm{fl}} \ll \frac{6\eta_s S L_2}{\rho g h_2^3 A}. \tag{7.18}$$

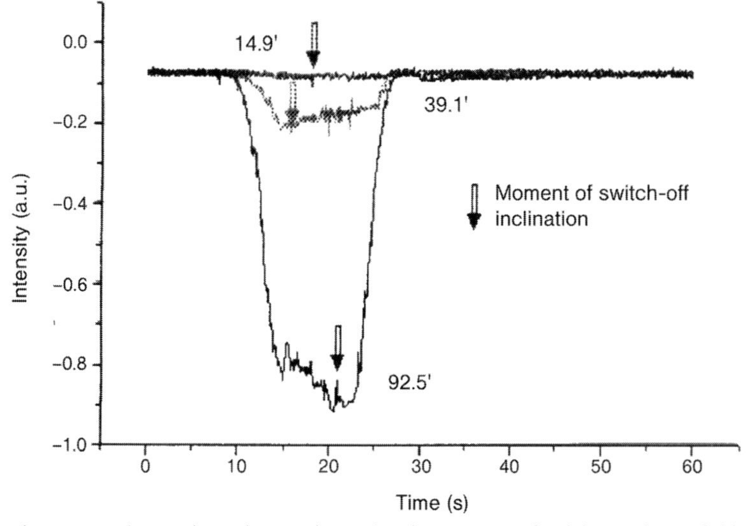

Figure 7.18 The $I(t)$ dependences obtained in the presence of stabilizing electric field ($U = 10$ V).

It results in $T_{fl} \ll 3.6 \times 10^3$ s at $h_2 = 100\,\mu$m, $A = 3$ mm, $S = h_1$, $A = 6$ mm^2, $\eta_s = 0.1$ Pa s, $\rho = 10^3$ kg/m^3. Such operation time is rather long enough for a number of possible applications.

At last, we would like to mention that the construction shown in Figure 7.16 is well suited for LC sensors of an angular acceleration ε. It is enough to totally fill the closed channel with a liquid crystal. In this case, the accelerated rotation of the sensor will induce the flow of LC relative to the channel. Contrary to the case of a linear acceleration and inclination sensors, this flow in the closed channel will not be change during rotation with a constant acceleration ε. The pressure gradient acting

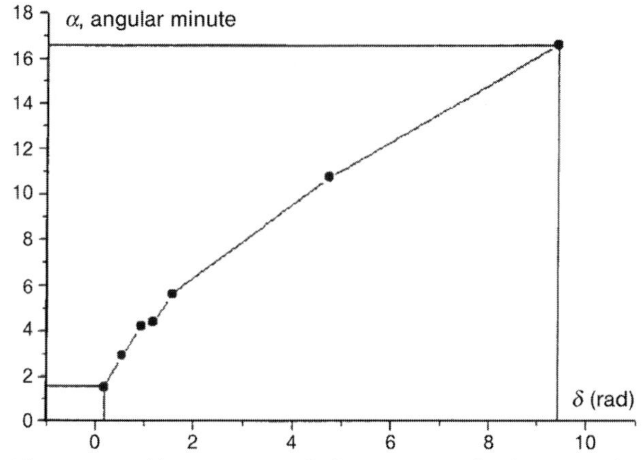

Figure 7.19 Calibration curve $\alpha(\delta)$ for LC sensor of inclination in the absence of electric field.

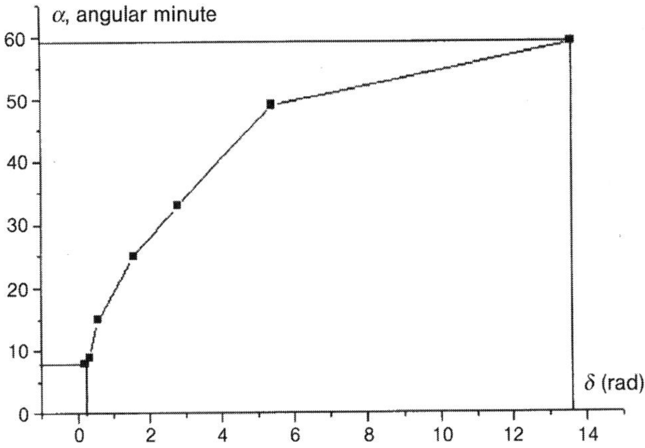

Figure 7.20 Calibration curve $\alpha(\delta)$ for LC sensor of inclination in the presence of low electric voltage ($U = 5$ V).

on channel 2 can be expressed in the simplest way for the circular closed channel 1 with a radius R:

$$G = \frac{2\pi R^2}{L_2}\rho\varepsilon. \tag{7.19}$$

The minimal value of registered angular acceleration strongly depends on R. For $R = 10$ cm and $L_2 = 3$ mm, it is of order $10^{-4}\,\text{c}^{-2}$ for the threshold sensitivity of the second channel $G_{min} \sim 1$ Pa/m.

7.4 Liquid Crystals for the Control of Liquid and Gas Flows

One of the prospective applications of liquid sensors described above is the control of liquid and gas flows. Over 40% of all liquid, gas, and steam measurements made in industry are still accomplished using common types of differential pressure flowmeters [32]. In most cases, it is needed to create an annular restriction in a pipeline with the help of orifice plate, Venturi tube, or nozzle to provide the pressure difference between the pressure upstream and pressure downstream of these elements. The increase in the velocity at the restriction causes a corresponding increase in the pressure difference. So, measurements of the latter parameter can be used to determine the instantaneous rate of flow and the overall consumption in pipelines.

One of the major advantages of the differential pressure flowmeters is that the measurement uncertainty can be predicted without the need for calibration if it is manufactured and installed in accordance with one of the international standards covering these devices. In addition, this type of differential pressure flowmeter is simple, has no moving parts, and is therefore reliable. The main disadvantages of these devices are their limited range (typically 3 : 1), the permanent pressure drop

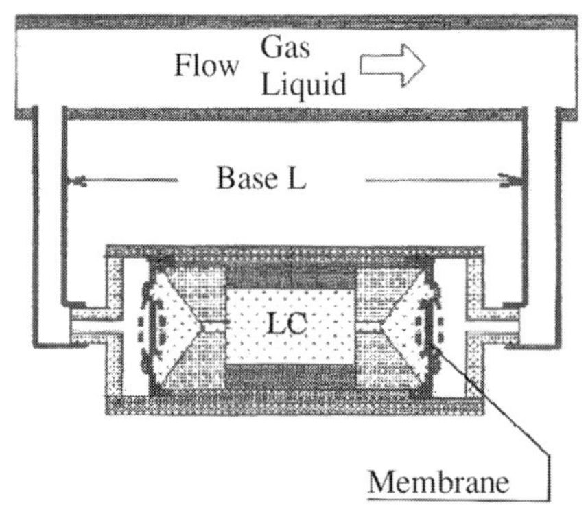

Figure 7.21 General scheme of application of LC differential sensor to control flows in a pipeline.

they produce in the pipeline (which can result in higher pumping costs), and their sensitivity to installation effects (which can be minimized using straight lengths of pipe before and after the flowmeter) [32].

Liquid crystal differential pressure sensors seem to be very promising for the control of flows due to high sensitivity. In principle, this makes it possible to avoid some disadvantages mentioned above.

A general scheme of the connection of the liquid crystal differential pressure sensor with a pipeline is shown in Figure 7.21.

The pressure difference ΔP applied to the sensor can be produced in two ways. In the case of laminar flows through the pipelines (of radius R) and of a constant cross section, the viscous losses ΔP_{vis} play a dominant role. This can be expressed by the Poiseuille formula

$$\Delta P = \frac{8\eta_s L}{\pi R^4} Q. \tag{7.20}$$

It is most important to estimate the viscous losses in flows of water as a liquid. At $R = 1$ cm, $L = 10$ cm, $\eta_s = 10^{-3}$ Pa s, we will obtain the value of ΔP_{vis} about 1 Pa for $Q = 2$ l/min. The corresponding Reynolds number Re equal to 2000 is close to the critical value $Re_c \approx 2300$ at which turbulence can arise in the flow of liquid through the pipeline with smooth walls. So, a very sensitive differential manometer has to be used to control the laminar flow of water in such pipeline. In principle, such device can be made on the basis of the liquid crystal differential pressure sensors described above. As LC pressure sensors have a limited operating time, a special decision has to be made for the control of one-directional flow of long duration. For example, it was proposed [33] to provide periodic changes in the direction of the pressure gradient via the use of special unit as shown in Figure 7.22.

The inversion of the direction of a pressure gradient is achieved with the help of a two-way valve.

As it follows from Equation 7.20 and estimates made, the control of laminar flows via registration of the viscous losses is suitable for pipelines of relatively small diameters (≤ 2 cm in the case of water). In this case, LC sensors show advantages in

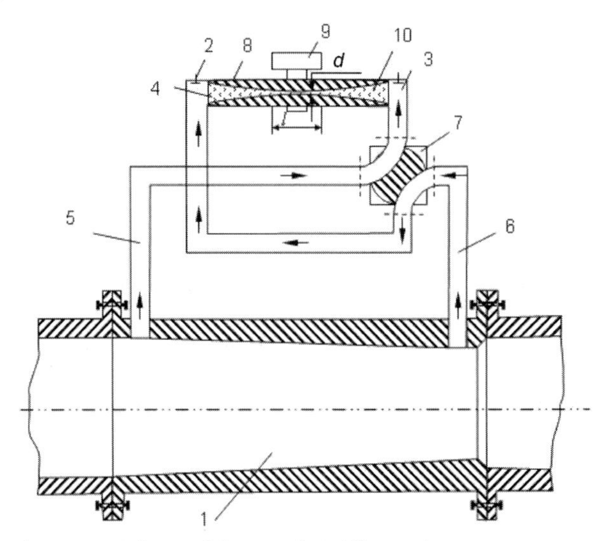

Figure 7.22 Scheme of the use of LC differential pressure sensor for the control of flows in pipelines: (1) Venturi tube; (2, 3) chambers with LC; (4) membranes; (5, 6) connected tubes; (7) two-way valve; (8) thin capillary with LC; (9) photon-coupled pair; (10) liquid crystal. (After Orlov *et al.* [32].)

comparison with analogous devices [2] due to high sensitivity to the pressure difference. In the case of turbulent flows through a pipeline of a large diameter, the special inserts (orifice plates or conical Venturi tubes) have to be used (the example of the conical insert is shown in Figure 7.22). The description and technical characteristics of such units can be found in special literature. As usual, there is no simple connection between the pressure losses ΔP_{turb} and the instant flow rate similar to (7.20). In the first approximation, $\Delta P_{\text{turb}} \sim Q^2$. So, a wide dynamic range of the registered pressure has to be realized to provide a relatively narrow range of an instant flow rate. For LC sensors, it can be achieved by using electric fields or (and) the cells of a variable gap.

An alternative technical decision was proposed in [34] for registration of flows. It is based on vortex separation at the flow of liquid or gases round the obstacle. This phenomenon is widely used in vortex shedding flowmeters, which shows a steady acceptance of flow measurement technique [35]. Such sensors have a number of important advantages in comparison with the alternative flow sensors. Namely, their moving parts produce a frequency output that varies linearly with a flow rate over a wide range of Reynolds numbers. The vortex meter has a very simple construction, provides accuracy (1% or better) comparable to higher priced techniques, and works equally well on liquids and gases. Industrial vortex shedding flowmeters are normally available in pipe sizes ranging from 15 to 300 mm (1/2–12 in.). A typical range 0.06–2.2 l/s for water flows in a 15 mm pipe [35] is achievable. A lower limit of this range corresponds to the mean velocity flow of about 0.5 m/s. Using highly sensitive LC cells makes it possible to decrease this value.

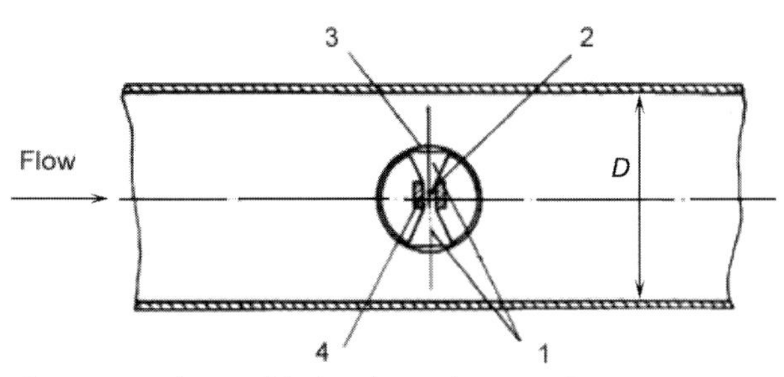

Figure 7.23 Scheme of the liquid crystal sensor of a vortex flowmeter: (1) a hollow cylinder, slits filled with LC; (2) LC cell; (3) separating limp diaphragm; (4) a capacity transducer (or photon-coupled pair).

The scheme of the liquid crystal sensor of a vortex flowmeter is shown in Figure 7.23.

The liquid crystal cell is placed into the cylinder and connected with its outer surface by two slits filled with a liquid crystal and is separated from the flowing media by a limp diaphragm. The flow around the cylinder results in vortex separation and periodic pressure difference acted on the diaphragms and transformed to a liquid crystal cell. The frequency of the vortex separation f is determined as

$$f = \frac{Sv}{D}, \tag{7.21}$$

where D is diameter of a cylinder, v is flow velocity, and S is the Strouhal number equal to 0.19 for Reynolds numbers $500 < Re < 10^5$. So, the measurement of the frequency f helps determine the flow velocity v. For $D = 1$ cm, $v = 0.1$ m/s, the frequency f is about 2 Hz at a pressure amplitude, $\Delta P \approx \rho v^2/2$ is equal to about 5 Pa for a water flow. Such parameters are suit well for the registration with the help of LC cell, as it was shown above. It does not hold for gas flows of the same velocity due to essentially small values of ρ and ΔP. It is worthwhile to note that the given decision makes it possible to delete the special unit shown in Figure 7.22 and needed to change the pressure gradient direction. Sensors of such type can be used both in pipelines and in open flows of liquids.

7.5 Application of Liquid Crystals for Detecting and Visualizing Acoustic Fields

As it was shown above, the linear response of a liquid crystal layer to the action of a pressure gradient that operates LC sensors decreases essentially with increasing frequency and becomes negligible at frequencies $f = 10^3$–10^4 Hz. Nevertheless, it is well known that liquid crystals can be used for the registration and visualization of acoustic fields of ultrasonic frequencies. One can find a number of original papers and some reviews (see, for example, Refs [20, 36, 37]) devoted to this topic. The substantial interest this problem has drawn is about different possible applications of the acoustooptical phenomena. In particular, it would be very desirable to find a simple and effective way to visualize acoustic fields passing through liquid-like nontransparent media with some objects inside to get an acoustic image of this object using a liquid crystal screen. If realized, such technique can be rather attractive not only for industry but also for ultrasonic diagnostics widely used in modern medicine [38]. In the latter case, a high sensitivity of the screen plays the key role as the intensity I_a of diagnostic ultrasound is strictly limited (not more than 10 mW/cm^2 at typical frequencies 3–15 MHz). Though this value is comparable to the sensitivity of LC cells with relatively thick (100–200 μm) homeotropic layers of a nematic phase [37, 39], it can be insufficient to register an ultrasonic wave passing through biological tissues due to the strong attenuation of ultrasound. It was found that the maximal sensitivity of LC cells to the ultrasound (as in the case of low-frequency vibrations) could be provided using polarized light passing through the homeotropic LC layer [37, 39]. The sensitivity can be essentially improved at oblique incidence of ultrasonic wave relative to the normal of the LC cell [40]. The analogous effect can be obtained via the additional destabilization of a nematic layer by electric field [39] or by mechanical vibrations [41]. The second problem arising in industrial and biomedical applications of acoustooptical phenomena in liquid crystals is related to the narrow dynamic range of LC cells relative to the registered ultrasonic intensities. Finally, a lateral resolution of acoustic images visualized through liquid crystals is not very high. In particular, stationary shear flows induced by ultrasound in a nematic layer are not strictly localized [37]. To some extent, this problem can be solved by dividing the entire LC panel into small separate subcells [42].

Now, we intend to show that the basic idea of the additional use of stabilizing and destabilizing electromagnetic fields can be successfully applied to control the technical parameters of LC acoustooptical sensors and screens applied for the registration, visualization, and mapping of acoustic fields of ultrasonic frequency range. Such an idea is applicable to the sensors based on nematic liquid crystals. So, the acoustooptical effects in cholesteric liquid crystals [36, 37] are out of scope of our consideration.

7.5.1 Acoustic Flows in Liquid Crystals

In spite of a long history of investigations of acoustooptical effects in liquid crystals, their theoretical description has not yet been considered [43–46]. The strict quantitative solution of the problem [45, 46] includes the calculation of propagation of an ultrasonic wave through the multilayer system (liquid–solid–liquid crystal–liquid). Results of such calculations strongly depend not only on viscoelastic parameters of a liquid crystal but also on mechanical properties and size of solid substrate that confine liquid crystals. A schematic representation of acoustic waves in a liquid crystal cell [45] is given in Figure 7.24.

Physical mechanisms of interaction of ultrasonic waves with a liquid crystal media resulting in steady changes in orientational structure and optical properties is of key importance to the problem under consideration. There were a number of attempts to

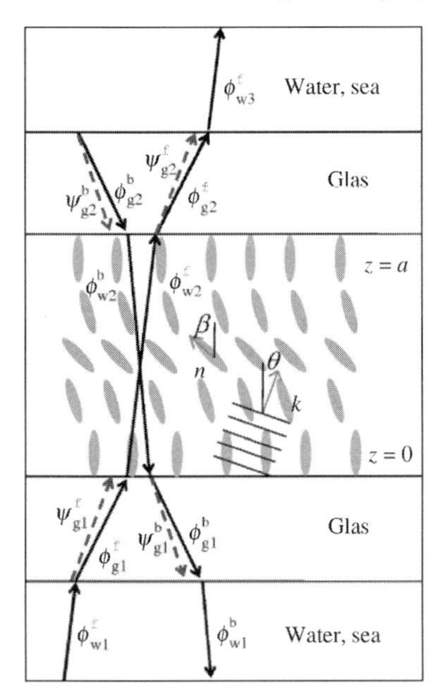

Figure 7.24 Schematic representation of a liquid crystal cell and sound waves traveling through it. Solid lines indicate the longitudinal sound waves that are produced as original sound wave propagates forward from the bottom/first layer represented with letter f in superscripts and the backward-propagating sound waves created by reflection at interfaces represented by letter b in superscripts. The change in the direction of the waves at each layer interface is required by Snell's law. In solid glass layers, the dashed lines represent the transverse sound waves. (After Malanoski *et al.* [45].)

explain acoustooptical phenomena in nematics [40, 47, 48]. At present, steady acoustic flows arising in LC layer are considered as the most probable reason of acoustooptical phenomena [20, 40, 49]. In a general case of oblique incidence of sound, the steady acoustic flows schematically shown in Figure 7.25 are created via the action of both longitudinal and overdamped shear waves propagating in z-direction [49].

Analogous flows are well known for isotropic liquids. In this case, they originate from nonlinear convective stresses $\sigma_{ij} \sim \langle \rho v_i v_j \rangle$ averaged in time that can be determined from Navier–Stokes equations.

The anisotropic character of viscoelastic properties of liquid crystals makes corresponding solution essentially more complicated. Moreover, some declinations from classic Leslie–Erickson theory at ultrasonic frequencies connected with relaxation phenomena (Chapter 4) can contribute to both linear and nonlinear acoustic phenomena. The mentioned factors prevent the strict quantitative calculations needed for comparison with the experimental data. Nevertheless, there are a number of similar properties of shear flows induced by a pressure gradient and ultrasonic waves important for practical applications. In particular, the homeotropic orientation turns out to be the most sensitive to the influence of ultrasonic waves, as it was mentioned above. In the case of oblique incidence of sound, a steady acoustic flow in a

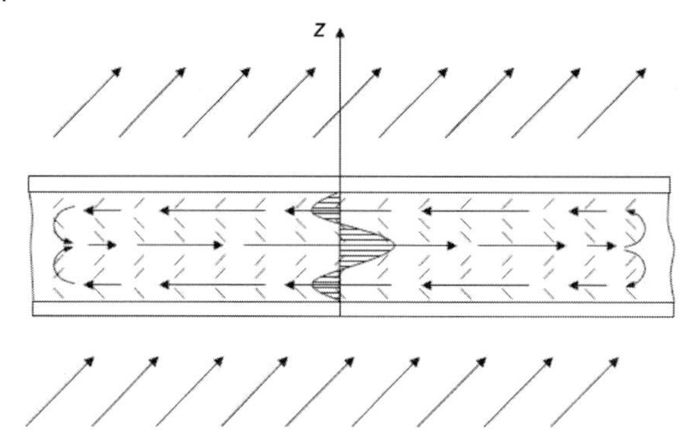

Figure 7.25 The structure of steady shear flows generated in a homeotropic LC cell under oblique incidence of sound. (After Kozhevnikov [49].)

homeotropic LC cell shown in Figure 7.25 is characterized by the velocity $v^{(2)}$ expressed from the next differential equation [49]

$$\eta_1 \frac{d^3 v_x^{(2)}}{dz^3} \cong \rho \frac{d^2 \langle v_x v_z \rangle}{dz^2}, \tag{7.22}$$

where $\langle v_x v_z \rangle$ is the average value of the combination of linear velocities in longitudinal (v_z) and shear (v_x) waves. So, there is no threshold in action of both the low-frequency pressure gradient and the ultrasonic wave on the initial homeotropic structure of a liquid crystal. The functional dependence of the intensity I of polarized light on the layer thickness $h(I \sim h^{14})$ is also the same as in the case of small linear deformations of a homeotropic layer under a steady Poiseuille flow.

The above-mentioned analogies confirm the possibility of controlling the optical response of a homeotropic layer on the ultrasonic wave via the variation of the layer thickness and application of electric field as in the case of Poiseuille flows. While the first possibility is well investigated [37], the action of electric (magnetic) fields on acoustooptical effects was studied in experiments only rarely [20, 39, 50–55].

7.5.2
Acoustooptical Effects on Liquid Crystals in the Presence of Electric and Magnetic Fields

The general geometry of experiments on the action of an ultrasound wave on a homeotropic LC layer in the presence of electric or (and) magnetic fields is shown in Figure 7.26.

In this geometry, an ultrasonic wave (with a wave vector $\boldsymbol{q}$) acted on the sealed or open edge of the LC cell under propagation in the plane of the layer. It essentially differs from the geometry shown in Figures 7.24 and 7.25 with a normal or oblique incidence of the sound. The acoustic contact was provided by gluing the piezo transducers directly to the sealed edge of LC cell [50, 51] or by using intermediate liquid media [52, 53]. In the latter case, the LC cell was immersed (partly or totally) in a liquid. The latter variant [53] is shown in Figure 7.27. It helped study of a combined

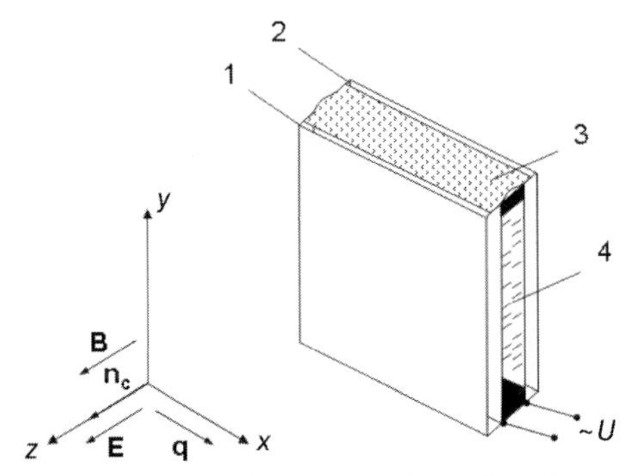

Figure 7.26 General geometry and LC cell used in acoustooptical experiments under action of electromagnetic fields: (1, 2) glass substrates; (3) sealed edges; (4) homeotropic layer of LC.

action of acoustic and electromagnetic fields on the orientation and the optical properties of a homeotropic layer. A peculiar feature of such construction is placing of the LC cell into the liquid crystal medium that transmits the liquid crystal layer hydrostatic pressure and ultrasonic vibrations without using any intermediate material.

In the described experiments, MBBA or the nematic mixture N37 (MBBA : EBBA) by Charkov production was used to fill the chamber and the LC cell (the layer thickness $h = 105 \pm 5\,\mu m$). The internal surface of one substrate of the cell was coated with a transparent conductive film. The internal surface of the other substrate had a conductive reflecting coating. An electric field directed along the x-axis could be formed by applying a voltage to the cell. The ultrasound vibrations were transmitted to the open side of the cell from a piezo transducer placed in the XZ-plane. The light

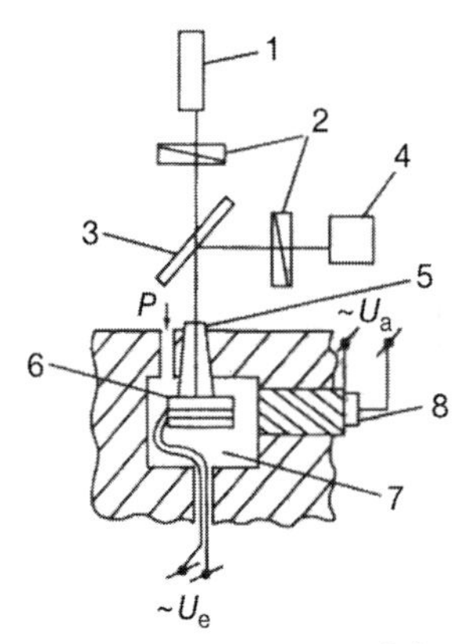

Figure 7.27 Acoustooptical chamber and optical scheme for the study of a combined action of acoustic and electromagnetic fields at variation of temperature and pressure. (1) He–Ne laser; (2) polarizers; (3) beam splitter; (4) glass cone; (5) LC cell; (6) liquid crystal; (7) photodiode; (8) piezo transducer. (After Pasechnik *et al.* [53].)

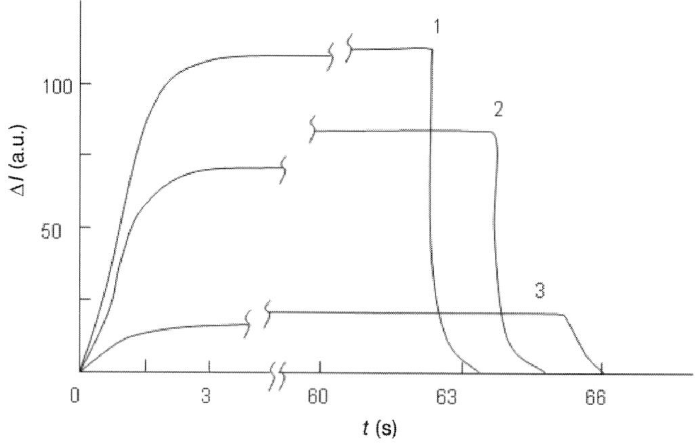

Figure 7.28 Time dependences of light intensity changes induced by turning on (off) ultrasonic vibrations at different values B of the magnetic field induction: (1) $B = 0$ T, (2) $B = 55$ mT, (3) $B = 145$ mT; N37, $T = 313$, 5 K, $P = 0.1$ MPa, $U_E = 0$. (After Pasechnik [54].)

propagated along the x-axis and passed twice (due to reflection) through the NLC layer. The chamber was placed between the electromagnet poles, so the combined action of acoustic and electromagnetic fields could be investigated.

In the experiment, the time dependence of light intensity changes $\Delta I(t)$ induced by turning on (off) ultrasonic vibrations of frequency in the presence of magnetic (electric) field were registered by means of an XY-recorder. In the case of relatively small amplitudes of voltage U_a applied to the piezo transducer, the deviation of an orientation from the homeotropic one was small enough and the light intensity increased for some seconds to the stationary value ΔI_s. Such behavior is demonstrated in Figure 7.28.

At higher values of the ultrasonic intensity proportional to U_a^2 or in the presence of destabilizing electric fields, a number of local extremes $\Delta I(t)$ were registered (Figure 7.29). Such behavior can be explained by relatively large value of the mean phase delay ($\delta > \pi$) as in the case of the action of low-frequency pressure gradient.

Results presented in Figures 7.28 and 7.29 show that both magnetic and electric fields modified static and dynamic characteristics of the acoustooptical response. Examples of the dependence of the ultrasonically induced stationary variations of light intensity changes on the magnetic field induction are shown in Figure 7.30.

Data presented are obtained in the vicinity of Fréedericksz transition, where relatively weak stabilizing magnetic field can essentially modify the acoustooptical response. In particular, it results in a decrease in LC cell sensitivity (Figure 7.29) and change in response times. Similar results were obtained via registration of acoustically induced optical diffraction in the presence of magnetic field [50, 51]. One can wait for the same effects from a stabilizing action of electric field (in the case of liquid crystals with a positive value of $\Delta\varepsilon$) that was confirmed previously for the case of the interaction of surface waves and electric field [55].

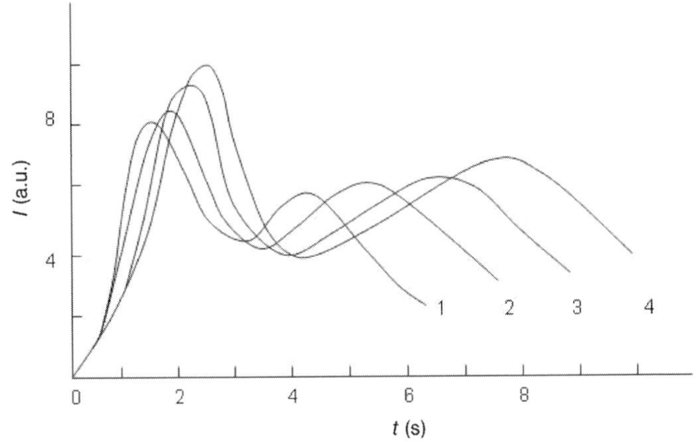

Figure 7.29 Light intensity changes $\Delta I(t)$ induced by turning on ultrasonic vibrations ($U_a = 20\,V$) at destabilizing electric field $U_E = 4.6$ V and stabilizing magnetic field of different values of induction; (1) $B = 0$; (2) 6.4 mT; (3) 18.4 mT; (4) 40.3 mT. N37, $T = 300$ K. (After Pasechnik [54].)

In contrast, destabilizing electric field at voltages U lower than Fréedericksz transition voltage U_F leads to an increase (up to some times) in sensitivity of LC cell [52, 54]. It is in accordance with the results [39] obtained previously for different experimental geometries. At the same time, the application of destabilizing electric field results in a slowdown of relaxation processes taking place at this orientational transformation (Fréedericksz transition). The critical increase in the director relax-

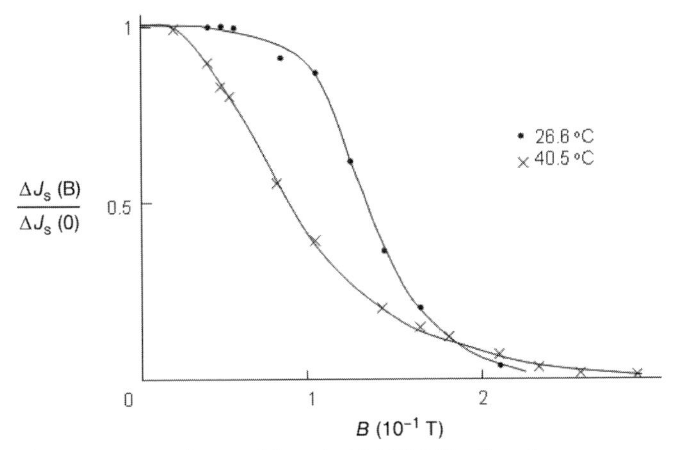

Figure 7.30 Dependences of the ratio $\Delta I_s(B)/\Delta I_s(B = 0)$ on the magnetic field induction in N37 at different temperatures; ΔI_s – ultrasonically induced stationary changes of light intensity at low values of a phase delay $\delta < \pi$. (After Pasechnik [54].)

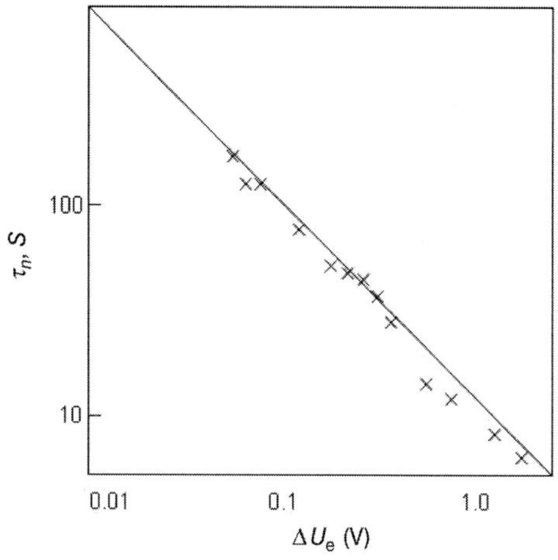

Figure 7.31 Dependence of the director relaxation time τ_n on $\Delta U = U_F - U$; in the vicinity of Fréedericksz transition. MBBA, $T = 300$ K, $P = 50$ MPa, $U_F = 4.28$ V. Straight line: the simple power law. (After Pasechnik [54].)

ation time τ_n, extracted via acoustooptical investigations is shown in Figure 7.31. In the vicinity of Fréedericksz transition, it is well described by a simple power law [54]

$$\tau_n \sim (\Delta U)^{-1},$$

which is valid for phase transition of the second order (see Chapter 4).

It is worthwhile to note that the influence of magnetic field on the dynamics of acoustooptical effects is most pronounced in the vicinity of the Fréedericksz transition, as it is shown in Figure 7.30.

Most of the presented results have found the explanation [53, 54] in the framework of a simple hydrodynamic model taking into account steady shear flows induced by ultrasound in a conventional isotropic liquid. It is based on the linearized equation of the director motion similar to that used above for the description of influence of a quasistationary Poiseuille flows on a homeotropic layer of LC (see, for example, Equation 5.93)

To summarize experimental results, one can conclude that electric (magnetic) fields are possible to modify the stationary and dynamic characteristics of acoustooptical effects. It is of interest for practical applications of liquid crystals for registration, visualization, and mapping of acoustic fields. Such possibility has not yet been realized.

We can summarize the main results presented in Chapter 7 as follows:

1. Extremely high sensitivity of nematic layers to the action of steady and low-frequency flows induced by a pressure gradient is very attractive for sensor applications.

2. Linear deformations of the initial homeotropic orientation under flows detected by polarizing light can be considered as the basic mechanooptical effect providing the highest threshold sensitivity to the sensors.

3. The use of stabilizing electric fields is very effective for optimization of technical characteristics (threshold sensitivity, dynamic range, operating times) of LC sensors.
4. Variation of the nematic layer thickness combined with electric fields and the use of a number of optical channels make it possible to propose liquid crystal sensors of the pressure gradient characterized by the high threshold sensitivity and wide dynamic range.
5. A number of particular liquid crystal sensors (differential pressure sensors, sensors of acceleration, vibrations, inclination, liquid, and gas flows) can be proposed on the basis of pressure gradient sensors.
6. The main advantages of such sensors compared to the analogous ones include high sensitivity and the possibility of the electric control of characteristics.
7. Liquid crystal sensors based on the flow phenomena are most effective for the registration of steady or low-frequency mechanical perturbations.
8. At the same time, they can be used for registration, visualization, and mapping of ultrasonic fields as the latter induce steady acoustic flows inside liquid crystal cells. In this case, the use of electric (magnetic) fields is also effective for the control of sensor characteristics.

References

1 Fraden, J. (2004) *Handbook of Modern Sensor: Physics, Designs and Applications*, Springer, New York.

2 Webster, J.G.(ed.) (2000) *Mechanical Variables Measurement: Solid, Fluid, and Thermal*, CRC Press LLC, Boca Raton.

3 Udd, E.(ed.) (2006) *Fiber Optic Sensors: An Introduction for Engineers and Scientists*, Wiley & Sons, Inc., New York.

4 Chandrasekhar, S. (1992) *Liquid Crystals*, Cambridge University Press, Cambridge.

5 Payne, P.(ed.) (1991) *Concise Encyclopedia of Biological & Biomedical Measurement Systems (Advances in Systems Control and Information Engineering)*, Pergamon Press, Oxford.

6 Biga, F.Y., Luks, F.I.,*et al.* (2007) Medical displays, in *Liquid Crystals: Frontiers in Biomedical Applications* (eds S.J. Woltman, G.R. Crawford, and G.D. Jay), World Scientific Publishing, Singapore, pp. 81–149.

7 Vukovic, D., Nagarajah, C.R., Toncich, D., and Chowdary, F. (2000) Thermochromic liquid crystal based sensing system for analysis of air flow. *Measurement*, **27**, 101.

8 Yang, J.S., Hong, C.H., and Choi, G.M. (2007) Heat transfer measurement using thermochromatic liquid crystal. *Curr. Appl. Phys.*, **7**, 413.

9 Lim, K.B., Lee, C.H., Sung, N.W., and Lee, S.H. (2007) An experimental study on the characteristics of heat transfer on the turbulent round impingement jet according to the inclined angle of convex surface using the liquid crystal transient method. *Exp. Therm. Fluid Sci.*, **31**, 711.

10 Woliński, T.R., Konopka, W., Bock, W.J., and Dąbrowski, R. (1998) Progress in liquid crystalline optical fiber systems for pressure monitoring. *SPIE*, **3318**, 338.

11 Yablonskii, S.V., Nakano, K., Mikhailov, A.S., Ozaki, M., and Yoshino, K. (2001)

New applications of FLC: sensor based on ferroelectric liquid crystalline freely suspended films. *J. Soc. Elect. Mater. Eng.*, **10**, 267.

12 Fernandes, P.R.G., Kimura, N.M., and Maki, J.N. (2004) Mechano-optical effect in isotropic phase of a lyotropic liquid crystal. *Mol. Cryst. Liq. Cryst.*, **421**, 243.

13 McCamley, M.K., Artenstein, A.W., and Crawford, G.P. (2007) Liquid crystal biosensors, in *Liquid Crystals: Frontiers in Biomedical Applications* (eds S.J. Woltman, G.R. Crawford, and G.D. Jay), World Scientific Publishing, Singapore, pp. 241–296.

14 Cadwell, K.D., Lockwood, N.A. *et al.* (2007) Detection of organophosphorous nerve agents using liquid crystals supported on chemically functionalized surfaces. *Sens. Actuator B-Chem.*, **128**, 91.

15 Bi, X., Huang, S., Hartono, D. *et al.* (2007) Liquid-crystal based optical sensors for simultaneous detection of multiple glycine oligomers with micromolar concentrations. *Sens. Actuator B-Chem.*, **127**, 406.

16 Shiyanovskii, S.V., Lavrentovich, O.D., Schneider, T., Ishikawa, T., Smalyukh, I.I., Woolverton, J., Niehaus, G.D., and Doane, K.J. (2005) Lyotropic chromonic liquid crystals for biological sensing applications. *Mol. Cryst. Liq. Cryst.*, **434**, 259.

17 Balandin, V.A., Gevorkyan, E.V., and Pasechnik, S.V. (1993) Increase of the sensitivity of a liquid-crystal cell to low-frequency pressure vibrations. *Tech. Phys. Lett.*, **19**, 154.

18 Balandin, V.A., Pasechnik, S.V., and Gevorkyan, E.V. (1995) Method of pressure measurement. Patent of Russia, RU2036447 (C1).

19 Balandin, V.A., Gevorkian, E.V., and Pasechnik, S.V. (1996) The optical response of nematics on the low-frequency variations of pressure in a tilted geometry under an electric field. *Mol. Mater.*, **6**, 45.

20 Kapustin, A.P. and Kapustina, O.A. (1986) *Akustika Zhidkih Krtistallov*, Nauka, Moscow (in Russian).

21 Pasechnik, S.V. and Torchinskaya, A.V. (1999) Behavior of nematic layer oriented by electric field and pressure gradient in the stripped liquid crystal cell. *Mol. Cryst. Liq Cryst.*, **331**, 341–347.

22 Pasechnik, S.V., Torchinskaya, A.V. *et al.* (2001) Nonlinear optical response of nematic liquid crystal on varying pressure difference in the presence of electric field. *Mol. Cryst. Liq. Cryst.*, **367**, 3515.

23 Shmeliova, D.V., Pasechnik, S.V. *et al.* (2002) The application of liquid crystal cells of variable thickness in electrically controlled sensors of low-frequency vibrations. Proceedings of 22nd International Display Research Conference, Nice, France, p. 457.

24 Pasechnik, S.V., Kravchuk, A.S. *et al.* (2003) Prospects for the application of liquid-crystal cells for the registration and visualization of mechanical oscillations in biomedical technologies. *J. Soc. Inform. Display*, **11**, 15.

25 Pasechnik, S.V., Shmeliova, D.V. *et al.* (2004) Orientational oscillations in homeotropic layers of liquid crystals induced by low frequency pressure gradient in the presence of stabilizing and destabilizing electric field. *Mol. Cryst. Liq. Cryst.*, **409**, 449.

26 Torchinskaya, A.V. (2002) Oscillating Poiseuille flow in a nematic liquid crystal, oriented by electric field. PhD thesis, Moscow State Academy of Instrument Engineering and Computer Science, Moscow.

27 Barnik, M.I., Belyaev, V.V., and Grebenkin, M.F. (1978) Elektricheskie, Opticheskie i vyzko-uprugie svo'istva ZhK smesi azoksisoedineni'i. *Krisstallografiya*, **23**, 805.

28 Grebenkin, M.F., Seliverstov, V.A. *et al.* (1975) Svoistva nematicheskih ghidkih kristallov s pologiteljnoi dielectricheskoi anizotropiei. *Krisstallografiya*, **20**, 984.

29 Balandin, V.A., Pasechnik, S.V., and Orlov, V.A. (1994) Differential pressure gauge. Patent of Russia, RU2008637 (C1)

30 Pasechnik, S.V., Balandin, V.A., and Tsvetkov, V.A. (1999) The meter of the pressure difference. Patent of Russia, RU2127876 (C1).

31 Shmeliova, D.V. (2005) Investigation of LC structures induced by shear flow and electric field. PhD thesis, Moscow State Academy of Instrument Engineering and Computer Science, Moscow.

32 Thorn, R. *et al.* (2000) Differential pressure flowmeters, Part 9.1, in *Mechanical Variables Measurement: Solid, Fluid, and Thermal* (ed. J.G. Webster), CRC Press LLC, Boca Raton.

33 Orlov, V.A., Balandin, V.A., Pasechnik, S.V., Zotkin, S.P., Solovjev, A.E., and Vasiljev, I.N. (1992) Pressure difference meter. Patent of Russia, SU1719944 (A1)

34 Balandin, V.A., Pasechnik, S.V. *et al.* (1996) Transducer of vortex flowmeter. Patent of Russia, RU2071035 (C1).

35 Thorn, R. *et al.* (2000) Vortex shedding flowmeters, Part 9.89, in *Mechanical Variables Measurement: Solid, Fluid, and Thermal* (ed. J.G. Webster), CRC Press LLC, Boca Raton.

36 Perbet, J.N., Hareng, M., Le Berre, S., and Mourey, B. (1979) Visualisation d'images acoustiques â l'aide d'un cristal liquidw nematique. *Rev. Tech. Thom.*, **4**, 837.

37 Kapustina, O.A. (1984) Acoustooptical phenomena in liquid crystals. *Mol. Cryst. Liq. Cryst.*, **112**, 1.

38 Hill, C.R., Bamber, J.C., and ter Haar, G.R.(eds) (2004) *Physical Principles of Medical Ultrasonics*, John Wiley & Sons, Ltd, Chichester, UK.

39 Kagawa, Y., Hatakeyama, T., and Tanaka, Y. (1974) Detection and visualization of ultrasonic fields and vibrations by means of liquid crystals. *J. Sound Vib.*, **36**, 407.

40 Candau, S., Ferre, A., Petters, A. *et al.* (1980) Acoustical streaming in a film of nematic liquid crystal. *Mol. Cryst. Liq. Cryst.*, **61**, 7.

41 Bocharov, Ju.V. and Kapustina, O.A. (1999) Peculiarities of the orientational distortion of a nematic liquid crystal under a two-wave action. *Acoust. Phys.*, **45**, 408.

42 Nagai, S. and Iizuka, K. (1978) An experiment on the resolution of acoustic image detector with a nematic liquid crystal. *Jpn. J. Appl. Phys.*, **17**, 723.

43 Selinger, J.V., Spector, M.S. *et al.* (2002) Acoustic realignment of nematic liquid crystals. *Phys. Rev. E*, **66**, 051708.

44 Greanya, V.A., Spector, M.S. *et al.* (2003) Acousto-optic response of nematic liquid crystals. *J. Appl. Phys.*, **94**, 7571.

45 Malanoski, A.P., Greanya, V.A., Weslowski, B.T. *et al.* (2004) Theory of the acoustic realignment of nematic liquid crystals. *Phys. Rev. E*, **69**, 021705.

46 Kozhevnikov, E.N. (2005) Deformation of a homeotropic nematic liquid crystal layer at oblique incidence of an ultrasonic beam. *Acoust. Phys.*, **51**, 688.

47 Dion, J.-L. (1979) Orienting action on liquid crystals related to the minimum entropy production. *J. Appl. Phys.*, **50**, 2965.

48 Kozhevnikov, E.N. (1980) Orientational instability of nematic liquid crystals under a sonic field in the absence of spreading. *Akust. Zh.*, **26**, 966 (in Russian).

49 Kozhevnikov, E.N. (1982) Acousto-optic effect induced by an incident ultrasonic beam in a normally oriented nematic liquid crystal layer. *Sov. Phys. JETP*, **55**, 96.

50 Popov, A.I., Pasechnik, S.V., Nozdrev, V.F., and Balandin, V.A. (1982) Dynamics of ultrasound-induced diffraction effects in a nematic liquid crystal oriented with a magnetic field. *Sov. Tech. Phys. Lett.*, **8**, 431.

51 Popov, A.I., Pasechnik, S.V., Balandin, V.A., and Nozdrev, V.F. (1984) Diffraction of light by periodic structures induced by ultrasound in a layer of nematic liquid crystal. *Sov. Phys. Crystallogr.*, **29**, 74.

52 Ezhov, S.G., Pasechnik, S.V., and Balandin, V.A. (1984) Effect of an electric field on the temporal characteristics of the acousto-optic effect in nematic liquid crystals. *Sov. Tech. Phys. Lett.*, **10**, 202.

53 Pasechnik, S.V., Kireev, V.I., and Balandin, V.A. (1996) Nematic orientation structure under the action of ultrasound and electric field at varying pressure. *Bull. Russ. Acad. Sci. Phys.*, **60**, 533.

54 Pasechnik, S.V. (1998) Ultrasound at phase transitions and structural transformations in liquid crystals. DSc thesis, Moscow State Academy of Instrument Engineering and Computer Science, Moscow.

55 Hakemi, H. (1982) The effect of thickness on the acousto-electro-optical study of a nematic liquid crystal. *J. Appl. Phys.*, **53**, 6137.

Index

Liquid Crystals: Viscous and Elastic Properties
S. V. Pasechnik, V. G. Chigrinov, and D. V. Shmeliova

ISBN: 978-3-527-40720-0